AF784791

Digital Audio Signal Processing

Third Edition

Udo Zölzer
Helmut Schmidt University Hamburg, Germany

with

Martin Holters, Etienne Gerat, Patrick Nowak, Purbaditya Bhattacharya, Lasse Köper, and Daniel Ahlers

WILEY

This third edition first published 2022
© 2022 John Wiley & Sons Ltd

Edition History: 1e (1997); 2e (2008)

All rights reserved. No part of this publication may be reproduced, stored in a retrieval system, or transmitted, in any form or by any means, electronic, mechanical, photocopying, recording or otherwise, except as permitted by law. Advice on how to obtain permission to reuse material from this title is available at http://www.wiley.com/go/permissions.

The right of Udo Zölzer to be identified as the author of this work has been asserted in accordance with law.

Registered Offices
John Wiley & Sons, Inc., 111 River Street, Hoboken, NJ 07030, USA
John Wiley & Sons Ltd, The Atrium, Southern Gate, Chichester, West Sussex, PO19 8SQ, UK

Editorial Office
The Atrium, Southern Gate, Chichester, West Sussex, PO19 8SQ, UK

For details of our global editorial offices, customer services, and more information about Wiley products visit us at www.wiley.com.

Wiley also publishes its books in a variety of electronic formats and by print-on-demand. Some content that appears in standard print versions of this book may not be available in other formats.

Limit of Liability/Disclaimer of Warranty

In view of ongoing research, equipment modifications, changes in governmental regulations, and the constant flow of information relating to the use of experimental reagents, equipment, and devices, the reader is urged to review and evaluate the information provided in the package insert or instructions for each chemical, piece of equipment, reagent, or device for, among other things, any changes in the instructions or indication of usage and for added warnings and precautions. While the publisher and author have used their best efforts in preparing this work, they make no representations or warranties with respect to the accuracy or completeness of the contents of this work and specifically disclaim all warranties, including without limitation any implied warranties of merchantability or fitness fora particular purpose. No warranty may be created or extended by sales representatives, written sales materials or promotional statements for this work. The fact that an organization, website, or product is referred to in this work as a citation and/or potential source of further information does not mean that the publisher and author endorse the information or services the organization, website, or product may provide or recommendations it may make. This work is sold with the understanding that the publisher is not engaged in rendering professional services. The advice and strategies contained herein may not be suitable for your situation. You should consult with a specialist where appropriate. Further, readers should be aware that websites listed in this work may have changed or disappeared between when this work was written and when it is read. Neither the publisher nor author shall be liable for any loss of profit or any other commercial damages, including but not limited to special, incidental, consequential, or other damages.

Library of Congress Cataloging-in-Publication data

Name: Zölzer, Udo, author.
Title: Digital audio signal processing / Udo Zölzer, Helmut Schmidt University Hamburg, Germany with Martin Holters, Etienne Gerat, Patrick Nowak, Purbaditya Bhattacharya, Lasse Köper, and Daniel Ahlers.
Other titles: Digitale Audiosignalverarbeitung. English
Description: Third edition. | Hoboken : Wiley, 2022. | Includes bibliographical references and index.
Identifiers: LCCN 2021059110 (print) | LCCN 2021059111 (ebook) | ISBN 9781119832676 (cloth) | ISBN 9781119832683 (adobe pdf) | ISBN 9781119832690 (epub)
Subjects: LCSH: Sound–Recording and reproducing–Digital techniques. | Signal processing–Digital techniques.
Classification: LCC TK7881.4 .Z6513 2022 (print) | LCC TK7881.4 (ebook) | DDC 621.389/3–dc23/eng/20220103
LC record available at https://lccn.loc.gov/2021059110
LC ebook record available at https://lccn.loc.gov/2021059111

Cover Design: Wiley
Cover Image: © whiteMocca/Shutterstock

Set in 10/12pt TimesLTStd by Straive, Chennai, India

10 9 8 7 6 5 4 3 2 1

M 03M000030000004881 040322

Contents

Preface for the Third Edition

This 3rd Edition is a revised and extended version of the book. It offers three new chapters (1, 10, and 11) and extensions on a further four chapters (5, 7, 8, and 9). The content of this book is the basis of a course on *Digital Audio Signal Processing* at the Hamburg University of Technology (TUHH) and a course on *Multimedia Signal Processing* at the Helmut Schmidt University Hamburg. For further study, you can find lecture slides, exercises, Matlab examples, and interactive audio demonstrations on the website.

`https://mydasp.com/dasp.`

My thanks go to Dr. Martin Holters for the interactive audio demonstrations and maintaining the website `http://www.dafx.de`. Thanks also to him and Lasse Köper for their contribution to the new Chapter 10 on "Nonlinear Processing". I would like to thank Daniel Ahlers for his contributions and updates to Chapter 5 on "Audio Processing Systems". Purbaditya Bhattacharya and Patrick Nowak made contributions to Chapter 7 on "Room Simulation" and are mainly responsible for the new Chapter 11 on "Machine Learning for Audio". Purbaditya Bhattacharya also made contributions to Chapter 9 on "Audio Coding". Thanks to both of them. Finally, I would like to thank Etienne Gerat for his refinements and extensions to Chapter 8 on "Dynamic Range Control". Last but not least, I would like to thank all participants of the lectures over all these years who have been my driving force:-).

Udo Zölzer
Hamburg, January 2022

Preface for the Second Edition

This 2nd Edition represents a revised and extended version, and offers an improved description as well as new issues and extended references. The content of this book is the basis of a course on *Digital Audio Signal Processing* at the Hamburg University of Technology (TU Hamburg-Harburg) and a course on *Multimedia Signal Processing* at the Helmut Schmidt University Hamburg. For further study, you can find interactive audio demonstrations, exercises, and Matlab examples on the website.

`http://ant.hsu-hh.de/dasp/.`

In addition to the basics of digital audio signal processing introduced in this 2nd edition, further advanced algorithms for digital audio effects can be found in the book *DAFX – Digital Audio Effects* (Ed. U. Zölzer) with a related website.

`http://www.dafx.de.`

My thanks go to Prof. Dieter Leckschat, Dr. Gerald Schuller, Udo Ahlvers, Mijail Guillemard, Christian Helmrich, Martin Holters, Dr. Florian Keiler, Stephan Möller, Francois-Xavier Nsabimana, Christian Ruwwe, Harald Schorr, Dr. Oomke Weikert, Catja Wilkens, and Christian Zimmermann.

Udo Zölzer
Hamburg, December 2007

Preface for the First Edition

Digital audio signal processing is employed in recording and storing music and speech signals, for sound mixing and production of digital programs, in digital transmission to broadcast receivers as well as in consumer products like CDs, DATs, and PCs. In the latter case, the audio signal is in a digital form all the way from the microphone up to the loudspeakers, enabling real-time processing with fast digital signal processors.

This book provides the basis of an advanced course in *Digital Audio Signal Processing* which I have been giving since 1992 at the Technical University Hamburg-Harburg. It is directed at students studying engineering, computer science, and physics but is also for professionals who are looking for solutions to problems in audio signal processing in the fields of studio engineering, consumer electronics, and multimedia. The mathematical and theoretical fundamentals of digital audio signal processing systems will be presented and typical applications with an emphasis on realization aspects will be discussed. Prior knowledge of systems theory, digital signal processing, and multirate signal processing are assumed as a prerequisite.

The book is divided into two parts. The first part (Chapters 1–4) presents a basis for hardware systems used in digital audio signal processing. The second part (Chapters 5–9) discusses algorithms for processing digital audio signals. Chapter 1 describes the course taken by an audio signal from its recording in a studio up to its reproduction at home. Chapter 2 contains a representation of signal quantization, dither techniques, and spectral shaping of quantization errors used for reducing the nonlinear effects of quantization. In the end, a comparison is made between the fixed-point and floating-point number representations as well as their associated effects on format conversion and algorithms. Chapter 3 describes methods for AD/DA conversion of signals, starting with Nyquist sampling, methods for oversampling techniques, and delta-sigma modulation. The chapter closes with a presentation of some circuit design of AD/DA converters. After an introduction to digital signal processors and digital audio interfaces, Chapter 4 describes simple hardware systems based on single- and multiprocessor solutions. The algorithms introduced in Chapters 5–9 are, to a great extent, implemented in real-time on the hardware platforms presented in Chapter 4. Chapter 5 describes digital audio equalizers. Apart from the implementation aspects of recursive audio filters, non-recursive linear phase filters based on fast convolution and filter banks are introduced. Filter designs, parametric filter structures, and precautions for reducing quantization errors in recursive filters are dealt with in detail. Chapter 6 deals with room simulation. Methods for simulation of artificial room impulse response and methods for approximation of measured impulse responses are discussed. In Chapter 7, the dynamic range control of audio signals is described. These methods are applied at several positions in the audio chain from the microphone to the loudspeakers to adapt to the dynamics of the recording, transmission, and listening environment. Chapter 8 contains a presentation of methods for synchronous and asynchronous sampling rate conversion. Efficient algorithms are described which are suitable for real-time processing as well as off-line processing. Both lossless and lossy audio coding are discussed in Chapter 9. Lossless audio coding is applied for the storage of higher word lengths. Lossy audio coding, however, plays a significant role in communication systems.

I would like to thank Prof. Fliege (University of Mannheim), Prof. Kammeyer (University of Bremen), and Prof. Heute (University of Kiel) for comments and support. I am also grateful to my colleagues at the TUHH and especially Dr. Alfred Mertins, Dr. Thomas Boltze, Dr. Bernd Redmer, Dr. Martin Schönle, Dr. Manfred Schusdziarra, Dr. Tanja Karp, Georg Dickmann, Werner Eckel, Thomas Scholz, Rüdiger Wolf, Jens Wohlers, Horst Zölzer, Bärbel Erdmann, Ursula Seifert, and Dieter Gödecke. Additionally, I would like to say a word of gratitude to all those students who helped me in carrying out this work successfully.

Special thanks go to Saeed Khawaja for his help during translation and to Dr. Anthony Macgrath for proof-reading the text. I also would like to thank Jenny Smith, Colin McKerracher, Ian Stoneham, and Christian Rauscher (Wiley).

My special thanks are directed to my wife Elke and my daughter Franziska.

Udo Zölzer
Hamburg, July 1997

Chapter 1

Introduction

U. Zölzer

In this first chapter, we will introduce the basics of signals and systems, and describe the transmission of signals through these systems [Opp14]. These fundamental concepts and the describing algorithms lay the foundation for digital audio signal processing. We will start with analog signals and analog systems, then we will sample the analog signals and perform digital signal processing, and finally reconstruct an analog output signal from the digital output signal. Figure 1.1 shows a typical audio application of capturing a vocalist and transmission to a loudspeaker via an amplifier for reproduction in another room for a listener or listening audience. The microphone delivers an electrical input signal $x(t)$ and the output signal $y(t)$ is the signal that will be received by the listener's ear. Both signals are continuous-time input and output signals. The entire chain of operations from microphone, amplifier, loudspeaker, and sound transmission through the listening room to the listener can be modeled by a system with a continuous-time impulse response $h(t)$. Such an impulse response can be acquired by an impulse response measurement approach. The entire continuous-time approach description can also be represented by a discrete-time approach through sampling the microphone signal $x(n)$, using the discrete-time impulse response $h(n)$, and then delivering the output signal $y(n)$. Both continuous-time and discrete-time signal-processing techniques [Opp10, Opp14] will be introduced in the following sections.

1.1 Continuous-time Signals and Convolution

Continuous-time signals $x(t)$, as shown in Fig. 1.2, can be used as test signals to analyze the behavior of the response of a physical system to an excitation signal. We need a few simple test signals that will allow for the derivation of all important relations to obtain the input/output

Digital Audio Signal Processing, Third Edition. Udo Zölzer.
© 2022 John Wiley & Sons Ltd. Published 2022 by John Wiley & Sons Ltd.

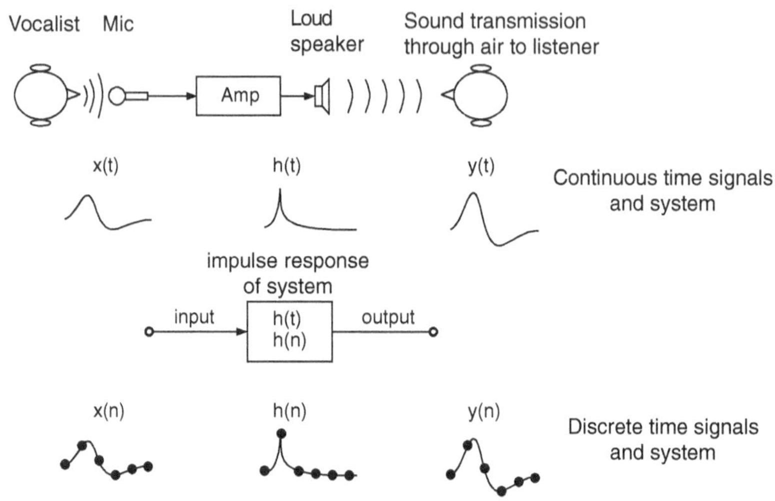

Figure 1.1 Audio capturing and reproduction for a listener, and representations of the operations by a signal and system model with input and output signals and by a system represented by an impulse response.

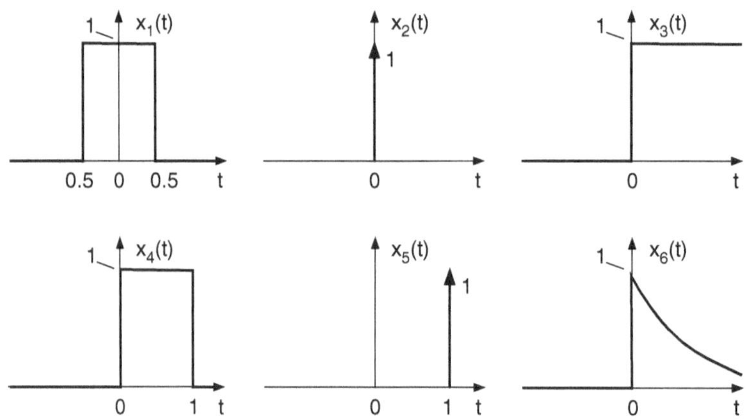

Figure 1.2 Continuous-time signals $x_1(t) = \text{rect}(t)$, $x_2(t) = \delta(t)$, $x_3(t) = \epsilon(t)$, $x_4(t) = x_2(t - 0.5)$, $x_5(t) = x_2(t - 1)$, and $x_6(t) = \exp(-at) \cdot \epsilon(t)$.

description of an input signal transformed to an output signal (handclap → acoustical transmission through room → received by human ear). The rectangular (rect) function is defined by

$$\text{rect}\left(\frac{t}{T}\right) = \begin{cases} T & \text{for} \quad |t| < \frac{T}{2}, \\ 0 & \text{for} \quad |t| > \frac{T}{2}. \end{cases} \tag{1.1}$$

The Dirac impulse $\delta(t)$ is defined by

$$\delta(t) = \lim_{T \to 0} \frac{1}{T} \text{rect}\left(\frac{t}{T}\right) \Rightarrow \int_{-\infty}^{\infty} \delta(t)\, dt = 1. \tag{1.2}$$

The step function is defined by

$$\epsilon(t) = \begin{cases} 0 & \text{for } t < 0, \\ 1 & \text{for } t > 0. \end{cases} \tag{1.3}$$

A general signal $x(t)$ can be written using the sampling property of the Dirac impulse as

$$x(t) = \int_{-\infty}^{\infty} x(\tau)\delta(t-\tau)\, d\tau. \tag{1.4}$$

Continuous-time systems transform the input $x(t)$ to the output $y(t) = \mathcal{T}\{x(t)\}$. A time-domain description can be given by the following signal flow graph: $x(t) \rightarrow \boxed{h(t)} \rightarrow y(t)$. The system parameter inside the box is called the *impulse response* $h(t) = \mathcal{T}\{\delta(t)\}$ of the system. It describes the output of the system $y(t)$ when the input is the Dirac $x(t) = \delta(t)$. Using Eq. (1.4), we can easily derive that the input/output relation of a system with impulse response $h(t)$ is given by the integral (sliding the folded impulse response $h(t-\tau)$ along the input and performing weighting and integration)

$$\boxed{y(t) = \mathcal{T}\{x(t)\} = \int_{-\infty}^{\infty} x(\tau)\cdot h(t-\tau)d\tau = \int_{-\infty}^{\infty} x(t-\tau)\cdot h(\tau)\, d\tau,} \tag{1.5}$$

which is called continuous-time convolution. The *convolution integral* describes a *filter operation* and is written as $y(t) = x(t) * h(t)$. *Causality* of a system implies $h(t) = 0$ for $t < 0$ and *stability* of a system is achieved if the integral of impulse response $\int_{-\infty}^{\infty} |h(t)|\, dt < M < \infty$. A simple example for continuous-time convolution is demonstrated in Fig. 1.3.

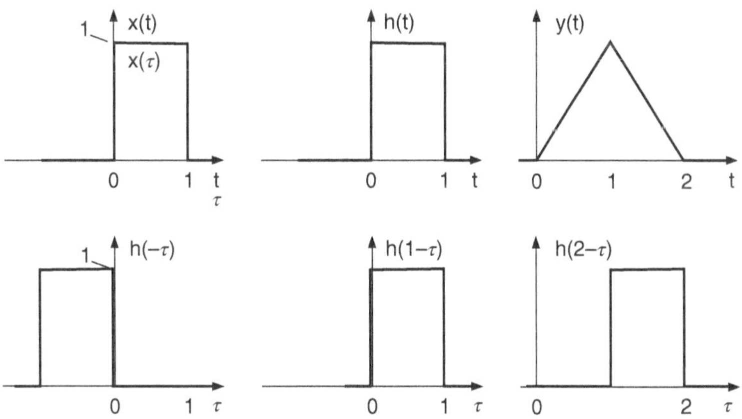

Figure 1.3 Continuous-time convolution $y(t) = \int x(\tau)h(t-\tau)\, d\tau$ showing the folded version of the impulse response $h(-\tau)$ and shifted versions $h(t-\tau)$ for $t = 0, 1, 2$.

Using the complex exponential $x(t) = e^{j\omega t}$ as input with $\omega = 2\pi f$, the output is given by the convolution integral as

$$y(t) = \int_{-\infty}^{\infty} e^{j\omega(t-\tau)}\cdot h(\tau)d\tau = e^{j\omega t}\cdot \underbrace{\int_{-\infty}^{\infty} h(\tau)\cdot e^{-j\omega\tau}d\tau}_{H(j\omega)\rightarrow\text{Fourier integral}} \tag{1.6}$$

$$= e^{j\omega t}\cdot H(j\omega) = e^{j\omega t}\cdot |H(j\omega)|e^{j\varphi(\omega)}. \tag{1.7}$$

This shows that for a exponential input $x(t)$, the output $y(t)$ is again an exponential signal where the input signal is weighted by the complex number $H(j\omega) = |H(j\omega)|e^{j\varphi(\omega)}$, which is the Fourier transform (integral) of the impulse response $h(t)$, and is also called the *frequency response* of a continuous-time system given by

$$\boxed{H(j\omega) = \int_{-\infty}^{\infty} h(t) \cdot e^{-j\omega t}\, dt.} \tag{1.8}$$

From $H(j\omega)$, we can compute the magnitude response

$$|H(j\omega)| = \sqrt{\mathrm{Re}^2\{H(j\omega)\} + \mathrm{Im}^2\{H(j\omega)\}} \tag{1.9}$$

and the phase response

$$\varphi(\omega) = \arctan\left(\frac{\mathrm{Im}\{H(j\omega)\}}{\mathrm{Re}\{H(j\omega)\}}\right), \mathrm{Re}\{H(j\omega)\} > 0 \tag{1.10}$$

of a continuous-time system. For a given signal $x(t)$, we can give its *continuous-time Fourier transform* as

$$\boxed{X(j\omega) = \int_{-\infty}^{\infty} x(t) \cdot e^{-j\omega t} dt \wedge X(f) = \int_{-\infty}^{\infty} x(t) \cdot e^{-j2\pi f t}\, dt.} \tag{1.11}$$

The *Fourier integral* describes a *spectral transform* from time domain $x(t)$ to frequency domain $X(j\omega)$, which is called the *Fourier spectrum* or *Fourier transform* of $x(t)$. The *inverse continuous-time Fourier transform* is given by

$$\boxed{x(t) = \frac{1}{2\pi}\int_{-\infty}^{\infty} X(j\omega) \cdot e^{j\omega t} d\omega \wedge x(t) = \int_{-\infty}^{\infty} X(f) \cdot e^{j2\pi f t}\, df,} \tag{1.12}$$

which takes the Fourier spectrum $X(j\omega)$ and reconstructs the input $x(t)$. In the following, useful Fourier transform pairs are listed in Eqs. (1.13)–(1.23). An important relation between the time domain $x(t) \rightarrow \boxed{h(t)} \rightarrow y(t)$, using $x(t)$ and $h(t)$ giving $y(t)$, and frequency domain $X(j\omega) \rightarrow \boxed{H(j\omega)} \rightarrow Y(j\omega)$, using $X(j\omega)$ and $H(j\omega)$ giving $Y(j\omega)$, shows that convolution in the time domain can be described by multiplication in the frequency domain.

Fourier Transform Pairs[1]

$$x(t) \circ\!\!-\!\!\bullet\, X(f) \tag{1.13}$$

$$x(t - t_0) \circ\!\!-\!\!\bullet\, X(f) \cdot e^{-j2\pi f t_0} \tag{1.14}$$

$$x(t) \cdot e^{j2\pi f_c t} \circ\!\!-\!\!\bullet\, X(f - f_c) \tag{1.15}$$

$$x(t) \cdot \cos(2\pi f_c t) \circ\!\!-\!\!\bullet\, \frac{1}{2}[X(f - f_c) + X(f + f_c)] \tag{1.16}$$

$$c_1 \cdot x_1(t) + c_2 \cdot x_2(t) \circ\!\!-\!\!\bullet\, c_1 \cdot X_1(f) + c_2 \cdot X_2(f) \tag{1.17}$$

$$x_1(t) * x_2(t) \circ\!\!-\!\!\bullet\, X_1(f) \cdot X_2(f) \tag{1.18}$$

[1] $\mathrm{sinc}(f) = \frac{\sin(\pi f)}{\pi f}$.

$$x(t) \rightarrow \boxed{h(t)} \rightarrow y(t) \circ\!\!-\!\!\bullet\; X(j\omega) \rightarrow \boxed{H(j\omega)} \rightarrow Y(j\omega) \tag{1.19}$$

$$\delta(t) \circ\!\!-\!\!\bullet\; 1 \tag{1.20}$$

$$\delta(t - t_0) \circ\!\!-\!\!\bullet\; e^{-j2\pi f t_0} \tag{1.21}$$

$$\mathrm{rect}\left(\frac{t}{T}\right) \circ\!\!-\!\!\bullet\; T \cdot \mathrm{sinc}(Tf) \tag{1.22}$$

$$f_S \cdot \mathrm{sinc}(f_S t) \circ\!\!-\!\!\bullet\; \mathrm{rect}\left(\frac{f}{f_S}\right) \tag{1.23}$$

Fourier Transforms of Even and Causal Signals

Figure 1.4 shows the Fourier transforms of even and causal rect signals and Fig. 1.5 shows two even sinc signals and their Fourier transforms. The ripple in the passband is based on the truncated length of the sinc signal.

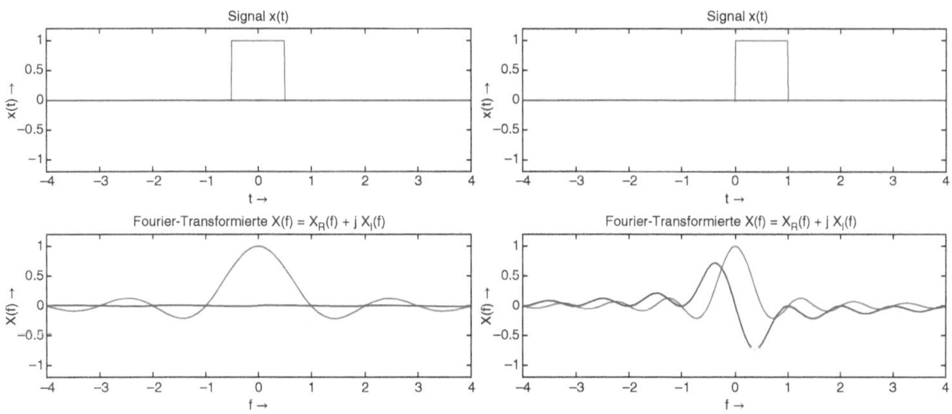

Figure 1.4 Fourier transforms of an even and a causal rect signal. The small imaginary part of the lower left plot arises from a small asymmetry of the rect signal in the upper left plot.

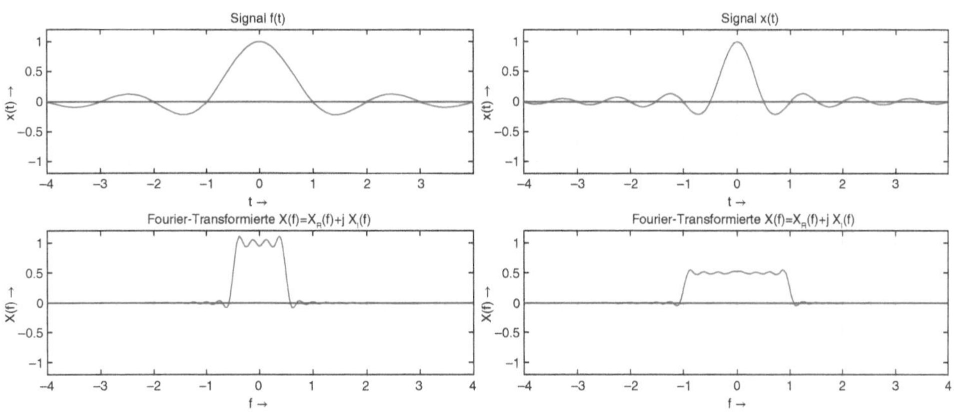

Figure 1.5 Fourier transforms of two even sinc signals.

1.2 Continuous-time Fourier Transform and Laplace Transform

The extension of the continuous-time Fourier transform to the Laplace transform allows for the transform of signals and impulse responses where the Fourier transform does not converge but the Laplace transform converges for a given convergence region. This extension of the continuous-time Fourier transform

$$\boxed{X(j\omega) = \int_{-\infty}^{\infty} x(t) \cdot e^{-j\omega t}\, dt} \tag{1.24}$$

is achieved by introducing a real part to the imaginary part according to a new complex variable $\boxed{s = \sigma + j\omega}$, which then gives

$$X(\sigma + j\omega) = \int_{-\infty}^{\infty} x(t) \cdot e^{-(\sigma+j\omega)t} dt \tag{1.25}$$

$$= \int_{-\infty}^{\infty} x(t) \cdot e^{-\sigma t} \cdot e^{-j\omega t}\, dt \tag{1.26}$$

and thus the Laplace transform

$$\boxed{X(s) = \int_{-\infty}^{\infty} x(t) \cdot e^{-st}\, dt.} \tag{1.27}$$

The Laplace transform of signals often leads to a rational function $X(s) = \frac{N(s)}{D(s)}$ with a numerator polynomial $N(s)$ and a denominator polynomial $D(s)$ in the variable s. The zeros of the numerator $N(s)$ are called the zeros of $X(s)$ and the zeros of $D(s)$ are called the poles of $X(s)$. The rational function $X(s)$ can be in given in polynomial, pole/zero, and partial expansion forms.

1.3 Sampling and Reconstruction

For digital signal processing, the sampling of $x(t)$ with a sampling rate $f_S = \frac{1}{T_S}$ and a sampling interval T_S is performed, which leads to a sequence of numbers $x(n)$ with time index n. According to the sampling theorem, the input signal $x(t)$ must be band limited to $f_S/2$. The sampling and the reconstruction of $x(t)$ from the number sequence $x(n)$ is achieved by the following sequence of operations: $x(t) \rightarrow \boxed{\text{ADC}} \rightarrow x(n) \rightarrow \boxed{\text{DAC}} \rightarrow x(t)$. Both operations are performed by an analog-to-digital converter (ADC) and a digital-to-analog converter (DAC). The converters can be considered as mixed continuous-time and discrete-time systems.

Sampling and quantization (analog-to-digital conversion) can be described by

$$x_d(t) = x(t) \cdot d(t) = x(t) \cdot \sum_{n=-\infty}^{\infty} \delta(t - nT_S) \tag{1.28}$$

$$= \sum_{n=-\infty}^{\infty} x_d(nT_S) \cdot \delta\left(t - nT_S\right), \tag{1.29}$$

$$x(n) = \left[x_d(nT_S)\right]_Q, \tag{1.30}$$

where the input $x(t)$ is sampled by multiplying it with a series of Dirac impulses $d(t) = \sum_{n=-\infty}^{\infty} \delta(t - nT_S)$ giving the ideal sampled $x_d(t)$ and then quantization of the samples $x_d(nT_S)$ to the sequence of numbers $x(n)$ with a finite number representation. Figure 1.6 shows in the left column the time-domain signals involved.

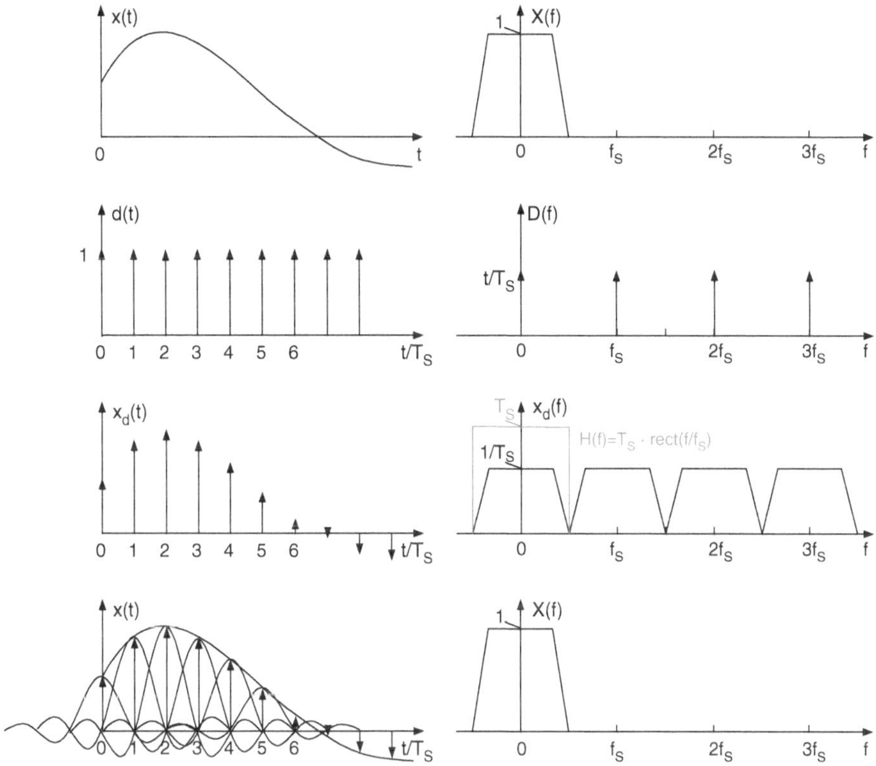

Figure 1.6 Sampling and reconstruction – Time-domain signals (left column) and corresponding Fourier spectra (right column).

Reconstruction (digital-to-analog conversion) of the continuous-time $x(t)$ from the sampled sequence $x(n)$ can be written as a convolution operation given by

$$x(t) = \sum_{n=-\infty}^{\infty} x(n) \cdot \operatorname{sinc}\left(f_S(t - nT_S)\right), \tag{1.31}$$

which is shown in the bottom left plot of Fig. 1.6. The **Fourier transforms** of the individual signals for sampling are given by

$$x(t) \circ\!\!-\!\!\bullet\, X(f) \tag{1.32}$$

$$d(t) = \sum_{n=-\infty}^{\infty} \delta(t - nT_S) \circ\!\!-\!\!\bullet\, D(f) = \frac{1}{T_S} \sum_{n=-\infty}^{\infty} \delta(f - nf_S) \tag{1.33}$$

$$x_d(t) = x(t) \cdot d(t) \circ\!\!-\!\!\bullet\, X_d(f) = X(f) * D(f) \tag{1.34}$$

$$= \frac{1}{T_S} \sum_{n=-\infty}^{\infty} X(f - nf_S) \tag{1.35}$$

and are shown in Fig. 1.6 (right column). Sampling leads to the periodic extension of the baseband spectrum at multiples of the sampling rate f_S and the scaling by $1/T_S$. The reconstruction of

$$x(t) = x_d(t) * h(t) \circ\!\!-\!\!\bullet X(f) = X_d(f) \cdot H(f) \tag{1.36}$$

$$= X_d(f) \cdot T_S \cdot \mathrm{rect}\left(\frac{f}{f_S}\right) \tag{1.37}$$

is achieved by convolution with a system impulse response $h(t)$, which acts an ideal lowpass filter with cutoff frequency $f_S/2$ and passband gain of T_S to compensate for the scaling of the sampling operation, as shown in the bottom plots of Fig. 1.6.

1.4 Discrete-time Signals and Convolution

With the discrete-time input signal $x(n)$, we can describe an input/output relation in the discrete-time domain, in signal flow graph form, as $x(n) \rightarrow \boxed{h(n)} \rightarrow y(n) = \mathcal{T}\{x(n)\}$, where the input is transformed into an output signal $y(n)$. Again we are using discrete-time test signals represented in Fig. 1.7, such as the unit pulse

$$\delta(n) = \begin{cases} 1 & \text{for} \quad n = 0, \\ 0 & \text{for} \quad n \neq 0, \end{cases} \tag{1.38}$$

and the step sequence

$$\epsilon(n) = \begin{cases} 1 & \text{for} \quad n \geq 0, \\ 0 & \text{for} \quad n < 0. \end{cases} \tag{1.39}$$

The lower row in Fig. 1.7 shows further discrete-time signals which can be formulated using delayed unit pulses $\delta(n-k)$ according to

$$x(n) = \sum_k x(k) \cdot \delta(n-k) \tag{1.40}$$

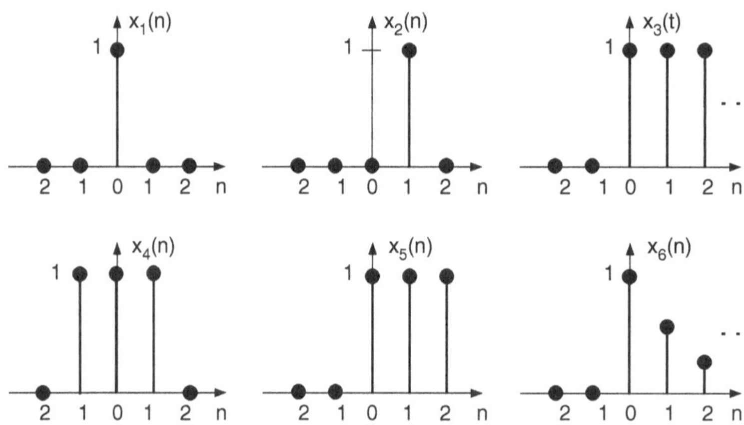

Figure 1.7 Discrete-time signals $x_1(n) = \delta(n)$, $x_2(n) = \delta(n-1)$, $x_3(t) = \epsilon(n)$, $x_4(n) = \delta(n+1) + \delta(n) + \delta(n-1)$, $x_5(n) = x_4(n-1)$, and $x_6(n) = \exp(-an) \cdot \epsilon(n)$.

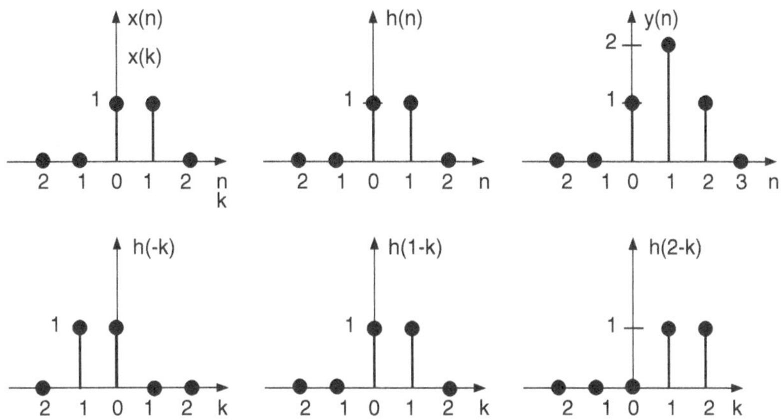

Figure 1.8 Discrete-time convolution $y(n) = \sum_k x(k)h(n-k)$ showing the folded version of the impulse response $h(-k)$ and shifted versions $h(n-k)$ for $n = 0, 1, 2$.

and explicitly as, for example, $x_6(n) = \exp(-an) \cdot \epsilon(n)$. The discrete-time system described by $y(n) = \mathcal{T}\{x(n)\}$ leads to the definition of the impulse response $h(n) = \mathcal{T}\{\delta(n)\}$ of the discrete-time system when the unit pulse $\delta(n)$ is used as an input signal $x(n)$. Using Eq. (1.40), we can derive the discrete-time convolution as

$$\boxed{y(n) = \mathcal{T}\{x(n)\} = \sum_{k=-\infty}^{\infty} x(k) \cdot h(n-k) = \sum_{k=-\infty}^{\infty} x(n-k) \cdot h(n).} \tag{1.41}$$

The *convolution sum* describes a *digital filter operation* and is written as $y(n) = x(n) * h(n)$. The discrete-time convolution is illustrated in Fig. 1.8. *Causality* of a discrete time system implies $h(n) = 0$ for $n < 0$ and *stability* of a discrete-time system is achieved if $\sum_n |h(n)| < M < \infty$.

Using the complex exponential $x(n) = e^{j\Omega n}$ as input with $\Omega = 2\pi \frac{f}{f_S}$, the output is given by the convolution sum as

$$y(n) = \sum_{k=-\infty}^{\infty} e^{j\Omega(n-k)} \cdot h(k) = e^{j\Omega n} \cdot \underbrace{\sum_k h(k) \cdot e^{-j\Omega k}}_{H(e^{j\Omega})}$$

$$= e^{j\Omega n} \cdot H(e^{j\Omega}) = e^{j\Omega n} \cdot |H(e^{j\Omega n})| e^{j\varphi(\Omega)}.$$

This shows that for an exponential input $x(n)$, the output $y(n)$ is again an exponential signal where the input signal is weighted by the complex number $H(e^{j\Omega}) = |H(e^{j\Omega})|e^{j\varphi(\Omega)}$, which is the discrete-time Fourier transform of the impulse response $h(n)$, and is also called the *frequency response*

$$\boxed{H(e^{j\Omega}) = \sum_{n=-\infty}^{\infty} h(n) \cdot e^{-j\Omega n}} \tag{1.42}$$

of a discrete-time system. From $H(e^{j\Omega})$, we can compute the magnitude response

$$|H(e^{j\Omega})| = \sqrt{\mathrm{Re}^2\{H(e^{j\Omega})\} + \mathrm{Im}^2\{H(e^{j\Omega})\}} \tag{1.43}$$

and the phase response

$$\varphi(\Omega) = \arctan\left(\frac{\mathrm{Im}\{H(e^{j\Omega})\}}{\mathrm{Re}\{H(e^{j\Omega})\}}\right), \mathrm{Re}\{H(e^{j\Omega})\} > 0 \tag{1.44}$$

of a discrete-time system. For a given signal $x(n)$, we can give its *discrete-time Fourier transform* as

$$\boxed{X(e^{j\Omega}) = \sum_{n=-\infty}^{\infty} x(n) \cdot e^{-j\Omega n}.} \tag{1.45}$$

The *discrete-time Fourier Transform* describes a *spectral transform* from the discrete-time domain $x(n)$ to the frequency domain $X(e^{j\Omega})$, which is called the *Fourier spectrum* or *Fourier transform* of $x(n)$. The *inverse discrete-time Fourier transform* is given by

$$\boxed{x(n) = \frac{1}{2\pi}\int_{-\pi}^{\pi} X(e^{j\Omega}) \cdot \cdot e^{j\Omega n} d\Omega,} \tag{1.46}$$

which takes the discrete-time Fourier spectrum $X(e^{j\Omega})$ and reconstructs the input $x(n)$. In the following, useful discrete-time Fourier transform pairs are listed in Eqs. (1.47)–(1.56). An important relation between the time domain $x(n) \rightarrow \boxed{h(n)} \rightarrow y(n)$, using $x(n)$ and $h(n)$ giving $y(n)$, and the frequency domain $X(j\omega) \rightarrow \boxed{H(j\omega)} \rightarrow Y(e^{j\Omega})$, using $X(e^{j\Omega})$ and $H(e^{j\Omega})$ giving $Y(e^{j\Omega})$, shows that convolution in the time domain can be described by multiplication in the frequency domain.

Discrete-time Fourier Transform Pairs

$$x(n) \circ\!\!-\!\!\bullet\, X(e^{j\Omega}) \tag{1.47}$$

$$x(n-m) \circ\!\!-\!\!\bullet\, X(e^{j\Omega})e^{-jm\Omega} \tag{1.48}$$

$$x(n)e^{j\Omega_0 n} \circ\!\!-\!\!\bullet\, X(e^{j(\Omega-\Omega_0)}) \tag{1.49}$$

$$x_1(n) * x_2(n) \circ\!\!-\!\!\bullet\, X_1(e^{j\Omega}) \cdot X_2(e^{j\Omega}) \tag{1.50}$$

$$x(n) \rightarrow \boxed{h(n)} \rightarrow y(n) \circ\!\!-\!\!\bullet\, X(e^{j\Omega}) \rightarrow \boxed{H(e^{j\Omega})} \rightarrow Y(e^{j\Omega}) \tag{1.51}$$

$$\delta(n) \circ\!\!-\!\!\bullet\, 1 \tag{1.52}$$

$$\delta(n-m) \circ\!\!-\!\!\bullet\, e^{-jm\Omega} \tag{1.53}$$

$$\delta(n) + \delta(n-1) \circ\!\!-\!\!\bullet\, 1 + e^{-j\Omega} = 1 + \cos\Omega - j\sin\Omega \tag{1.54}$$

$$\sum_{k=0}^{N-1} \delta(n-k) \circ\!\!-\!\!\bullet\, \sum_{k=0}^{N-1} e^{-jk\Omega} \tag{1.55}$$

$$\sum_{k=0}^{N-1} h(k)\delta(n-k) \circ\!\!-\!\!\bullet\, \sum_{k=0}^{N-1} h(k)e^{-jk\Omega} \tag{1.56}$$

1.5 Discrete-time Fourier Transform and Z-Transform

The extension of the discrete-time Fourier transform to the Z-transform allows for the transform of signals and impulse responses where the discrete-time Fourier transform does not converge but the Z-transform converges for a given convergence region. This extension of the discrete-time Fourier transform

$$\boxed{X(e^{j\Omega}) = \sum_{n=-\infty}^{\infty} x(n) \cdot e^{-j\Omega n}} \tag{1.57}$$

is achieved by introducing a radius r to the complex exponential $e^{j\Omega}$ according to a new complex variable $\boxed{z = re^{j\Omega}}$, which then gives

$$X(re^{j\Omega}) = \sum_{n=-\infty}^{\infty} x(n) \cdot (re^{j\Omega})^{-n} \tag{1.58}$$

$$= \sum_{n=-\infty}^{\infty} x(n) \cdot r^{-n} \cdot e^{-j\Omega n} \tag{1.59}$$

and thus the Z-transform

$$\boxed{X(z) = \sum_{n=-\infty}^{\infty} x(n) \cdot z^{-n}.} \tag{1.60}$$

The Z-transform of many signals leads to a rational function $X(z) = \frac{N(z)}{D(z)}$, with a numerator polynomial $N(z)$ and a denominator polynomial $D(z)$ in the variable z. The zeros of the numerator $N(z)$ are called the zeros of $X(z)$ and the zeros of $D(z)$ are called the poles of $X(z)$. The rational function $X(z)$ can be given in polynomial, pole/zero, and partial expansion forms.

1.6 Discrete Fourier Transform

To reduce the the length of the input signal $x(n)$ to N samples and to compute the Fourier spectrum only at a finite number of N frequency samples, we can discretize the unit circle $\Omega_k = \frac{2\pi}{N}k$ at N frequency samples, and using N samples, $x(n)$ gives

$$X(e^{j\Omega_k}) = \sum_{n=0}^{N-1} x(n)\, e^{-j\Omega_k n}, \quad \Omega_k = \frac{2\pi}{N}k, \quad k = 0,1,\ \dots\ ,N-1 \tag{1.61}$$

and the discrete Fourier transform (DFT)

$$\boxed{\begin{array}{c} X(k) = \sum_{n=0}^{N-1} x(n)\, e^{-j\frac{2\pi}{N}kn} \\ k = 0,1,\ \dots\ ,N-1 \end{array}} \tag{1.62}$$

and the inverse discrete Fourier transform (IDFT)

$$\boxed{\begin{array}{c} x(n) = \frac{1}{N}\sum_{k=0}^{N-1} X(k)\, e^{j\frac{2\pi}{N}kn} \\ n = 0,1,\ \dots\ ,N-1. \end{array}} \tag{1.63}$$

The DFT of a causal audio signal is shown in Fig. 1.9.

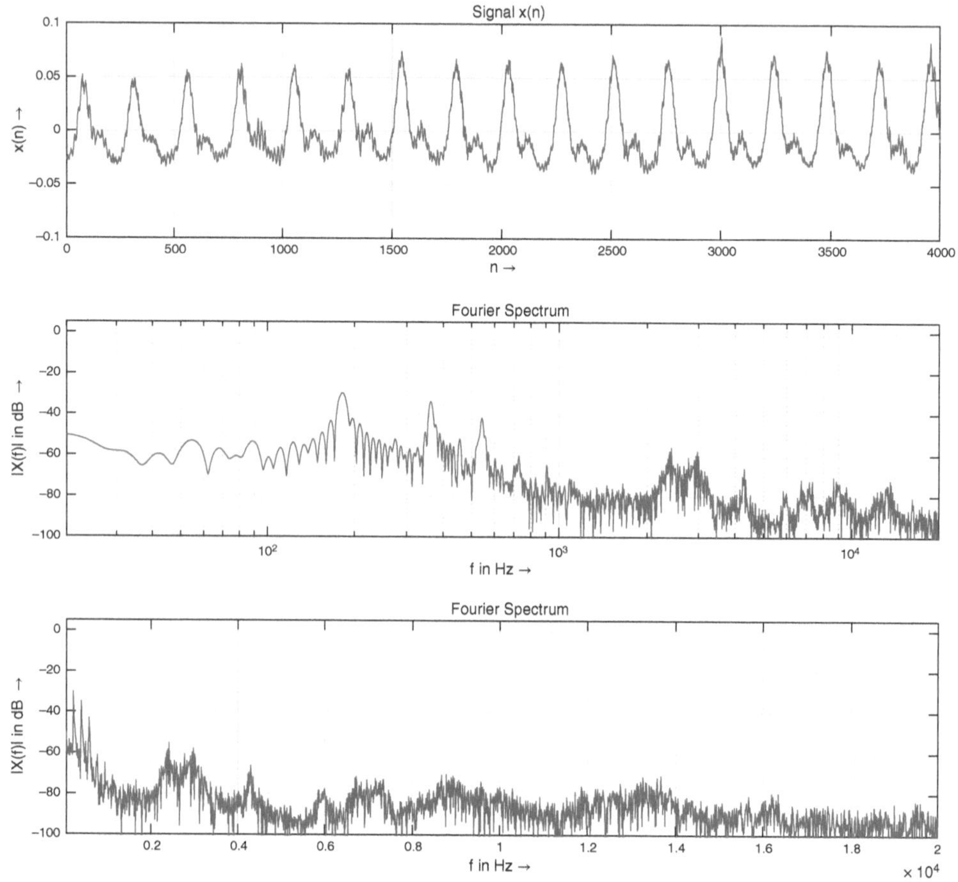

Figure 1.9 Fourier transform of an audio signal $x(n)$ (top plot) and its magnitude spectrum $|X(f)|$ in dB versus logarithmic (middle plot) and linear frequency axis (bottom plot).

1.7 FIR and IIR Filters

For causal signals $x(n)$ and causal impulse responses $h(n)$, the convolution sum is given by $y(n) = \sum_{k=0}^{\infty} h(k) \cdot x(n-k) = \sum_{k=0}^{\infty} x(k) \cdot h(n-k)$. A further simplification is possible if the impulse response is decaying very rapidly and can be truncated in length after N samples, which then leads to a **finite impulse response (FIR)** filter with N taps (Matlab file y=filter(b,1,x)) and the discrete-time convolution

$$\boxed{y(n) = \sum_{k=0}^{N-1} h(k) \cdot x(n-k) = \sum_{k=0}^{N-1} b(k) \cdot x(n-k),} \tag{1.64}$$

which is also called a difference equation. The difference equation can be represented by an FIR signal flow graph (block diagram), as shown in Fig. 1.10. The coefficients $b(k)$ correspond to the impulse response $h(k)$. Deriving the Z-transform of Eq. (1.64) gives

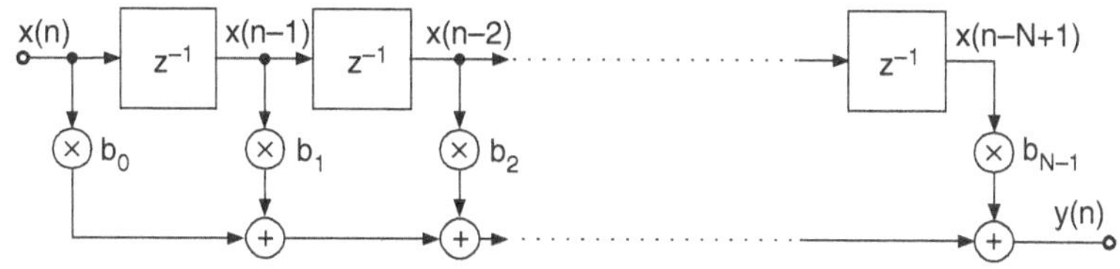

Figure 1.10 Block diagram of FIR filter.

$$Y(z) = \sum_{k=0}^{N-1} b(k) \cdot z^{-k} \cdot X(z), \tag{1.65}$$

from which we can derive the system transfer function $H(z) = \frac{Y(z)}{X(z)}$, which gives

$$H(z) = \sum_{k=0}^{N-1} b(k) \cdot z^{-k} = \frac{\sum_{k=0}^{N-1} b(k) \cdot z^{N-1-k}}{z^{N-1}}. \tag{1.66}$$

The frequency response of the FIR filter (Matlab file freqz(b,1)) can be derived by taking the discrete-time Fourier transform of the impulse response or evaluating the system transfer function on the unit circle by replacing $z \to e^{j\Omega}$, which gives

$$H_{\text{FIR}}(e^{j\Omega}) = \sum_{k=0}^{N-1} b(k) \cdot e^{-j\Omega k}, \quad \Omega = 2\pi \frac{f}{f_S}. \tag{1.67}$$

Figure 1.11 shows a symmetric FIR impulse response with five taps, the magnitude response, phase response, and the pole/zero plot.

If the impulse response is decaying infinitely long, we have an **infinite impulse response (IIR)** filter of order N (Matlab file y=filter(b,[1 a],x)) which is dependent on its input $x(n)$ and N delayed versions $x(n-k)$, weighted by the coefficients $b(k)$, and N delayed versions $y(n-k)$ of the output $y(n)$, weighted by the coefficients $a(k)$ according to

$$\boxed{y(n) = \sum_{k=0}^{N} b(k) \cdot x(n-k) - \sum_{k=1}^{N} a(k) \cdot y(n-k)}, \tag{1.68}$$

which is also called a difference equation. The difference equation can be represented by an IIR signal flow graph (block diagram), as shown in Fig. 1.12. Performing the Z-transform of the recursive difference equation in Eq. (1.68) gives

$$Y(z) = \sum_{k=0}^{N} b(k) \cdot z^{-k} \cdot X(z) - \sum_{k=1}^{N} a(k) \cdot z^{-k} \cdot Y(z), \tag{1.69}$$

from which we can derive the system transfer function $H(z) = \frac{Y(z)}{X(z)}$ by deriving

$$H(z) = \frac{\sum_{k=0}^{N} b(k) \cdot z^{-k}}{1 + \sum_{k=1}^{N} a(k) \cdot z^{-k}}. \tag{1.70}$$

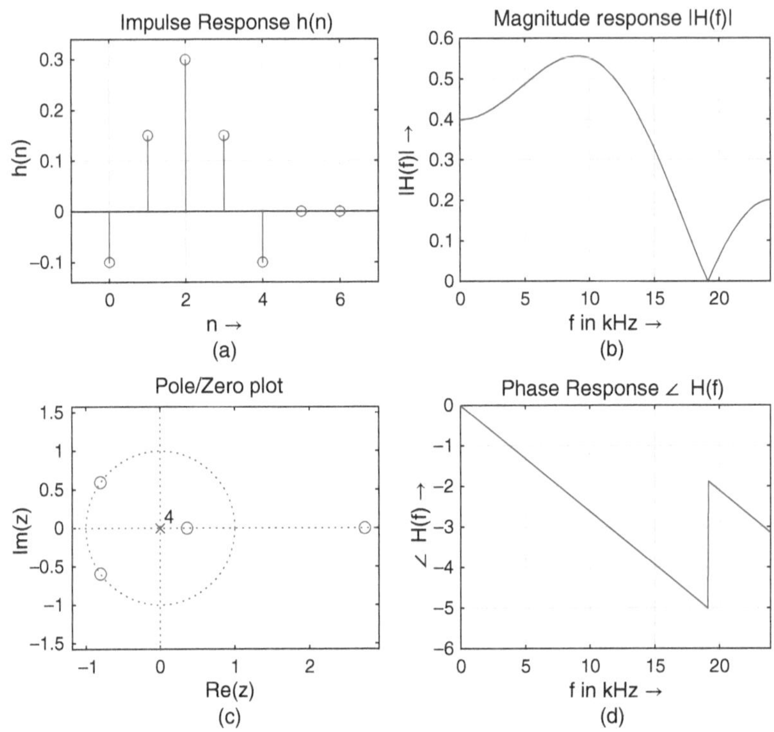

Figure 1.11 FIR filter impulse response, magnitude response, phase response, and pole/zero plot.

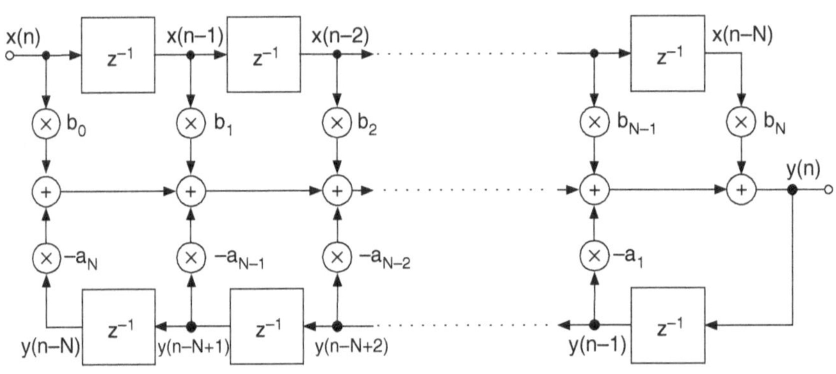

Figure 1.12 Block diagram of an IIR filter.

Typically, $N = 1$ or $N = 2$, and for higher order, we can use a cascade of first- and second-order sub-filters. Figure 1.13 shows a signal flow graph for an IIR filter cascade. The frequency response of the IIR filter (freqz(b,[1 a])) can be derived by evaluating the system transfer function on the unit circle by replacing $z \rightarrow e^{j\Omega}$, which gives the frequency response of an IIR filter of order N

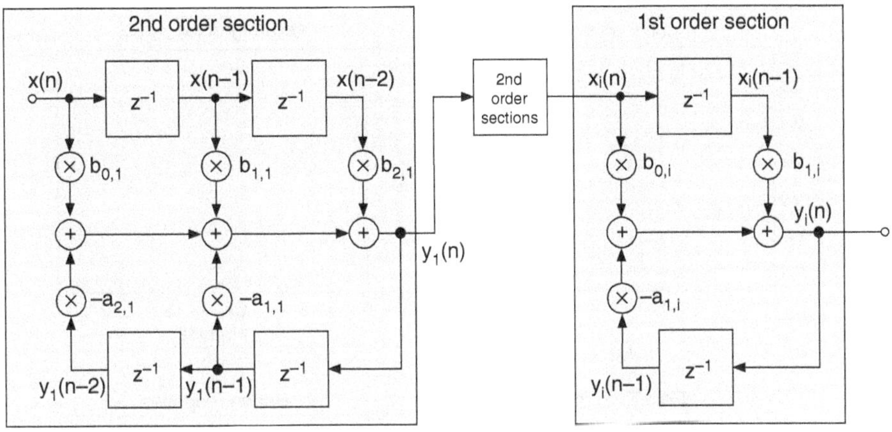

Figure 1.13 IIR filter cascade.

(freqz(b,[1 a])) according to

$$H_{\mathrm{IIR}}(e^{j\Omega}) = \frac{\sum_{k=0}^{N} b(k)\cdot e^{-j\Omega k}}{1+\sum_{k=1}^{N} a(k)\cdot e^{-j\Omega k}}, \qquad \Omega = 2\pi\frac{f}{f_S}. \tag{1.71}$$

Figure 1.14 shows an IIR impulse response with infinite decay, the magnitude response, phase response, and the pole/zero plot.

Some specific FIR filters are the *N-tap moving average filter*,

$$h_{\mathrm{MA}}(n) = \frac{1}{N}\sum_{k=0}^{N-1}\delta(n-k), \tag{1.72}$$

the *triangular filter*,

$$h_{\mathrm{TRI}}(n) = h_{\mathrm{MA}}(n) * h_{\mathrm{MA}}(n), \tag{1.73}$$

and the equi-ripple *Parks–McClellan filter*,

$$h_{\mathrm{PM}}(n) = \sum_{k=0}^{N-1} h(k)\delta(n-k), \tag{1.74}$$

where the impulse response is computed based on the Parks–McClellan [Opp14] design approach. *Complementary filters*, according to

$$h_{\mathrm{C}}(n) = \delta\left(n-\frac{N-1}{2}\right) - \sum_{k=0}^{N-1} h(k)\delta(n-k), \tag{1.75}$$

allow for the use of a lowpass FIR filter of odd length N to perform lowpass-to-highpass and bandpass-to-bandstop transformations. Some frequency responses of these FIR filters are shown in Fig. 1.15.

Some simple but effective **first- and second-order IIR filters** for lowpass (LP), highpass (HP), and bandpass (BP) filtering are given by

$$H_{\mathrm{LP}}(z) = \frac{1-a}{2}\,\frac{1+z^{-1}}{1-az^{-1}}, \tag{1.76}$$

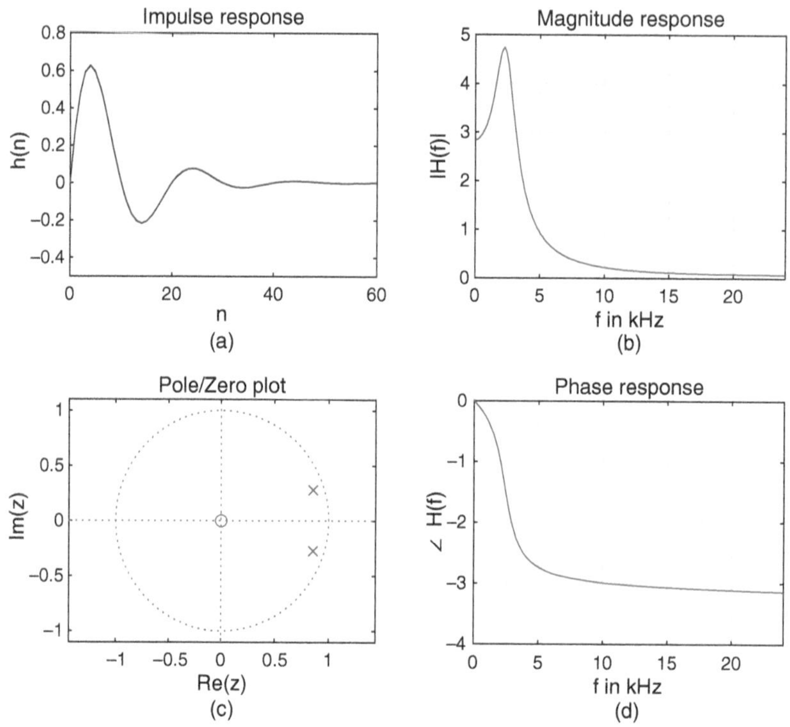

Figure 1.14 IIR filter and impulse response, magnitude response, phase response, and pole/zero plot.

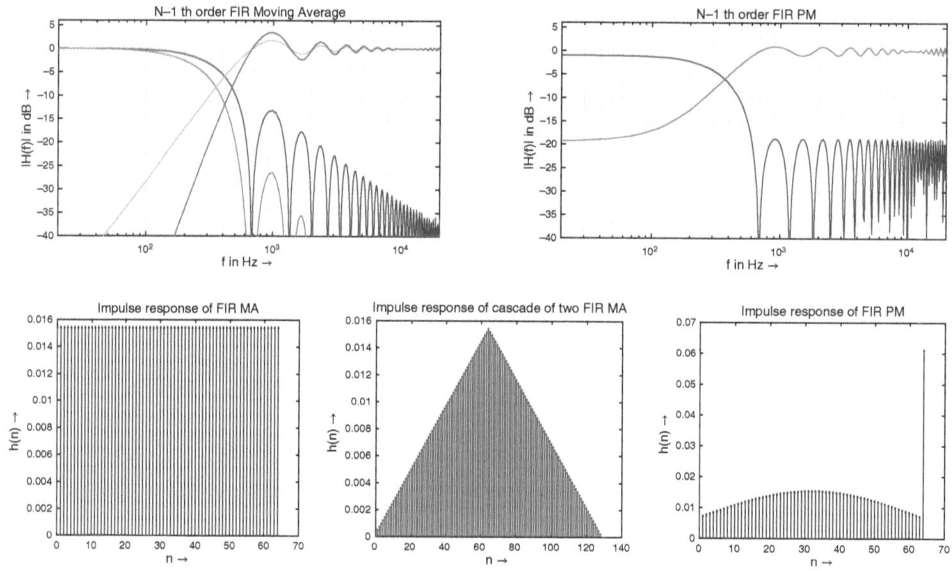

Figure 1.15 Frequency responses and impulse responses of a moving average FIR and a Parks–McClellan FIR filter. In the two frequency response plots, a complementary frequency response is also shown.

$$a = \frac{1 - \sin(2\pi f_c/f_S)}{\cos(2\pi f_c/f_S)}. \tag{1.77}$$

$$H_{HP}(z) = \frac{1+a}{2}\frac{1 - z^{-1}}{1 - az^{-1}}, \tag{1.78}$$

$$a = \frac{1 - \sin(2\pi f_c/f_S)}{\cos(2\pi f_c/f_S)}. \tag{1.79}$$

$$H_{BP}(z) = \frac{1-a}{2}\frac{1 - z^{-2}}{1 - b(1+a)z^{-1} + az^{-2}}, \tag{1.80}$$

$$a = \frac{1 - \sin(2\pi f_b/f_S)}{\cos(2\pi f_b/f_S)}, \tag{1.81}$$

$$b = \cos(2\pi f_c/f_S). \tag{1.82}$$

They all have one (LP, HP) or two coefficients (BP) for controlling their cutoff frequency f_c in Hz (−3-dB magnitude of LP or HP frequency response) and the BP bandwidth f_b in Hz (lower and higher than −3-dB magnitude of BP frequency response). Their frequency responses are shown in Fig. 1.16. These simple first- and second-order IIR filters (LP, HP, and BP) can be further implemented using first- and second-order allpass filters $H_{AP_1}(z)$ and $H_{AP_2}(z)$, as shown in the

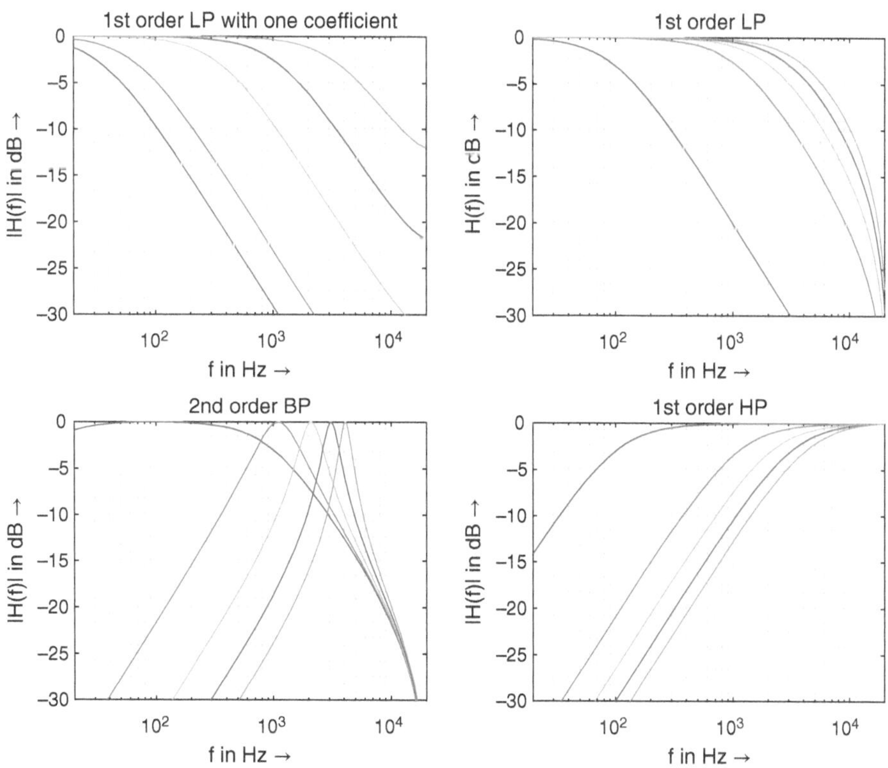

Figure 1.16 Frequency response of IIR filter.

sequence of system transfer functions given by

$$H_{AP_1}(z) = \frac{-a + z^{-1}}{1 - az^{-1}}, \tag{1.83}$$

$$H_{AP_2}(z) = \frac{a - b(1+a)z^{-1} + z^{-2}}{1 - b(1+a)z^{-1} + az^{-2}}, \tag{1.84}$$

$$H_{LP}(z) = \frac{1}{2}(1 + H_{AP_1}(z)), \tag{1.85}$$

$$H_{HP}(z) = \frac{1}{2}(1 - H_{AP_1}(z)), \tag{1.86}$$

$$H_{BP}(z) = \frac{1}{2}(1 - H_{AP_2}(z)), \tag{1.87}$$

where the coefficients a and b are given by Eq. (1.81) and Eq. (1.82), respectively. The lowpass transfer function in Eq. (1.85) is an additive parallel connection of direct path and a first-order allpass of Eq. (1.83). The highpass transfer function in Eq. (1.86) is a subtractive parallel connection of direct path and a first-order allpass of Eq. (1.83). Finally, the bandpass transfer function in Eq. (1.87) is a subtractive parallel connection of direct path and a first-order allpass of Eq. (1.84).

1.8 Adaptive Filters

Adaptive filters change their filter characteristic according to a predefined error criterion. A generic structure is shown in Fig. 1.17 where a desired signal $d(n)$ is approximated by a reference signal $x(n)$ passing through an adaptive filter giving an output signal $y(n)$, which should be made close to $d(n)$ such that the error signal $e(n)$ between the desired signal $d(n)$ and the output signal $y(n)$ gets smaller with time. The coefficients of the adaptive filter are derived by minimizing the error energy. Based on the generic structure shown in Fig. 1.17, special adaptive filter approaches for system identification, inverse filtering, echo cancellation, and linear prediction can be achieved. We will show how to perform linear prediction as an adaptive FIR filter technique and derive the adaption of the filter coefficients by the autocorrelation method and the least-mean-square (LMS) method.

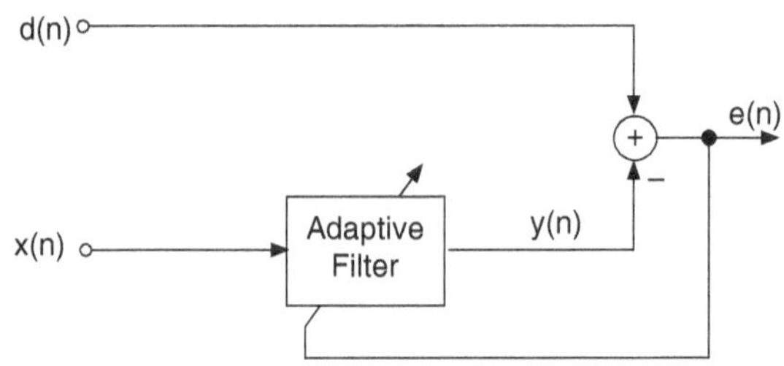

Figure 1.17 Adaptive signal approximation.

Adaptive Linear Prediction using the Autocorrelation Method

For the task of adaptive linear prediction, the generic structure is simplified because the desired signal $d(n)$ is equal to the input $x(n)$, which leads to the signal flow graph of linear prediction

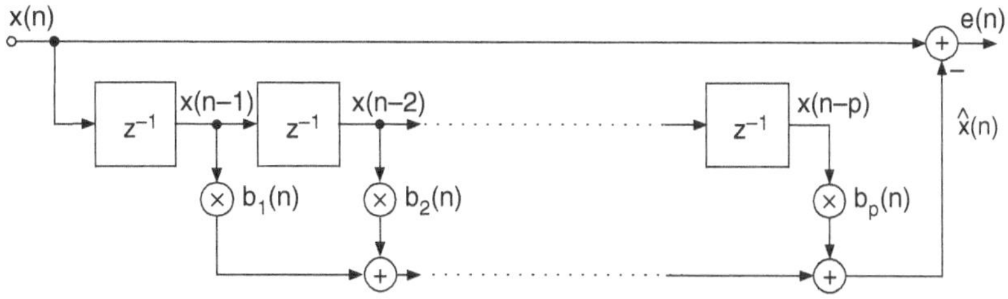

Figure 1.18 Linear prediction.

shown in Fig. 1.18. The adaptive filter is, in that case, an FIR filter with p coefficients. Linear prediction computes an estimate

$$\hat{x}(n) = \sum_{k=1}^{p} b(k)x(n-k) = \sum_{k=1}^{p} b_k x(n-k) \tag{1.88}$$

of the actual sample $x(n)$ by a linear combination of past weighted input samples. Here p is the prediction order and b_k are the prediction coefficients. The difference between $x(n)$ and its prediction $\hat{x}(n)$ is called the *prediction error* and is given by

$$e(n) = x(n) - \hat{x}(n) = x(n) - \sum_{k=1}^{p} b_k x(n-k). \tag{1.89}$$

The filter coefficients b_k are computed by minimizing a cost function

$$J = E\{e^2(n)\} \tag{1.90}$$

and thus minimizing the quadratic error[2]. The partial derivatives of the cost function with respect to the filter coefficients b_k (for $k = 1, \ldots p$) lead to

$$\frac{\partial J}{\partial b_k} = 2E\left\{e(n) \cdot \frac{\partial e(n)}{\partial b_k}\right\} = -2E\{e(n)x(n-k)\} \tag{1.91}$$

$$= -2E\left\{\left[x(n) - \sum_{i=1}^{p} b_i x(n-i)\right] x(n-k)\right\} \tag{1.92}$$

$$= -2E\{x(n)x(n-k)\} + 2\sum_{i=1}^{p} b_i E\{x(n-i)x(n-k)\}. \tag{1.93}$$

Computing the filter coefficients b_k by setting the partial derivative to zero gives, for $k = 1, \ldots, p$,

$$\frac{\partial J}{\partial b_k} = 0 \tag{1.94}$$

$$\Leftrightarrow \sum_{i=1}^{p} b_i E\{x(n-i)x(n-k)\} = E\{x(n)x(n-k)\}. \tag{1.95}$$

By using the autocorrelation sequence

$$r_{xx}(k) = E\{x(n)x(n-k)\} = \sum_{n=-\infty}^{\infty} x(n)x(n-k) \tag{1.96}$$

[2] Expectation $E\{e^2(n)\} = \sum_n e^2(n)$.

and the properties

$$r_{xx}(k-i) = E\{x(n)x(n-(k-i))\}$$
$$= E\{x(n-i)x(n-k))\} \tag{1.97}$$

$$\text{and} \quad r_{xx}(k) = r_{xx}(-k) \quad \text{(even sequence)} \tag{1.98}$$

leads to the set of equations (1.95) in matrix form as

$$\begin{pmatrix} r_{xx}(0) & r_{xx}(1) & r_{xx}(2) & \cdots & r_{xx}(p-1) \\ r_{xx}(1) & r_{xx}(0) & r_{xx}(1) & \cdots & r_{xx}(p-2) \\ r_{xx}(2) & r_{xx}(1) & r_{xx}(0) & \ddots & \vdots \\ \vdots & \vdots & \ddots & \ddots & r_{xx}(1) \\ r_{xx}(p-1) & r_{xx}(p-2) & \cdots & r_{xx}(1) & r_{xx}(0) \end{pmatrix} \begin{pmatrix} b_1 \\ b_2 \\ \vdots \\ b_p \end{pmatrix} = \begin{pmatrix} r_{xx}(1) \\ r_{xx}(2) \\ \vdots \\ r_{xx}(p) \end{pmatrix}. \tag{1.99}$$

These equations are called *Wiener–Hopf equations* or *normal equations*. The so-called autocorrelation matrix **R** is symmetric and has identical values on each diagonal. It is a *Toeplitz matrix* and the set of equations can be solved efficiently by the *Levinson–Durbin recursion*. The described procedure is called the autocorrelation method for updating the prediction coefficients.

Adaptive Linear Prediction using the LMS Method

In this method, a recursive algorithm for minimizing a simplified cost function $\hat{J}$ is applied by minimizing the LMS error. All filter coefficients will be updated every sampling cycle based on previous values. This recursive approach for the filter coefficients uses the gradient descent formula

$$b_k(n+1) = b_k(n) - \frac{1}{2}\mu(n)\frac{\partial J}{\partial b_k(n)} \tag{1.100}$$

with gradient weight $\mu(n)$. Using the simplified cost function (instantaneous value of the quadratic error)

$$\hat{J} = e^2(n) \quad \text{and} \quad e(n) = x(n) - \sum_{k=1}^{p} b_k(n)x(n-k) \tag{1.101}$$

leads to

$$\frac{\partial \hat{J}}{\partial b_k(n)} = \frac{\partial e^2(n)}{\partial b_k(n)} = 2e(n)\cdot\frac{\partial e(n)}{\partial b_k} = -2e(n)x(n-k) \tag{1.102}$$

and thus, for $k = 1, \ldots, p$, the recursive LMS algorithm

$$\boxed{b_k(n+1) = b_k(n) + \mu(n)e(n)x(n-k).} \tag{1.103}$$

The gradient weight μ can be adjusted using the different LMS approaches listed in Table 1.1.

Linear Prediction for Coding and Source-filter Processing

Linear prediction can be found in coding and source-filter processing which have the same signal flow graph of operation, as shown in Fig. 1.19. The first part is the coding operation and consists of a forward prediction and coefficient calculation using the autocorrelation or LMS method. The coder transfer function is given by

$$H_C(z) = 1 - P(z). \tag{1.104}$$

Table 1.1 Computing the gradient weight $\mu(n)$.

Standard LMS	$\mu(n) = \text{const.}$
Normalized LMS (NLMS)	$\mu(n) = \dfrac{\alpha}{\beta + \sum_{k=0}^{N-1} x^2(n-k)}$
	β takes care of large $\mu(n)$
Power Normalized (PNLMS)	$\mu(n) = \dfrac{\alpha}{N \cdot \sigma_x^2(n)}$
- Limitation:	$\mu(n) = \min\{\mu(n), \mu_{max}\}$
- Recursive computation:	$\sigma_x^2(n) = \lambda\sigma_x^2(n-1) + (1-\lambda)x^2(n)$,
	$0 < \lambda < 1$

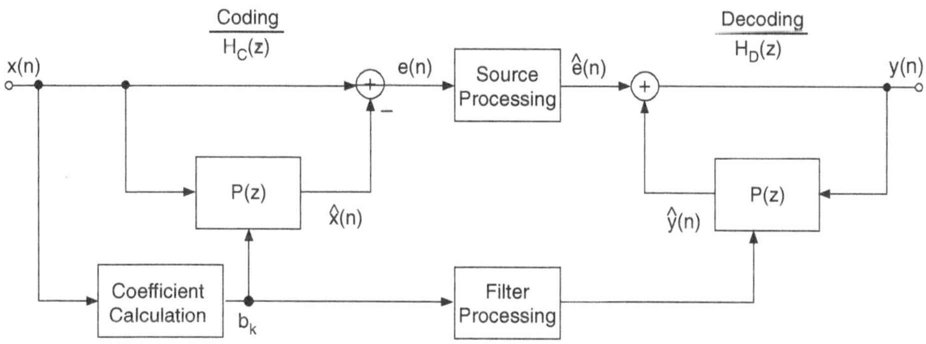

Figure 1.19 Linear prediction for coding and source-filter processing.

Then the error signal $e(n)$ is quantized for the coding application and further processed for source-filter modifications. The receiver performs a decoding operation by inverse filtering using the prediction filter and the corresponding coefficients from the coder or processed and modified coefficients. The decoder transfer function is given by

$$H_D(z) = \frac{1}{1 - P(z)}. \tag{1.105}$$

Source-filter processing extracts, in the time domain, the source signal $e(n)$ through linear prediction and simultaneously the filter coefficients b_k of the prediction filter $P(z)$. The source signal is a whitened version of the input and the filter transfer function $H_D(z)$ represents the spectral envelope of the input $x(n)$. Both the source signal $e(n)$ and the prediction coefficients b_k can be modified according to specific tasks like sound morphing (keeping the source signal and modifying the spectral envelope).

1.9 Exercises

1. Signals and Systems

1. Sketch the impulse responses:

   (a) $h(n) = a \cdot \delta(n)$ (amplifier);

(b) $h(n) = \delta(n - M)$ (delay by M samples);

(c) $h(n) = 0{,}5\,[\delta(n) + \delta(n-1)]$ (2-tap moving average, simple lowpass filter);

(d) $h(n) = \frac{1}{M}\sum_{k=0}^{M}\delta(n-k)$ (M-tap moving average, simple lowpass filter);

(e) $h(n) = a^n \cdot \epsilon(n)$, $|a| < 1$ (lowpass filter);

(f) $h(n) = a^n \cdot \sin\left(2\pi\frac{1}{20}n\right) \cdot \epsilon(n)$, $|a| < 1$ (lowpass filter);

and compute $y(n)$ for $x(n) = \delta(n) + \delta(n-1)$ using convolution (folding the impulse response and folding the input signal. Write a Matlab script for the computation of $y(n)$.

2. Compute the output signals:

(a) $y(n) = A\cos(\Omega n) \star h(n)$ (eigen signal passing through system);

(b) $y(n) = x(n) + ay(n-1)$, $|a| < 1$ (first-order lowpass filter);

(c) $y(n) = x(n) + y(n-1)$ (integrator);

(d) $y(n) = x(n) - x(n-1)$ (differentiator);

for $x(n) = \delta(n)$ and plot the output signals for $x(n) = \sum_{k=0}^{7}\delta(n-k)$ using Matlab.

2. Discrete-time Fourier Transform

1. Compute the discrete-time Fourier transform of:

(a) $x(n) = a \cdot \delta(n)$;

(b) $x(n) = \delta(n - M)$;

(c) $x(n) = 0{,}5\,[\delta(n) + \delta(n-1)]$;

(d) $x(n) = a^n \cdot \epsilon(n)$, $|a| < 1$;

and plot the results using Matlab.

2. Using the difference equation

$$y(n) = a\sin(\Omega_0 n) \cdot x(n-1) + 2a\cos(\Omega_0 n) \cdot y(n-1) - a^2 \cdot y(n-2):$$

(a) sketch the signal flow graph;

(b) compute the frequency response $H(e^{j\Omega}) = \frac{Y(e^{j\Omega})}{X(e^{j\Omega})}$;

(c) plot the magnitude response and phase response using Matlab;

(d) program the difference equation and plot the impulse response.

3. FIR and IIR Filters

1. FIR filters with linear phase (even and odd symmetry, linear phase, constant group delay):

   (a) $h(n) = 0.5 \cdot \delta(n) + 0.5 \cdot \delta(n-1)$;

   (b) $h(n) = 0.25 \cdot \delta(n) + 0.5 \cdot \delta(n-1) + 0.25 \cdot \delta(n)$;

   (c) $h(n) = 0.5 \cdot \delta(n) - 0.5 \cdot \delta(n-1)$;

   (d) $h(n) = 0.5 \cdot \delta(n) - 0.5 \cdot \delta(n-2)$.

   Plot the impulse responses, magnitude and phase responses (freqz.m), pole/zero plots (zplane.m), and group delay (grpdelay.m). Discuss the results.

2. Use FIR filters to approximate a given impulse response by truncating the length of the impulse response to N taps.

3. Do experimentation with the Parks–McClellan design (firpm.m) and design several lowpass filters with linear phase. Use the Matlab files freqz.m and zplane.m for your evaluation.

4. Show the IIR signal flow graphs of first-order LP and HP filters, and second-order BP filters. Verify the frequency responses.

4. Adaptive Filters

1. Write a small script for linear prediction using the LMS algorithm. Use an audio signal and experiment with the prediction order and monitor the error.

2. Modify the linear prediction approach by replacing $d(n) = x(n) \star h(n)$, where $h(n)$ is the unknown system impulse response. Modify the code to perform system identification using the LMS algorithm.

References

[Opp14] A.V. Oppenheim, A.S. Willsky, with H. Nawab: *Signals and Systems*, Pearson New International Edition, 2nd Edition, 2014.

[Opp10] A.V. Oppenheim, R.W. Schafer, with J.R. Buck: *Discrete-time Signal Processing*, Prentice Hall Signal Processing Series, 3rd Edition, 2010.

Chapter 2

Quantization

U. Zölzer

Basic operations for analog-to-digital (AD) conversion of a continuous-time signal $x(t)$ are the sampling and quantization of $x(n)$ yielding the quantized sequence $x_Q(n)$ (see Fig. 2.1). Before discussing AD/digital-to-analog (DA) conversion techniques and the choice of the sampling frequency $f_S = \frac{1}{T_S}$ in Chapter 3, we will introduce the quantization of the samples $x(n)$ with a finite number of bits. The digitization of a sampled signal with continuous amplitude is called quantization. The effects of quantization, starting with the classical quantization model, are discussed in Section 2.1. In Section 2.2, dither techniques are presented which, for low-level signals, linearize the process of quantization. In Section 2.3, spectral shaping of quantization errors is described. Section 2.4 deals with number representation for digital audio signals and their effects on algorithms.

2.1 Signal Quantization

2.1.1 Classical Quantization Model

Quantization is described by *Widrow's quantization theorem* [Wid61]. It says that a quantizer can be modeled (see Fig. 2.2) as the addition of a uniformly distributed random signal $e(n)$ to the original signal $x(n)$ (see Fig. 2.2, [Wid61]). This additive model given by

$$x_Q(n) = x(n) + e(n) \tag{2.1}$$

is based on the difference between quantized output and input according to the error signal

$$e(n) = x_Q(n) - x(n). \tag{2.2}$$

This linear model of the output $x_Q(n)$ is only then valid when the input amplitude has a wide dynamic range and the quantization error $e(n)$ is not correlated with the signal $x(n)$. Owing to the

Digital Audio Signal Processing, Third Edition. Udo Zölzer.
© 2022 John Wiley & Sons Ltd. Published 2022 by John Wiley & Sons Ltd.

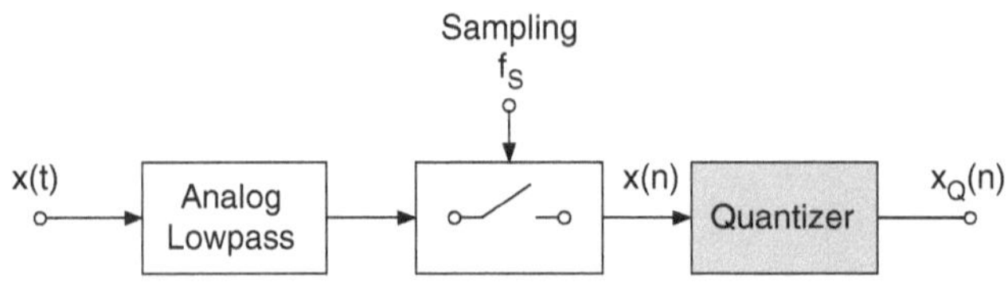

Figure 2.1 AD conversion and quantization.

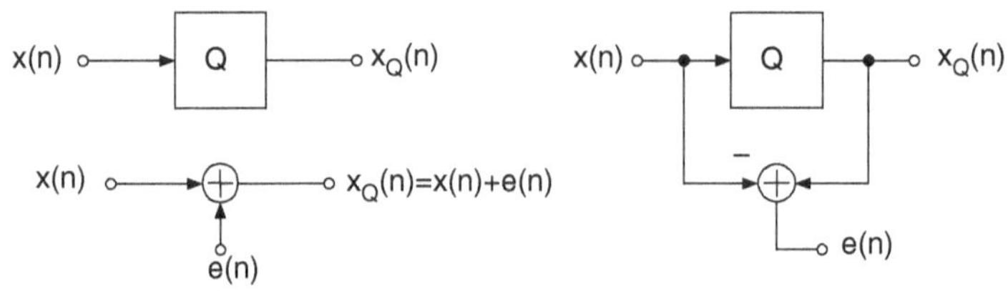

Figure 2.2 Quantization.

statistical independence of consecutive quantization errors, the autocorrelation of the error signal is given by $r_{EE}(m) = \sigma_E^2 \cdot \delta(m)$, which yields a power density spectrum $S_{EE}(e^{j\Omega}) = \sigma_E^2$.

The nonlinear process of quantization is described by a nonlinear characteristic curve, as shown in Fig. 2.3a, where Q denotes the quantization step. The difference between output and input of the quantizer provides the quantization error $e(n) = x_Q(n) - x(n)$, which is shown in Fig. 2.3b. The uniform probability density function (PDF) of the quantization error is given (see Fig. 2.3b) by

$$p_E(e) = \frac{1}{Q}\,\mathrm{rect}\left(\frac{e}{Q}\right). \tag{2.3}$$

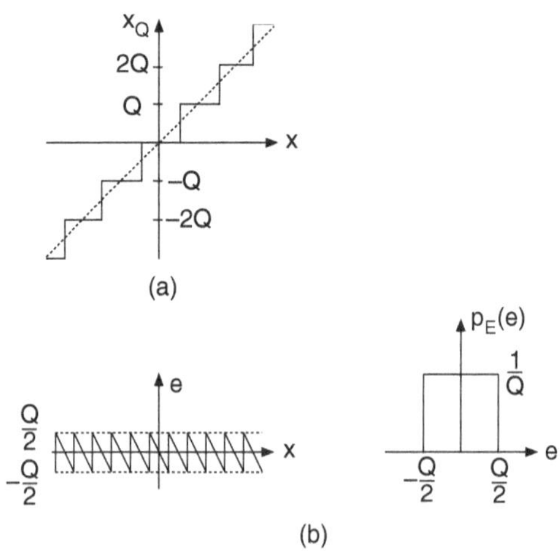

Figure 2.3 (a) Nonlinear characteristic curve of a quantizer. (b) Quantization error e and its probability density function (PDF) $p_E(e)$.

The mth moment of a random variable E with a PDF $p_E(e)$ is defined as the expected value of E^m:

$$E\{E^m\} = \int_{-\infty}^{\infty} e^m p_E(e)\, de. \tag{2.4}$$

For a uniformly distributed random process, as in Eq. (2.3), the first two moments are given by

$$m_E = E\{E\} = 0 \qquad \text{mean value,} \tag{2.5}$$

$$\sigma_E^2 = E\{E^2\} = \frac{Q^2}{12} \qquad \text{variance.} \tag{2.6}$$

The signal-to-noise ratio (SNR)

$$\text{SNR} = 10 \log_{10}\left(\frac{\sigma_X^2}{\sigma_E^2}\right) \qquad \text{dB} \tag{2.7}$$

is defined as the ratio of signal power σ_X^2 to error power σ_E^2.

For a quantizer with input range $\pm x_{max}$ and word length w, the quantization step size can be expressed as

$$Q = 2x_{max}/2^w. \tag{2.8}$$

By defining a peak factor

$$P_F = \frac{x_{max}}{\sigma_X} = \frac{2^{w-1}Q}{\sigma_X}, \tag{2.9}$$

the variances of the input and the quantization error can be written as

$$\sigma_X^2 = \frac{x_{max}^2}{P_F^2} \qquad \text{and} \tag{2.10}$$

$$\sigma_E^2 = \frac{Q^2}{12} = \frac{1}{12}\frac{x_{max}^2}{2^{2w}}2^2 = \frac{1}{3}x_{max}^2 2^{-2w}. \tag{2.11}$$

The SNR is then given by

$$\begin{aligned} \text{SNR} &= 10 \log_{10}\left(\frac{x_{max}^2/P_F^2}{\frac{1}{3}x_{max}^2 2^{-2w}}\right) = 10 \log_{10}\left(2^{2w}\frac{3}{P_F^2}\right) \\ &= 6.02\, w - 10 \log_{10}(P_F^2/3) \qquad \text{dB.} \end{aligned} \tag{2.12}$$

A sinusoidal signal (PDF as in Fig. 2.4) with $P_F = \sqrt{2}$ gives

$$\text{SNR} = 6.02\, w + 1.76 \qquad \text{dB.} \tag{2.13}$$

For a signal with uniform PDF (see Fig. 2.4) and $P_F = \sqrt{3}$, we can write

$$\text{SNR} = 6.02\, w \qquad \text{dB,} \tag{2.14}$$

and for a Gaussian distributed signal (probability of overload $< 10^{-5}$ leads to $P_F = 4.61$, see Fig. 2.5), it follows that

$$\text{SNR} = 6.02\, w - 8.5 \qquad \text{dB.} \tag{2.15}$$

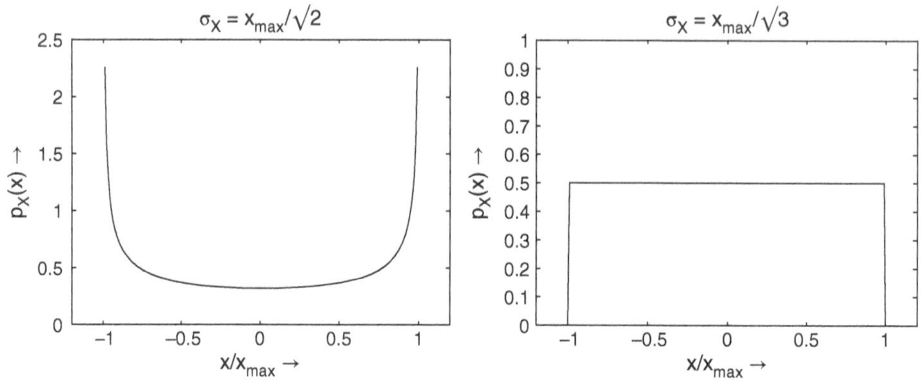

Figure 2.4 Probability density function (sinusoidal signal and signal with uniform PDF).

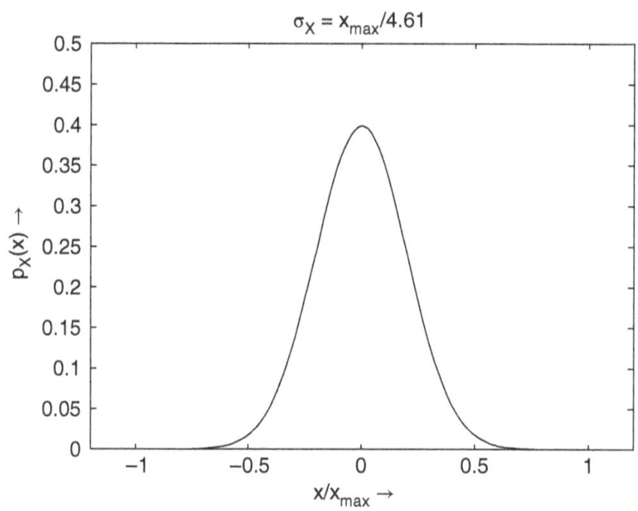

Figure 2.5 Probability density function (signal with Gaussian PDF).

It is obvious that the SNR depends on the PDF of the input. For digital audio signals that exhibit nearly Gaussian distribution, the maximum SNR for a given word length w is 8.5 dB lower than the *rule of thumb* formula (2.14) for the SNR.

2.1.2 Quantization Theorem

The statement of the *quantization theorem* for amplitude sampling (digitizing the amplitude) of signals has been given by Widrow [Wid61]. The analogy for digitizing the time axis is the *sampling theorem* given by Shannon [Sha48]. Figure 2.6 shows the amplitude quantization and the time quantization. First of all, the PDF of the output signal of a quantizer is determined in

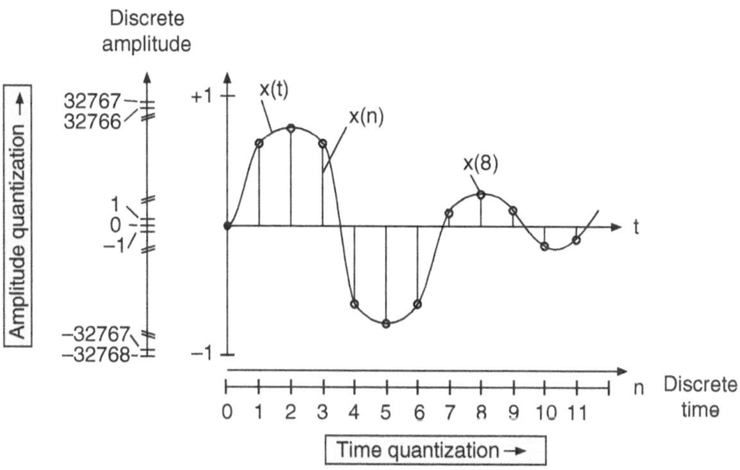

Figure 2.6 Amplitude and time quantization.

terms of the PDF of the input signal. Both PDFs are shown in Fig. 2.7. The respective characteristic functions (Fourier transform of a PDF) of the input and output signals form the basis for Widrow's *quantization theorem*.

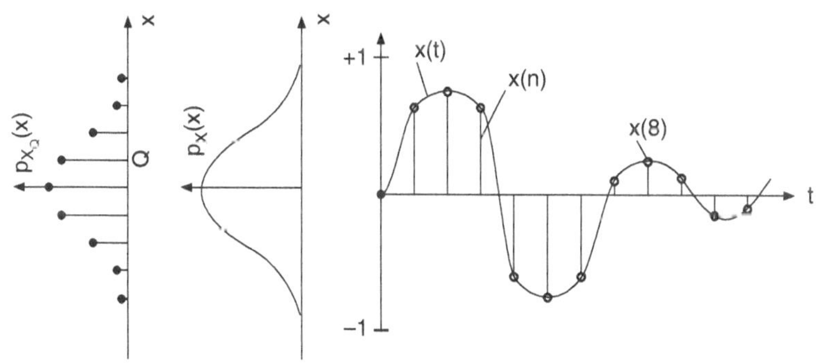

Figure 2.7 Probability density function of signal $x(n)$ and quantized signal $x_Q(n)$.

First-order Statistics of the Quantizer Output

Quantization of a continuous-amplitude signal x with PDF $p_X(x)$ leads to a discrete-amplitude signal y with PDF $p_Y(y)$ (see Fig. 2.8). The continuous PDF of the input is sampled by integrating over all quantization intervals (zone sampling). This leads to a discrete PDF of the output.

In the quantization intervals, the discrete PDF of the output is determined by the probability

$$W[kQ] = W\left[-\frac{Q}{2} + kQ \leq x < \frac{Q}{2} + kQ\right] = \int_{-\frac{Q}{2}+kQ}^{\frac{Q}{2}+kQ} p_X(x)\, dx. \tag{2.16}$$

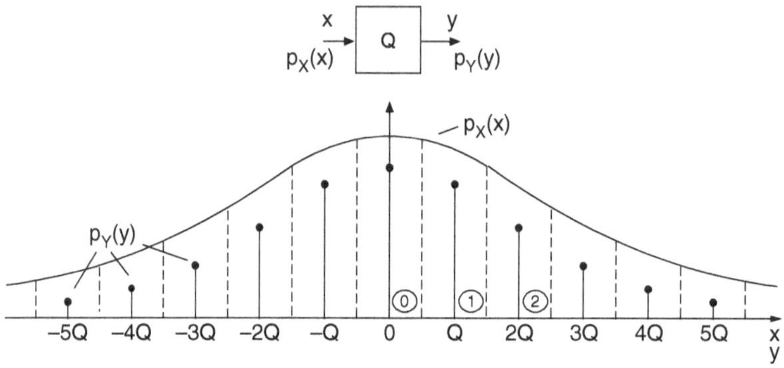

Figure 2.8 Zone sampling of the PDF.

For the intervals $k = 0, 1, 2$, it follows that

$$\begin{aligned} p_Y(y) &= \delta(0) \int_{-\frac{Q}{2}}^{\frac{Q}{2}} p_X(x)\, dx & -\tfrac{Q}{2} \le y < \tfrac{Q}{2}, \\ &= \delta(y-Q) \int_{-\frac{Q}{2}+Q}^{\frac{Q}{2}+Q} p_X(x)\, dx & -\tfrac{Q}{2} + Q \le y < \tfrac{Q}{2} + Q, \\ &= \delta(y-2Q) \int_{-\frac{Q}{2}+2Q}^{\frac{Q}{2}+2Q} p_X(x)\, dx & -\tfrac{Q}{2} + 2Q \le y < \tfrac{Q}{2} + 2Q. \end{aligned}$$

The summation over all intervals gives the PDF of the output

$$p_Y(y) = \sum_{k=-\infty}^{\infty} \delta(y - kQ) W(kQ) \tag{2.17}$$

$$= \sum_{k=-\infty}^{\infty} \delta(y - kQ) W(y), \tag{2.18}$$

where

$$W(kQ) = \int_{-\frac{Q}{2}+kQ}^{\frac{Q}{2}+kQ} p_X(x)\, dx, \tag{2.19}$$

$$W(y) = \int_{-\infty}^{\infty} \text{rect}\left(\frac{y-x}{Q}\right) p_X(x)\, dx \tag{2.20}$$

$$= \text{rect}\left(\frac{y}{Q}\right) * p_X(y). \tag{2.21}$$

Using

$$\delta_Q(y) = \sum_{k=-\infty}^{\infty} \delta(y - kQ), \tag{2.22}$$

the PDF of the output is given by

$$p_Y(y) = \delta_Q(y) \left[\text{rect}\left(\frac{y}{Q}\right) * p_X(y)\right]. \tag{2.23}$$

Hence, the PDF of the output can be determined by convolution of a rect function [Lip92] with the PDF of the input. This is followed by an *amplitude sampling* with resolution Q as described in Eq. (2.23) (see Fig. 2.9).

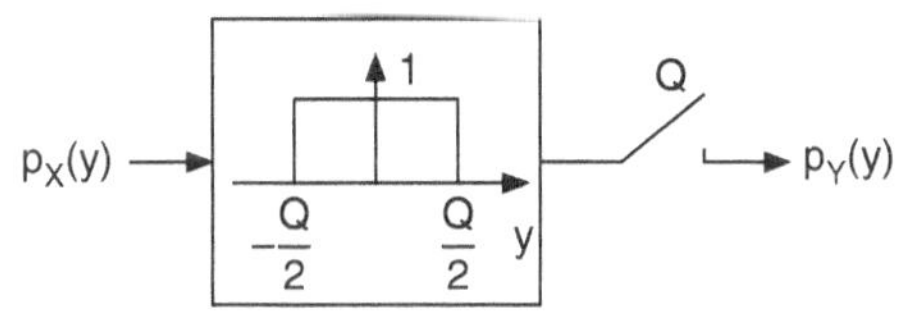

Figure 2.9 Determining the PDF of the output.

Using $FT\{f_1(t) \cdot f_2(t)\} = \frac{1}{2\pi} F_1(j\omega) * F_2(j\omega)$, the characteristic function (Fourier transform of $p_Y(y)$) can be written as

$$P_Y(ju) = \frac{1}{2\pi} u_o \sum_{k=-\infty}^{\infty} \delta(u - ku_o) * \left[Q \frac{\sin\left(u\frac{Q}{2}\right)}{u\frac{Q}{2}} \cdot P_X(ju)\right], \tag{2.24}$$

$$\text{with} \qquad u_o = \frac{2\pi}{Q},$$

$$= \sum_{k=-\infty}^{\infty} \delta(u - ku_o) * \left[\frac{\sin\left(u\frac{Q}{2}\right)}{u\frac{Q}{2}} \cdot P_X(ju)\right]. \tag{2.25}$$

$$\boxed{P_Y(ju) = \sum_{k=-\infty}^{\infty} P_X(ju - jku_o) \frac{\sin\left[(u - ku_o)\frac{Q}{2}\right]}{(u - ku_o)\frac{Q}{2}}}. \tag{2.26}$$

Equation (2.26) describes the sampling of the continuous PDF of the input. If the quantization frequency $u_o = 2\pi/Q$ is twice the highest frequency of the characteristic function $P_X(ju)$, then periodically recurring spectra do not overlap. Hence, a reconstruction of the PDF of the input $p_X(x)$ from the quantized PDF of the output $p_Y(y)$ is possible (see Fig. 2.10). This is known as the *quantization theorem* of Widrow. In contrast to the first sampling theorem (Shannon's sampling theorem, ideal amplitude sampling in the time domain) $F^A(j\omega) = \frac{1}{T} \sum_{k=-\infty}^{\infty} F(j\omega - jk\omega_o)$, it can be observed that there is an additional multiplication of the periodically characteristic function with $\frac{\sin\left[(u-ku_o)\frac{Q}{2}\right]}{(u-ku_o)\frac{Q}{2}}$ (see Eq. (2.26)).

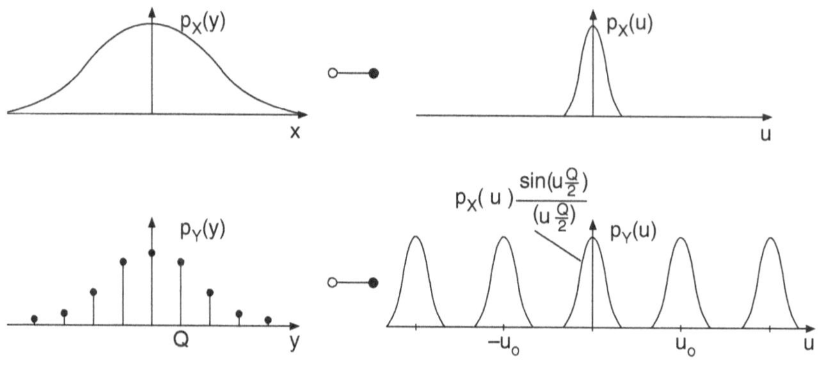

Figure 2.10 Spectral representation.

If the baseband of the characteristic function ($k = 0$)

$$P_Y(ju) = P_X(ju)\underbrace{\frac{\sin\left(u\frac{Q}{2}\right)}{u\frac{Q}{2}}}_{P_E(ju)} \tag{2.27}$$

is considered, it is observed that it is a product of two characteristic functions. The multiplication of characteristic functions leads to the convolution of PDFs, from which the addition of two statistically independent signals can be concluded. Hence, the characteristic function of the quantization error is

$$P_E(ju) = \frac{\sin\left(u\frac{Q}{2}\right)}{u\frac{Q}{2}} \tag{2.28}$$

and the PDF

$$p_E(e) = \frac{1}{Q}\,\text{rect}\left(\frac{e}{Q}\right) \tag{2.29}$$

(see Fig. 2.11).

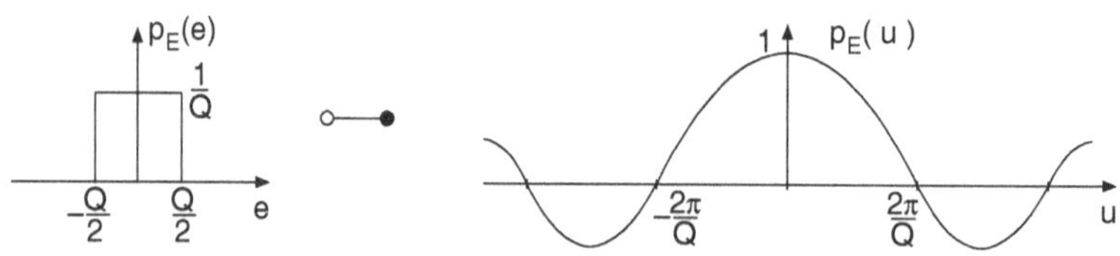

Figure 2.11 PDF and characteristic function of quantization error.

The modeling of the quantization process as an addition of a statistically independent noise signal to the input signal leads to a continuous PDF of the output (see Fig. 2.12, convolution of PDFs and sampling in the interval Q gives the discrete PDF of the output). The PDF of the

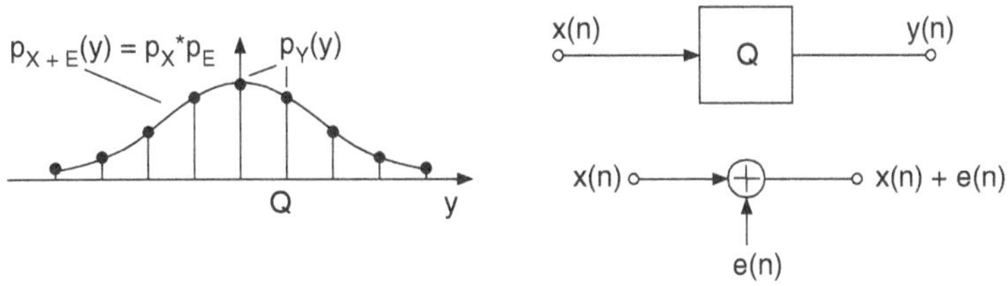

Figure 2.12 PDF of the model.

discrete-valued output comprises Dirac pulses at distance Q with values equal to the continuous PDF (see Eq. (2.23)). Only if the *quantization theorem* is valid, the continuous PDF can be reconstructed from the discrete PDF.

In many cases, it is not necessary to reconstruct the PDF of the input. It is sufficient to calculate the moments of the input from the output. The mth moment can be expressed in terms of the PDF or the characteristic function:

$$E\{Y^m\} = \int_{-\infty}^{\infty} y^m p_Y(y)dy \tag{2.30}$$

$$= (-j)^m \left. \frac{d^m P_Y(ju)}{du^m} \right|_{u=0}. \tag{2.31}$$

If the *quantization theorem* is satisfied, then the periodic terms in Eq. (2.26) do not overlap and the mth derivative of $P_Y(ju)$ is solely determined by the baseband[1] so that with Eq. (2.26), it can be written

$$E\{Y^m\} = (-j)^m \frac{d^m}{du^m} P_X(ju) \left. \frac{\sin\left(u\frac{Q}{2}\right)}{u\frac{Q}{2}} \right|_{u=0}. \tag{2.32}$$

With Eq. (2.32), the first two moments can be determined as

$$m_Y = E\{Y\} = E\{X\}, \tag{2.33}$$

$$\sigma_Y^2 = E\{Y^2\} = \underbrace{E\{X^2\}}_{\sigma_X^2} + \underbrace{\frac{Q^2}{12}}_{\sigma_E^2}. \tag{2.34}$$

Second-order Statistics of Quantizer Output

To describe the properties of the output in the frequency domain, two output values Y_1 (at time n_1) and Y_2 (at time n_2) are considered [Lip92]. For the joint density function,

$$p_{Y_1Y_2}(y_1, y_2) = \delta_{QQ}(y_1, y_2)\left[\text{rect}\left(\frac{y_1}{Q}, \frac{y_2}{Q}\right) * p_{X_1X_2}(y_1, y_2)\right] \tag{2.35}$$

[1] This is also valid owing to the weaker condition of Sripad and Snyder [Sri77] discussed in the next section.

with

$$\delta_{QQ}(y_1, y_2) = \delta_Q(y_1) \cdot \delta_Q(y_2) \tag{2.36}$$

and

$$\mathrm{rect}\left(\frac{y_1}{Q}, \frac{y_2}{Q}\right) = \mathrm{rect}\left(\frac{y_1}{Q}\right) \cdot \mathrm{rect}\left(\frac{y_2}{Q}\right). \tag{2.37}$$

For the two-dimensional Fourier transform, it follows that

$$P_{Y_1Y_2}(ju_1, ju_2) = \sum_{k=-\infty}^{\infty} \sum_{l=-\infty}^{\infty} \delta(u_1 - ku_o)\delta(u_2 - lu_o)$$

$$* \left[\frac{\sin\left(u_1 \frac{Q}{2}\right)}{u_1 \frac{Q}{2}} \cdot \frac{\sin\left(u_2 \frac{Q}{2}\right)}{u_2 \frac{Q}{2}} \cdot P_{X_1X_2}(ju_1, ju_2) \right] \tag{2.38}$$

$$= \sum_{k=-\infty}^{\infty} \sum_{l=-\infty}^{\infty} P_{X_1X_2}(ju_1 - jku_o, ju_2 - jlu_o)$$

$$\frac{\sin\left[(u_1 - ku_o)\frac{Q}{2}\right]}{(u_1 - ku_o)\frac{Q}{2}} \cdot \frac{\sin\left[(u_2 - lu_o)\frac{Q}{2}\right]}{(u_2 - lu_o)\frac{Q}{2}}. \tag{2.39}$$

Similar to the one-dimensional *quantization theorem*, a two-dimensional theorem [Wid61] can be formulated: the joint density function of the input can be reconstructed from the joint density function of the output, if $P_{X_1X_2}(ju_1, ju_2) = 0$ for $u_1 \geq u_o/2$ and $u_2 \geq u_o/2$. Here again, the moments of the joint density function can be calculated as follows:

$$E\{Y_1^m Y_2^n\} = (-j)^{m+n} \frac{\partial^{m+n}}{\partial u_1^m \partial u_2^n} P_{X_1X_2}(ju_1, ju_2) \frac{\sin\left(u_1 \frac{Q}{2}\right)}{u_1 \frac{Q}{2}} \frac{\sin\left(u_2 \frac{Q}{2}\right)}{u_2 \frac{Q}{2}} \Bigg|_{u_1=0,\, u_2=0}. \tag{2.40}$$

From this, the autocorrelation function with $m = n_2 - n_1$ can be written as

$$r_{YY}(m) = E\{Y_1 Y_2\}(m) = \begin{cases} E\{X^2\} + \frac{Q^2}{12} & \textit{form} = 0, \\ E\{X_1 X_2\}(m) & \text{elsewhere}, \end{cases} \tag{2.41}$$

(for $m = 0$, we obtain Eq. (2.34)).

2.1.3 Statistics of Quantization Error

First-order Statistics of Quantization Error

The PDF of the quantization error depends on the PDF of the input and is dealt with in the following. The quantization error $e = x_Q - x$ is restricted to the interval $[-\frac{Q}{2}, \frac{Q}{2}]$. It depends linearly on the input (see Fig.2.13). If the input value lies in the interval $[-\frac{Q}{2}, \frac{Q}{2}]$, then the error is $e = 0 - x$. For the PDF, we obtain $p_E(e) = p_X(e)$. If the input value lies in the interval $[-\frac{Q}{2} + Q, \frac{Q}{2} + Q]$, then the quantization error is $e = Q\lfloor Q^{-1}x + 0.5\rfloor - x$ and is again restricted to

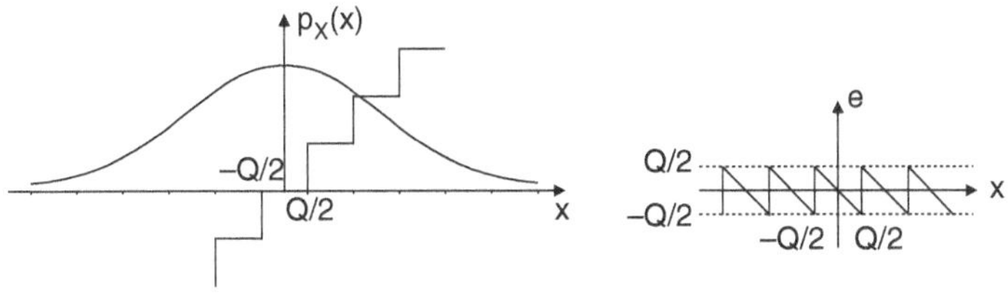

Figure 2.13 Probability density function and quantization error.

$[-\frac{Q}{2}, \frac{Q}{2}]$. The PDF of the quantization error is consequently $p_E(e) = p_X(e + Q)$ and is added to the first term. For the sum over all intervals, we can write

$$p_E(e) = \begin{cases} \sum\limits_{k=-\infty}^{\infty} p_X(e - kQ) & \mathit{for}\ -\frac{Q}{2} \le e < \frac{Q}{2}, \\ 0 & \text{elsewhere.} \end{cases} \tag{2.42}$$

Because of the restricted values of the variable of the PDF, we can write

$$p_E(e) = \mathrm{rect}\left(\frac{e}{Q}\right) \sum_{k=-\infty}^{\infty} p_X(e - kQ) \tag{2.43}$$

$$= \mathrm{rect}\left(\frac{e}{Q}\right) [p_X(e) * \delta_Q(e)]. \tag{2.44}$$

The PDF of the quantization error is determined by the PDF of the input and can be computed by shifting and windowing a zone. All individual zones are summed up for calculating the PDF of the quantization error [Lip92]. A simple graphical interpretation of this overlapping is shown in Fig. 2.14. The overlapping leads to a uniform distribution of the quantization error if the input PDF $p_X(x)$ is spread over a sufficient number of quantization intervals.

For the Fourier transform of the PDF from Eq. (2.44) follows

$$P_E(ju) = \frac{1}{2\pi} Q \frac{\sin\left(u\frac{Q}{2}\right)}{u\frac{Q}{2}} * \left[P_X(ju)\frac{2\pi}{Q} \sum_{k=-\infty}^{\infty} \delta(u - ku_o)\right] \tag{2.45}$$

$$= \frac{\sin\left(u\frac{Q}{2}\right)}{u\frac{Q}{2}} * \left[\sum_{k=-\infty}^{\infty} P_X(jku_o)\delta(u - ku_o)\right] \tag{2.46}$$

$$= \sum_{k=-\infty}^{\infty} P_X(jku_o)\left[\frac{\sin\left(u\frac{Q}{2}\right)}{u\frac{Q}{2}} * \delta(u - ku_o)\right], \tag{2.47}$$

$$\boxed{P_E(ju) = \sum_{k=-\infty}^{\infty} P_X(jku_o) \frac{\sin\left[(u - ku_o)\frac{Q}{2}\right]}{(u - ku_o)\frac{Q}{2}}.} \tag{2.48}$$

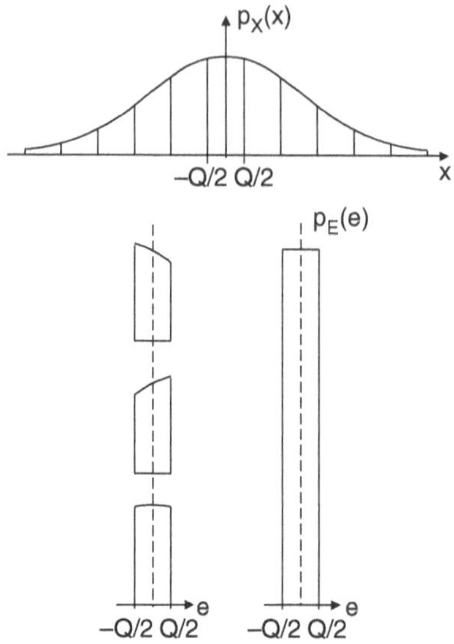

Figure 2.14 Probability density function of the quantization error.

If the quantization theorem is satisfied, i.e. if $P_X(ju) = 0$ for $u > u_o/2$, then there is only one non-zero term ($k = 0$ in Eq. (2.48)). The characteristic function of the quantization error is reduced, with $P_X(0) = 1$, to

$$P_E(ju) = \frac{\sin\left(u\frac{Q}{2}\right)}{u\frac{Q}{2}}. \tag{2.49}$$

Hence, the PDF of the quantization error is

$$p_E(e) = \frac{1}{Q}\,\text{rect}\left(\frac{e}{Q}\right). \tag{2.50}$$

Sripad and Snyder [Sri77] have modified the sufficient condition of Widrow (band-limited characteristic function of input) for a quantization error of uniform PDF by the weaker condition

$$\boxed{P_X(jku_o) = P_X\left(j\frac{2\pi k}{Q}\right) = 0 \qquad \text{for all } k \neq 0.} \tag{2.51}$$

The uniform distribution of the input PDF

$$p_X(x) = \frac{1}{Q}\,\text{rect}\left(\frac{x}{Q}\right) \tag{2.52}$$

with characteristic function

$$P_X(ju) = \frac{\sin\left(u\frac{Q}{2}\right)}{u\frac{Q}{2}} \tag{2.53}$$

does not satisfy Widrow's condition for a band-limited characteristic function, but instead the weaker condition

$$P_X\left(j\frac{2\pi k}{Q}\right) = \frac{\sin(\pi k)}{\pi k} = 0 \qquad \text{for all } k \neq 0 \tag{2.54}$$

is fulfilled. From this follows the uniform PDF in Eq. (2.49) of the quantization error. The weaker condition from Sripad and Snyder extends the class of input signals for which a uniform PDF of the quantization error can be assumed.

To show the deviation from the uniform PDF of the quantization error as a function of the PDF of the input, Eq. (2.48) can be written as

$$\begin{aligned} P_E(ju) &= P_X(0)\frac{\sin\left[u\frac{Q}{2}\right]}{u\frac{Q}{2}} + \sum_{k=-\infty,\,k\neq 0}^{\infty} P_X\left(j\frac{2\pi k}{Q}\right)\frac{\sin\left[(u-ku_o)\frac{Q}{2}\right]}{(u-ku_o)\frac{Q}{2}} \\ &= \frac{\sin\left[u\frac{Q}{2}\right]}{u\frac{Q}{2}} + \sum_{k=-\infty,\,k\neq 0}^{\infty} P_X\left(j\frac{2\pi k}{Q}\right)\frac{\sin\left[u\frac{Q}{2}\right]}{u\frac{Q}{2}} * \delta(u-ku_0). \end{aligned} \tag{2.55}$$

The inverse Fourier transform yields

$$p_E(e) = \frac{1}{Q}\,\mathrm{rect}\left(\frac{e}{Q}\right)\left[1 + \sum_{k=-\infty,\,k\neq 0}^{\infty} P_X\left(j\frac{2\pi k}{Q}\right)\exp\left(j\frac{2\pi k e}{Q}\right)\right] \tag{2.56}$$

$$= \begin{cases} \frac{1}{Q}\left[1 + \sum_{k\neq 0}^{\infty} P_X\left(j\frac{2\pi k}{Q}\right)\exp\left(j\frac{2\pi k e}{Q}\right)\right] & \textit{for } -\frac{Q}{2} \le e < \frac{Q}{2}, \\ 0 & \text{elsewhere.} \end{cases} \tag{2.57}$$

Equation (2.56) shows the effect of the input PDF on the deviation from a uniform PDF.

Second-order Statistics of Quantization Error

For describing the spectral properties of the error signal, two values E_1 (at time n_1) and E_2 (at time n_2) are considered [Lip92]. The joint PDF is given by

$$p_{E_1E_2}(e_1, e_2) = \mathrm{rect}\left(\frac{e_1}{Q}, \frac{e_2}{Q}\right)\left[p_{X_1X_2}(e_1, e_2) * \delta_{QQ}(e_1, e_2)\right]. \tag{2.58}$$

Here $\delta_{QQ}(e_1, e_2) = \delta_Q(e_1)\cdot\delta_Q(e_2)$ and $\mathrm{rect}(\frac{e_1}{Q}, \frac{e_2}{Q}) = \mathrm{rect}(\frac{e_1}{Q})\cdot\mathrm{rect}(\frac{e_2}{Q})$. For the Fourier transform of the joint PDF, a similar procedure to that shown by Eqs. (2.45)–(2.48) leads to

$$\begin{aligned} P_{E_1E_2}(ju_1, ju_2) = \sum_{k_1=-\infty}^{\infty}\sum_{k_2=-\infty}^{\infty} & P_{X_1X_2}(jk_1u_o, jk_2u_o) \\ & \frac{\sin\left[(u_1-k_1u_o)\frac{Q}{2}\right]}{(u_1-k_1u_o)\frac{Q}{2}}\,\frac{\sin\left[(u_2-k_2u_o)\frac{Q}{2}\right]}{(u_2-k_2u_o)\frac{Q}{2}}. \end{aligned} \tag{2.59}$$

If the quantization theorem and/or the Sripad–Snyder condition

$$\boxed{P_{X_1X_2}(jk_1u_o,\, jk_2u_o) = 0 \qquad \text{for all } k_1, k_2 \neq 0} \tag{2.60}$$

are satisfied, then

$$P_{E_1E_2}(ju_1, ju_2) = \frac{\sin\left[u_1 \frac{Q}{2}\right]}{u_1 \frac{Q}{2}} \frac{\sin\left[u_2 \frac{Q}{2}\right]}{u_2 \frac{Q}{2}}. \tag{2.61}$$

For the joint PDF of the quantization error, it then holds that

$$p_{E_1E_2}(e_1, e_2) = \frac{1}{Q} \operatorname{rect}\left(\frac{e_1}{Q}\right) \cdot \frac{1}{Q} \operatorname{rect}\left(\frac{e_2}{Q}\right) \qquad -\frac{Q}{2} \leq e_1, e_2 < \frac{Q}{2} \tag{2.62}$$

$$= p_{E_1}(e_1) \cdot p_{E_2}(e_2). \tag{2.63}$$

Owing to the statistical independence of quantization errors (Eq. (2.63)),

$$E\{E_1^m E_2^n\} = E\{E_1^m\} \cdot E\{E_2^n\}. \tag{2.64}$$

For the moments of the joint PDF,

$$E\{E_1^m E_2^n\} = (-j)^{m+n} \frac{\partial^{m+n}}{\partial u_1^m \partial u_2^n} P_{E_1E_2}(u_1, u_2)\Bigg|_{u_1=0,\, u_2=0}. \tag{2.65}$$

From this, it follows for the autocorrelation function with $m = n_2 - n_1$

$$r_{EE}(m) = \mathrm{E}\{E_1E_2\} = \begin{cases} \mathrm{E}\{E^2\} & form = 0, \\ \mathrm{E}\{E_1E_2\} & \text{elsewhere}, \end{cases} \tag{2.66}$$

$$= \begin{cases} \frac{Q^2}{12} & form = 0, \\ 0 & \text{elsewhere}, \end{cases} \tag{2.67}$$

$$= \underbrace{\frac{Q^2}{12}}_{\sigma_E^2} \delta(m). \tag{2.68}$$

The power density spectrum of the quantization error is then given by

$$S_{EE}(e^{j\Omega}) = \sum_{m=-\infty}^{+\infty} r_{EE}(m) e^{-j\Omega m} = \frac{Q^2}{12}, \tag{2.69}$$

which is equal to the variance $\sigma_E^2 = \frac{Q^2}{12}$ of the quantization error (see Fig. 2.15).

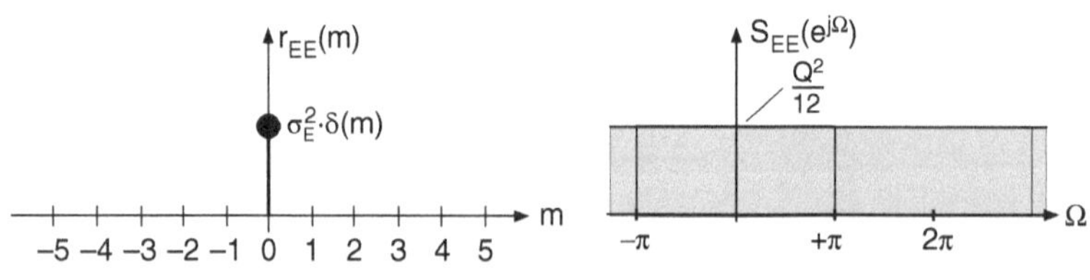

Figure 2.15 Autocorrelation $r_{EE}(m)$ and power density spectrum $S_{EE}(e^{j\Omega})$ of quantization error $e(n)$.

Correlation of Signal and Quantization Error

For describing the correlation of the signal and the quantization error [Sri77], the second moment of the output with Eq. (2.26) is derived as follows:

$$E\{Y^2\} = (-j)^2 \frac{d^2 P_Y(ju)}{du^2}\bigg|_{u=0} \tag{2.70}$$

$$\begin{aligned} &= (-j)^2 \sum_{k=-\infty}^{\infty} \left[\ddot{P}_X\left(-\frac{2\pi k}{Q}\right) \frac{\sin(\pi k)}{\pi k} \right. \\ &+ Q\dot{P}_X\left(-\frac{2\pi k}{Q}\right) \frac{\sin(\pi k) - \pi k \cos(\pi k)}{\pi^2 k^2} \\ &\left. + \frac{Q^2}{4} P_X\left(-\frac{2\pi k}{Q}\right) \frac{(2 - \pi^2 k^2)\sin(\pi k) - 2\pi k \cos(\pi k)}{\pi^3 k^3} \right] \end{aligned} \tag{2.71}$$

$$= E\{X^2\} + \frac{Q}{\pi} \sum_{k=-\infty,\, k\neq 0}^{\infty} \frac{(-1)^k}{k} \dot{P}_X\left(-\frac{2\pi k}{Q}\right) + E\{E^2\}. \tag{2.72}$$

With the quantization error $e(n) = y(n) - x(n)$,

$$E\{Y^2\} = E\{X^2\} + 2E\{X \cdot E\} + E\{E^2\}, \tag{2.73}$$

where the term $E\{X \cdot E\}$, with Eq. (2.72), is written as

$$E\{X \cdot E\} = \frac{Q}{2\pi} \sum_{k=-\infty,\, k\neq 0}^{\infty} \frac{(-1)^k}{k} \dot{P}_X\left(-\frac{2\pi k}{Q}\right). \tag{2.74}$$

With the assumption of a Gaussian PDF of the input, we obtain

$$p_X(x) = \frac{1}{\sqrt{2\pi}\sigma} \exp\left(\frac{-x^2}{2\sigma^2}\right) \tag{2.75}$$

with the characteristic function

$$P_X(ju) = \exp\left(\frac{-u^2\sigma^2}{2}\right). \tag{2.76}$$

Using Eq. (2.57), the PDF of the quantization error is then given by

$$p_E(e) = \begin{cases} \frac{1}{Q}\left[1 + 2\sum_{k=1}^{\infty} \cos\left(\frac{2\pi k e}{Q}\right) \exp\left(-\frac{2\pi^2 k^2 \sigma^2}{Q^2}\right)\right] & for \quad -\frac{Q}{2} \leq e < \frac{Q}{2}, \\ 0 & \text{elsewhere.} \end{cases} \tag{2.77}$$

Figure 2.16a shows the PDF in Eq. (2.77) of the quantization error for different variances of the input.

For the mean value and the variance of a quantization error, it follows from Eq. (2.77) that $E\{E\} = 0$ and

$$E\{E^2\} = \int_{-\infty}^{\infty} e^2 p_E(e)\, de = \frac{Q^2}{12}\left[1 + \frac{12}{\pi^2}\sum_{k=1}^{\infty} \frac{(-1)^k}{k^2} \exp\left(-\frac{2\pi^2 k^2 \sigma^2}{Q^2}\right)\right]. \tag{2.78}$$

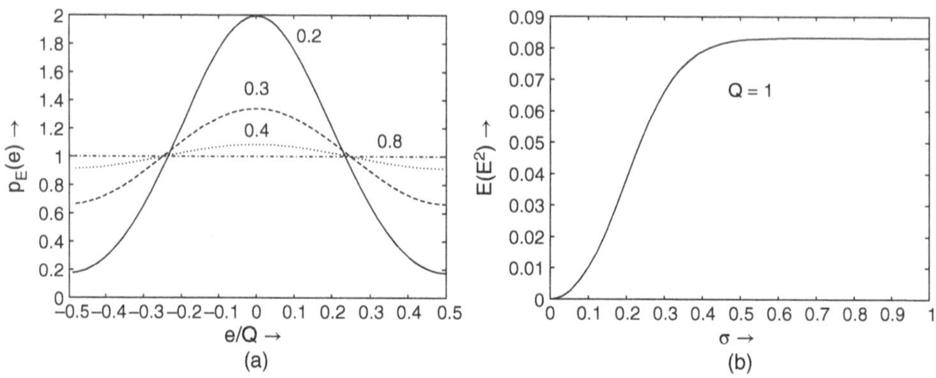

Figure 2.16 (a) Probability density function of quantization error for different standard deviations of a Gaussian PDF input. (b) Variance of quantization error for different standard deviations of a Gaussian PDF input.

Figure 2.16b shows the variance of the quantization error in Eq. (2.78) for different variances of the input.

For a Gaussian PDF input, as given by Eqs. (2.75) and (2.76), the correlation (see Eq. (2.74)) between input and quantization error is expressed as

$$E\{X \cdot E\} = 2\sigma^2 \sum_{k=1}^{\infty} (-1)^k \exp\left(-\frac{2\pi^2 k^2 \sigma^2}{Q^2}\right). \tag{2.79}$$

The correlation is negligible for large values of $\frac{\sigma}{Q}$.

2.2 Dither

2.2.1 Basics

The *requantization* (renewed quantization of already quantized signals) to limited word lengths occurs repeatedly during storage, format conversion, and signal processing algorithms. Here, small signal levels lead to error signals which depend on the input. Owing to quantization, nonlinear distortion occurs for low-level signals. The conditions for the classical quantization model are not satisfied anymore. To reduce these effects for signals of small amplitude, a linearization of the nonlinear characteristic curve of the quantizer is performed. This is done by adding a random sequence $d(n)$ to the quantized signal $x(n)$ (see Fig. 2.17) before the actual quantization process. The specification of the word length is shown in Fig. 2.18. This random signal is called dither.

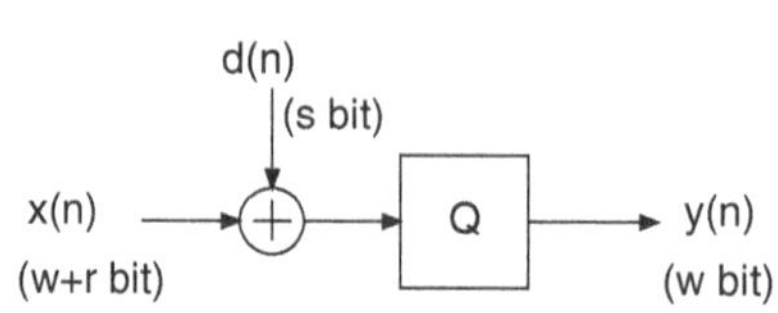

Figure 2.17 Addition of a random sequence before a quantizer.

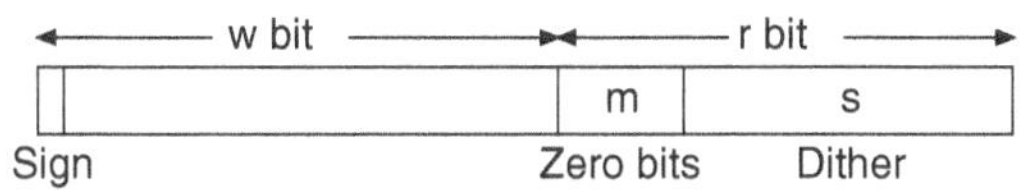

Figure 2.18 Specification of the word length.

The statistical independence of the error signal from the input is not achieved, but the conditional moments of the error signal can be affected [Lip92, Ger89, Wan92, Wan00].

The sequence $d(n)$, with amplitude range $(-\frac{Q}{2} \leq d(n) \leq \frac{Q}{2})$, is generated with the help of a random number generator and is added to the input. For a dither value with $Q = 2^{-(w-1)}$:

$$d_k = k2^{-r}Q \qquad -2^{s-1} \leq k \leq 2^{s-1} - 1. \tag{2.80}$$

The index k of the random number d_k characterizes the value from the set of $N = 2^s$ possible numbers with the probability

$$P(d_k) = \begin{cases} 2^{-s} & \text{for } -2^{s-1} \leq k \leq 2^{s-1} - 1, \\ 0 & \text{elsewhere.} \end{cases} \tag{2.81}$$

With the mean value $\overline{d} = \sum_k d_k P(d_k)$, the variance $\sigma_D^2 = \sum_k [d_k - \overline{d}]^2 P(d_k)$, and the quadratic mean $\overline{d^2} = \sum_k d_k^2 P(d_k)$, we can rewrite the variance as $\sigma_d^2 = \overline{d^2} - \overline{d}^2$.

For a static input amplitude V and the dither value d_k, the rounding operation [Lip86] is expressed as

$$g(V + d_k) = Q\left\lfloor \frac{V + d_k}{Q} + 0.5 \right\rfloor. \tag{2.82}$$

For the mean of the output $\overline{g}(V)$ as a function of the input V, we can write

$$\overline{g}(V) = \sum_k g(V + d_k)P(d_k). \tag{2.83}$$

The quadratic mean of the output $\overline{g^2}(V)$ for input V is given by

$$\overline{g^2}(V) = \sum_k g^2(V + d_k)P(d_k). \tag{2.84}$$

For the variance $d_R^2(V)$ for input V,

$$d_R^2(V) = \sum_k \{g(V + d_k) - \overline{g}(V)\}^2 P(d_k) = \overline{g^2}(V) - \{\overline{g}(V)\}^2. \tag{2.85}$$

The above-mentioned equations have the input V as a parameter. Figures 2.19 and 2.20 illustrate the mean output $\overline{g}(V)$ and the standard deviation $d_R(V)$ within a quantization step size, which are given by Eqs. (2.83), (2.84), and (2.85). The examples of rounding and truncation demonstrate the linearization of the characteristic curve of the quantizer. The coarse step size is replaced by a finer one. The quadratic deviation from the mean output $d_R^2(V)$ is termed *noise modulation*. For a uniform PDF dither, this noise modulation depends on the amplitude (see Figs. 2.19 and 2.20). It is maximum in the middle of the quantization step size and approaches zero towards the end. The linearization and the suppression of the noise modulation can be achieved by a triangular PDF dither with bipolar characteristic [Van89] and rounding operation (see Fig. 2.20). Triangular

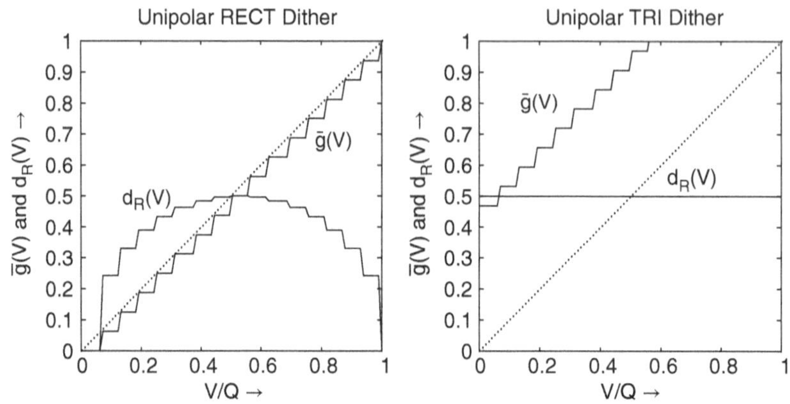

Figure 2.19 Truncation – linearizing and suppression of noise modulation ($s = 4$, $m = 0$).

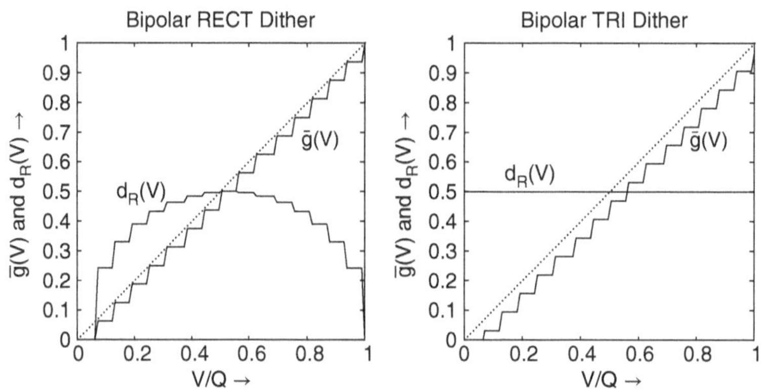

Figure 2.20 Rounding – linearizing and suppression of noise modulation ($s = 4$, $m = 1$).

PDF dither is obtained by adding two statistically independent dither signals with uniform PDF (convolution of PDFs). A dither signal with a higher-order PDF is not necessary for audio signals [Lip92, Wan00].

The total noise power for this quantization technique consists of the dither power and the power of the quantization error [Lip86]. The following noise powers are obtained by integration with respect to V as follows.

1. Mean dither power d^2:

$$d^2 = \frac{1}{Q}\int_0^Q d_R^2(V)dV \tag{2.86}$$

$$= \frac{1}{Q}\int_0^Q \sum_k \{g(V+d_k) - \overline{g}(V)\}^2 P(d_k)\, dV. \tag{2.87}$$

(This is equal to the deviation from mean output in accordance with Eq. (2.83).)

2. Mean of total noise power d_{tot}^2:

$$d_{tot}^2 = \frac{1}{Q}\int_0^Q \sum_k \{g(V + d_k) - V\}^2 P(d_k)\, dV. \tag{2.88}$$

(This is equal to the deviation from an ideal straight line.)

To derive a relationship between d_{tot}^2 and d^2, the quantization error given by

$$Q(V + d_k) = g(V + d_k) - (V + d_k) \tag{2.89}$$

is used to rewrite Eq. (2.88) as

$$d_{tot}^2 = \sum_k P(d_k)\frac{1}{Q}\int_0^Q (Q^2(V + d_k) + 2d_k Q(V + d_k) + d_k^2)\, dV \tag{2.90}$$

$$\begin{aligned} &= \sum_k P(d_k)\frac{1}{Q}\int_0^Q Q^2(V + d_k)\, dV \\ &\quad + 2\sum_k d_k P(d_k)\frac{1}{Q}\int_0^Q Q(V + d_k)\, dV \\ &\quad + \sum_k d_k^2 P(d_k)\frac{1}{Q}\int_0^Q dV. \end{aligned} \tag{2.91}$$

The integrals in Eq. (2.91) are independent of d_k. Moreover, $\sum_k P(d_k) = 1$. With the mean value of the quantization error

$$\overline{e} = \frac{1}{Q}\int_0^Q Q(V)\, dV \tag{2.92}$$

and the quadratic mean error

$$\overline{e^2} = \frac{1}{Q}\int_0^Q Q^2(V)\, dV, \tag{2.93}$$

it is possible to rewrite Eq. (2.91) as

$$d_{tot}^2 = \overline{e^2} + 2\overline{d}\overline{e} + \overline{d^2}. \tag{2.94}$$

With $\sigma_E^2 = \overline{e^2} - \overline{e}^2$ and $\sigma_D^2 = \overline{d^2} - \overline{d}^2$, Eq. (2.94) can be written as

$$\boxed{d_{tot}^2 = \sigma_E^2 + (\overline{d} + \overline{e})^2 + \sigma_D^2}\ . \tag{2.95}$$

Equations (2.94) and (2.95) describe the total noise power as a function of the quantization $(\overline{e}, \overline{e^2}, \sigma_E^2)$ and the dither addition $(\overline{d}, \overline{d^2}, \sigma_D^2)$. It can be seen that for zero-mean quantization, the middle term in Eq. (2.95) results in $\overline{d} + \overline{e} = 0$. The acoustically perceptible part of the total error power is represented by σ_E^2 and σ_D^2.

2.2.2 Implementation

The random sequence $d(n)$ is generated with the help of a random number generator with uniform PDF. For generating a triangular PDF random sequence, two independent uniform PDF random sequences $d_1(n)$ and $d_2(n)$ can be added. To generate a triangular highpass dither, the dither value $d_1(n)$ is added to $-d_1(n-1)$. Thus, only one random number generator is required. In conclusion, the following dither sequences can be implemented:

$$d_{RECT}(n) = d_1(n); \tag{2.96}$$

$$d_{TRI}(n) = d_1(n) + d_2(n); \tag{2.97}$$

$$d_{HP}(n) = d_1(n) - d_1(n-1). \tag{2.98}$$

The power density spectra of triangular PDF dither and triangular PDF HP dither are shown in Fig. 2.21. Figure 2.22 shows histograms of a uniform PDF dither and a triangular PDF highpass dither together with their respective power density spectra. The amplitude range of a uniform PDF dither lies between $\pm Q/2$, whereas it lies between $\pm Q$ for triangular PDF dither. The total noise power for triangular PDF dither is doubled.

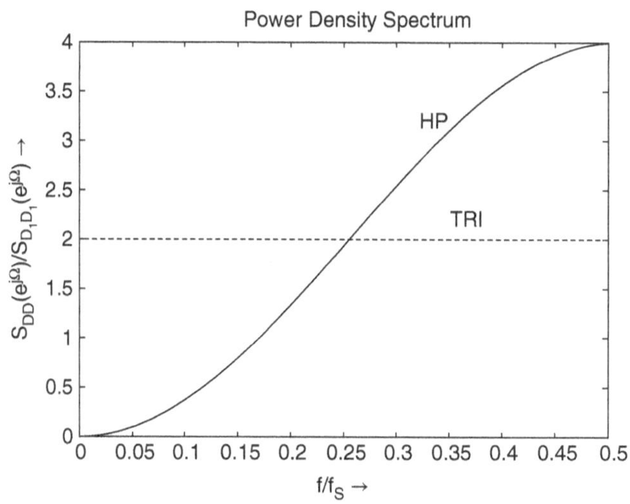

Figure 2.21 Normalized power density spectrum for triangular PDF dither (TRI) with $d_1(n) + d_2(n)$ and triangular PDF highpass dither (HP) with $d_1(n) - d_1(n-1)$.

2.2.3 Examples

The effect of the input amplitude of the quantizer is shown in Fig. 2.23 for a 16-bit quantizer ($Q = 2^{-15}$). A quantized sinusoidal signal with amplitude 2^{-15} (one-bit amplitude) and frequency $f/f_S = 64/1024$ is shown in Fig. 2.23a,b for rounding and truncation. Figure 2.23c,d shows their corresponding spectra. For truncation, Fig. 2.23c shows the spectral line of the signal and the spectral distribution of the quantization error with the harmonics of the input signal. For rounding, Fig. 2.23d shows that, with special signal frequency $f/f_S = 64/1024$, the quantization error is concentrated in uneven harmonics.

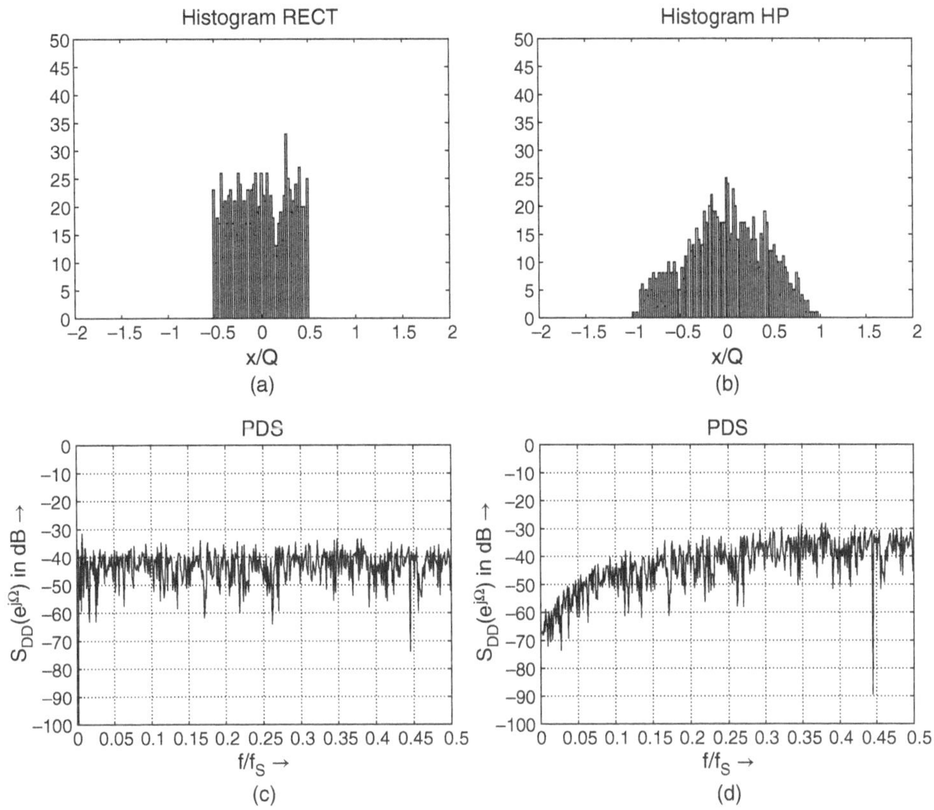

Figure 2.22 (a,d) Histogram and (c,d) power density spectrum of uniform PDF dither (RECT) with $d_1(n)$ and triangular PDF highpass dither (HP) with $d_1(n) - d_1(n-1)$.

In the following, only the rounding operation is used. By adding a uniform PDF random signal to the actual signal before quantization, the quantized signal shown in Fig. 2.24a results. The corresponding power density spectrum is illustrated in Fig. 2.24c. In the time domain, it is observed that the one-bit amplitudes approach zero so that the regular pattern of the quantized signal is affected. The resulting power density spectrum in Fig. 2.24c shows that the harmonics do not occur anymore and the noise power is uniformly distributed over the frequencies. For triangular PDF dither, the quantized signal is shown in Fig. 2.24b. Owing to triangular PDF, amplitudes of $\pm 2Q$ occur in addition to the signal values of $\pm Q$ and zero. Figure 2.24d shows the increase of the total noise power.

To illustrate the noise modulation for uniform PDF dither, the amplitude of the input is reduced to $A = 2^{-18}$ and the frequency is chosen as $f/f_S = 14/1024$. This means that the input amplitude to the quantizer is 0.25 bit. For a quantizer without additive dither, the quantized output signal is zero. For RECT dither, the quantized signal is shown in Fig. 2.25a. The input signal with amplitude $0.25Q$ is also shown. The power density spectrum of the quantized signal is shown in Fig. 2.25c. The spectral line of the signal and the uniform distribution of the quantization error can be seen. However, in the time domain, a correlation between positive and negative amplitudes of the input and the quantized positive and negative values of the output can be observed. In hearing tests, this noise modulation occurs if the amplitude of the input is decreased continuously and falls below

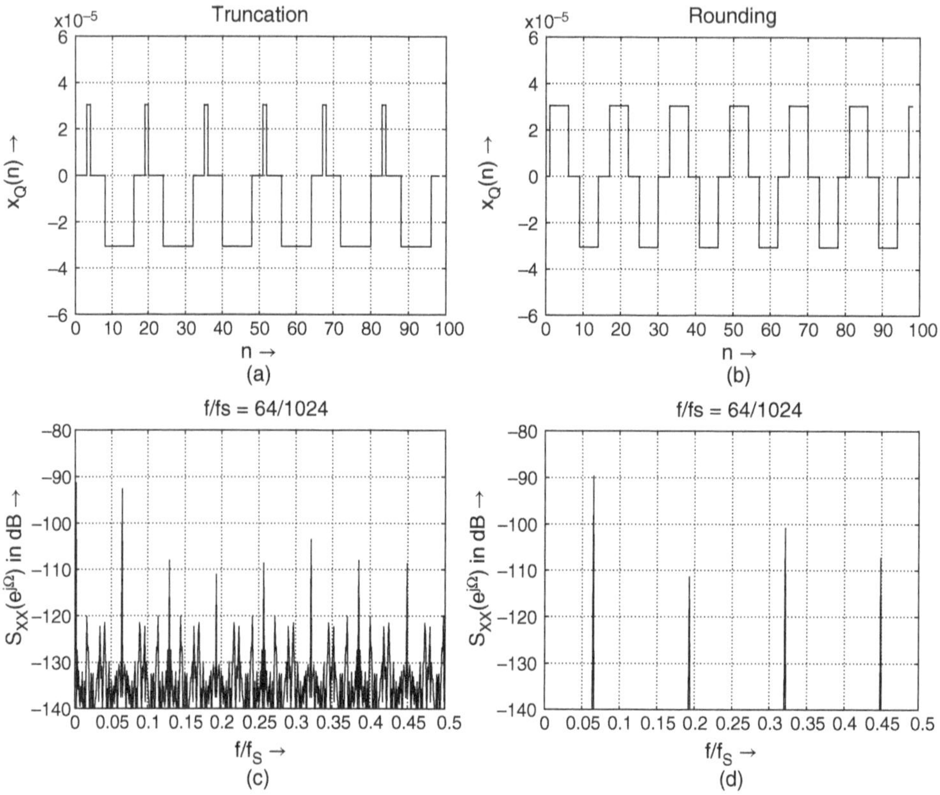

Figure 2.23 One-bit amplitude – quantizer with truncation (a,c) and rounding (b,d).

the amplitude of the quantization step. This process occurs for all fade-out processes that occur in speech and music signals. For positive low-amplitude signals, two output states zero and Q occur, and for negative low-amplitude signals, the output states zero and -Q occur. This is observed as a disturbing rattle which is overlapped on the actual signal. If the input level is further reduced, the quantized output approaches zero.

To reduce this noise modulation at low levels, a triangular PDF dither is used. Figure 2.25b shows the quantized signal and Fig. 2.25d shows the power density spectrum. It can be observed that the quantized signal has an irregular pattern. Hence, a direct association of positive half-waves with the positive output values, as well as vice versa, is not possible. The power density spectrum shows the spectral line of the signal along with an increase in noise power owing to triangular PDF dither. In acoustic hearing tests, the use of triangular PDF dither results in a constant noise floor even if the input level is reduced to zero.

2.3 Spectrum Shaping of Quantization – Noise Shaping

Using the linear model of a quantizer in Fig. 2.26 and the relations

$$e(n) = y(n) - x(n), \tag{2.99}$$

$$y(n) = [x(n)]_Q \tag{2.100}$$

$$= x(n) + e(n), \tag{2.101}$$

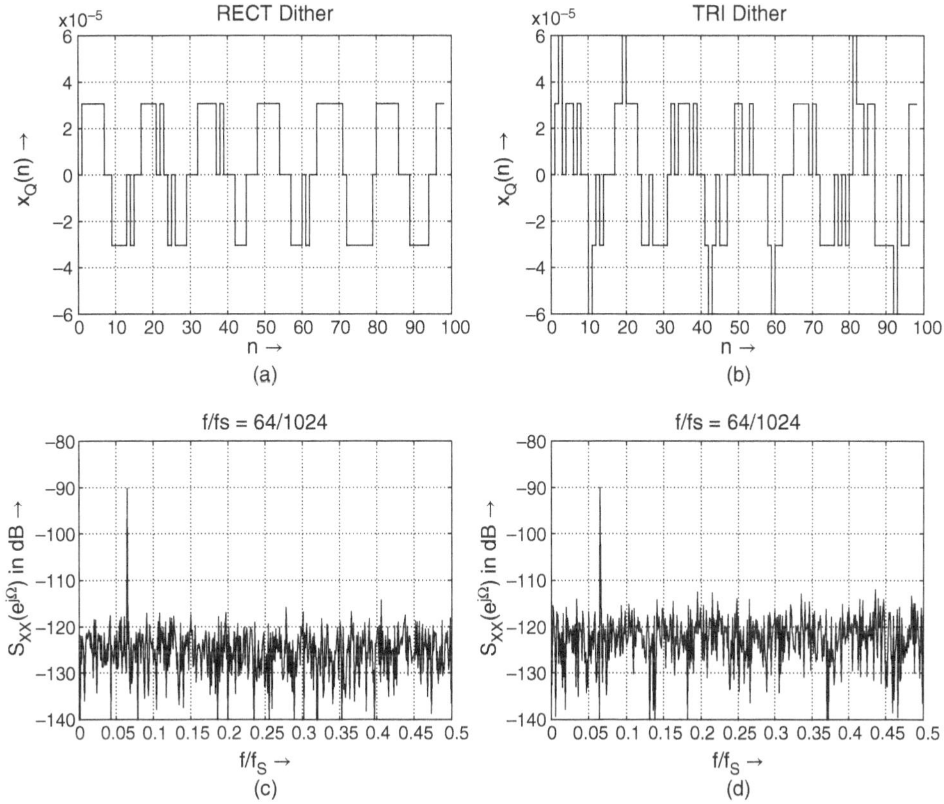

Figure 2.24 One-bit amplitude – rounding with RECT dither (a,c) and TRI dither (b,d).

the quantization error $e(n)$ may be isolated and fed back through a transfer function $H(z)$, as shown in Fig. 2.27. This leads to the spectral shaping of the quantization error as given by

$$y(n) = [x(n) - e(n) * h(n)]_Q \tag{2.102}$$

$$= x(n) + e(n) - e(n) * h(n), \tag{2.103}$$

$$e_1(n) = y(n) - x(n) \tag{2.104}$$

$$= e(n) * (\delta(n) - h(n)), \tag{2.105}$$

and the corresponding Z-transforms

$$Y(z) = X(z) + E(z)(1 - H(z)), \tag{2.106}$$

$$E_1(z) = E(z)(1 - H(z)). \tag{2.107}$$

A simple spectrum shaping of the quantization error $e(n)$ is achieved by feeding back with $H(z) = z^{-1}$, as shown in Fig. 2.28, and leads to

$$y(n) = [x(n) - e(n-1)]_Q \tag{2.108}$$

$$= x(n) - e(n-1) + e(n), \tag{2.109}$$

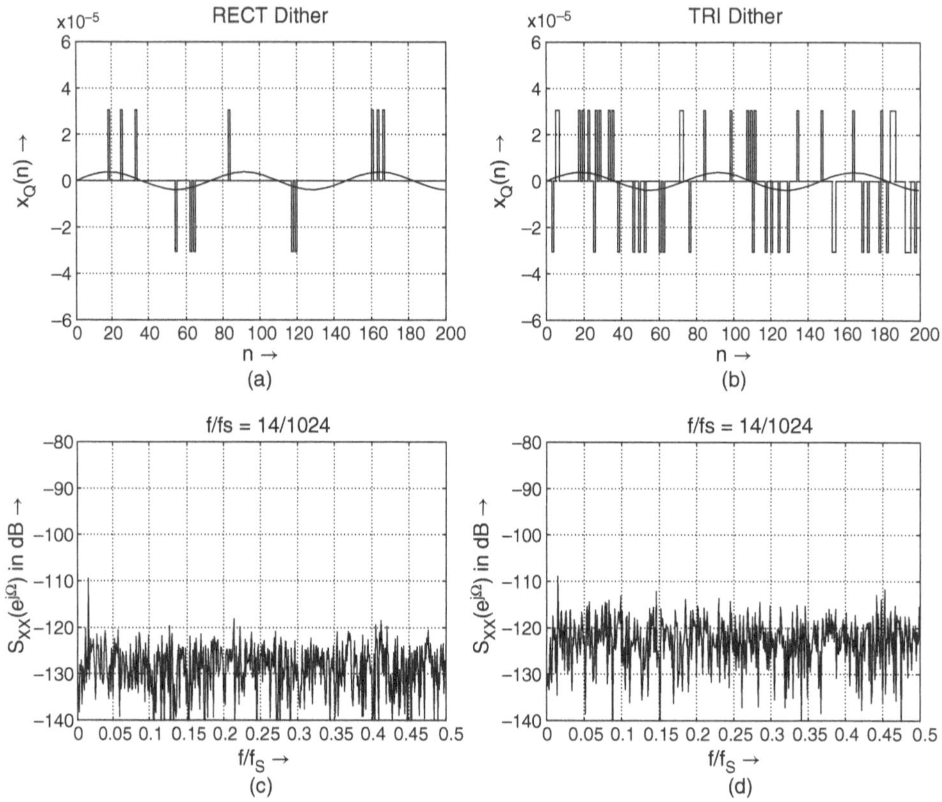

Figure 2.25 Noise modulation at 0.25-bit amplitude – rounding with RECT dither (a,c) and TRI dither (b,d).

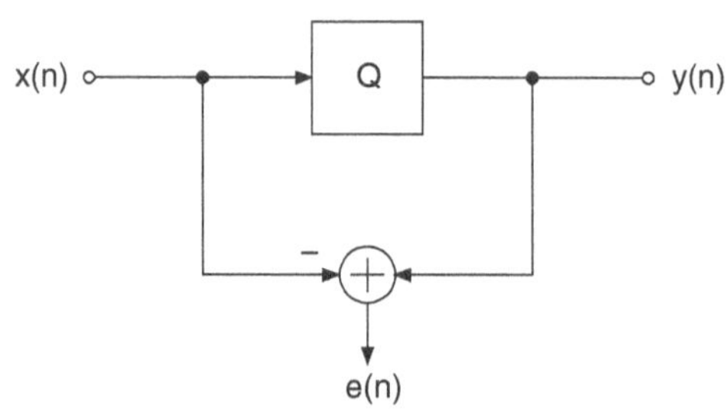

Figure 2.26 Linear model of a quantizer.

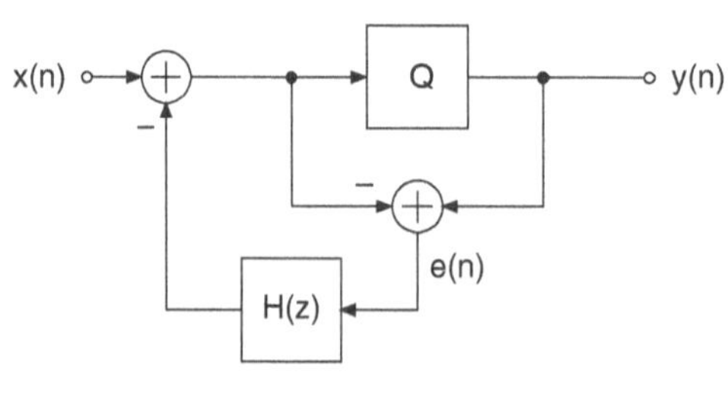

Figure 2.27 Spectrum shaping of quantization error.

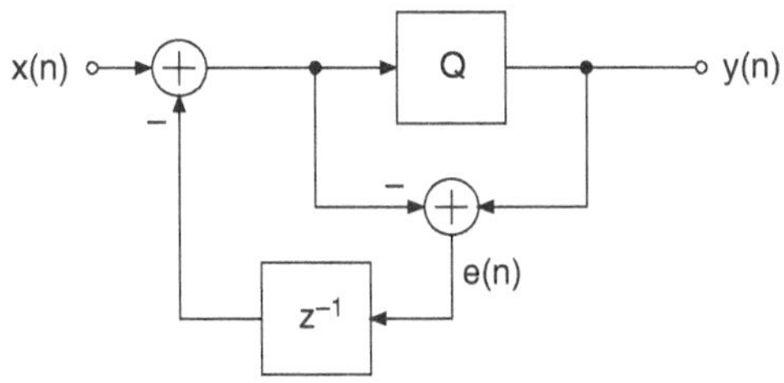

Figure 2.28 Highpass spectrum shaping of quantization error.

$$e_1(n) = y(n) - x(n) \tag{2.110}$$

$$= e(n) - e(n-1), \tag{2.111}$$

and the Z-transforms

$$Y(z) = X(z) + E(z)(1 - z^{-1}), \tag{2.112}$$

$$E_1(z) = E(z)(1 - z^{-1}). \tag{2.113}$$

Equation (2.113) shows a highpass weighting of the original error signal $e(n)$. By choosing $H(z) = z^{-1}(-2 + z^{-1})$, second-order highpass weighting given by

$$E_2(z) = E(z)(1 - 2z^{-1} + z^{-2}) \tag{2.114}$$

can be achieved. The power density spectrum of the error signal for the two cases is given by

$$S_{E_1E_1}(e^{j\Omega}) = |1 - e^{-j\Omega}|^2 S_{EE}(e^{j\Omega}), \tag{2.115}$$

$$S_{E_2E_2}(e^{j\Omega}) = |1 - 2e^{-j\Omega} + e^{-j2\Omega}|^2 S_{EE}(e^{j\Omega}). \tag{2.116}$$

Figure 2.29 shows the weighting of the power density spectrum by this noise shaping technique.

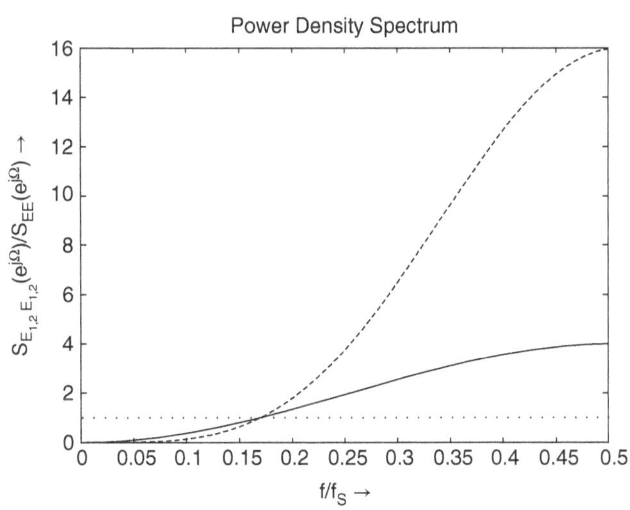

Figure 2.29 Spectrum shaping ($S_{EE}(e^{j\Omega})$, $\cdots$; $S_{E_1E_1}(e^{j\Omega})$, —; $S_{E_2E_2}(e^{j\Omega})$, - - -).

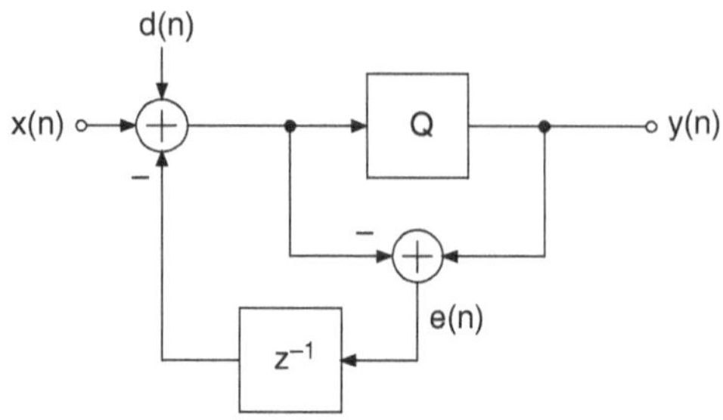

Figure 2.30 Dither and spectrum shaping.

By adding a dither signal $d(n)$ (see Fig. 2.30), the output and the error are given by

$$y(n) = [x(n) + d(n) - e(n-1)]_Q \tag{2.117}$$

$$= x(n) + d(n) - e(n-1) + e(n) \tag{2.118}$$

and

$$e_1(n) = y(n) - x(n) \tag{2.119}$$

$$= d(n) + e(n) - e(n-1). \tag{2.120}$$

For the Z-transforms, we write

$$Y(z) = X(z) + E(z)(1 - z^{-1}) + D(z), \tag{2.121}$$

$$E_1(z) = E(z)(1 - z^{-1}) + D(z). \tag{2.122}$$

The modified error signal $e_1(n)$ consists of the dither and the highpass shaped quantization error.

By moving the addition (Fig. 2.31) of the dither directly before the quantizer, a highpass spectrum shaping is obtained for both the error signal and the dither. Here, the following relationships hold:

$$y(n) = [x(n) + d(n) - e_0(n-1)]_Q \tag{2.123}$$

$$= x(n) + d(n) - e_0(n-1) + e(n); \tag{2.124}$$

$$e_0(n) = y(n) - (x(n) - e_0(n-1)) \tag{2.125}$$

$$= d(n) + e(n); \tag{2.126}$$

$$y(n) = x(n) + d(n) - d(n-1) + e(n) - e(n-1); \tag{2.127}$$

$$e_1(n) = d(n) - d(n-1) + e(n) - e(n-1), \tag{2.128}$$

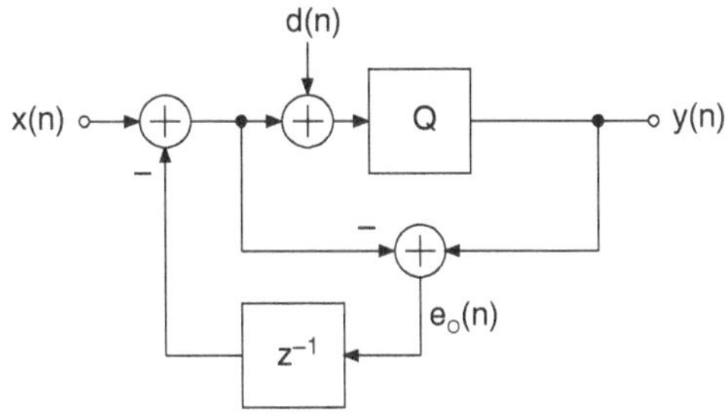

Figure 2.31 Modified dither and spectrum shaping.

with the Z-transforms given by

$$Y(z) = X(z) + E(z)(1 - z^{-1}) + D(z)(1 - z^{-1}), \quad (2.129)$$

$$E_1(z) = E(z)(1 - z^{-1}) + D(z)(1 - z^{-1}). \quad (2.130)$$

Apart from the discussed feedback structures which are easy to implement on a digital signal processor and which lead to highpass noise shaping, there are psychoacoustic-based noise-shaping methods that have been proposed in the literature [Ger89, Wan92, Hel07]. These methods use special approximations of the hearing threshold (threshold in quiet, absolute threshold) for the feedback structure $1 - H(z)$. Figure 2.32a shows several hearing threshold models as a function of frequency [ISO389, Ter79, Wan92]. It can be seen that the sensitivity of human hearing is high for frequencies between 2 and 6 kHz and sharply decreases for high and low frequencies. Figure 2.32b also shows the inverse ISO 389-7 threshold curve, which represents an approximation of the filtering operation in our perception. The feedback filter of the noise shaper should affect the quantization error with the inverse ISO389 weighting curve. Hence, the noise power in the frequency range with high sensitivity should be reduced and shifted toward lower and higher frequencies. Figure 2.33a shows the unweighted power density spectra of the quantization error for three special filters $H(z)$ [Wan92, Hel07]. Figure 2.33b depicts the same three power density spectra, weighted by the inverse ISO 389 threshold of Fig. 2.32b. These weighted power density spectra (PDS) show that the perceived noise power is reduced by all three noise shapers versus the frequency axis. Figure 2.34 shows a sinusoid with amplitude $Q = 2^{-15}$, which is quantized to $w = 16$ bit with psychoacoustic noise shaping. The quantized signal $x_Q(n)$ consists of different amplitudes reflecting the low-level signal. The power density spectrum of the quantized signal reflects the psychoacoustic weighting of the noise shaper with a fixed filter. A time-variant psychoacoustic noise shaping is described in [DeK03, Hel07], where the instantaneous masking threshold is used for adaptation of a time-variant filter.

2.4 Number Representation

The different applications in digital signal processing and transmission of audio signals lead to the question of the type of number representation for digital audio signals. In this section, basic properties of fixed-point and floating-point number representation in the context of digital audio signal processing are presented.

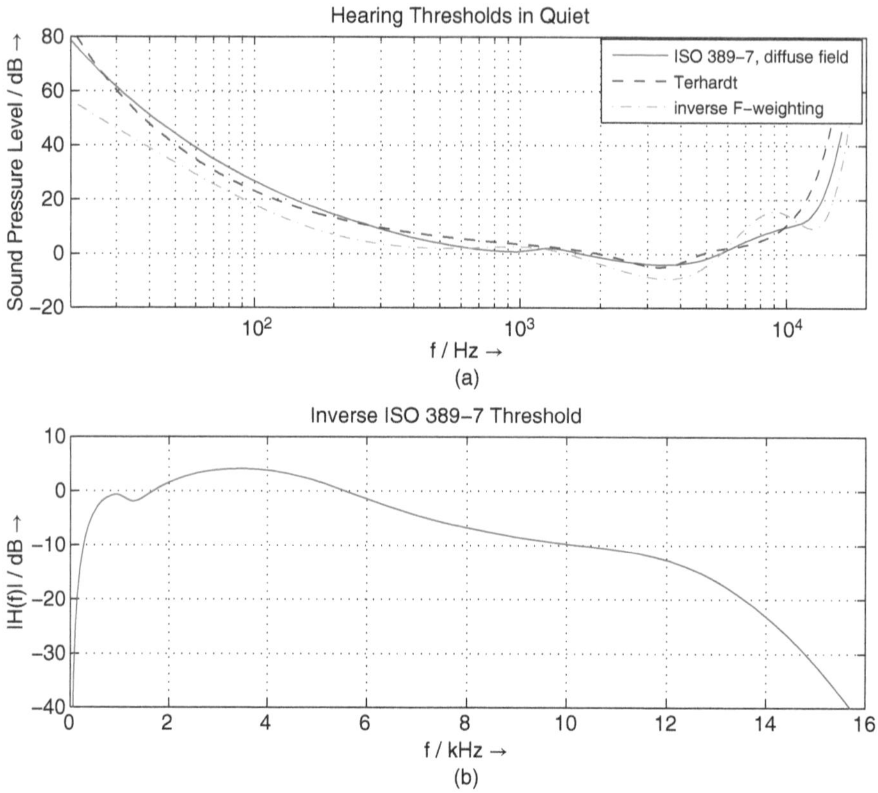

Figure 2.32 (a) Hearing thresholds in quiet. (b) Inverse ISO 389-7 threshold curve.

2.4.1 Fixed-point Number Representation

In general, an arbitrary real number x can be approximated by a finite summation

$$x_Q = \sum_{i=0}^{w-1} b_i 2^i, \tag{2.131}$$

where the possible values for b_i are 0 and 1.

The fixed-point number representation with a finite number w of binary places leads to four different interpretations of the number range (see Table 2.1 and Fig. 2.35).

The signed fractional representation (2s complement) is the usual format for digital audio signals and for algorithms in fixed-point arithmetic. For address and modulo operation, the unsigned integer is used. Owing to finite word length w, overflow occurs, as shown in Fig. 2.36. These curves have to be taken into consideration while carrying out operations, especially additions in 2s complement arithmetic.

Quantization is carried out with techniques, as shown in Table 2.2, for rounding and truncation. The quantization step size is characterized by $Q = 2^{-(w-1)}$ and the symbol $\lfloor x \rfloor$ denotes the biggest integer smaller than or equal to x. Figure 2.37 shows the rounding and truncation curves for 2s complement number representation. The absolute error shown in Fig. 2.37 is given by $e = x_Q - x$.

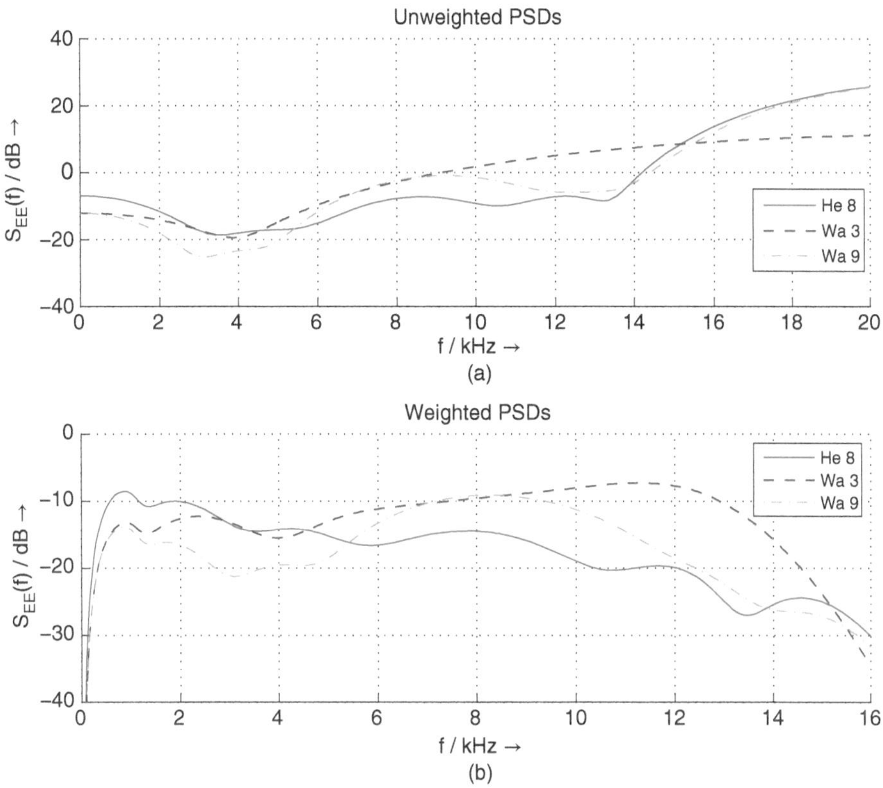

Figure 2.33 Power density spectra of three filter approximations (Wa3, third order filter; Wa9, ninth-order filter; He8, eighth-order filter [Wan92, Hel07]): (a) unweighted power spectral densities (PSDs); (b) inverse ISO 389-7 weighted PSDs.

Table 2.1 Bit location and range of values.

Type	Bit location	Range of values		
Signed 2s c.	$x_Q = -b_0 + \sum_{i=1}^{w-1} b_{-i}2^{-i}$	-1	$\leq x_Q \leq$	$1 - 2^{-(w-1)}$
Unsigned 2s c.	$x_Q = \sum_{i=1}^{w} b_{-i}2^{-i}$	0	$\leq x_Q \leq$	$1 - 2^{-w}$
Signed int.	$x_Q = -b_{w-1}2^{w-1} + \sum_{i=0}^{w-2} b_i 2^i$	-2^{w-1}	$\leq x_Q \leq$	$2^{w-1} - 1$
Unsigned int.	$x_Q = \sum_{i=0}^{w-1} b_i 2^i$	0	$\leq x_Q \leq$	$2^w - 1$

Digital audio signals are coded in the 2s complement number representation. For 2s complement representation, the range of values from $-X_{max}$ to $+X_{max}$ is normalized to the range -1 to $+1$ and is represented by the weighted finite sum $x_Q = -b_0 + b_1 \cdot 0.5 + b_2 \cdot 0.25 + b_3 \cdot 0.125 + \cdots + b_{w-1} \cdot 2^{-(w-1)}$. The variables b_0 to b_{w-1} are called bits and can take the values 1 or 0. The bit b_0 is called MSB (most significant bit) and b_{w-1} is called LSB (least significant bit). For positive numbers, b_0 is equal to 0 and for negative numbers, b_0 equals 1. For a three-bit quantization (see Fig. 2.38), a quantized value can be represented by $x_Q = -b_0 + b_1 \cdot 0.5 + b_2 \cdot 0.25$. The smallest

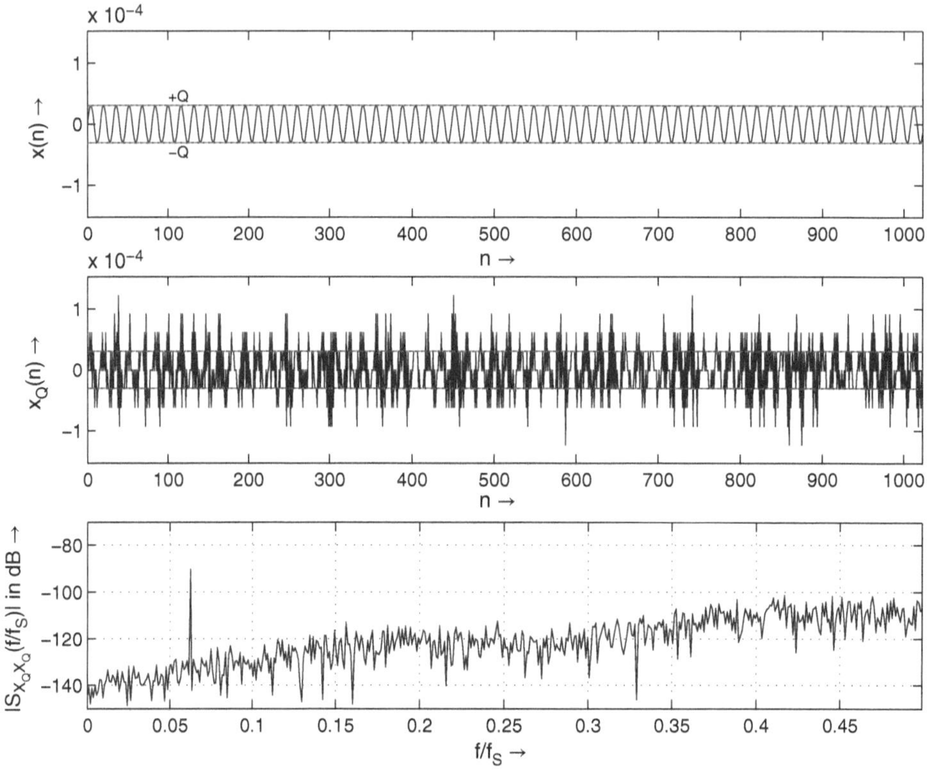

Figure 2.34 Psychoacoustic noise shaping: signal $x(n)$; quantized signal $x_Q(n)$; and power density spectrum of quantized signal.

Bit	31	30	29		2	1	0	
	-2^{0}	2^{-1}	2^{-2}	• • •	2^{-29}	2^{-30}	2^{-31}	Signed 2′ Complement
	2^{-1}	2^{-2}	2^{-3}	• • •	2^{-30}	2^{-31}	2^{-32}	Unsigned 2′ Complement
	-2^{31}	2^{30}	2^{29}	• • •	2^{2}	2^{1}	2^{0}	Signed Integer
	2^{31}	2^{30}	2^{29}	• • •	2^{2}	2^{1}	2^{0}	Unsigned Integer

Figure 2.35 Fixed-point formats.

quantization step size is 0.25. For a positive number 0.75, it follows that $0.75 = -0 + 1 \cdot 0.5 + 1 \cdot 0.25$. The binary coding for 0.75 is 011.

Dynamic Range. The dynamic range of a number representation is defined as the ratio of maximum to minimum number. For fixed-point representation with

$$x_{Qmax} = (1 - 2^{-(w-1)}), \tag{2.132}$$

$$x_{Qmin} = 2^{-(w-1)}, \tag{2.133}$$

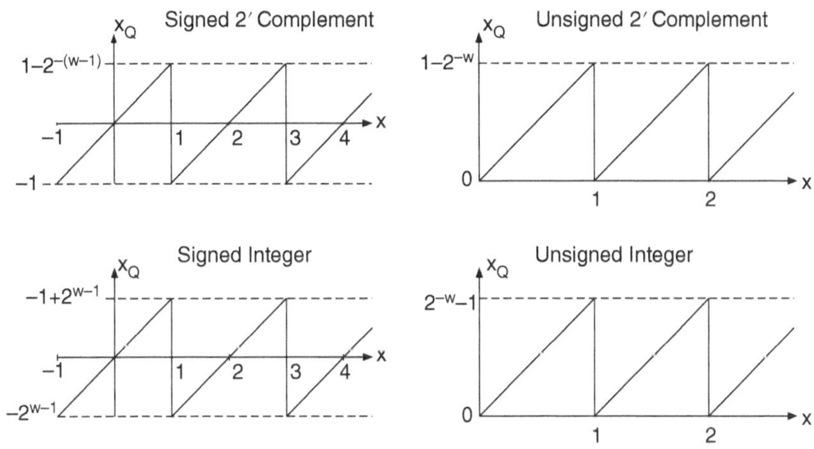

Figure 2.36 Number range.

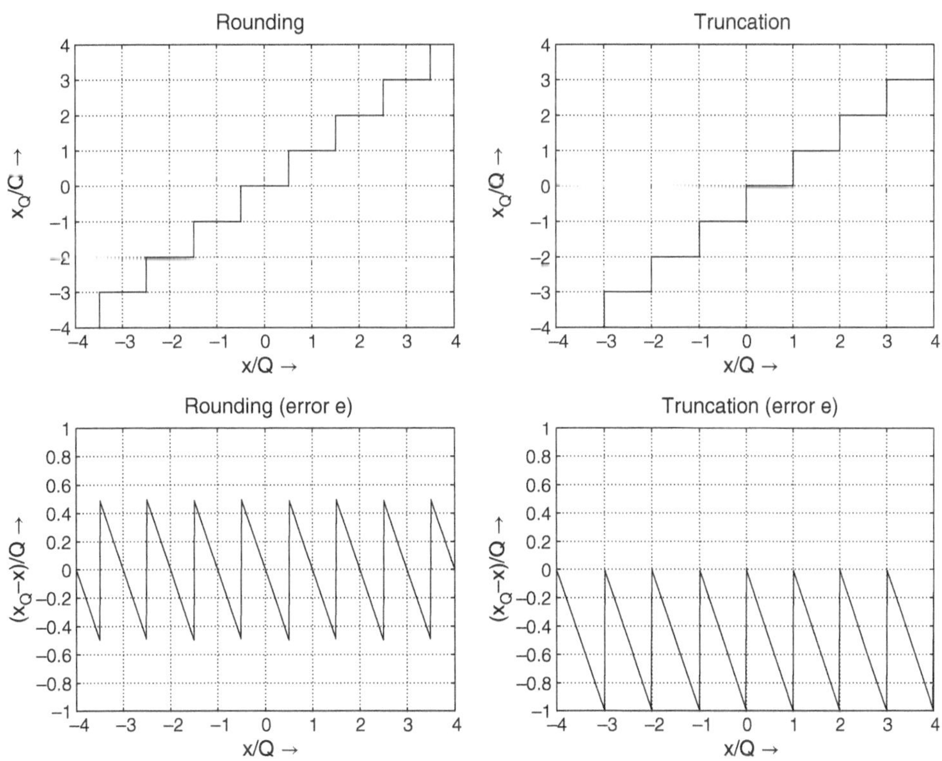

Figure 2.37 Rounding and truncation curves.

Table 2.2 Rounding and truncation of 2s complement numbers.

Type	Quantization	Error limits		
2s c. (r)	$x_Q = Q\lfloor Q^{-1}x + 0.5\rfloor$	$-Q/2$	$\leq x_Q - x \leq$	$Q/2$
2s c. (t)	$x_Q = Q\lfloor Q^{-1}x\rfloor$	$-Q$	$\leq x_Q - x \leq$	0

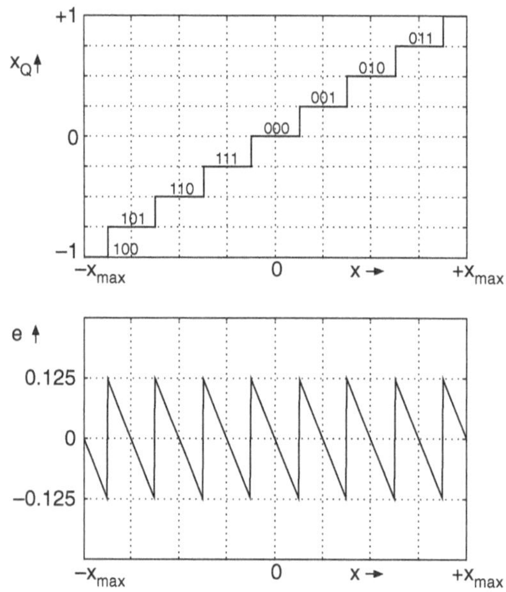

Figure 2.38 Rounding curve and error signal for $w = 3$ bit.

the dynamic range is given by

$$\mathrm{DR}_F = 20\,\log_{10}\left(\frac{x_{Qmax}}{x_{Qmin}}\right) = 20\,\log_{10}\left(\frac{1-Q}{Q}\right)$$
$$= 20\,\log_{10}(2^{w-1} - 1) \quad \text{in dB.} \tag{2.134}$$

Multiplication and Addition of Fixed-point Numbers. For the multiplication of two fixed-point numbers in the range from -1 to $+1$, the result is always less than 1. For the addition of two fixed-point numbers, care must be taken for the result to remain in the range from -1 to $+1$. An addition of $0.6 + 0.7 = 1.3$ must be carried out in the form $0.5(0.6 + 0.7) = 0.65$. This multiplication by the factor 0.5 or generally 2^{-s} is called scaling. An integer in the range from one to, for instance, eight is chosen for the scaling coefficient s.

Error Model. The quantization process for fixed-point numbers can be approximated as an addition of an error signal $e(n)$ to the signal $x(n)$ (see Fig. 2.39). The error signal is a random signal with white power density spectrum.

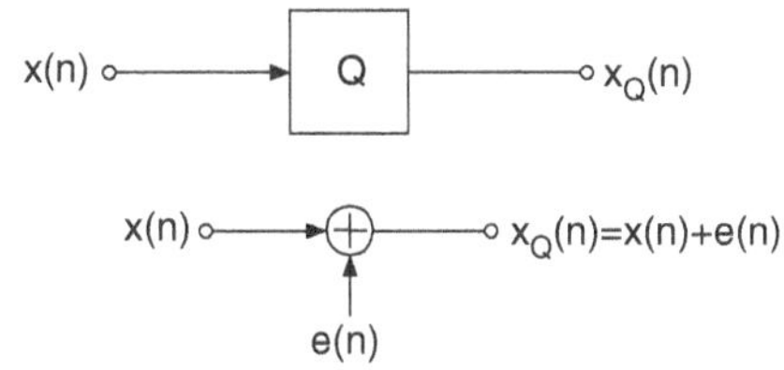

Figure 2.39 Model of a fixed-point quantizer.

Signal-to-noise Ratio. The SNR for a fixed-point quantizer is defined by

$$\text{SNR} = 10 \log_{10} \left(\frac{\sigma_X^2}{\sigma_E^2} \right), \tag{2.135}$$

where σ_X^2 is the signal power and σ_E^2 is the noise power.

2.4.2 Floating-point Number Representation

The representation of a floating-point number is given by

$$x_Q = M_G \, 2^{E_G} \tag{2.136}$$

with

$$0.5 \leq M_G < 1, \tag{2.137}$$

where M_G denotes the normalized mantissa and E_G the exponent. The normalized standard format (IEEE) is shown in Fig. 2.40 and special cases are given in Table 2.3. The mantissa M is implemented with a word length of w_M bits and is in fixed-point number representation. The exponent E is implemented with a word length of w_E bits and is an integer in the range from $-2^{w_E-1}+2$ to $2^{w_E-1}-1$. For an exponent word length of $w_E = 8$ bits, its range of values is between -126 and $+127$. The range of values of the mantissa is between 0.5 and 1. This is denoted as the normalized mantissa and is responsible for a unique representation of a number. For a fixed-point number in the range between 0.5 and 1, it follows that the exponent of the floating-point number representation is $E = 0$. For representing a fixed-point number in the range between 0.25 and 0.5 in floating-point representation, the range of values of the normalized mantissa M lies between 0.5 and 1, and for the exponent it follows $E = -1$. As an example, for a fixed-point number 0.75,

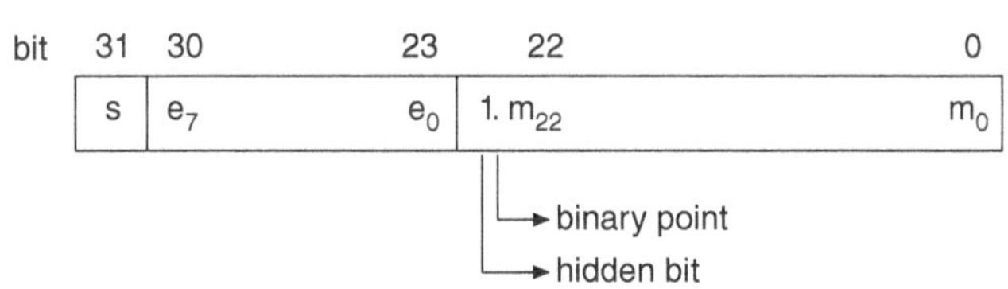

Figure 2.40 Floating-point number representation.

Table 2.3 Special cases of floating-point number representation.

Type	Exponent	Mantissa	Value
NAN	255	$\neq 0$	undefined
Infinity	255	0	$(-1)^s$ infinity
Normal	$1 \leq e \leq 254$	any	$(-1)^s(0.m)2^{e-127}$
Zero	0	0	$(-1)^s \cdot 0$

the floating-point number $0.75 \cdot 2^0$ results. The fixed-point number 0.375 is not represented as the floating-point number $0.375 \cdot 2^0$. With the normalized mantissa, the floating-point number is expressed as $0.75 \cdot 2^{-1}$. Owing to normalization, ambiguity of the floating-point number representation is avoided. Numbers > 1 can be represented. For example, 1.5 becomes $0.75 \cdot 2^1$ in floating-point number representation.

Figure 2.41 shows the rounding and truncation curves for floating-point representation and the absolute error $e = x_Q - x$. The curves for floating-point quantization show that for small amplitudes, small quantization steps sizes occur. In contrast to fixed-point representation, the absolute error is dependent on the input signal.

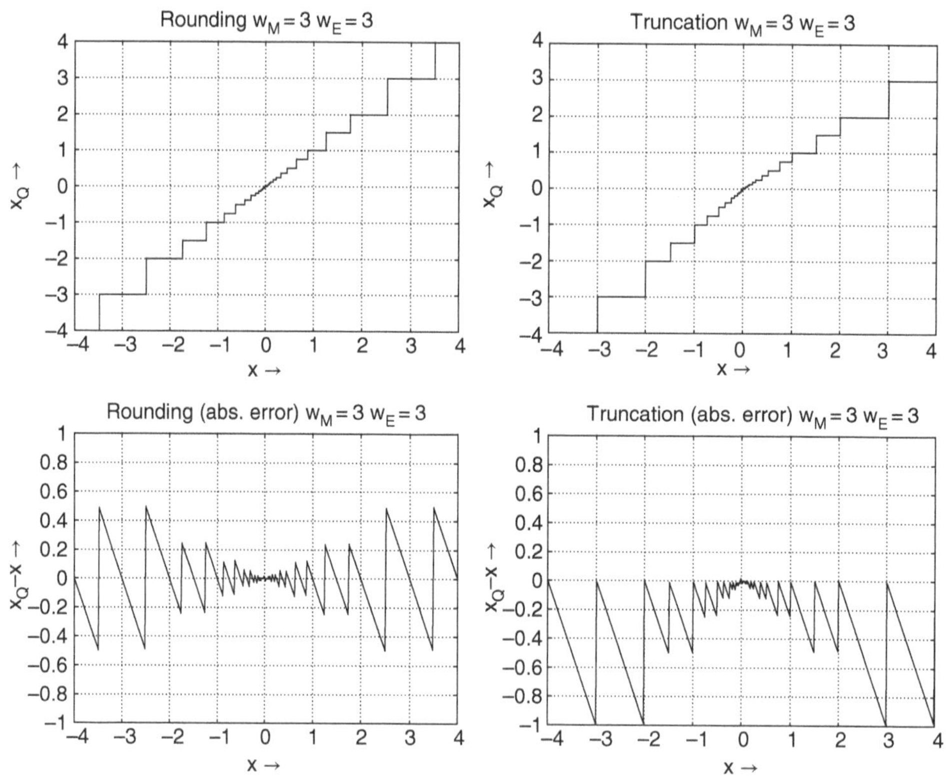

Figure 2.41 Rounding and truncation curves for floating-point representation.

In the interval

$$2^{E_G} \le x < 2^{E_G+1}, \tag{2.138}$$

the quantization step is given by

$$Q_G = 2^{-(w_M-1)} 2^{E_G}. \tag{2.139}$$

For the relative error

$$e_r = \frac{x_Q - x}{x} \tag{2.140}$$

of the floating-point representation, a constant upper limit can be stated as

$$|e_r| \le 2^{-(w_M-1)}. \tag{2.141}$$

Dynamic Range. With the maximum and minimum numbers given by

$$x_{Qmax} = \left(1 - 2^{-(w_M-1)}\right) 2^{E_{Gmax}}, \tag{2.142}$$

$$x_{Qmin} = 0.5\, 2^{E_{Gmin}}, \tag{2.143}$$

and

$$E_{Gmax} = 2^{w_E-1} - 1, \tag{2.144}$$

$$E_{Gmin} = -2^{w_E-1} + 2, \tag{2.145}$$

the dynamic range for floating-point representation is given by

$$\begin{aligned} \mathrm{DR}_G &= 20\log_{10}\left(\frac{(1-2^{-(w_M-1)})2^{E_{Gmax}}}{0.5\, 2^{E_{Gmin}}}\right) \\ &- 20\log_{10}(1-2^{-(w_M-1)})2^{E_{Gmax}-E_{Gmin}+1} \\ &= 20\log_{10}(1-2^{-(w_M-1)})2^{2^{w_E}-2}\ \mathrm{dB}. \end{aligned} \tag{2.146}$$

Multiplication and Addition of Floating-point Numbers. For multiplications with floating-point numbers, the exponents of both numbers $x_{Q1} = M_1 2^{E_1}$ and $x_{Q2} = M_2 2^{E_2}$ are added and the mantissas are multiplied. The resulting exponent $E_G = E_1 + E_2$ is adjusted so that $M_G = M_1 M_2$ lies in the interval $0.5 \le M_G < 1$. For additions, the smaller number is denormalized to get the same exponent. Then both mantissa are added and the result is normalized.

Error Model. With the definition of the relative error $e_r(n) = \frac{x_Q(n)-x(n)}{x(n)}$, the quantized signal can be written as

$$x_Q(n) = x(n)\cdot(1+e_r(n)) = x(n) + x(n)\cdot e_r(n). \tag{2.147}$$

Floating-point quantization can be modeled as an additive error signal $e(n) = x(n)\cdot e_r(n)$ to the signal $x(n)$ (see Fig. 2.42).

Signal-to-noise Ratio. Under the assumption that the relative error is independent of the input x, the noise power of the floating-point quantizer can be written as

$$\sigma_E^2 = \sigma_X^2 \cdot \sigma_{E_r}^2. \tag{2.148}$$

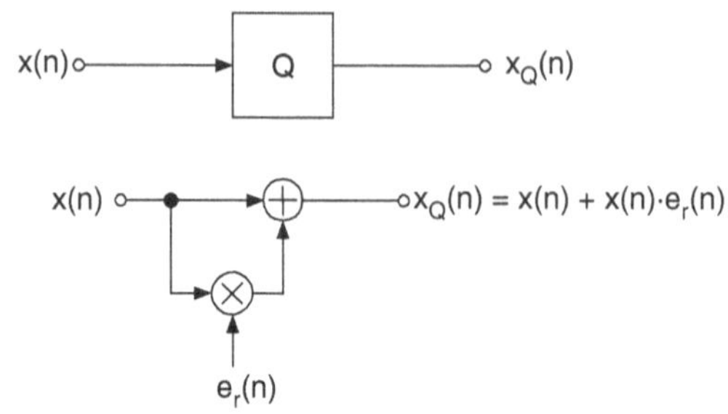

Figure 2.42 Model of a floating-point quantizer.

For the SNR, we can derive

$$\begin{aligned} \text{SNR} &= 10\log_{10}\left(\frac{\sigma_X^2}{\sigma_E^2}\right) \\ &= 10\log_{10}\left(\frac{\sigma_X^2}{\sigma_X^2\cdot\sigma_{E_r}^2}\right) \\ &= 10\log_{10}\left(\frac{1}{\sigma_{E_r}^2}\right). \end{aligned} \tag{2.149}$$

Equation (2.149) shows that the SNR is independent of the level of the input. It is only dependent on the noise power $\sigma_{E_r}^2$ which, in turn, is only dependent on the word length w_M of the mantissa of the floating-point representation.

2.4.3 Effects on Format Conversion and Algorithms

First, a comparison of the SNR is made for the fixed-point and floating-point number representations. Figure 2.43 shows the SNR as a function of input level for both number representations. The fixed-point word length is $w = 16$ bits. The word length of the mantissa in floating-point representation is also $w_M = 16$ bits whereas that of the exponent is $w_E = 4$ bits The SNR for the floating-point representation shows that it is independent of input level and varies as a saw-tooth curve in a 6-dB grid. If the input level is so low that a normalization of the mantissa arising from finite number representation is not possible, then the SNR is comparable to the fixed-point representation. While using the full range, both fixed-point and floating-point result in the same SNR. It can be noticed that the SNR for the fixed-point representation depends on the input level. This SNR in the digital domain is an exact image of the level-dependent SNR of an analog signal in the analog domain. A floating-point representation cannot improve this SNR. Rather, the floating-point curve is vertically shifted downwards to the value of the SNR of an analog signal.

AD/DA Conversion. Before processing, storing, and transmission of audio signals, the analog audio signal is converted into a digital signal. The precision of this conversion depends on the word length w of the AD converter. The resulting SNR is $6w$ dB for uniform PDF inputs. The SNR in the analog domain depends on the level. This linear dependence of the SNR on the level is preserved after AD conversion with subsequent fixed-point representation.

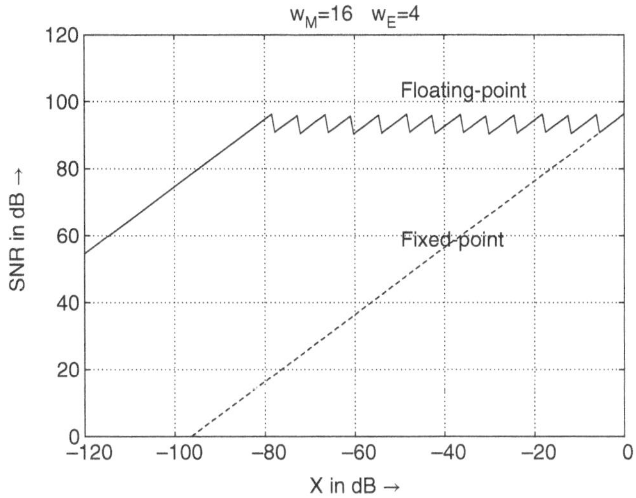

Figure 2.43 Signal-to-noise ratio for an input level.

Equalizers. While implementing equalizers with recursive digital filters, the SNR depends on the choice of the recursive filter structure. By a suitable choice of a filter structure and methods to spectrally shape the quantization errors, optimal SNRs are obtained for a given word length. The SNR for a fixed-point representation depends on the word length, and for a floating-point representation on the word length of the mantissa. For filter implementations with fixed-point arithmetic, boost filters have to be implemented with a scaling within the filter algorithm. The properties of a floating-point representation take care of automatic scaling in boost filters. If an insert inpit/output (I/O) in fixed-point representation follows a boost filter in floating-point representation, then the same scaling as in fixed-point arithmetics has to be done.

Dynamic Range Control. Dynamic range control is performed by a simple multiplicative weighting of the input signal with a control factor. The latter follows from calculating the peak and root-mean-square (RMS) value of the input signal. The number representation of the signal has no influence on the properties of the algorithm. Owing to the normalized mantissa in the floating-point representation, some simplifications are produced while determining the control factor.

Mixing/Summation. While mixing signals to a stereo image, only multiplications and additions occur. Under the assumption of incoherent signals, an overload reserve can be estimated. This implies a reserve of 20/30 dB for 48/96 sources. For fixed-point representation, the overload reserve is provided by a number of overflow bits in the accumulator of a digital signal processor (DSP). The properties of automatic scaling in floating-point arithmetic provide for overload reserves. For both number representations, the summation signal must be matched with the number representation of the output. While dealing with AES/EBU outputs or MADI outputs, both number representations are adjusted to a fixed-point format. Similarly, within heterogeneous system solutions, it is logical to make heterogeneous use of both number representations though corresponding number representations have to be converted.

Because the SNR in fixed-point representation depends on the input level, a conversion from fixed-point to floating-point representation does not lead to a change of the SNR, i.e. the conversion does not improve the SNR. Further signal processing with floating-point or fixed-point arithmetic does not alter the SNR as long as the algorithms are chosen and programmed accordingly. Reconversion from floating-point to fixed-point representation again leads to a level-dependent SNR.

As a consequence, for two-channel DSP systems which operate with AES/EBU or with analog inputs and outputs, and which are used for equalization, dynamic range control, room simulation etc., the above-mentioned holds. These conclusions are also valid for digital mixing consoles for which digital inputs from AD converters or from multitrack machines are represented in fixed-point format (AES/EBU or MADI). The number representation for inserts and auxiliaries is specific to a system. Digital AES/EBU (or MADI) inputs and outputs are realized in fixed-point number representation.

2.5 JS Applet – Quantization, Dither, and Noise Shaping

This applet shown in Fig. 2.44 demonstrates audio effects resulting from quantization. It is designed as a first insight into the perceptual effects of quantizing an audio signal.

The following functions can be selected on the lower right of the graphical user interface:

- Quantizer
  - word length w leads to a quantization step size $Q = 2^{-(w-1)}$;
- Dither
  - rect dither – uniform probability density function,
  - tri dither – triangular probability density function,
  - highpass dither – triangular probability density function and highpass power spectral density;
- Noise shaping
  - first-order $H(z) = z^{-1}$,
  - second-order $H(z) = -2z^{-1} + z^{-2}$,
  - psychoacoustic noise shaping.

You can choose between two predefined audio files from our web server (*audio1.wav* or *audio2.wav*) or your own local WAV file to be processed [Gui05].

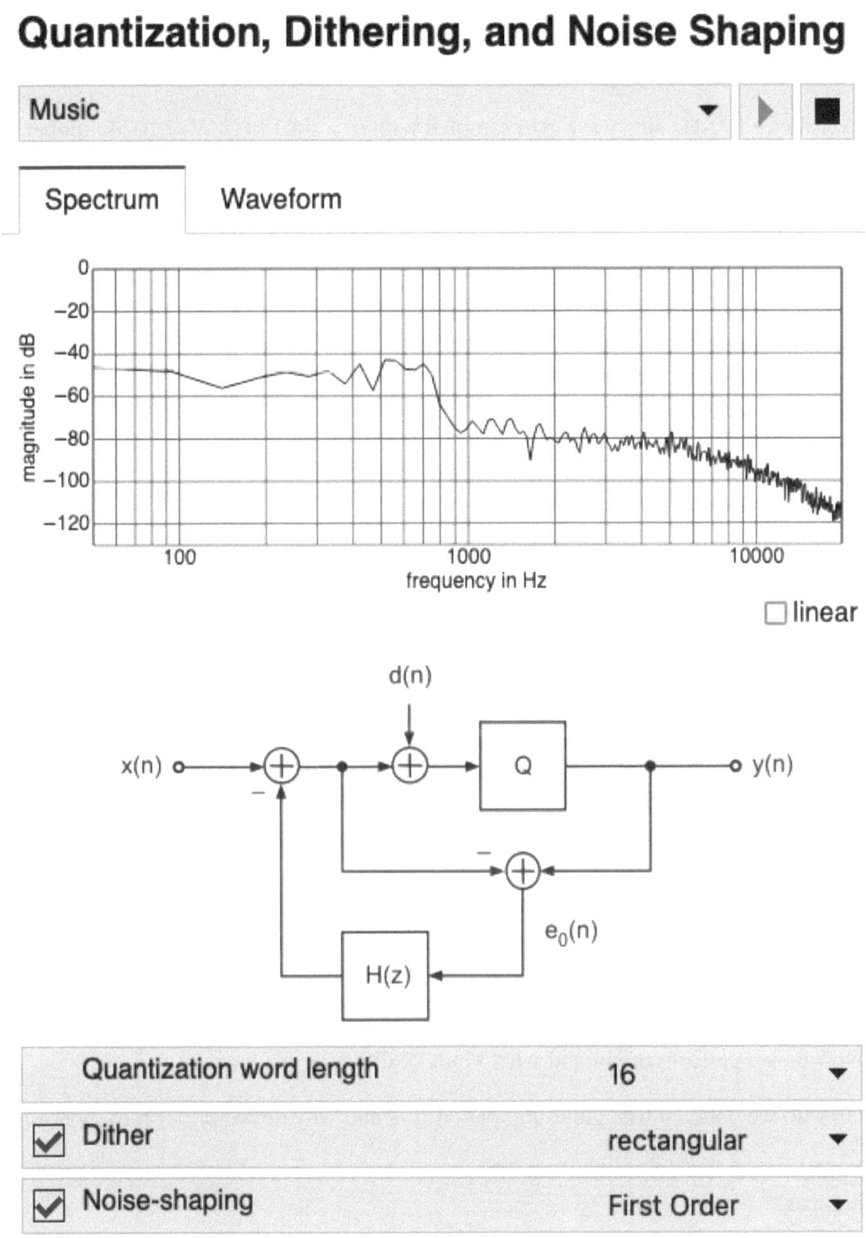

Figure 2.44 JS applet – quantization, dither, and noise shaping.

2.6 Exercises

1. Quantization

1. Consider a 100-Hz sine wave $x(n)$ sampled with $f_S = 44.1$ kHz, $N = 1024$ number of samples, and $w = 3$ bit (word length). What is the number of quantization levels? What is the quantization step Q when the signal is normalized to $-1 \leq x(n) < 1$. Show graphically how quantization is performed. What is the maximum error for this 3-bit quantizer? Write a Matlab code for quantization with rounding and truncation.

2. Derive the mean value, the variance, and the peak factor P_F of sequence $e(n)$ if the signal has a uniform probability density function (PDF) in the range $-\frac{Q}{2} < e(n) < -\frac{Q}{2}$. Derive the signal-to-noise ratio (SNR) for this case. What will happen if we increase our word length by one bit?

3. As the input signal level decreases from maximum amplitude to very low amplitudes, the error signal becomes more audible. Describe the error calculated above when w decreases to 1 bit? Is the classical quantization model still valid? What can be done to avoid this distortion?

4. Write a Matlab code for a quantizer with $w = 16$ bit with rounding and truncation.

   - Plot the nonlinear transfer characteristic and the error signal when the input signal covers the range $3Q < x(n) < 3Q$.

   - Consider the sine wave $x(n) = A\sin(2\pi \frac{f}{f_S} n), n = 0, \ldots, N-1$ with $A = Q, \frac{f}{f_S} = 64/N$ and $N = 1024$. Plot the output signal ($n = 0, \ldots, 99$) of a quantizer with rounding and truncation in the time domain and the frequency domain.

   - Compute for both quantization types the quantization error and the SNR.

2. Dither

1. What is dither and when do we have to use it?

2. How do we perform dither and what kinds of dither do we have?

3. How do we obtain a triangular highpass dither and why do we prefer it to other dithers?

4. Matlab: Generate corresponding dither signals for rectangular, triangular, and triangular highpass.

5. Plot the amplitude distribution and the spectrum of the output $x_Q(n)$ of a quantizer for every dither type.

3. Noise Shaping

1. What is noise shaping and when do we do it?

2. Why is it necessary to dither during noise shaping and how do we do this?

3. Matlab: The first noise shaper used is without dither and assumes that the transfer function in the feedback structure can be first-order $H(z) = z^{-1}$ or second-order $H(z) = -2z^{-1} + z^{-2}$. Plot the output $x_Q(n)$ and the error signal $e(n)$ and its spectrum. Show with a plot how the error signal will be shaped.

4. The same noise shaper is now used with a dither signal. Is it really necessary to dither with noise shaping? Where would you add your dither in the flow graph to achieve better results?

5. In the feedback structure, we now use a psychoacoustic-based noise shaper which uses the Wannamaker filter coefficients:

$$h_3 = [1.623, -0.982, 0.109];$$

$$h_5 = [2.033, -2.165, 1.959, -1.590, 0.6149];$$

$$h_9 = [2.412, -3.370, 3.937, -4.174, 3.353, -2.205, 1.281, -0.569, 0.0847].$$

Show with a Matlab plot how the error is shaped by this filter.

References

[DeK03] D. De Koning, W. Verhelst: *On Psychoacoustic Noise Shaping for Audio Requantization*, Proc. ICASSP-03, Vol. 5, pp. 453–456, April 2003.

[Ger89] M.A. Gerzon, P.G. Craven: *Optimal Noise Shaping and Dither of Digital Signals*, Proc. 87th AES Convention, New York, Preprint No. 2822, October 1989.

[Gui05] M. Guillemard, C. Ruwwe, U. Zölzer: *J-DAFx - Digital Audio Effects in Java*, Proc. 8th Int. Conference on Digital Audio Effects (DAFx-05), Madrid, Spain, pp.161–166, 2005.

[Hel07] C.R. Helmrich, M. Holters, and U. Zölzer, *Improved Psychoacoustic Noise Shaping for Requantization of High-Resolution Digital Audio*, AES 31st International Conference, London, UK, June 2007.

[ISO389] ISO 389-7:2005, *Acoustics – Reference zero for the calibration of audiometric equipment – Part 7: Reference threshold of hearing under free-field and diffuse-field listening conditions*, Geneva, Switzerland, 2005.

[Lip86] S.P. Lipshitz, J. Vanderkoy: *Digital Dither*, Proc. 81st AES Convention, Los Angeles, Preprint No. 2412, November 1986.

[Lip92] S.P. Lipshitz, R.A. Wannamaker, J. Vanderkoy: *Quantization and Dither: A Theoretical Survey*, J. Audio Eng. Soc., Vol. 40, No. 5, pp. 355–375, May 1992.

[Sha48] C.E. Shannon: *A Mathematical Theory of Communication*, Bell Systems, Techn. J., pp. 379–423, pp. 623–656, 1948.

[Sri77] A.B. Sripad, D.L. Snyder: *A Necessary and Sufficient Condition for Quantization Errors to be Uniform and White*, IEEE Trans. ASSP, Vol. 25, pp. 442–448, Oct. 1977.

[Ter79] E. Terhardt, *Calculating Virtual Pitch*, Hearing Res., Vol. 1, pp. 155–182, 1979.

[Van89] J. Vanderkooy, S.P. Lipshitz: *Digital Dither: Signal Processing with Resolution Far below the Least Significant Bit*, Proc. AES Int. Conf. on Audio in Digital Times, pp. 87–96, May 1989.

[Wan92] R.A. Wannamaker: *Psychoacoustically Optimal Noise Shaping*, J. Audio Eng. Soc., Vol. 40, No. 7/8, pp. 611–620, July/August 1992.

[Wan00] R.A. Wannamaker, S.P. Lipshitz, J. Vanderkooy, J.N. Wright: *A Theory of Nonsubtractive Dither*, IEEE Trans. Signal Processing, Vol. 48, No. 2, pp. 499–516, 2000.

[Wid61] B. Widrow: *Statistical Analysis of Amplitude-Quantized Sampled-Data Systems*, Trans. AIEE, Pt. II, Vol. 79, pp. 555–568, Jan. 1961.

Chapter 3

Sampling Rate Conversion

U. Zölzer

Several different sampling rates are established for digital audio applications. For broadcasting, professional, and consumer audio, sampling rates of 32, 48, and 44.1 kHz are used, respectively. Moreover, other sampling rates are derived from different frame rates for film and video. In connecting systems with different uncoupled sampling rates, there is a need for sampling rate conversion. In this chapter, synchronous sampling rate conversion with rational factor L/M for coupled clock rates and asynchronous sampling rate conversion will be discussed where the different sampling rates are not synchronized with each other.

3.1 Basics

Sampling rate conversion consists of upsampling and downsampling and anti-imaging and antialiasing filtering [Cro83, Vai93, Fli00, Opp99]. The discrete-time Fourier transform of the sampled signal $x(n)$ with sampling frequency $f_S = 1/T$ ($\omega_S = 2\pi f_S$) is given by

$$X(e^{j\Omega}) = \frac{1}{T}\sum_{k=-\infty}^{\infty} X_a\left(j\omega + jk\underbrace{\frac{2\pi}{T}}_{\omega_S}\right) \quad \text{with } \Omega = \omega T \tag{3.1}$$

with the Fourier transform $X_a(j\omega)$ of the continuous-time signal $x(t)$. For ideal sampling, the condition

$$X(e^{j\Omega}) = \frac{1}{T}X_a(j\omega), \quad |\Omega| \leq \pi \tag{3.2}$$

holds.

Digital Audio Signal Processing, Third Edition. Udo Zölzer.
© 2022 John Wiley & Sons Ltd. Published 2022 by John Wiley & Sons Ltd.

3.1.1 Upsampling and Anti-Imaging Filtering

For upsampling, the signal

$$x(n) \circ\!\!-\!\!\bullet\, X(e^{j\Omega}) \tag{3.3}$$

by factor L between consecutive samples $L-1$, zero samples will be included (see Fig. 3.1). This leads to the upsampled signal

$$w(m) = \begin{cases} x\left(\frac{m}{L}\right) & m = 0, \pm L, \pm 2L, ..., \\ 0 & \text{otherwise,} \end{cases} \tag{3.4}$$

with sampling frequency $f_S' = 1/T' = L \cdot f_S = L/T$ ($\Omega' = \Omega/L$) and the corresponding Fourier transform

$$W(e^{j\Omega'}) = \sum_{m=-\infty}^{\infty} w(m)e^{-jm\Omega'} = \sum_{m=-\infty}^{\infty} x(m)e^{-jmL\Omega'} = X(e^{jL\Omega'}). \tag{3.5}$$

The suppression of the image spectra is achieved by anti-imaging filtering of $w(m)$ with $h(m)$, such that the output signal is given by

$$y(m) = w(m) * h(m), \tag{3.6}$$

$$Y(e^{j\Omega'}) = H(e^{j\Omega'}) \cdot X(e^{j\Omega' L}). \tag{3.7}$$

For adjusting the signal power in the baseband, the Fourier transform of the impulse response

$$H(e^{j\Omega'}) = \begin{cases} L & |\Omega'| \leq \pi/L, \\ 0 & \text{otherwise,} \end{cases} \tag{3.8}$$

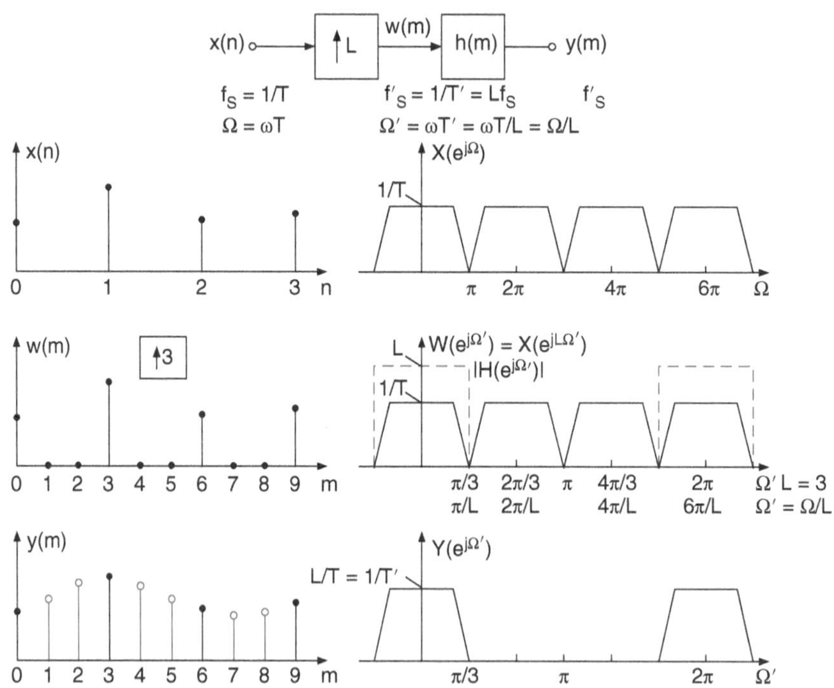

Figure 3.1 Upsampling by L and anti-imaging filtering in the time and frequency domains.

needs a gain factor L in the passband, such that the output signal $y(m)$ has the Fourier transform given by

$$Y(e^{j\Omega'}) = LX(e^{j\Omega' L}) \tag{3.9}$$

$$= \underbrace{L\frac{1}{T}\sum_{k=-\infty}^{\infty} X_a\left(j\omega + jLk\frac{2\pi}{T}\right)}_{\text{with Eqs. (3.1) and (3.5)}} \tag{3.10}$$

$$= L\frac{1}{LT'}\sum_{k=-\infty}^{\infty} X_a\left(j\omega + jLk\frac{2\pi}{LT'}\right) \tag{3.11}$$

$$= \underbrace{\frac{1}{T'}\sum_{k=-\infty}^{\infty} X_a\left(j\omega + jk\frac{2\pi}{T'}\right)}_{\text{spectrum of signal with } f_S'=Lf_S}. \tag{3.12}$$

The output signal represents the sampling of the input $x(t)$ with sampling frequency $f_S' = Lf_S$.

3.1.2 Downsampling and Antialiasing Filtering

For downsampling a signal $x(n)$ by M, the signal has to be band limited to π/M to avoid aliasing after the downsampling operation (see Fig. 3.2). Band limiting is achieved by filtering with $H(e^{j\Omega})$ according to

$$w(m) = x(m) * h(m), \tag{3.13}$$

$$W(e^{j\Omega}) = X(e^{j\Omega}) \cdot H(e^{j\Omega}), \tag{3.14}$$

$$H(e^{j\Omega}) = \begin{cases} 1 & |\Omega| \leq \pi/M, \\ 0 & \text{otherwise.} \end{cases} \tag{3.15}$$

Downsampling of $w(m)$ is performed by taking every Mth sample, which leads to the output signal

$$y(n) = w(Mn) \tag{3.16}$$

with the Fourier transform

$$Y(e^{j\Omega'}) = \frac{1}{M}\sum_{l=0}^{M-1} W(e^{j(\Omega' - 2\pi l)/M)}). \tag{3.17}$$

For the baseband spectrum ($|\Omega'| \leq \pi$ and $l = 0$), we get

$$Y(e^{j\Omega'}) = \frac{1}{M}H(e^{j\Omega'/M}) \cdot X(e^{j\Omega'/M}) = \frac{1}{M}X(e^{j\Omega'/M}) \quad |\Omega'| \leq \pi \tag{3.18}$$

and for the Fourier transform of the output signal, we can derive

$$Y(e^{j\Omega'}) = \frac{1}{M}X(e^{j\Omega'/M}) = \underbrace{\frac{1}{M}\frac{1}{T}\sum_{k=-\infty}^{\infty} X_a\left(j\omega + jk\frac{2\pi}{MT}\right)}_{\text{with Eq. (3.1)}} \tag{3.19}$$

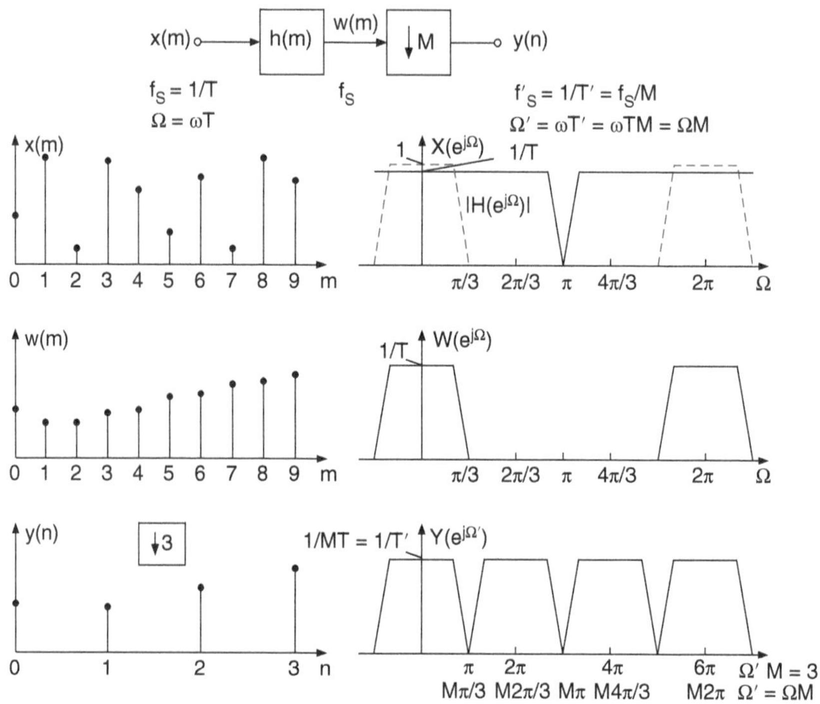

Figure 3.2 Antialiasing filtering and downsampling by M in the time and frequency domains.

$$= \underbrace{\frac{1}{T'} \sum_{k=-\infty}^{\infty} X_a\left(j\omega + jk\frac{2\pi}{T'}\right)}_{\text{spectrum of signal with } f'_S = f_S/M}, \tag{3.20}$$

which represents a sampled signal $y(n)$ with $f'_S = f_S/M$.

3.2 Synchronous Conversion

Sampling rate conversion for coupled sampling rates by a rational factor L/M can be performed by the system shown in Fig. 3.3. After upsampling by a factor L, anti-imaging filtering at Lf_S is done followed by downsampling by factor M. Because after upsampling and filtering only every Mth sample is used, it is possible to develop efficient algorithms that reduce the complexity. In this respect, two methods are in use; one is based on a time-domain interpretation [Cro83] and

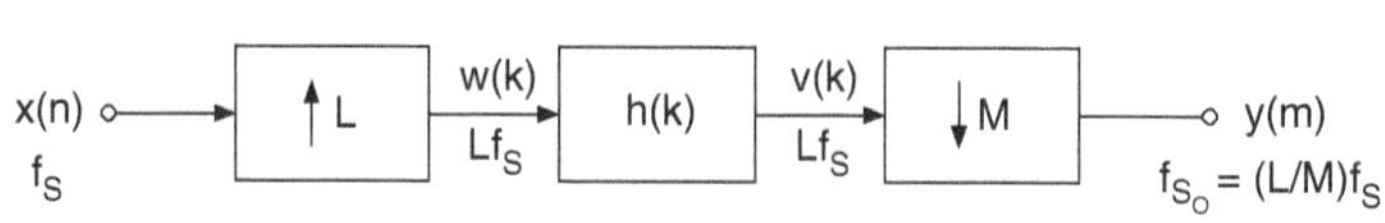

Figure 3.3 Sampling rate conversion by factor L/M.

the other [Hsi87] uses Z-domain fundamentals. Owing to its computational efficiency, only the method in the Z-domain will be considered.

Starting with the finite impulse response $h(n)$ of length N and its Z-transform

$$H(z) = \sum_{n=0}^{N-1} h(n)z^{-n}, \tag{3.21}$$

the polyphase representation [Cro83, Vai93, Fli00] with M components can be expressed as

$$H(z) = \sum_{k=0}^{M-1} z^{-k} E_k(z^M) \qquad \text{type 1} \tag{3.22}$$

$$\text{with} \qquad e_k(n) = h(nM + k), \quad k = 0,1,...,M-1 \tag{3.23}$$

or

$$H(z) = \sum_{k=0}^{M-1} z^{-(M-1-k)} R_k(z^M) \qquad \text{type 2} \tag{3.24}$$

$$\text{with} \qquad r_k(n) = h(nM - k), \quad k = 0,1,...,M-1. \tag{3.25}$$

The polyphase decomposition as given in Eqs. (3.22) and (3.24) is denoted as type 1 and 2, respectively. The type-1 polyphase decomposition corresponds to a commutator model in the anti-clockwise direction, whereas the type 2 is in the clockwise direction. The relationship between $R(z)$ and $E(z)$ is described by

$$R_k(z) = E_{M-1-k}(z). \tag{3.26}$$

With the help of the identities [Vai93] shown in Fig. 3.4 and the decomposition (Euclid's theorem)

$$z^{-1} = z^{-pL} z^{qM}, \tag{3.27}$$

it is possible to move the inner delay elements of Fig. 3.5. Equation (3.27) is valid if M and L are prime numbers. In a cascade of upsampling and downsampling, the order of functional blocks can be exchanged (see Fig. 3.5b).

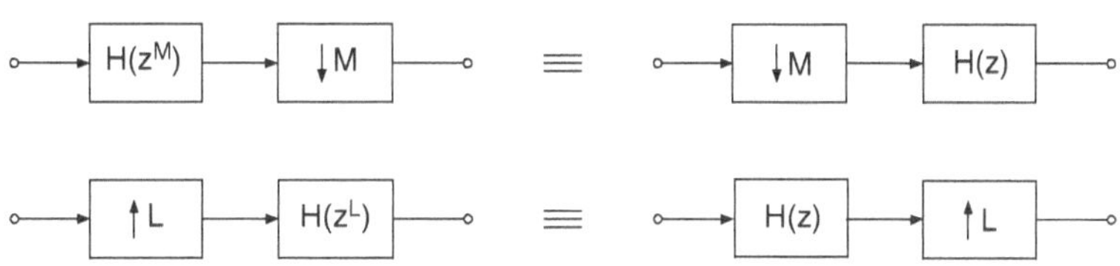

Figure 3.4 Identities for sampling rate conversion.

The use of polyphase decomposition can be demonstrated with the help of an example for $L = 2$ and $M = 3$. This implies a sampling rate conversion from 48 kHz to 32 kHz. Figures 3.6 and 3.7 show two different solutions for polyphase decomposition of sampling rate conversion by 2/3. Further decompositions of the upsampling decomposition of Fig. 3.7 are demonstrated in Fig. 3.8. First, interpolation is implemented with a polyphase decomposition and the delay z^{-1} is decomposed to $z^{-1} = z^{-2}z^3$. Then, the downsampler of factor 3 is moved through the adder into the two paths (Fig. 3.8b) and the delays are moved according to the identities of Fig. 3.4. In Fig. 3.8c, the upsampler is exchanged with the downsampler and in a last step (Fig. 3.8d), another polyphase decomposition of $E_0(z)$ and $E_1(z)$ is carried out. The actual filter operations $E_{0k}(z)$ and $E_{1k}(z)$ with $k = 0,1,2$ are performed at $\frac{1}{3}$ of the input sampling rate.

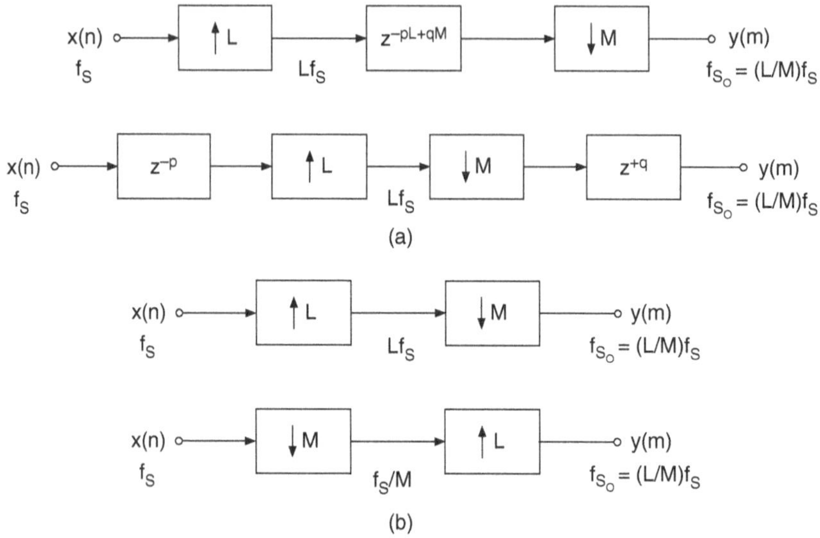

Figure 3.5 Decomposition in accordance with Euclid's theorem.

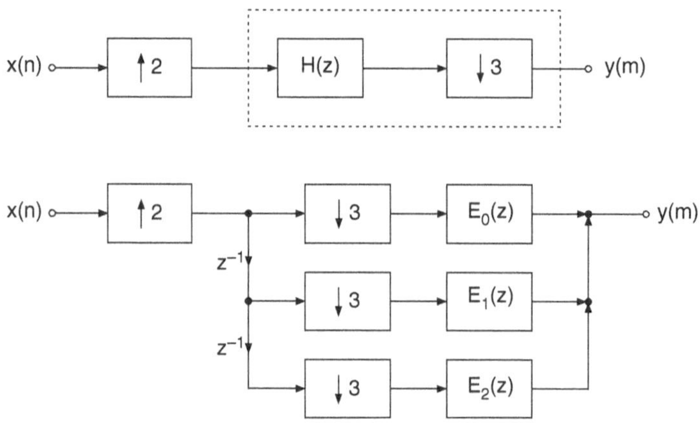

Figure 3.6 Polyphase decomposition for downsampling L/M = 2/3.

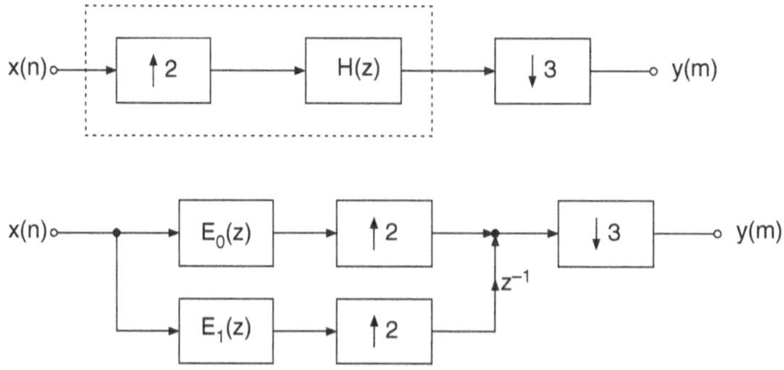

Figure 3.7 Polyphase decomposition for upsampling L/M = 2/3.

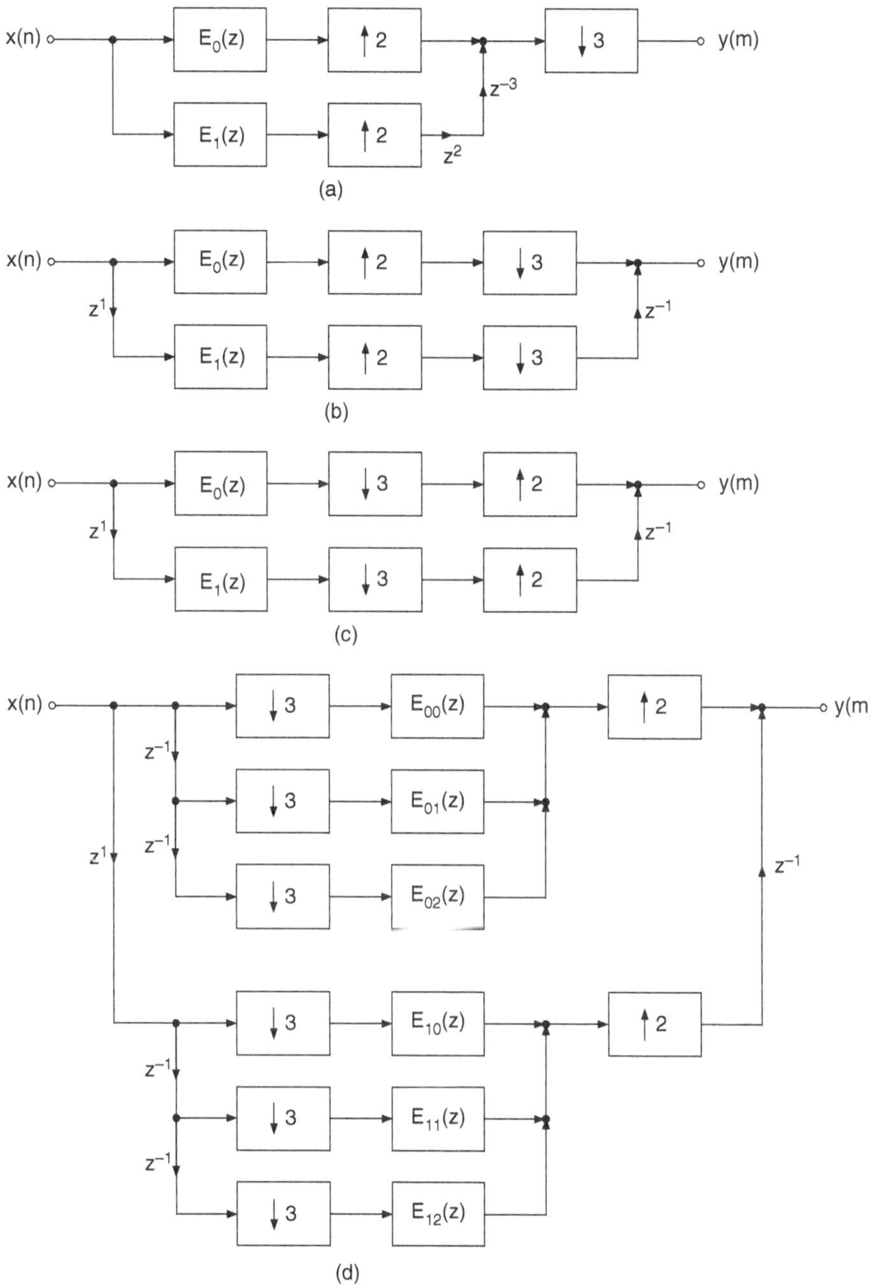

Figure 3.8 Sampling rate conversion by factor 2/3.

3.3 Asynchronous Conversion

Plesiochronous systems consist of partial systems with different and uncoupled sampling rates. Sampling rate conversion between such systems can be achieved through a DA conversion with the sampling rate of the first system followed by an AD conversion with the sampling rate of the second system. A digital approximation of this approach is made with a multi-rate system [Lag81, Lag82a, Lag82b, Lag82c, Lag83, Ram82, Ram84]. Figure 3.9a shows a system for increasing the sampling rate by a factor L followed by an anti-imaging filter $H(z)$ and a resampling of the interpolated signal $y(k)$. The samples $y(k)$ are held for a clock period (see Fig. 3.9c) and then sampled with output clock period $T_{S_O} = 1/f_{S_O}$. The interpolation sampling rate must be increased so that the difference of two consecutive samples $y(k)$ is smaller than the quantization step Q. The sample-and-hold function applied to $y(k)$ suppresses the spectral images at multiples of Lf_S (see Fig. 3.9b). The obtained signal is a band-limited continuous-time signal which can be sampled with output sampling rate f_{S_O}.

For the calculation of the necessary oversampling rate, the problem is considered in the frequency domain. The sinc function of a sample-and-hold system (see Fig. 3.9b) at frequency $\tilde{f} = \left(L - \frac{1}{2}\right) f_S$ is given by

$$E(\tilde{f}) = \frac{\sin\left(\frac{\pi\tilde{f}}{Lf_S}\right)}{\frac{\pi\tilde{f}}{Lf_S}} = \frac{\sin\left(\frac{\pi\left(L-\frac{1}{2}\right)f_S}{Lf_S}\right)}{\frac{\pi\left(L-\frac{1}{2}\right)f_S}{Lf_S}} = \frac{\sin\left(\pi - \frac{\pi}{2L}\right)}{\pi - \frac{\pi}{2L}}. \tag{3.28}$$

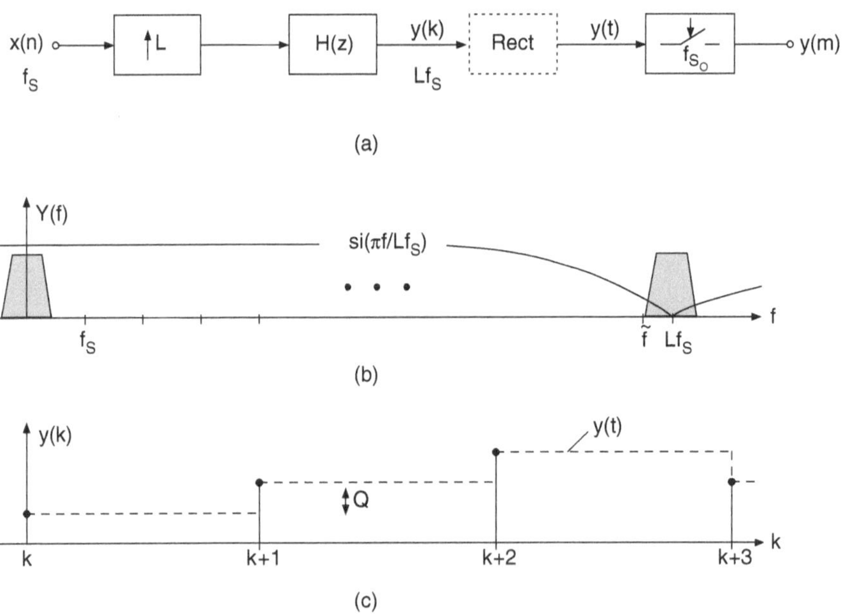

Figure 3.9 Approximation of DA/AD conversions.

With $\sin(\alpha - \beta) = \sin(\alpha)\cos(\beta) - \cos(\alpha)\sin(\beta)$, we derive

$$E(\tilde{f}) = \frac{\sin\left(\frac{\pi}{2L}\right)}{\pi\left(1 - \frac{1}{2L}\right)} \approx \frac{\pi/2L}{\pi\left(1 - \frac{1}{2L}\right)} \approx \frac{1}{2L-1} \approx \frac{1}{2L}. \tag{3.29}$$

This value of Eq. (3.29) should be lower than $\frac{Q}{2}$ and allows the computation of the interpolation factor L. For a given word length w and quantization step Q, the necessary interpolation rate L is calculated by

$$\frac{Q}{2} \geq \frac{1}{2L} \tag{3.30}$$

$$\frac{2^{-(w-1)}}{2} \geq \frac{1}{2L} \tag{3.31}$$

$$\hookrightarrow L \geq 2^{w-1}. \tag{3.32}$$

For a linear interpolation between upsampled samples $y(k)$, we can derive

$$E(\tilde{f}) = \frac{\sin^2\left(\frac{\pi\tilde{f}}{Lf_S}\right)}{\left(\frac{\pi\tilde{f}}{Lf_S}\right)^2} \tag{3.33}$$

$$= \frac{\sin^2\left(\frac{\pi\left(L-\frac{1}{2}\right)f_S}{Lf_S}\right)}{\left(\frac{\pi\left(L-\frac{1}{2}\right)f_S}{Lf_S}\right)^2} \tag{3.34}$$

$$\approx \frac{1}{(2L)^2}. \tag{3.35}$$

With this, it is possible to reduce the necessary interpolation rate to

$$L_1 \geq 2^{\frac{w}{2}-1}. \tag{3.36}$$

Figure (3.10) demonstrates this with a two-stage block diagram. First, interpolation up to a sampling rate $L_1 f_S$ is performed by conventional filtering. In the second stage, upsampling by factor L_2 is done by linear interpolation. The two-stage approach must satisfy the sampling rate $Lf_S = (L_1 L_2) f_S$.

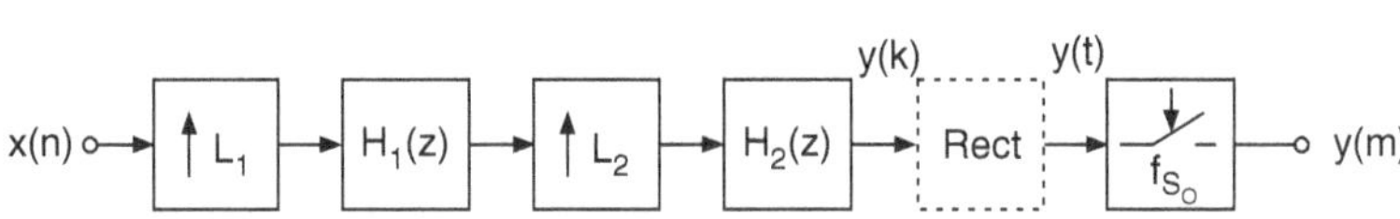

Figure 3.10 Linear interpolation before virtual sample-and-hold function.

The choice of the interpolation algorithm in the second stage enables the reduction of the first oversampling factor.

3.3.1 Single-stage Methods

Direct conversion methods implement the block diagram [Lag83, Smi84, Par90, Par91a, Par91b, Ada92, Ada93] shown in Fig. 3.9a. The calculation of a discrete sample on an output grid of sampling rate f_{S_O} from samples $x(n)$ at sampling rate f_{S_I} can be written as

$$\begin{aligned} \mathrm{DFT}[x(n-\alpha)] &= X(e^{j\Omega})e^{-j\alpha\Omega} \\ &= X(e^{j\Omega})H_\alpha(e^{j\Omega}), \end{aligned} \tag{3.37}$$

where $0 \leq \alpha < 1$. With the transfer function

$$H_\alpha(e^{j\Omega}) = e^{-j\alpha\Omega} \tag{3.38}$$

and the properties

$$H(e^{j\Omega}) = \begin{cases} 1 & 0 \leq |\Omega| \leq \Omega_c, \\ 0 & \Omega_c < |\Omega| < \pi, \end{cases} \tag{3.39}$$

the impulse response is given by

$$h_\alpha(n) = h(n-\alpha) = \frac{\Omega_c}{\pi}\frac{\sin[\Omega_c(n-\alpha)]}{\Omega_c(n-\alpha)}. \tag{3.40}$$

From Eq. (3.37), we can express the delayed signal

$$x(n-\alpha) = \sum_{m=-\infty}^{\infty} x(m)h(n-\alpha-m) \tag{3.41}$$

$$= \sum_{m=-\infty}^{\infty} x(m)\frac{\Omega_c}{\pi}\frac{\sin[\Omega_c(n-\alpha-m)]}{\Omega_c(n-\alpha-m)} \tag{3.42}$$

as the convolution between $x(n)$ and $h(n-\alpha)$. Figure 3.11 illustrates this convolution in the time domain for a fixed α. Fig. 3.12 shows the coefficients $h(n-\alpha_i)$ for discrete α_i ($i = 0, \ldots, 3$) which are obtained from the intersection of the sinc function with the discrete samples $x(n)$.

To limit the convolution sum, the impulse response is windowed, which gives

$$h_W(n-\alpha_i) = w(n)\frac{\Omega_c}{\pi}\frac{\sin[\Omega_c(n-\alpha_i)]}{\Omega_c(n-\alpha_i)} \qquad n = 0, \ldots, 2M. \tag{3.43}$$

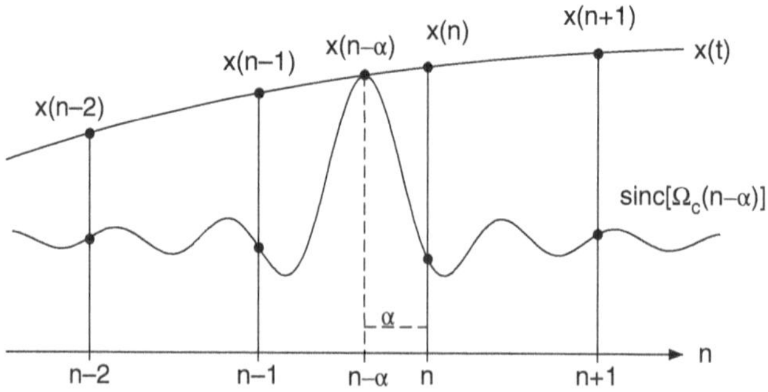

Figure 3.11 Convolution sum (3.42) in the time domain.

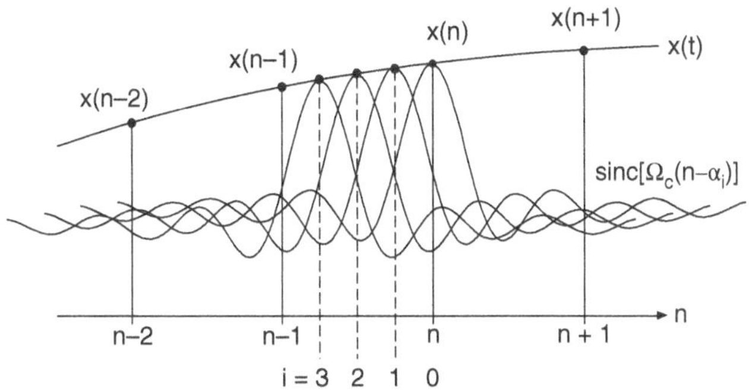

Figure 3.12 Convolution sum (3.42) for different α_i.

From this, the sample estimate

$$\hat{x}(n - \alpha_i) = \sum_{m=-M}^{M} x(m) h_W(n - \alpha_i - m) \tag{3.44}$$

results. A graphical interpretation of the time-variant impulse response which depends on α_i is shown in Fig. 3.13. The discrete segmentation between two input samples into N intervals leads to N partial impulse responses of length $2M + 1$.

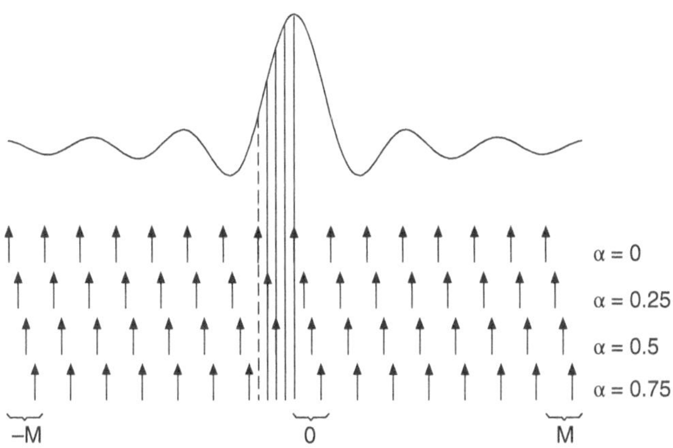

Figure 3.13 Sinc function and different impulse responses.

If the output sampling rate is smaller than the input sampling rate ($f_{S_O} < f_{S_I}$), band limiting (antialiasing) to the output sampling rate has to be done. This can be achieved with factor $\beta = \frac{f_{S_O}}{f_{S_I}}$ and leads, with the the scaling theorem of the Fourier transform, to

$$h(n - \alpha) = \frac{\beta\Omega_c}{\pi} \frac{\sin[\beta\Omega_c(n - \alpha)]}{\beta\Omega_c(n - \alpha)}. \tag{3.45}$$

This time scaling of the impulse response has the consequence that the number of coefficients of the time-variant partial impulse responses is increased. The number of required states also increases. Figure 3.14 shows the time-scaled impulse response and elucidates the increase of the number M of the coefficients.

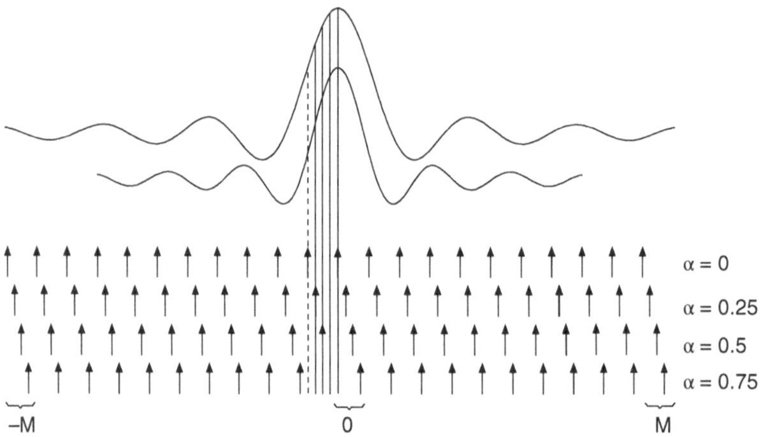

Figure 3.14 Time-scaled impulse response.

3.3.2 Multistage Methods

The basis of a multistage conversion method [Lag81, Lag82a, Lag82b, Lag82c, Kat85, Kat86] is shown in Fig. 3.15a and will be described in the frequency domain, as shown in Fig. 3.15b–d. The increase of the sampling rate up to the rate Lf_S before the sample-and-hold function is done in four stages. In the first two stages, the sampling rate is increased by a factor of two followed by an anti-imaging filter (see Fig. 3.15b,c), which leads to a four-times oversampled spectrum (Fig. 3.15d). In the third stage, the signal is upsampled by a factor of 32 and the image spectra are suppressed (see Fig. 3.15d,e). In the fourth stage (Fig. 3.15e), the signal is upsampled to a sampling rate of Lf_S by a factor of 256 and a linear interpolator. The sinc^2-function of the linear interpolator suppresses the images at multiples of $128f_S$ up to the spectrum at Lf_S. The virtual sample-and-hold function is shown in Fig. 3.15f, where resampling at the output sampling rate is performed. A direct conversion of this kind of cascaded interpolation structure requires anti-imaging filtering after every upsampling with the corresponding sampling rate. Although the necessary filter order decreases owing to a decrease of requirements for filter design, an implementation of the filters in the third and fourth stages is not possible directly. After a suggestion by Lagadec [Lag82c], the measurement of the ratio of input to output rate is used to control the polyphase filters in the third and fourth stages (see Fig. 3.16a, CON = control) to reduce complexity. Figure 3.16b–d illustrate an interpretation in the time domain. Figure 3.16b shows the interpolation of three samples between two input samples $x(n)$ with the help of the first and second interpolation stages. The abscissa represents the intervals of the input sampling rate and the sampling rate is increased by a factor of four. In Fig. 3.16c, the four-times oversampled signal is shown. The abscissa shows the four- times oversampled output grid. It is assumed that the output sample $y(m = 0)$ and input sample $x(n = 0)$ are identical. The output sample $y(m = 1)$ is now determined in such a form that, with the interpolator in the third stage, only two polyphase filters just before and after the output sample need to be calculated. Hence, only two out of a total of 31 possible polyphase filters are calculated in the third stage. Fig. 3.16d shows these two polyphase

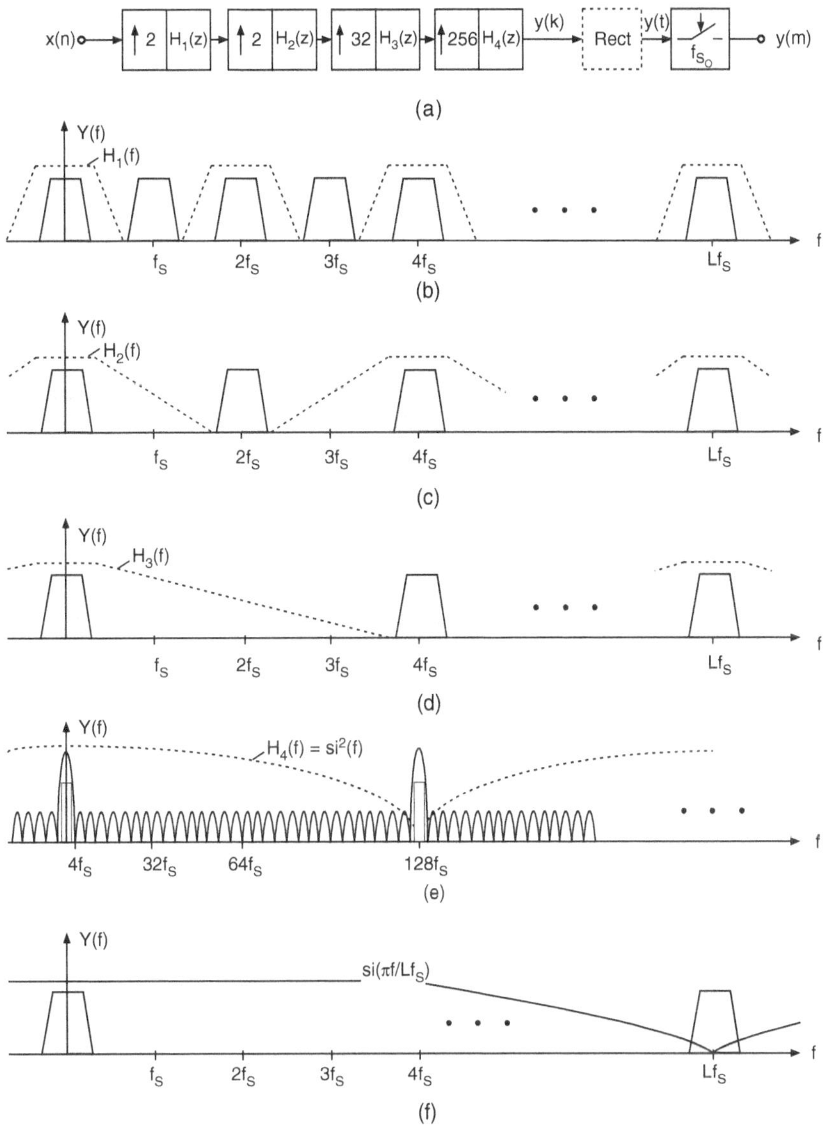

Figure 3.15 Multistage conversion – frequency-domain interpretation.

output samples. Between these two samples, the output sample $y(m = 1)$ is obtained with a linear interpolation on a grid of 255 values.

Instead of the third and fourth stages, special interpolation methods can be used to calculate the output $y(m)$ directly from the four-times oversampled input signal (see Fig. 3.17) [Sti91, Cuc91, Liu92]. The upsampling factor $L_3 = 2^{w-3}$ for the last stage is calculated according to $L = 2^{w-1} = L_1 L_2 L_3 = 2^2 L_3$. Section 3.4 is devoted to different interpolation methods which allow a real-time calculation of filter coefficients. This can be interpreted as time-variant filters in which the filter coefficients are derived from the ratio of sampling rates. The calculation of one filter coefficient set

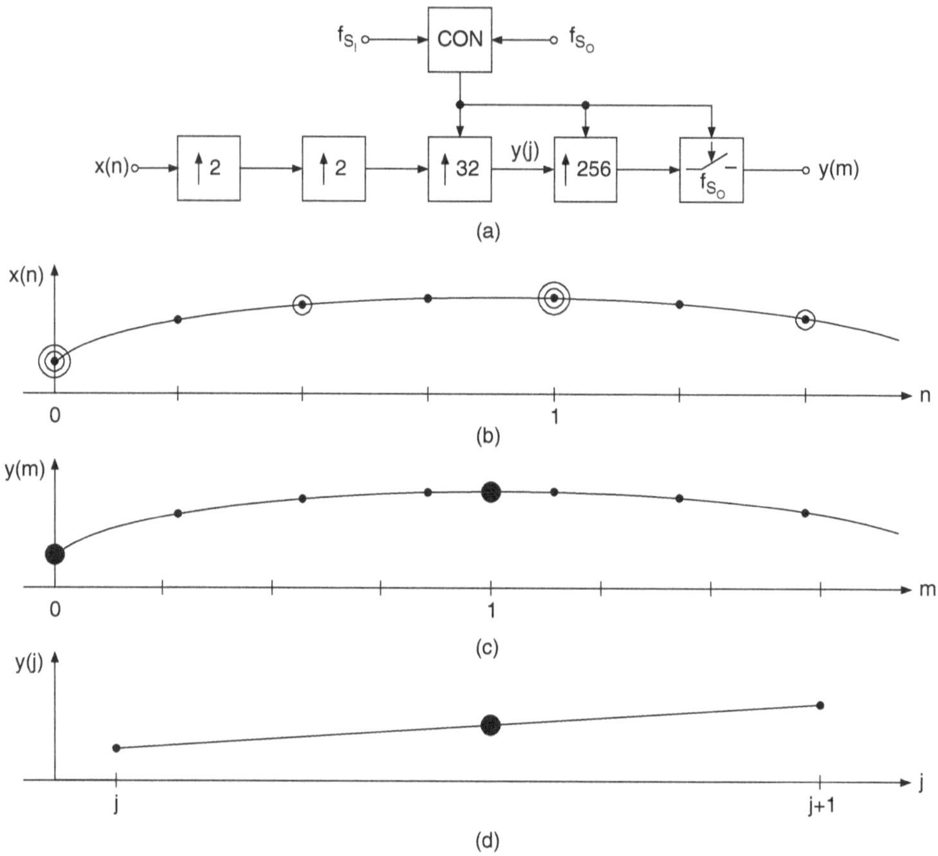

Figure 3.16 Time-domain interpretation.

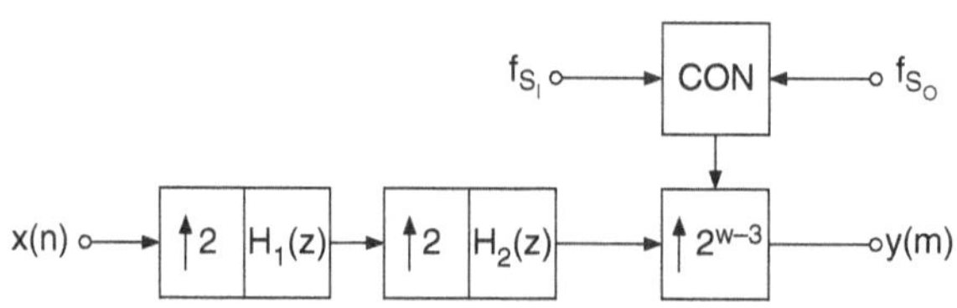

Figure 3.17 Sampling rate conversion with interpolation for calculating coefficients of a time-variant interpolation filter.

for the output sample at the output rate is done by measuring the ratio of input to output sampling rate as described in the next section.

3.3.3 Control of Interpolation Filters

The measurement of the ratio of input and output sampling rates is used to control the interpolation filters [Lag82a]. By increasing the sampling rate by a factor of L, the input sampling period is divided into $L = 2^{w-1} = 2^{15}$ parts for a signal word-length of $w = 16$ bits. The time instant of the output sample is calculated on this grid with the help of the measured ratio of sampling periods T_{S_O}/T_{S_I} as follows.

A counter is clocked with Lf_{S_I} and reset by every new input sampling clock. A sawtooth curve of the counter output versus time is obtained, as shown in Fig. 3.18. The counter runs from 0 to $L-1$ during one input sampling period. At time t_{i-2}, which corresponds to counter output z_{i-2}, the output sampling period T_{S_O} starts, and stops at time t_{i-1} with counter output z_{i-1}. The difference between both counter measurements allows for the calculation of the output sampling period T_{S_O} with a resolution of Lf_{S_I}.

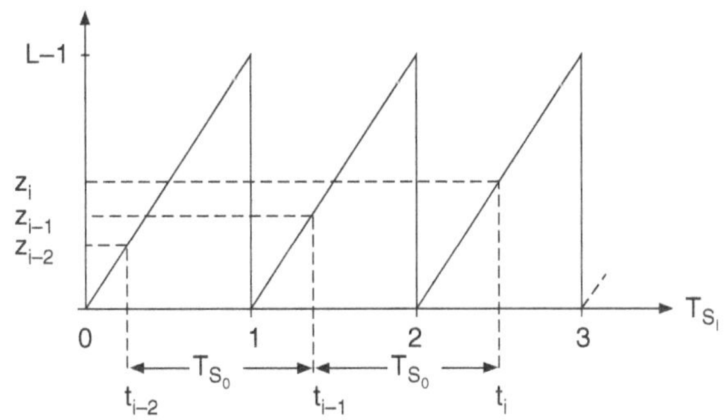

Figure 3.18 Calculation of t_i.

The new counter measurement is added to the difference of the previous counter measurements. As a result, the new counter measurement is obtained as

$$t_i = (t_{i-1} + T_{S_O}) \oplus T_{S_I}. \tag{3.46}$$

The modulo operation can be carried out with an accumulator of word length $w - 1 = 15$. The resulting time t_i determines the time instant of the output sample at the output sampling rate and therefore the choice of the polyphase filter in a single-stage conversion or the time instant for a multistage conversion.

The measurement of T_{S_O}/T_{S_I} is illustrated in Fig. 3.19.

- The input sampling rate f_{S_I} is increased to $M_Z f_{S_I}$ using a frequency multiplier, where $M_Z = 2^w$. This increased input clock by the factor M_Z triggers a w-bit counter. The counter output z is evaluated every M_O output sampling periods.
- Counting of M_O output sampling periods.
- Simultaneous counting of the M_I input sampling periods.

The time intervals d_1 and d_2 (see Fig. 3.19) are given by

$$d_1 = M_I T_{S_I} + \frac{z - z_0}{M_Z} T_{S_I} = \left(M_I + \frac{z - z_0}{M_Z}\right) T_{S_I}, \tag{3.47}$$

$$d_2 = M_O T_{S_O}, \tag{3.48}$$

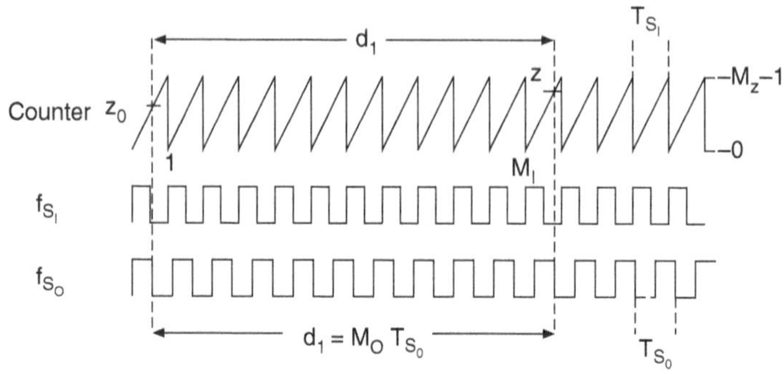

Figure 3.19 Measurement of T_{S_O}/T_{S_I}.

and with the requirement $d_1 = d_2$, we can write

$$M_O T_{S_O} = \left(M_I + \frac{z - z_0}{M_Z} \right) T_{S_I}$$

$$\frac{T_{S_O}}{T_{S_I}} = \frac{M_I + (z - z_0)/M_Z}{M_O} = \frac{M_Z M_I + (z - z_0)}{M_Z M_O}. \tag{3.49}$$

- Example 1: $w = 0 \rightarrow M_Z = 1$,

$$\frac{T_{S_O}}{T_{S_I}} = \frac{M_I}{2^{15}}. \tag{3.50}$$

With a precision of 15 bits, the averaging number is chosen as $M_O = 2^{15}$ and the number M_I has to be determined.

- Example 2: $w = 8 \rightarrow M_Z = 2^8$,

$$\frac{T_{S_O}}{T_{S_I}} = \frac{2^8 M_I + (z - z_0)}{2^8 2^7}. \tag{3.51}$$

With a precision of 15 bits, the averaging number is chosen as $M_O = 2^7$ and the number M_I, as well as the counter outputs, has to be determined.

The sampling rates at the input and output of a sampling rate converter can be calculated by evaluating the 8-bit increment of the counter for each output clock with

$$z = \frac{T_{S_O}}{T_{S_I}} M_Z = \frac{f_{S_I}}{f_{S_O}} 256, \tag{3.52}$$

as seen from Table 3.1.

Table 3.1 Counter increments for different sampling rate conversions.

Conversion/kHz	8-bit counter increment
32 → 48	170
44.1 → 48	235
32 → 44.1	185
48 → 44.1	278
48 → 32	384
44.1 → 32	352

3.4 Interpolation Methods

In the following sections, special interpolation methods are discussed. These methods enable the calculation of time-variant filter coefficients for sampling rate conversion and need an oversampled input sequence as well as the time instant of the output sample. A convolution of the oversampled input sequence with time-variant filter coefficients gives the output sample at the output sampling rate. This real-time computation of filter coefficients is not based on popular filter design methods. On the contrary, methods are presented for calculating filter coefficient sets for every input clock cycle where the filter coefficients are derived from the distance of output samples to the time grid of the oversampled input sequence.

3.4.1 Polynomial Interpolation

The aim of a polynomial interpolation [Liu92] is to determine a polynomial

$$p_N(x) = \sum_{i=0}^{N} a_i x^i \tag{3.53}$$

of Nth order representing exactly a function $f(x)$ at $N+1$ uniformly spaced x_i, i.e. $p_N(x_i) = f(x_i) = y_i$ for $i = 0, \dots, N$. This can be written as a set of linear equations:

$$\begin{bmatrix} 1 & x_0 & x_0^2 & \cdots & x_0^N \\ 1 & x_1 & x_1^2 & \cdots & x_1^N \\ \vdots & \vdots & \vdots & & \vdots \\ 1 & x_N & x_N^N & \cdots & x_N^N \end{bmatrix} \begin{bmatrix} a_0 \\ a_1 \\ \vdots \\ a_N \end{bmatrix} = \begin{bmatrix} y_0 \\ y_1 \\ \vdots \\ y_N \end{bmatrix}. \tag{3.54}$$

The polynomial coefficients a_i as functions of $y_0 \dots y_N$ are obtained with the help of Cramer's Rule according to

$$a_i = \frac{\overset{i\text{th column}}{\begin{vmatrix} 1 & x_0 & x_0^2 & \cdots & y_0 & \cdots & x_0^N \\ 1 & x_1 & x_1^2 & \cdots & y_1 & \cdots & x_1^N \\ \vdots & \vdots & \vdots & & \vdots & \cdots & \vdots \\ 1 & x_N & x_N^2 & \cdots & y_N & \cdots & x_N^N \end{vmatrix}}}{\begin{vmatrix} 1 & x_0 & x_0^2 & \cdots & x_0^N \\ 1 & x_1 & x_1^2 & \cdots & x_1^N \\ \vdots & \vdots & \vdots & & \vdots \\ 1 & x_N & x_N^2 & \cdots & x_N^N \end{vmatrix}}, \qquad i = 0,1, \dots, N. \tag{3.55}$$

For uniformly spaced $x_i = i$ with $i = 0,1, \ldots ,N$, the interpolation of an output sample with distance α gives

$$y(n+\alpha) = \sum_{i=0}^{N} a_i(n+\alpha)^i. \tag{3.56}$$

To determine the relationship between the output sample $y(n+\alpha)$ and y_i, a set of time-variant coefficients c_i needs to be determined such that

$$\boxed{y(n+\alpha) = \sum_{i=-N/2}^{N/2} c_i(\alpha)y(n+i).} \tag{3.57}$$

The calculation of time-variant coefficients $c_i(\alpha)$ will be illustrated by an example.

Example: Figure 3.20 shows the interpolation of an output sample of distance α with $N = 2$ and using three samples, which can be written as

$$y(n+\alpha) = \sum_{i=0}^{2} a_i(n+\alpha)^i. \tag{3.58}$$

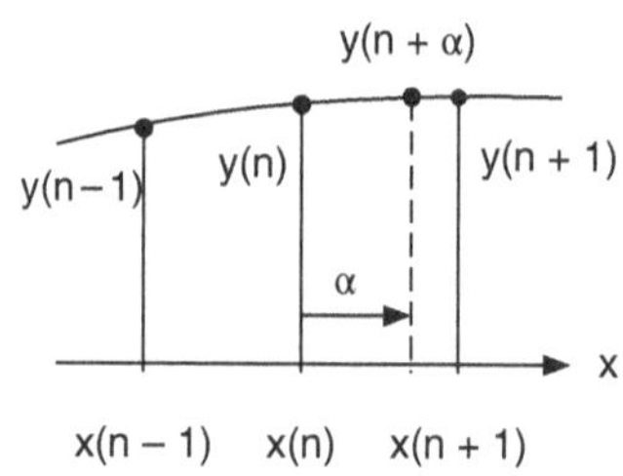

Figure 3.20 Polynomial interpolation with three samples.

The samples $y(n+i)$, with $i = -1,0,1$, can be expressed as

$$y(n+1) = \sum_{i=0}^{2} a_i(n+1)^i \qquad \alpha = 1,$$

$$y(n) = \sum_{i=0}^{2} a_i n^i \qquad \alpha = 0,$$

$$y(n-1) = \sum_{i=0}^{2} a_i(n-1)^i \qquad \alpha = -1, \tag{3.59}$$

or in matrix notation

$$\begin{bmatrix} 1 & (n+1) & (n+1)^2 \\ 1 & n & n^2 \\ 1 & (n-1) & (n-1)^2 \end{bmatrix} \begin{bmatrix} a_0 \\ a_1 \\ a_2 \end{bmatrix} = \begin{bmatrix} y(n+1) \\ y(n) \\ y(n-1) \end{bmatrix}. \tag{3.60}$$

The coefficients a_i as functions of y_i are then given by

$$\begin{bmatrix} a_0 \\ a_1 \\ a_2 \end{bmatrix} = \begin{bmatrix} \frac{n(n-1)}{2} & 1-n^2 & \frac{n(n+1)}{2} \\ -\frac{2n-1}{2} & 2n & -\frac{2n+1}{2} \\ \frac{1}{2} & -1 & \frac{1}{2} \end{bmatrix} \begin{bmatrix} y(n+1) \\ y(n) \\ y(n-1) \end{bmatrix}, \tag{3.61}$$

such that

$$y(n+\alpha) = a_0 + a_1(n+\alpha) + a_2(n+\alpha)^2 \tag{3.62}$$

is valid. The output sample $y(n+\alpha)$ can be written as

$$\begin{aligned} y(n+\alpha) &= \sum_{i=-1}^{1} c_i(\alpha) y(n+i) \\ &= c_{-1}y(n-1) + c_0 y(n) + c_1 y(n+1). \end{aligned} \tag{3.63}$$

Equation (3.62) with a_i from Eq. (3.61) leads to

$$\begin{aligned} y(n+\alpha) &= \left[\frac{1}{2}y(n+1) - y(n) + \frac{1}{2}y(n-1)\right](n+\alpha)^2 \\ &+ \left[-\frac{2n-1}{2}y(n+1) + 2ny(n) - \frac{2n+1}{2}y(n-1)\right](n+\alpha) \\ &+ \frac{n(n-1)}{2}y(n+1) + (1-n^2)y(n) + \frac{n(n+1)}{2}y(n-1). \end{aligned} \tag{3.64}$$

Comparing the coefficients from Eqs. (3.63) and (3.64) for $n = 0$ gives the coefficients

$$\begin{aligned} c_{-1} &= \frac{1}{2}\alpha(\alpha-1), \\ c_0 &= -(\alpha-1)(\alpha+1) = 1-\alpha^2, \\ c_1 &= \frac{1}{2}\alpha(\alpha+1). \end{aligned}$$

3.4.2 Lagrange Interpolation

Lagrange interpolation for $N+1$ samples makes use of the polynomials $l_i(x)$ which have the following properties (see Fig. 3.21):

$$l_i(x_k) = \delta_{ik} = \begin{cases} 1 & i = k, \\ 0 & \text{elsewhere}. \end{cases} \tag{3.65}$$

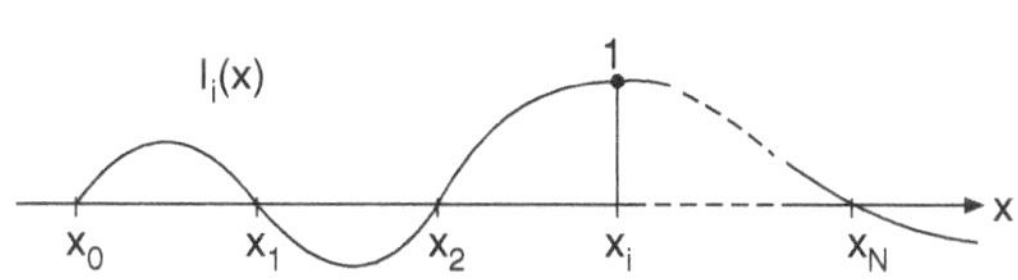

Figure 3.21 Lagrange polynomial.

Based on the zeros of the polynomial $l_i(x)$, it follows that

$$l_i(x) = a_i(x - x_0) \dots (x - x_{i-1})(x - x_{i+1}) \dots (x - x_N). \tag{3.66}$$

With $l_i(x_i) = 1$, the coefficients are given by

$$a_i(x_i) = \frac{1}{(x_i - x_0) \dots (x_i - x_{i-1})(x_i - x_{i+1}) \dots (x_i - x_N)}. \tag{3.67}$$

The interpolation polynomial is expressed as

$$\begin{aligned} p_N(x) &= \sum_{i=0}^{N} l_i(x) y_i \\ &= l_0(x) y_0 + \dots + l_N(x) y_N. \end{aligned} \tag{3.68}$$

With $a = \prod_{j=0}^{N}(x - x_j)$, Eq. (3.66) can be written as

$$\begin{aligned} l_i(x) = a_i \frac{a}{x - x_i} &= \frac{1}{\prod_{j=0,\, j \neq i}^{N} x_i - x_j} \frac{\prod_{j=0}^{N} x - x_j}{x - x_i} \\ &= \prod_{j=0,\, j \neq i}^{N} \frac{x - x_j}{x_i - x_j}. \end{aligned} \tag{3.69}$$

For uniformly spaced samples,

$$x_i = x_0 + ih, \tag{3.70}$$

and with the new variable α, as given by

$$x = x_0 + \alpha h, \tag{3.71}$$

we get

$$\frac{x - x_j}{x_i - x_j} = \frac{(x_0 + \alpha h) - (x_0 + jh)}{(x_0 + ih) - (x_0 + jh)} = \frac{\alpha - j}{i - j} \tag{3.72}$$

and hence

$$l_i(x(\alpha)) = \prod_{j=0,\, j \neq i}^{N} \frac{\alpha - j}{i - j}. \tag{3.73}$$

For even N, we can write

$$l_i(x(\alpha)) = \prod_{j=-\frac{N}{2},\, j \neq i}^{\frac{N}{2}} \frac{\alpha - j}{i - j} \tag{3.74}$$

and for odd N,

$$l_i(x(\alpha)) = \prod_{j=-\frac{N-1}{2},\, j \neq i}^{\frac{N+1}{2}} \frac{\alpha - j}{i - j}. \tag{3.75}$$

The interpolation of an output sample is given by

$$\boxed{y(n+\alpha) = \sum_{i=-N/2}^{N/2} l_i(\alpha) y(n+i).} \tag{3.76}$$

Example: $N = 2$, three samples.

$$l_{-1}(x(\alpha)) = \prod_{j=-1,\, j\neq -1}^{1} \frac{\alpha - j}{-1 - j} = \frac{1}{2}\alpha(\alpha - 1),$$

$$l_0(x(\alpha)) = \prod_{j=-1,\, j\neq 0}^{1} \frac{\alpha - j}{0 - j} = -(\alpha - 1)(\alpha + 1) = 1 - \alpha^2,$$

$$l_1(x(\alpha)) = \prod_{j=-1,\, j\neq 1}^{1} \frac{\alpha - j}{1 - j} = \frac{1}{2}\alpha(\alpha + 1).$$

3.4.3 Spline Interpolation

The interpolation using piecewise-defined functions that only exist over finite intervals is called spline interpolation [Cuc91]. The goal is to compute the sample $y(n+\alpha) = \sum_{i=-N/2}^{N/2} b_i^N(\alpha) y(n+i)$ from weighted samples $y(n+i)$.

A B-spline $M_k^N(x)$ of Nth order using $m+1$ samples is defined in the interval $[x_k, \ldots, x_{k+m}]$ by

$$M_k^N(x) = \sum_{i=k}^{k+m} a_i \phi_i(x) \tag{3.77}$$

with the truncated power functions

$$\phi_i(x) = (x - x_i)_+^N = \begin{cases} 0 & x < x_i, \\ (x - x_i)^N & x \geq x_i. \end{cases} \tag{3.78}$$

In the following, $M_0^N(x) = \sum_{i=0}^{m} a_i \phi_i(x)$ will be considered for $k = 0$, where $M_0^N(x) = 0$ for $x < x_0$ and $M_0^N(x) = 0$ for $x \geq x_m$. Figure 3.22 shows the truncated power functions and the B-spline of Nth order. With the definition of the truncated power functions, we can write

$$\begin{aligned} M_0^N(x) &= a_0\phi_0(x) + a_1\phi_1(x) + \ldots + a_m\phi_m(x) \\ &= a_0(x - x_0)_+^N + a_1(x - x_1)_+^N + \ldots + a_m(x - x_m)_+^N \end{aligned} \tag{3.79}$$

and after some calculations, we get

$$\begin{aligned} M_0^N(x) = {} & a_0(x_0^N + c_1 x_0^{N-1} x + \ldots + c_{N-1} x_0 x^{N-1} + x^N) \\ & + a_1(x_1^N + c_1 x_1^{N-1} x + \ldots + c_{N-1} x_1 x^{N-1} + x^N) \\ & \vdots \\ & + a_m(x_m^N + c_1 x_m^{N-1} x + \ldots + c_{N-1} x_m x^{N-1} + x^N). \end{aligned} \tag{3.80}$$

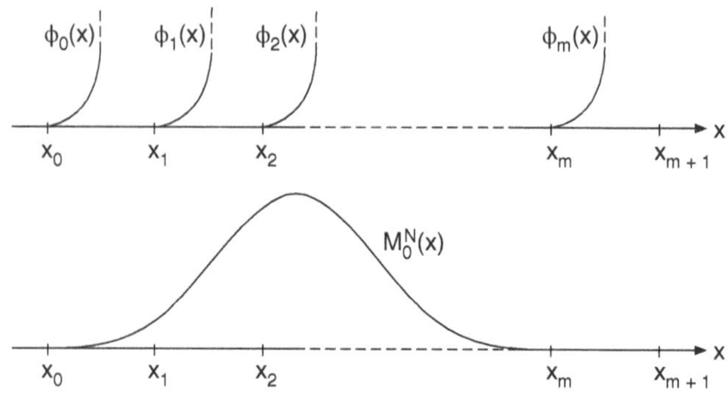

Figure 3.22 Truncated power functions and the B-spline of Nth order.

With the condition $M_0^N(x) = 0$ for $x \geq x_m$, the following set of linear equations can be written with Eq. (3.80) and the coefficients of the powers of x:

$$\begin{bmatrix} 1 & 1 & \cdots & 1 \\ x_0 & x_1 & \cdots & x_m \\ x_0^2 & x_1^2 & \cdots & x_m^2 \\ \vdots & \vdots & & \vdots \\ x_0^N & x_1^N & \cdots & x_m^N \end{bmatrix} \begin{bmatrix} a_0 \\ a_1 \\ a_2 \\ \vdots \\ a_m \end{bmatrix} = \begin{bmatrix} 0 \\ 0 \\ 0 \\ \vdots \\ 0 \end{bmatrix}. \tag{3.81}$$

The homogeneous set of linear equations has nontrivial solutions for $m > N$. The minimum requirement results in $m = N + 1$. For $m = N + 1$, the coefficients [Boe93] can be obtained as follows:

$$a_i = \frac{\overset{i\text{th column}}{\begin{vmatrix} 1 & 1 & 1 & \cdots & 0 & \cdots & 1 \\ x_0 & x_1 & x_2 & \cdots & 0 & \cdots & x_{N+1} \\ \vdots & \vdots & \vdots & & \vdots & & \vdots \\ x_0^N & x_1^N & x_2^N & \cdots & 0 & \cdots & x_{N+1}^N \end{vmatrix}}}{\begin{vmatrix} 1 & 1 & 1 & \cdots & 1 \\ x_0 & x_1 & x_2 & \cdots & x_{N+1} \\ \vdots & \vdots & \vdots & & \vdots \\ x_0^{N+1} & x_1^{N+1} & x_2^{N+1} & \cdots & x_{N+1}^{N+1} \end{vmatrix}}, \qquad i = 0,1, \ldots, N+1. \tag{3.82}$$

Setting the ith column of the determinant in the numerator of Eq. (3.82) equal to zero corresponds to deleting the column. Computing both determinants of *Vandermonde* matrices [Bar90] and division leads to the coefficients

$$a_i = \frac{1}{\prod_{j=0,i\neq j}^{N+1}(x_i - x_j)} \tag{3.83}$$

and hence

$$M_0^N(x) = \sum_{i=0}^{N+1} \frac{(x - x_i)_+^N}{\prod_{j=0,i\neq j}^{N+1}(x_i - x_j)}. \tag{3.84}$$

For some k, we obtain

$$M_k^N(x) = \sum_{i=k}^{k+N+1} \frac{(x - x_i)_+^N}{\prod_{j=0,i\neq j}^{N+1}(x_i - x_j)}. \tag{3.85}$$

Because the functions $M_k^N(x)$ decrease with increasing N, a normalization of the form $N_k^N(x) = (x_{k+N+1} - x_k)M_k^N$ is done, such that for equidistant samples, we get

$$N_k^N(x) = (N + 1) \cdot M_k^N(x). \tag{3.86}$$

The next example illustrates the computation of B-splines.

Example: For $N = 3$, $m = 4$, and five samples, the coefficients according to Eq. (3.83) are given by

$$a_0 = \frac{1}{(x_0 - x_4)(x_0 - x_3)(x_0 - x_2)(x_0 - x_1)},$$
$$a_1 = \frac{1}{(x_1 - x_4)(x_1 - x_3)(x_1 - x_2)(x_1 - x_0)},$$
$$a_2 = \frac{1}{(x_2 - x_4)(x_2 - x_3)(x_2 - x_1)(x_2 - x_0)},$$
$$a_3 = \frac{1}{(x_3 - x_4)(x_3 - x_2)(x_3 - x_1)(x_3 - x_0)},$$
$$a_4 = \frac{1}{(x_4 - x_3)(x_4 - x_2)(x_2 - x_1)(x_3 - x_0)}.$$

Figure 3.23a,b shows the truncated power functions and their summation for calculating $N_0^3(x)$. In Fig. 3.23c, the horizontally shifted $N_i^3(x)$ are depicted.

A linear combination of B-splines is called a spline. Figure 3.24 shows the interpolation of sample $y(n + \alpha)$ for splines of second and third orders. The shifted B-splines $N_i^N(x)$ are evaluated at the vertical line representing the distance α. With sample $y(n)$ and the normalized B-splines $N_i^N(x)$, the second- and third-order splines are expressed as

$$y(n + \alpha) = \sum_{i=-1}^{1} y(n + i)N_{n-1+i}^2(\alpha), \tag{3.87}$$

$$y(n + \alpha) = \sum_{i=-1}^{2} y(n + i)N_{n-2+i}^3(\alpha). \tag{3.88}$$

The computation of a second-order B-spline at the sample index α is based on the symmetry properties of the B-spline, which is depicted in Fig. 3.25. With Eqs. (3.77) and (3.86), and with the symmetry properties shown in Fig. 3.25, the B-splines can be written in the form of

$$N_2^2(\alpha) = N_0^2(\alpha) = 3\sum_{i=0}^{3} a_i(\alpha - x_i)_+^2$$

$$N_1^2(1 + \alpha) = N_0^2(1 + \alpha) = 3\sum_{i=0}^{3} a_i(1 + \alpha - x_i)_+^2$$

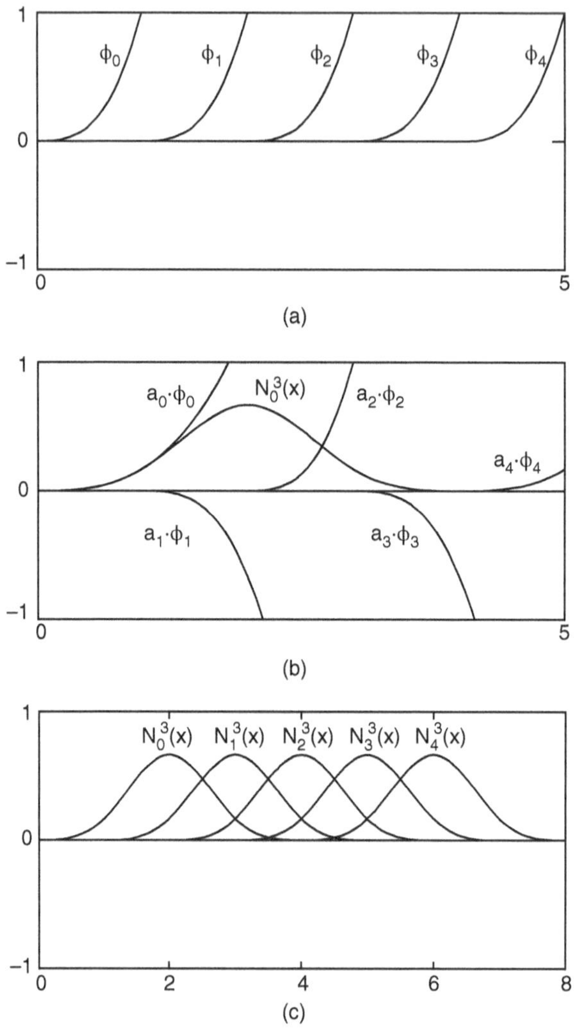

Figure 3.23 Third-order B-spline (N = 3, m = 4, five samples).

$$N_0^2(2+\alpha) = N_0^2(1-\alpha) = 3\sum_{i=0}^{3} a_i(2+\alpha-x_i)_+^2$$
$$= 3\sum_{i=0}^{3} a_i(1-\alpha-x_i)_+^2. \tag{3.89}$$

With Eq. (3.83), we get the coefficients

$$a_0 = \frac{1}{(0-1)(-2)(-3)} = -\frac{1}{6},$$
$$a_1 = \frac{1}{(1-0)(1-2)(1-3)} = \frac{1}{2},$$

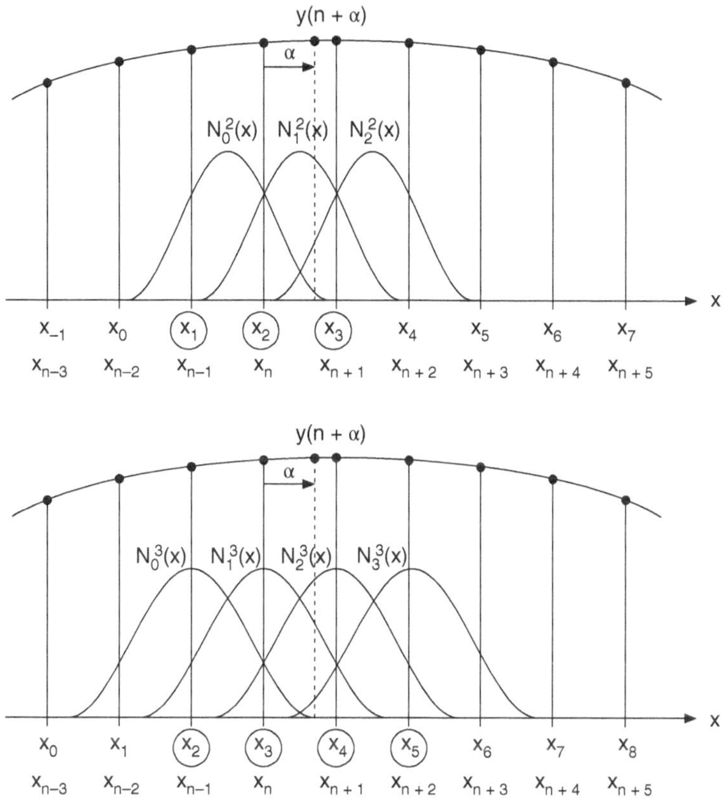

Figure 3.24 Interpolation with B-splines of second and third orders.

$$a_2 = \frac{1}{(2-0)(2-1)(2-3)} = -\frac{1}{2}, \tag{3.90}$$

and thus

$$\begin{aligned} N_2^2(\alpha) &= 3[a_0\alpha^2] = -\frac{1}{2}\alpha^2, \\ N_1^2(\alpha) &= 3[a_0(1+\alpha)^2 + a_1\alpha^2] = -\frac{1}{2}(1+\alpha)^2 + \frac{3}{2}\alpha^2, \\ N_0^2(\alpha) &= 3[a_0(1-\alpha)^2] = -\frac{1}{2}(1-\alpha)^2. \end{aligned} \tag{3.91}$$

Owing to the symmetrical properties of the B-splines, the time-variant coefficients of the second-order B-spline can be derived as

$$N_2^2(\alpha) = h(1) = -\frac{1}{2}\alpha^2, \tag{3.92}$$

$$N_1^2(\alpha) = h(2) = -\frac{1}{2}(1+\alpha)^2 + \frac{3}{2}\alpha^2, \tag{3.93}$$

$$N_0^2(\alpha) = h(3) = -\frac{1}{2}(1-\alpha)^2. \tag{3.94}$$

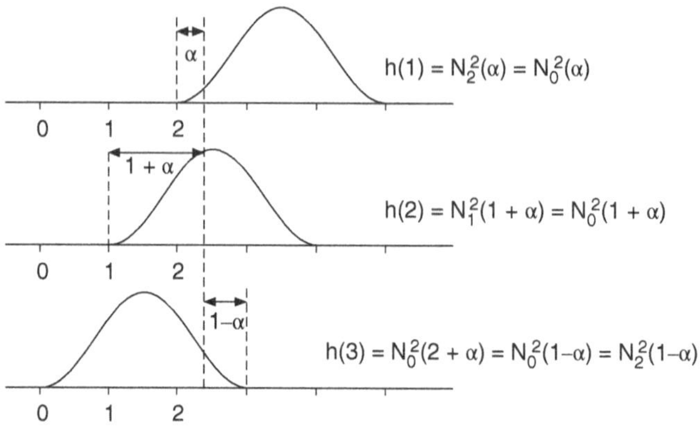

Figure 3.25 Exploiting the symmetry properties of a second-order B-spline.

In the same way, the time-variant coefficients of a third-order B-spline are given by

$$N_3^3(\alpha) = h(1) = \frac{1}{6}\alpha^3, \tag{3.95}$$

$$N_2^3(\alpha) = h(2) = \frac{1}{6}(1+\alpha)^3 - \frac{2}{3}\alpha^3, \tag{3.96}$$

$$N_1^3(\alpha) = h(3) = \frac{1}{6}(2-\alpha)^3 - \frac{2}{3}(1-\alpha)^3, \tag{3.97}$$

$$N_0^3(\alpha) = h(4) = \frac{1}{6}(1-\alpha)^3. \tag{3.98}$$

Higher-order B-splines are given by

$$y(n+\alpha) = \sum_{i=-2}^{2} y(n+i) N_{n-2+i}^4(\alpha), \tag{3.99}$$

$$y(n+\alpha) = \sum_{i=-2}^{3} y(n+i) N_{n-3+i}^5(\alpha), \tag{3.100}$$

$$y(n+\alpha) = \sum_{i=-3}^{3} y(n+i) N_{n-3+i}^6(\alpha). \tag{3.101}$$

Similar sets of coefficients can be derived here as well. Figure 3.26 illustrates this for fourth- and sixth-order B-splines.

Generally, for even orders, we get

$$\boxed{y(n+\alpha) = \sum_{i=-N/2}^{N/2} N_{N/2+i}^N(\alpha) y(n+i),} \tag{3.102}$$

and for odd orders,

$$\boxed{y(n+\alpha) = \sum_{i=-(N-1)/2}^{(N+1)/2} N_{(N-1)/2+i}^N(\alpha) y(n+i).} \tag{3.103}$$

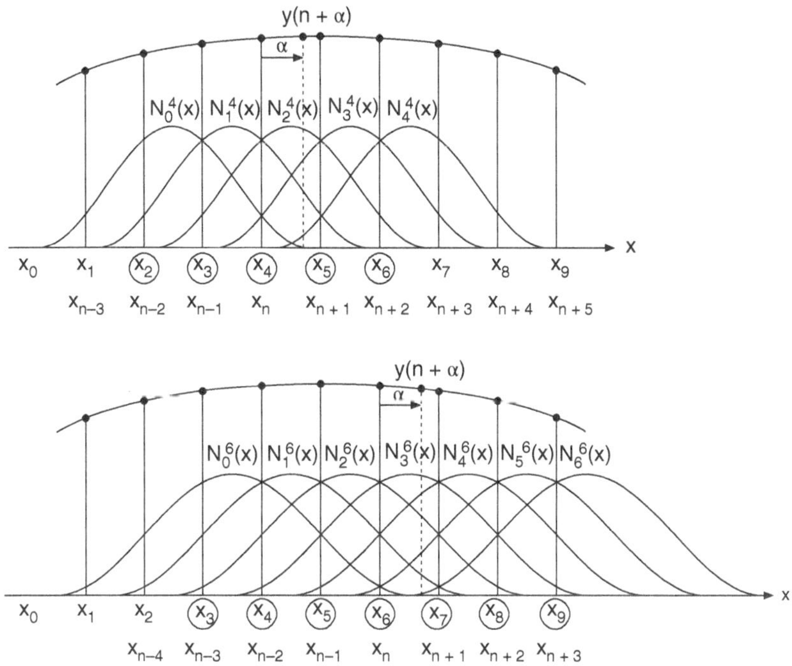

Figure 3.26 Interpolation with B-splines of fourth and sixth order.

For the application of interpolation, the properties in the frequency domain are important. The zero-order B-spline is given by

$$N_0^0(x) = \sum_{i=0}^{1} a_i \phi_i(x) = \begin{cases} 0 & x < 0, \\ 1 & 0 \leq x < 1, \\ 0 & x \geq 1, \end{cases} \tag{3.104}$$

and the Fourier transform gives the sinc function in the frequency domain. The first-order B-spline, given by

$$N_0^1(x) = 2\sum_{i=0}^{2} a_i \phi_i(x) = \begin{cases} 0 & x < 0, \\ \frac{1}{2}x & 0 \leq x < 1, \\ 1 - \frac{1}{2}x & 1 \leq x < 2, \\ 0 & x \geq 2, \end{cases} \tag{3.105}$$

leads to a sinc^2 function in the frequency domain. Higher-order B-splines can be derived by repeated convolution [Chu92] as given by

$$N^N(x) = N^0(x) * N^{N-1}(x). \tag{3.106}$$

Thus, the Fourier transform leads to

$$\text{FT}[N^N(x)] = \text{sinc}^{N+1}(f). \tag{3.107}$$

With the help of the properties in the frequency domain, the necessary order of the spline interpolation can be determined. Owing to the attenuation properties of the $\text{sinc}^{N+1}(f)$ function and the simple real-time calculation of the coefficients, spline interpolation is well suited to time-variant conversion in the last stage of a multistage sampling rate conversion system [Zöl94].

3.5 Exercises

1. Basics

Consider a simple sampling rate conversion system with a conversion rate of $\frac{4}{3}$. The system consists of two upsampling blocks, each by a factor of 2, and one downsampling block by a factor of 3.

1. What are anti-imaging and anti-aliasing filters and where do we need them in our system?
2. Sketch the block diagram.
3. Sketch the input, intermediate, and output spectra in the frequency domain.
4. How is the amplitude affected by the up- and downsampling and where does it come from?
5. Sketch the frequency response of the anti-aliasing and anti-imaging filters needed for this upsampling system.

2. Synchronous Conversion

Our system will now be upsampled directly by a factor of 4 and again downsampled by the factor of 3, but with linear interpolation and decimation methods. The input signal is $x(n) = \sin(\frac{n\pi}{6}), n = 0, \ldots, 48$.

1. What are the impulse responses of the two interpolation filters? Sketch their magnitude responses.
2. Plot the signals (input, intermediate, and output signal) in the time domain using Matlab.
3. What is the delay resulting from the causal interpolation/decimation filters?
4. Show the error introduced by this interpolation/decimation method in the frequency domain.

3. Polyphase Representation

Now we extend our system using a polyphase decomposition of the interpolation/decimation filters.

1. Sketch the idea of polyphase decomposition using a block diagram. What is the benefit of such decomposition?
2. Calculate the polyphase filters for up- and downsampling (using interpolation and decimation).
3. Plot with Matlab all resulting signals in the time and frequency domains.

4. Asynchronous Conversion

1. What is the basic concept of asynchronous sampling rate conversion?
2. Sketch the block diagram and discuss the individual operations.
3. What is the necessary oversampling factor L for a 20-bit resolution?
4. How can we simplify the oversampling operations?
5. How can we make use of polyphase filtering?
6. Why are half-band filters an efficient choice for the upsampling operation?
7. Which parameters determine the interpolation algorithms in the last stage of the conversion?

References

[Ada92] R. Adams, T. Kwan: *VLSI Architectures for Asynchronous Sample-Rate Conversion*, Proc. 93rd AES Convention, San Francisco, Preprint No. 3355, October 1992.

[Ada93] R. Adams, T. Kwan: *Theory and VLSI Implementations for Asynchronous Sample-Rate Conversion*, Proc. 94th AES Convention, Berlin, Preprint No. 3570, March 1993.

[Bar90] S. Barnett: *Matrices - Methods and Applications*, Oxford University Press, 1990.

[Boe93] W. Boehm, H. Prautzsch: *Numerical Methods*, AK Peters/Vieweg, 1993.

[Chu92] C.K. Chui (ed.): *Wavelets: A Tutorial in Theory and Applications*, Volume 2, Academic Press, Boston, 1992.

[Cro83] R.E. Crochiere, L.R. Rabiner: *Multirate Digital Signal Processing*, Prentice-Hall, Englewood Cliffs, 1983.

[Cuc91] S. Cucchi, F. Desinan, G. Parladori, G. Sicuranza: *DSP Implementation of Arbitrary Sampling Frequency Conversion for High Quality Sound Application*, Proc. IEEE ICASSP-91, Toronto, pp. 3609–3612, May 1991.

[Fli00] N. Fliege: *Multirate Digital Signal Processing*, J. Wiley & Sons, Chichester, 2000.

[Hsi87] C.-C. Hsiao: *Polyphase Filter Matrix for Rational Sampling Rate Conversions*, Proc. IEEE ICASSP-87, Dallas, pp. 2173–2176, April 1987.

[Kat85] Y. Katsumata, O. Hamada: *A Digital Audio Sampling Frequency Converter Employing New Digital Signal Processors*, Proc. 79th AES Convention, New York, Preprint No. 2272, October 1985.

[Kat86] Y. Katsumata, O. Hamada: *An Audio Sampling Frequency Conversion Using Digital Signal Processors*, Proc. IEEE ICASSP-86, Tokyo, pp. 33–36, 1986.

[Lag81] R. Lagadec, H.O. Kunz: *A Universal, Digital Sampling Frequency Converter for Digital Audio*, Proc. IEEE ICASSP-81, Atlanta, pp. 595–598, April 1981.

[Lag82a] R. Lagadec, D. Pelloni, D. Weiss: *A Two-Channel Professional Digital Audio Sampling Frequency Converter*, Proc. 71st AES Convention, Montreux, Preprint No. 1882, March 1982.

[Lag82b] D. Lagadec, D. Pelloni, D. Weiss: *A 2-Channel, 16-Bit Digital Sampling Frequency Converter for Professional Digital Audio*, Proc. IEEE ICASSP-82, Paris, pp. 93–96, May 1982.

[Lag82c] R. Lagadec: *Digital Sampling Frequency Conversion*, Digital Audio, Collected Papers from the AES Premier Conference, pp. 90–96, June 1982.

[Lag83] R. Lagadec, D. Pelloni, A. Koch: *Single-Stage Sampling Frequency Conversion*, Proc. 74th AES Convention, New York, Preprint No. 2039, October 1983.

[Liu92] G.-S. Liu, C.-H. Wei: *A New Variable Fractional Delay Filter with Nonlinear Interpolation*, IEEE Trans. Circuits and Systems-II: Analog and Digital Signal Processing, Vol. 39, No.2, pp. 123–126, February 1992.

[Opp99] A.V. Oppenheim, R.W. Schafer, J.R. Buck: *Discret-time Signal Processing*, Prentice-Hall, 2nd Edition, 1999.

[Par90] S. Park, R. Robles: *A Real-Time Method for Sample-Rate Conversion from CD to DAT*, Proc. IEEE Int. Conf. Consumer Electronics, Chicago, pp. 360–361, June 1990.

[Par91a] S. Park: *Low Cost Sample Rate Converters*, Proc. NAB Broadcast Engineering Conference, Las Vegas, April 1991.

[Par91b] S. Park, R. Robles: *A Novel Structure for Real-Time Digital Sample-Rate Converters with Finite Precision Error Analysis*, Proc. IEEE ICASSP-91, Toronto, pp. 3613–3616, May 1991.

[Ram82] T.A. Ramstad: *Sample-Rate Conversion by Arbitrary Ratios*, Proc. IEEE ICASSP-82, Paris, pp. 101–104, May 1982.

[Ram84] T.A. Ramstad: *Digital Methods for Conversion Between Arbitrary Sampling Frequencies*, IEEE Transactions on Acoustics, Speech and Signal Processing, vol. ASSP-32, no. 3, pp. 577–591, June 1984.

[Smi84] J.O. Smith, P. Gossett: *A Flexible Sampling-Rate Conversion Method*, Proc. IEEE ICASSP-84, pp. 19.4.1–19.4.4, 1984.

[Sti91] E.F. Stikvoort: *Digital Sampling Rate Converter with Interpolation in Continuous Time*, Proc. 90th AES Convention, Paris, Preprint No. 3018, Feb. 1991.

[Vai93] P.P. Vaidyanathan: *Multirate Systems and Filter Banks*, Prentice-Hall, Englewood Cliffs, 1993.

[Zöl94] U. Zölzer, T. Boltze: *Interpolation Algorithms: Theory and Application*, Proc. 97th AES Convention, San Francisco, Preprint No. 3898, November 1994.

Chapter 4

AD/DA Conversion

U. Zölzer

The conversion of a continuous-time signal $x(t)$ (voltage, current) into a sequence of numbers $x(n)$ is called analog-to-digital conversion (AD conversion). The reverse process is known as digital-to-analog conversion (DA conversion). The time sampling of a signal $x(t)$ is described by *Shannon's sampling theorem*. It states that a continuous-time signal with bandwidth f_B can be sampled with a sampling rate $f_S > 2f_B$ without changing the content of information in the signal. The original analog signal is reconstructed by lowpass filtering with bandwidth f_B. In addition to time sampling, the nonlinear procedure of digitizing the continuous-valued amplitude (quantization) of the sampled signal occurs. In Section 4.1, the basic concepts of Nyquist sampling, oversampling, and delta-sigma modulation are presented. In Sections 4.2 and 4.3, the principles of AD and DA converter circuits are discussed.

4.1 Methods

4.1.1 Nyquist Sampling

The sampling of a signal with sampling rate $f_S > 2f_B$ is called Nyquist sampling. The schematic diagram in Fig. 4.1 shows the procedure. The band limiting of the input at $f_S/2$ is carried out by an analog lowpass filter (Fig. 4.1a). The following sample-and-hold circuit samples the band-limited input at a sampling rate f_S. The constant amplitude of the time signal over the sampling interval $T_S = 1/f_S$ is converted to a number sequence $x(n)$ by a quantizer (Fig. 4.1b). This number sequence is fed to a digital signal processor (DSP) which performs signal processing algorithms. The output sequence $y(n)$ is delivered to a DA converter which gives a staircase as its output (Fig.4.1c). Following this, a lowpass filter gives the analog output $y(t)$ (Fig. 4.1d). Figure 4.2 demonstrates each step of AD/DA conversion in the frequency domain. The individual spectra in Fig. 4.2a–d correspond to the outputs in Fig. 4.1a–d.

Digital Audio Signal Processing, Third Edition. Udo Zölzer.
© 2022 John Wiley & Sons Ltd. Published 2022 by John Wiley & Sons Ltd.

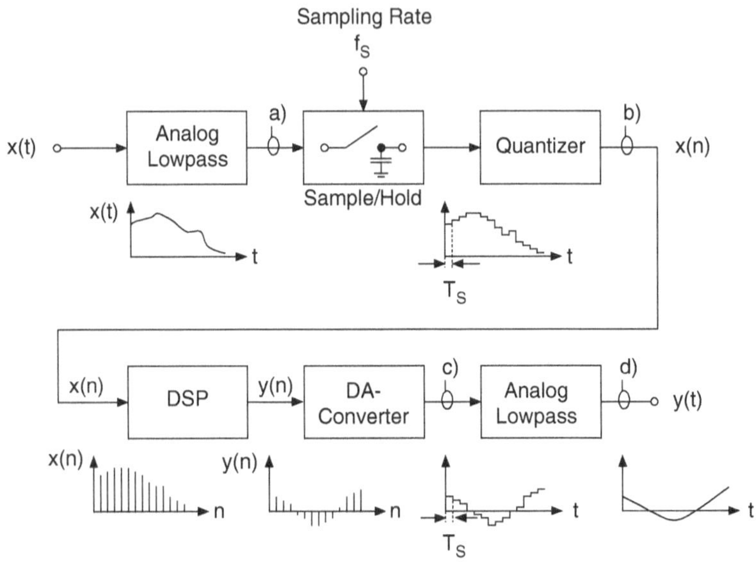

Figure 4.1 Schematic diagram of Nyquist sampling.

After band limiting (Fig. 4.2a) and sampling, a periodic spectrum with sampling rate f_S of the sampled signal is obtained, as shown in Fig. 4.2b. Assuming that consecutive quantization errors $e(n)$ are statistically independent of each other, the noise power has a spectral uniform distribution in the frequency domain $0 \leq f \leq f_S$. The output of the DA converter still has a periodic spectrum. However, this is weighted with the sinc function (sinc$=\frac{\sin(x)}{x}$) of the sample-and-hold circuit (Fig. 4.2c). The zeros of the sinc function are at multiples of the sampling rate f_S. To reconstruct the output (Fig. 4.2d), the image spectra are eliminated by an analog lowpass of sufficient stopband attenuation (see Fig. 4.2c).

The problems of Nyquist sampling lie in the steep band-limiting filter characteristic (anti-aliasing filter) of the analog input filter and the analog reconstruction filter (anti-imaging filter) of similar filter characteristic and sufficient stopband attenuation. Further, the sinc distortion owing to the sample-and-hold circuit needs to be compensated.

4.1.2 Oversampling

To increase the resolution of the conversion process and reduce the complexity of the analog filters, oversampling techniques are employed. Owing to the spectral uniform distribution of quantization error between 0 and f_S (see Fig. 4.3a), it is possible to reduce the power spectral density in the passband $0 \leq f \leq f_B$ through oversampling by a factor L, i.e. with the new sampling rate Lf_S (see Fig. 4.3b). For identical quantization step size Q, the dark-shaded areas (quantization error power σ_E^2) in Fig. 4.3a and Fig. 4.3b are equal. The increase in the SNR is also noticed from Fig. 4.3.

It follows that in the passband at a sampling rate of $f_S = 2f_B$, the power spectral density given by

$$S_{EE}(f) = \frac{Q^2}{12 f_S} \tag{4.1}$$

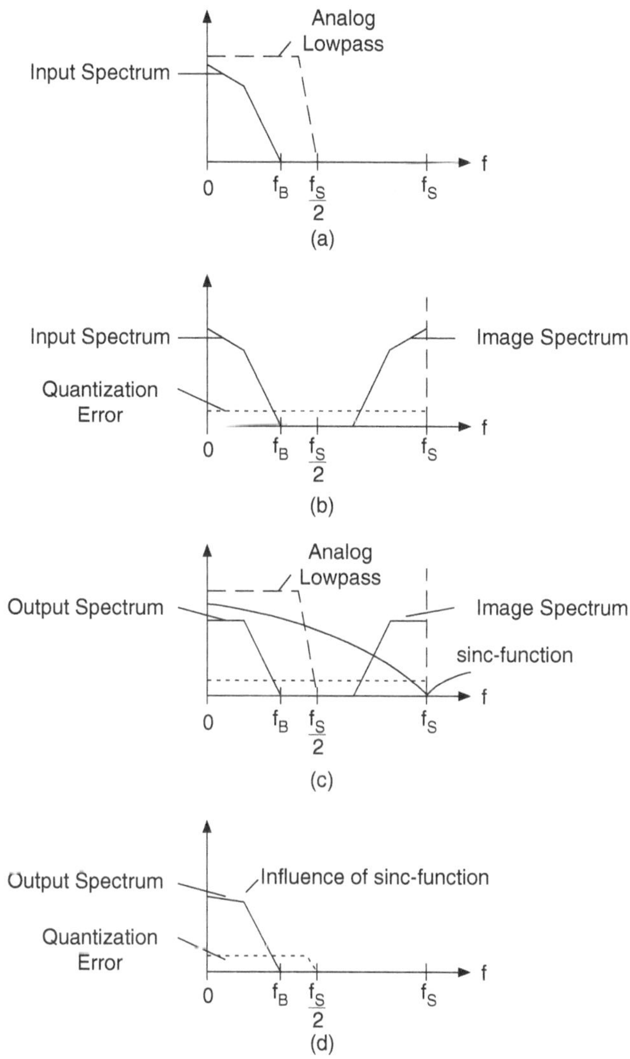

Figure 4.2 Nyquist sampling – interpretation in the frequency domain.

leads to the noise power

$$N_B^2 = \sigma_E^2 = 2\int_0^{f_B} S_{EE}(f)df = \frac{Q^2}{12}. \tag{4.2}$$

Owing to oversampling by a factor of L, a reduction of the power spectral density given by

$$S_{EE}(f) = \frac{Q^2}{12Lf_S} \tag{4.3}$$

is obtained (see Fig. 4.3b). With $f_S = 2f_B$, the error power in the audio band is given by

$$N_B^2 = 2f_B\frac{Q^2}{12Lf_S} = \frac{Q^2}{12}\frac{1}{L}. \tag{4.4}$$

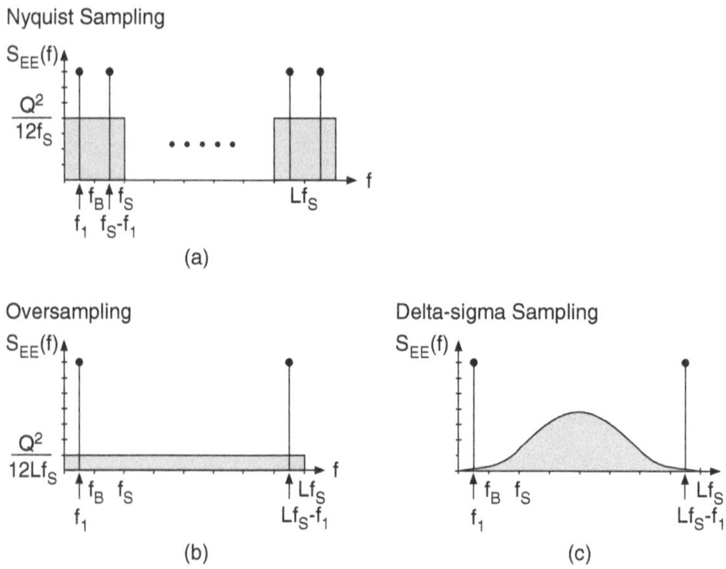

Figure 4.3 Influence of oversampling and delta-sigma technique on power spectral density of quantization error and on input sinusoid with frequency f_1.

The SNR (with $P_F = \sqrt{3}$) owing to oversampling can now be expressed as

$$\mathrm{SNR} = 6.02 \cdot w + 10 \log_{10}(L) \text{ dB}. \tag{4.5}$$

Figure 4.4a shows a schematic diagram of an oversampling AD converter. Owing to oversampling, the analog band-limiting lowpass filter can have a wider transition bandwidth, as shown in Fig. 4.4b. The quantization error power is distributed between 0 and the sampling rate Lf_S. To reduce the sampling rate, it is necessary to limit the bandwidth with a digital lowpass filter (see Fig. 4.4c). After this, the sampling rate is reduced by a factor L (see Fig. 4.4d) by taking every Lth output sample of the digital lowpass filter [Cro83, Vai93, Fli00].

Figure 4.5a shows the schematic diagram of an oversampling DA converter. The sampling rate is first increased by a factor of L. For this, $L - 1$ zeros are introduced between two consecutive input values [Cro83, Vai93, Fli00]. The following digital filter eliminates all image spectra (Fig. 4.5b) except the baseband spectrum and spectra at multiples of Lf_S (Fig. 4.5c). It interpolates $L - 1$ samples between two input samples. The w-bit DA converter operates at a sampling rate Lf_S. Its output is fed to an analog reconstruction filter which eliminates the image spectra at multiples of Lf_S.

4.1.3 Delta-sigma Modulation

Delta-sigma modulation using oversampling is a conversion strategy derived from delta modulation. In delta modulation (Fig. 4.6a), the difference between the input $x(t)$ and signal $x_1(t)$ is converted into a one-bit signal $y(n)$ at a very high sampling rate Lf_S. The sampling rate is higher than the necessary Nyquist rate f_S. The quantized signal $y(n)$ gives the signal $x_1(t)$ via an analog integrator. The demodulator consists of an integrator and a reconstruction lowpass filter.

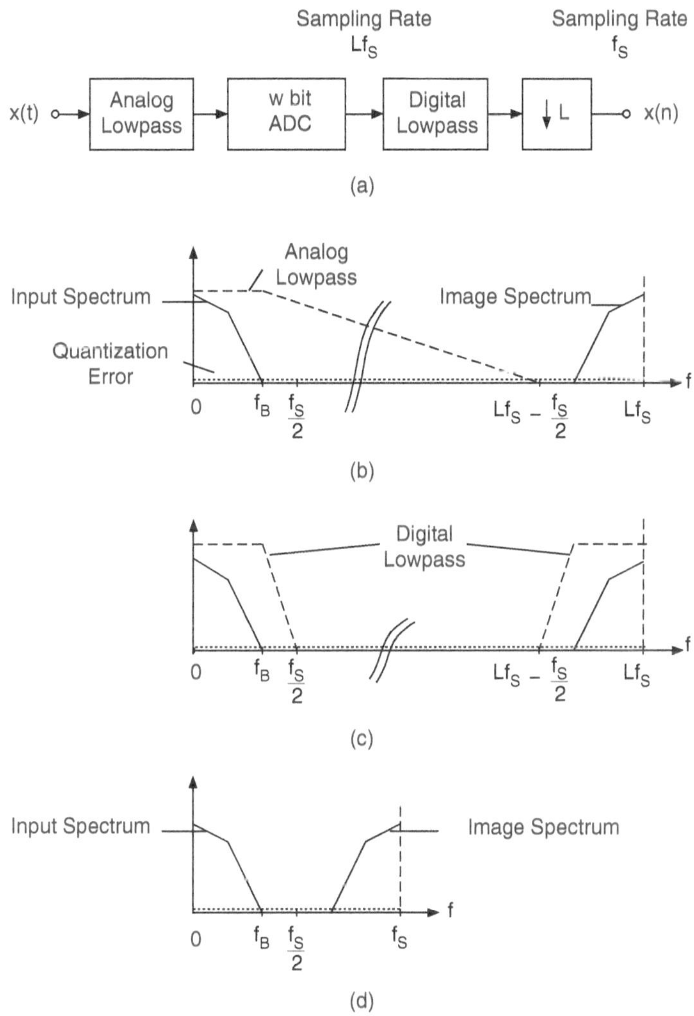

Figure 4.4 Oversampling AD converter and sampling rate reduction.

The extension to delta-sigma modulation [Ino63] consists of shifting the integrator from the demodulator to the input of the modulator (see Fig. 4.6b). With this, it is possible to combine the two integrators as a single integrator after addition (see Fig. 4.7a). The corresponding signals are shown in Fig. 4.8.

A time-discrete model of the delta-sigma modulator is given in Fig. 4.7b. The Z-transform of the output signal $y(n)$ is given by

$$\begin{aligned} Y(z) &= \frac{H(z)}{1+H(z)}X(z) + \frac{1}{1+H(z)}E(z) \\ &\approx X(z) + \frac{1}{1+H(z)}E(z). \end{aligned} \tag{4.6}$$

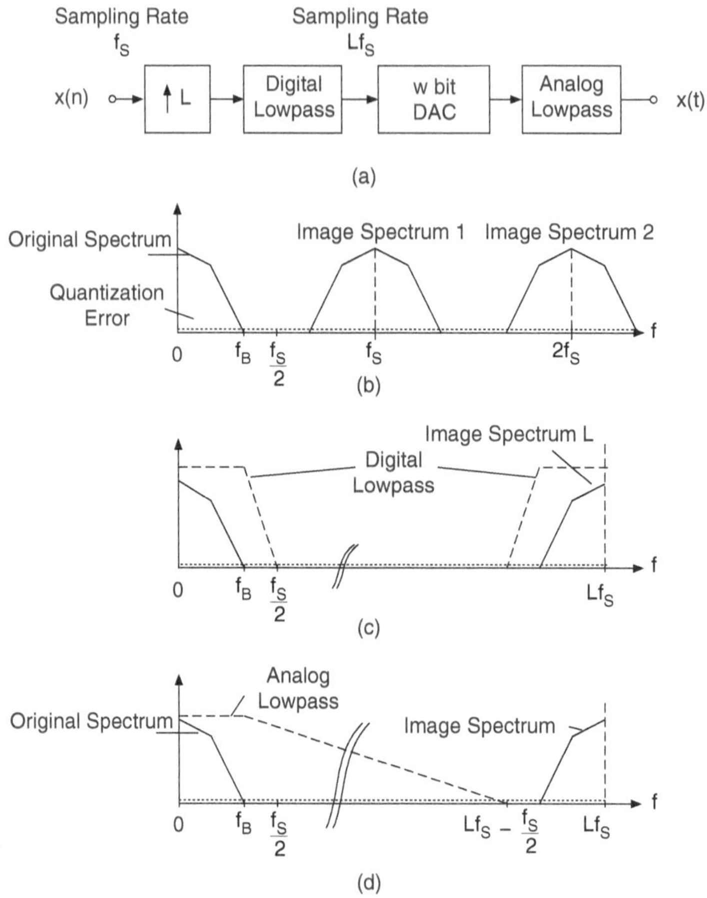

Figure 4.5 Oversampling and DA conversion.

For a large gain factor of the system $H(z)$, the input signal will not be affected. In contrast, the quantization error is shaped by the filter term $1/[(1 + H(z)]$.

The schematic diagrams of delta-sigma AD/DA conversion are shown in Figs. 4.9 and 4.10. For delta-sigma AD converters, a digital lowpass filter and a downsampler with factor L are used to reduce the sampling rate Lf_S down to f_S. The one-bit input to the digital lowpass filter leads to a w-bit output $x(n)$ at a sampling rate f_S. The delta-sigma DA converter consists of an upsampler with factor L, a digital lowpass filter to eliminate the mirror spectra, and a delta-sigma modulator followed by an analog reconstruction lowpass filter. To illustrate noise shaping in delta-sigma modulation in detail, first- and second-order systems as well as multistage techniques are investigated in the following sections.

First-order Delta-sigma Modulator

A time-discrete model of a first-order delta-sigma modulator is shown in Fig. 4.11. The difference equation for the output $y(n)$ is given by

$$y(n) = x(n-1) + e(n) - e(n-1). \tag{4.7}$$

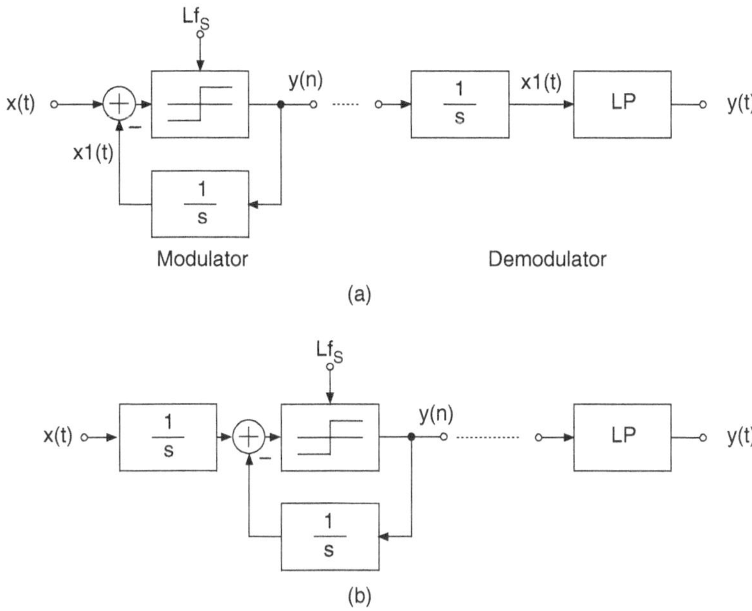

Figure 4.6 Delta modulation and displacement of integrator.

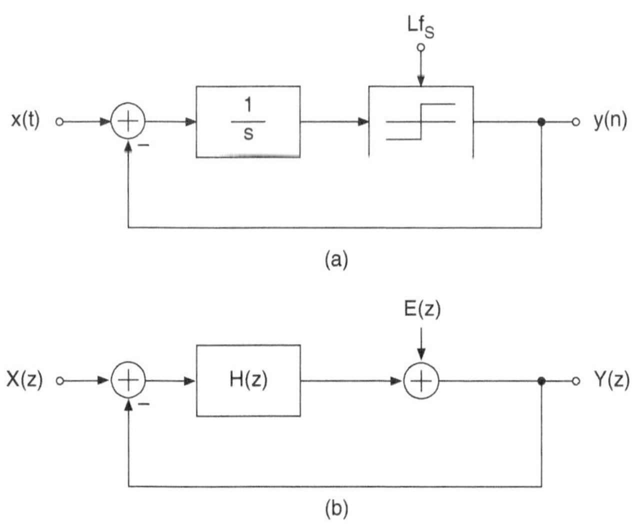

Figure 4.7 Delta-sigma modulation and time-discrete model.

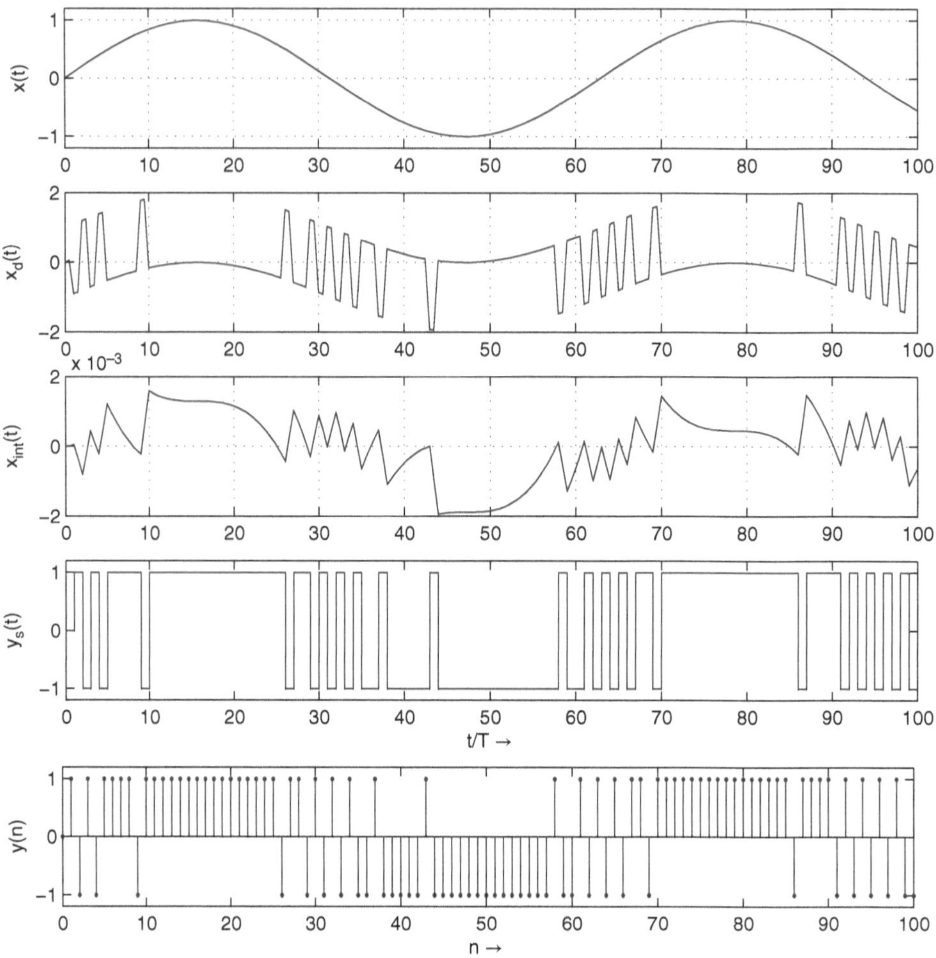

Figure 4.8 Signals in delta-sigma modulation.

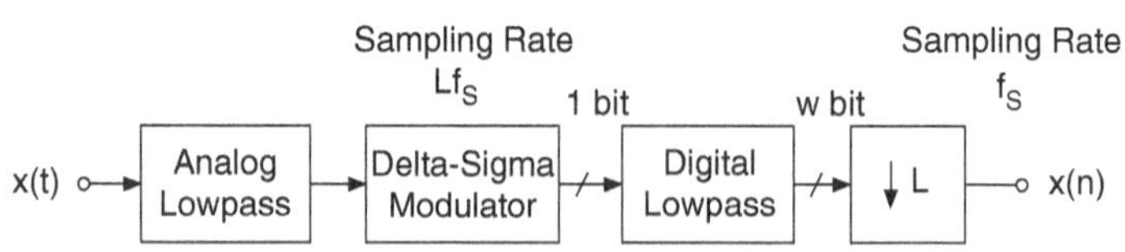

Figure 4.9 Oversampling delta-sigma AD converter.

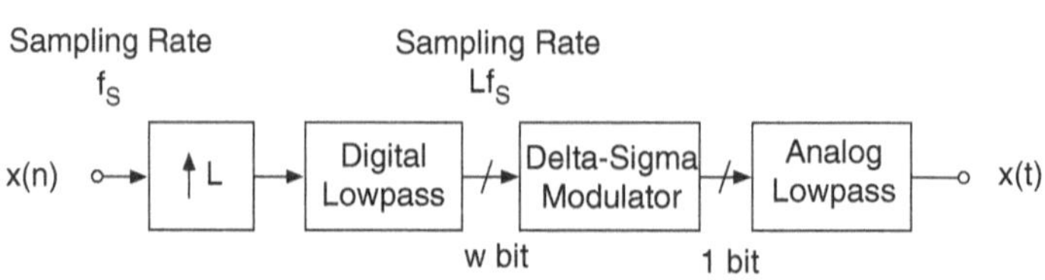

Figure 4.10 Oversampling delta-sigma DA converter.

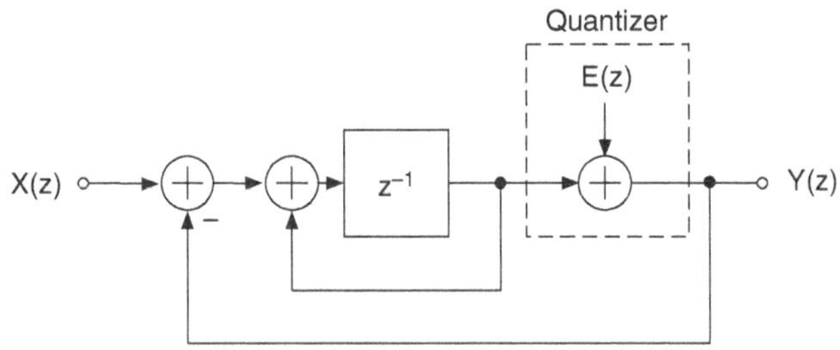

Figure 4.11 Time-discrete model of a first-order delta-sigma modulator.

The corresponding Z-transform leads to

$$Y(z) = z^{-1}X(z) + E(z)\underbrace{(1 - z^{-1})}_{H_E(z)}. \tag{4.8}$$

The power density spectrum of the error signal $e_1(n) = e(n) - e(n-1)$ is

$$S_{E_1E_1}(e^{j\Omega}) = S_{EE}(e^{j\Omega})\left|1 - e^{-j\Omega}\right|^2$$

$$= S_{EE}(e^{j\Omega})4\sin^2\left(\frac{\Omega}{2}\right), \tag{4.9}$$

where $S_{EE}(e^{j\Omega})$ denotes the power density spectrum of the quantization error $e(n)$. The error power in the frequency band $[-f_B, f_B]$, with $S_{EE}(f) = \frac{Q^2}{12Lf_S}$, can be written as

$$N_B^2 = S_{EE}(f)2\int_0^{f_B} 4\sin^2\left(\pi\frac{f}{Lf_S}\right)df \tag{4.10}$$

$$\simeq \frac{Q^2}{12}\frac{\pi^2}{3}\left(\frac{2f_B}{Lf_S}\right)^3. \tag{4.11}$$

With $f_S = 2f_B$, we get

$$N_B^2 = \frac{Q^2}{12}\frac{\pi^2}{3}\left(\frac{1}{L}\right)^3. \tag{4.12}$$

Second-order Delta-sigma Modulator

For the second-order delta-sigma modulator [Can85, Can92] shown in Fig. 4.12, the difference equation is expressed as

$$y(n) = x(n-1) + e(n) - 2e(n-1) + e(n-2) \tag{4.13}$$

and the Z-transform is given by

$$Y(z) = z^{-1}X(z) + E(z)\underbrace{(1 - 2z^{-1} + z^{-2})}_{H_E(z)=(1-z^{-1})^2}. \tag{4.14}$$

The power density spectrum of the error signal $e_1(n) = e(n) - 2e(n-1) + e(n-2)$ can be written as

$$S_{E_1E_1}(e^{j\Omega}) = S_{EE}(e^{j\Omega})|1 - e^{-j\Omega}|^4$$

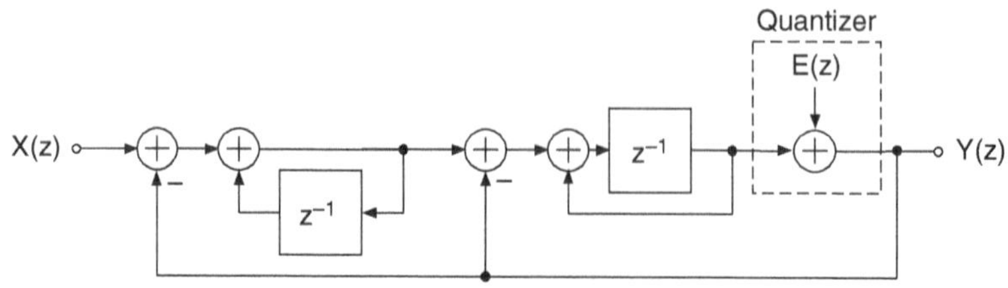

Figure 4.12 Time-discrete model of a second-order delta-sigma modulator.

$$= S_{EE}(e^{j\Omega})\left[4\sin^2\left(\frac{\Omega}{2}\right)\right]^2$$
$$= S_{EE}(e^{j\Omega})4[1-\cos(\Omega)]^2. \tag{4.15}$$

The error power in the frequency band $[-f_B, f_B]$ is given by

$$N_B^2 = S_{EE}(f)2\int_0^{f_B} 4[1-\cos(\Omega)]^2\, df \tag{4.16}$$

$$\simeq \frac{Q^2}{12}\frac{\pi^4}{5}\left(\frac{2f_B}{Lf_S}\right)^5, \tag{4.17}$$

and with $f_S = 2f_B$, we obtain

$$N_B^2 = \frac{Q^2}{12}\frac{\pi^4}{5}\left(\frac{1}{L}\right)^5. \tag{4.18}$$

Multistage Delta-sigma Modulator

A multistage delta-sigma modulator (MASH, [Mat87]) is shown in Fig. 4.13.

The Z-transforms of the output signals $y_{1..3}(n)$ are given by

$$Y_1(z) = X(z) + (1-z^{-1})E_1(z), \tag{4.19}$$

$$Y_2(z) = -E_1(z) + (1-z^{-1})E_2(z), \tag{4.20}$$

$$Y_3(z) = -E_2(z) + (1-z^{-1})E_3(z). \tag{4.21}$$

The Z-transform of the output obtained by addition and filtering leads to

$$\begin{aligned} Y(z) &= Y_1(z) + (1-z^{-1})Y_2(z) + (1-z^{-1})^2Y_3(z) \\ &= X(z) + (1-z^{-1})E_1(z) - (1-z^{-1})E_1(z) \\ &\quad +(1-z^{-1})^2E_2(z) - (1-z^{-1})^2E_2(z) + (1-z^{-1})^3E_3(z) \\ &= X(z) + \underbrace{(1-z^{-1})^3}_{H_E(z)}E_3(z). \end{aligned} \tag{4.22}$$

The error power in the frequency band $[-f_B, f_B]$

$$N_B^2 = \frac{Q^2}{12}\frac{\pi^6}{7}\left(\frac{2f_B}{Lf_S}\right)^7, \tag{4.23}$$

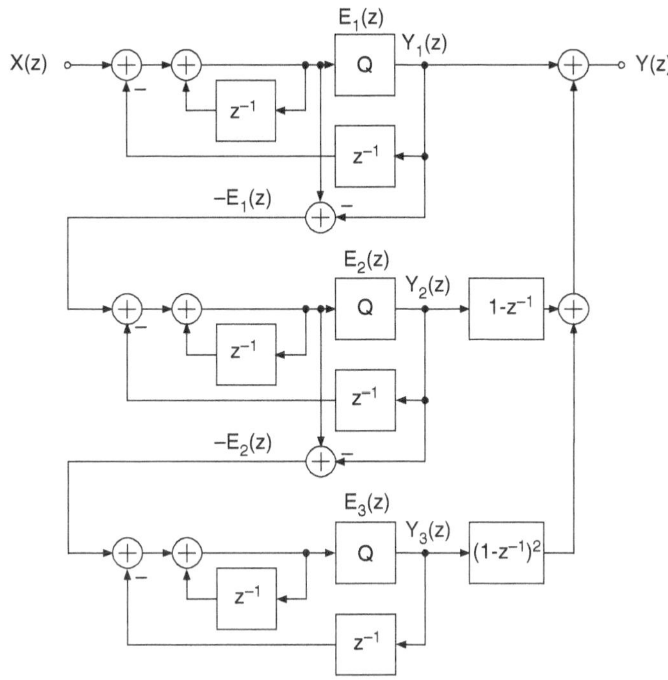

Figure 4.13 Time-discrete model of a multistage delta-sigma modulator.

with $f_S = 2f_B$, gives the following total noise power

$$N_B^2 = \frac{Q^2}{12}\frac{\pi^6}{7}\left(\frac{1}{L}\right)^7. \tag{4.24}$$

The error transfer functions in Fig. 4.14 show the noise shaping for three types of delta-sigma modulations as discussed before. The error power is shifted toward higher frequencies.

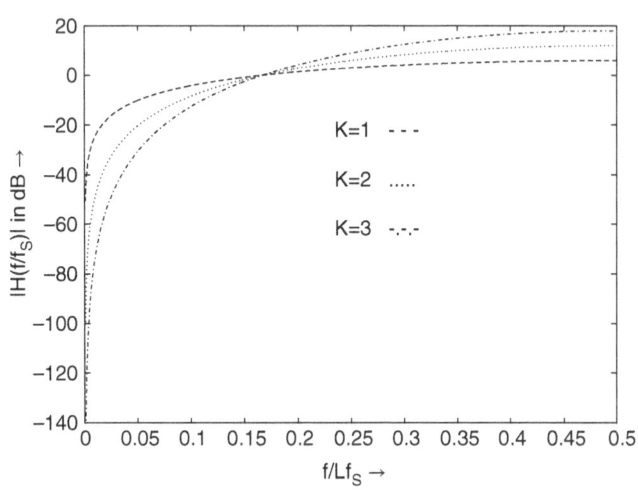

Figure 4.14 $H_E(z) = (1 - z^{-1})^K$ with $K = 1,2,3$.

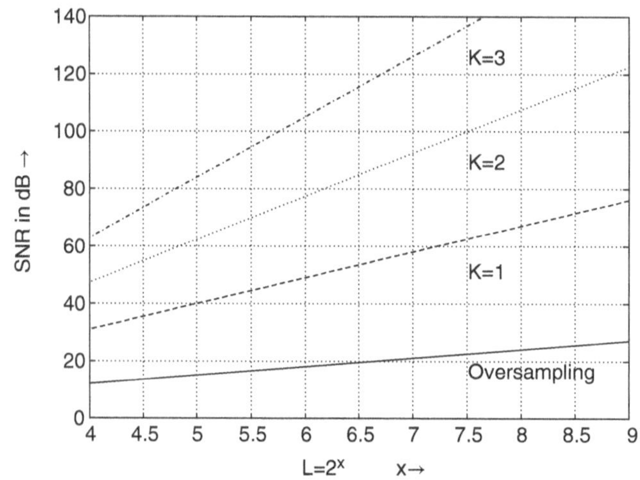

Figure 4.15 Improvement of signal-to-noise ratio as a function of oversampling and noise shaping ($L = 2^x$).

The improvement of the SNR by pure oversampling and delta-sigma modulation (first-, second-, and third-order) is shown in Fig. 4.15. For the general case of a kth order delta-sigma conversion with oversampling factor L, one can derive the SNR according to

$$\mathrm{SNR} = 6.02 \cdot w - 10\log_{10}\left(\frac{\pi^{2k}}{2k+1}\right) + (2k+1)10\log_{10}(L) \text{ in dB.} \tag{4.25}$$

In Eq. (4.25), w denotes the quantizer word-length of the delta-sigma modulator. The signal quantization after digital lowpass filtering and downsampling by L can can be performed with Eq. (4.25) according to the relation $w = \mathrm{SNR}/6$.

Higher-order Delta-sigma Modulator

A widening of the stopband for the highpass transfer function of the quantization error is achieved with higher-order delta-sigma modulation [Cha90]. As well as the zeros at $z = 1$, additional zeros are placed on the unit circle. Additionally, poles are integrated into the transfer function. A time-discrete model of a higher-order delta-sigma modulator is shown in Fig. 4.16.

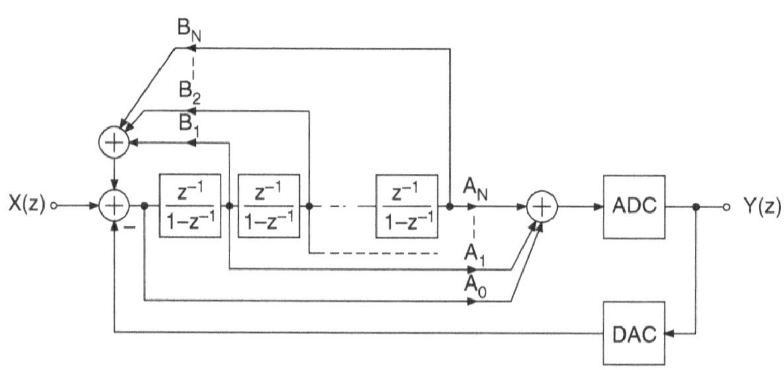

Figure 4.16 Higher-order delta-sigma modulator.

The transfer function in Fig. 4.16 can be written as

$$H(z) = \frac{A_0 + A_1\frac{z^{-1}}{1-z^{-1}} + A_2\left(\frac{z^{-1}}{1-z^{-1}}\right)^2 + \dots}{1 - B_1\frac{z^{-1}}{1-z^{-1}} - B_2\left(\frac{z^{-1}}{1-z^{-1}}\right)^2 + \dots}$$

$$= \frac{A_0(z-1)^N + A_1(z-1)^{N-1} + \dots + A_N}{(z-1)^N - B_1(z-1)^{N-1} - \dots - B_N}$$

$$= \frac{\sum_{i=0}^{N} A_i(z-1)^{N-i}}{(z-1)^N - \sum_{i=1}^{N} B_i(z-1)^{N-i}}. \quad (4.26)$$

The Z-transform of the output is given by

$$Y(z) = \frac{H(z)}{1+H(z)}X(z) + \frac{1}{1+H(z)}E(z) \quad (4.27)$$

$$= H_X(z)X(z) + H_E(Z)E(z). \quad (4.28)$$

The transfer function for the input is

$$H_X(z) = \frac{\sum_{i=0}^{N} A_i(z-1)^{N-i}}{(z-1)^N - \sum_{i=1}^{N} B_i(z-1)^{N-i} + \sum_{i=0}^{N} A_i(z-1)^{N-i}}, \quad (4.29)$$

and the transfer function for the error signal is given by

$$H_E(z) = \frac{(z-1)^N - \sum_{i=1}^{N} B_i(z-1)^{N-i}}{(z-1)^N - \sum_{i=1}^{N} B_i(z-1)^{N-i} + \sum_{i=0}^{N} A_i(z-1)^{N-i}}. \quad (4.30)$$

For Butterworth or Chebychev filter designs, the frequency responses, as shown in Fig. 4.17, are obtained for the error transfer functions. As a comparison, the frequency responses of first-, second-, and third-order delta-sigma modulation are shown. The widening of the stopband for Butterworth and Chebychev filters can be observed from Fig. 4.18.

Decimation Filter

The implementation of decimation filters for AD conversion and interpolation filters for DA conversion are performed with multirate systems [Fli00]. The necessary downsampler and upsampler are simple systems. For the former, every nth sample is taken out of the input sequence. For the latter, $(n-1)$ zeros are inserted between two input samples. For decimation, band limiting is performed by $H(z)$ followed by sampling rate reduction by a factor L. This procedure can be implemented in stages (see Fig. 4.19). The use of easy-to-implement filter structures at high sampling rates, like comb filters with transfer function

$$H_1(z) = \frac{1}{L}\frac{1-z^{-L}}{1-z^{-1}} \quad (4.31)$$

(shown in Fig. 4.20), allows simple implementation needing only delay systems and additions. To increase the stopband attenuation, a series of comb filters is used so that

$$H_1^M(z) = \left[\frac{1}{L}\frac{1-z^{-L}}{1-z^{-1}}\right]^M \quad (4.32)$$

is obtained.

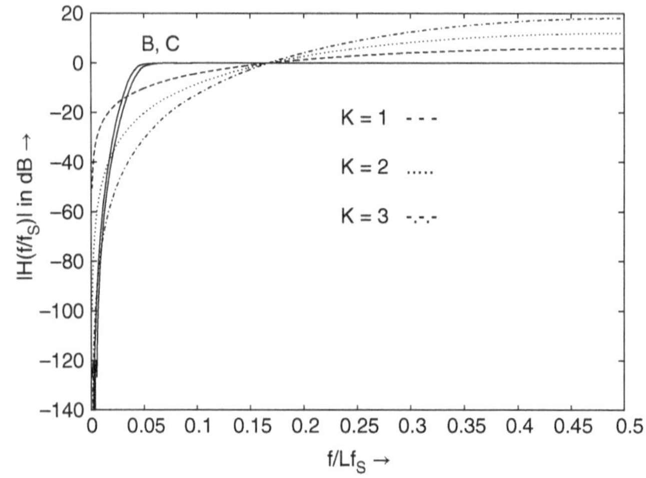

Figure 4.17 Comparison of different transfer functions of the error signal.

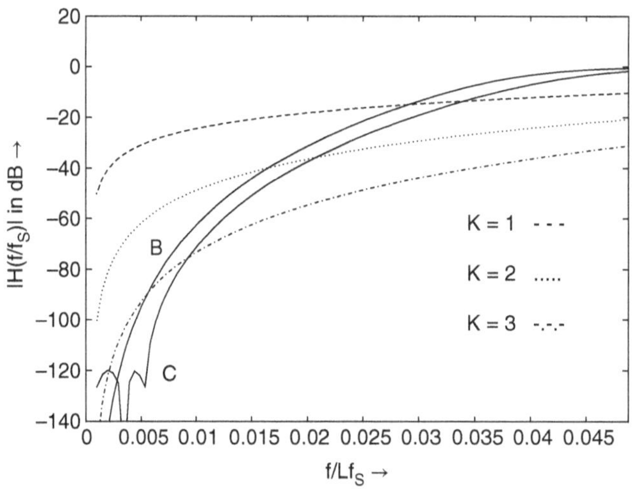

Figure 4.18 Transfer function of the error signal in the stopband.

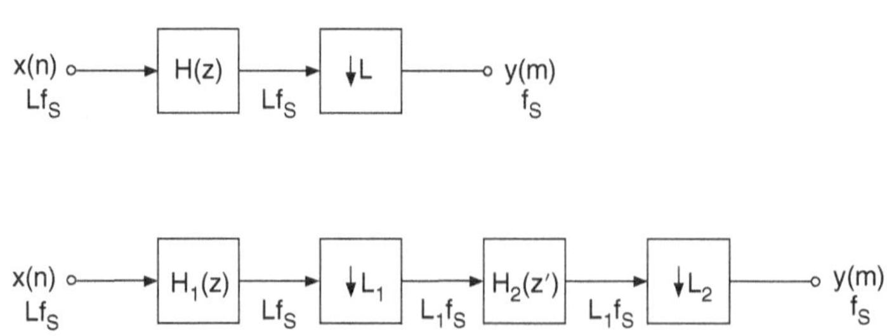

Figure 4.19 Several stages for sampling rate reduction.

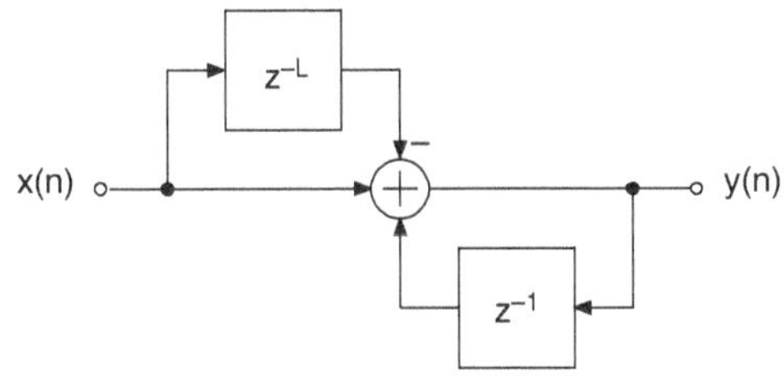

Figure 4.20 Signal flow diagram of a comb filter.

As well as additions at high sampling rates, complexity can be reduced further. Owing to sampling rate reduction by a factor of L, the numerator $(1 - z^{-L})$ can be moved so that it is placed after the downsampler (see Fig. 4.21). For a series of comb filters, the structure in Fig. 4.22 results. Now, M simple recursive accumulators have to be performed at the high sampling rate Lf_S. After this, downsampling by a factor L is carried out. The M nonrecursive systems are calculated with the output sampling rate f_S.

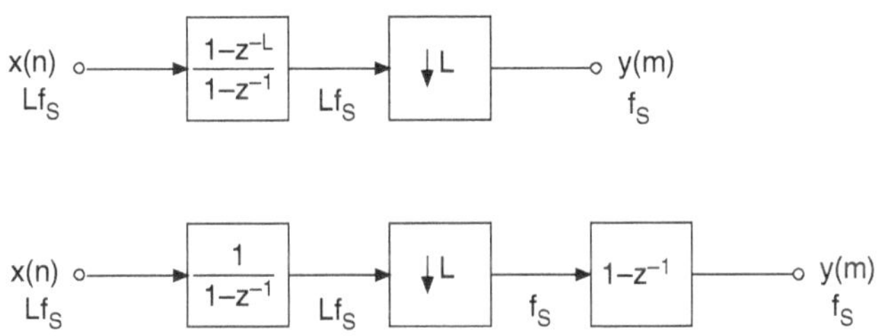

Figure 4.21 Comb filter for sampling rate reduction.

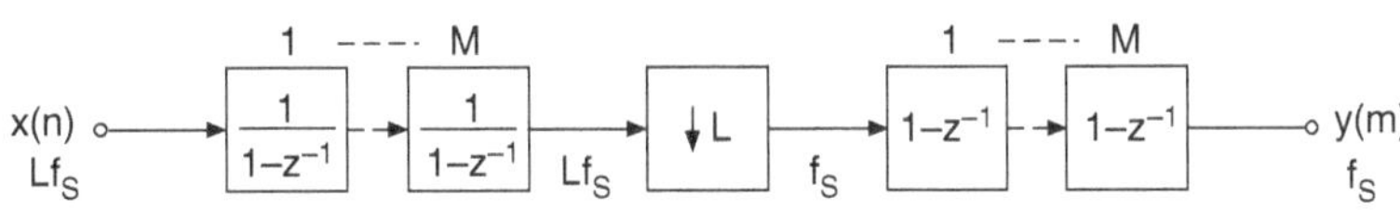

Figure 4.22 Series of comb filters for sampling rate reduction.

Figure 4.23a shows the frequency responses of a series of comb filters ($L = 16$). Figure 4.23b shows the resulting frequency response for the quantization error of a third-order delta-sigma modulator connected in series with a comb filter $H_1^4(z)$. The system delay owing to filtering and sampling rate reduction is given by

$$t_D = \frac{N-1}{2} \cdot \frac{1}{Lf_S}. \tag{4.33}$$

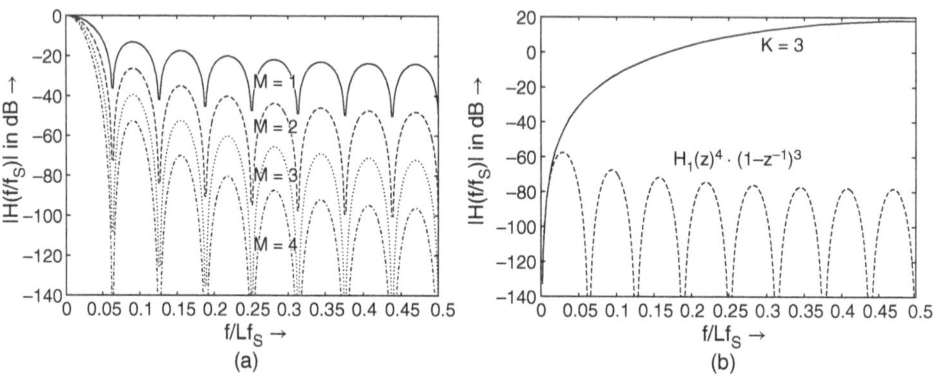

Figure 4.23 (a) Transfer function $H_1^M(z) = \left[\frac{1}{16}\frac{1-z^{-16}}{1-z^{-1}}\right]^M$ with $M = 1 \ldots 4$. (b) Third-order delta-sigma modulation and in series with $H_1^4(z)$.

Example: Delay time of conversion process (latency time)

1. Nyquist conversion

$$f_S = 48 \text{ kHz},$$
$$t_D = \frac{1}{f_S} = 20.83\ \mu s.$$

2. Delta-sigma modulation with single-stage downsampling

$$L = 64,$$
$$f_S = 48 \text{ kHz},$$
$$N = 4096,$$
$$t_D = 665\ \mu s.$$

3. Delta-sigma modulation with two-stage downsampling

$$L = 64,$$
$$f_S = 48 \text{ kHz}$$
$$L_1 = 16,$$
$$L_2 = 4,$$
$$N_1 = 61,$$
$$N_2 = 255,$$
$$t_{D_1} = 9.76\ \mu s,$$
$$t_{D_2} = 662\ \mu s.$$

4.2 AD Converters

The choice of an AD converter for a certain application is influenced by a number of factors. It mainly depends on the necessary resolution for a given conversion time. Both of these depend upon each other and are decisively influenced by the architecture of the AD converter [She86]. For this reason, the specifications of an AD converter are first discussed. This is followed by circuit principles which influence the mutual dependence of resolution and conversion time.

4.2.1 Specifications

In the following, the most important specifications for AD conversion are presented.

Resolution. The resolution for a given word length w of an AD converter determines the smallest amplitude

$$x_{min} = Q = x_{max}\ 2^{-(w-1)}, \tag{4.34}$$

which is equal to the quantization step Q.

Conversion Time. The minimum sampling interval $T_S = 1/f_S$ between two samples is called the conversion time.

Sample-and-hold Circuit. Before quantization, the time-continuous signal is sampled with the help of a sample-and-hold circuit, as shown in Fig. 4.24a. The sampling interval T_S is divided into sampling time t_S, in which the output voltage U_2 follows the input voltage U_1, and hold time t_H. During the hold time, the output voltage U_2 is constant and is converted into a binary word by quantization.

Aperture Delay. The time t_{AD} elapsed between the start of hold and actual hold modes (see Fig. 4.24b) is called the aperture delay.

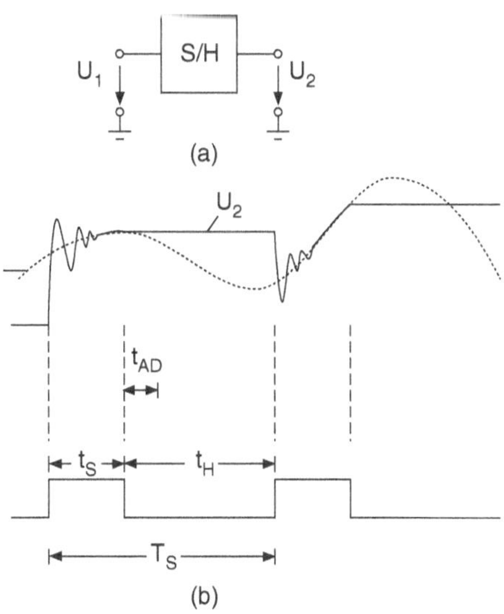

Figure 4.24 (a) Sample-and-hold circuit. (b) Input and output with clock signal. (t_S = samplingtime, t_H = holdtime, t_{AD} = aperturedelay.)

Aperture Jitter. The variation of aperture delay from sample to sample is called aperture jitter t_{ADJ}. The influence of aperture jitter limits the useful bandwidth of the sampled signal. This is because at high frequency, a deterioration of the SNR occurs. Assuming a Gaussian PDF aperture jitter, the SNR owing to aperture jitter as a function of frequency f can be written as

$$\text{SNR}_J = -20 \log_{10}\left(2\pi f t_{ADJ}\right) \text{ dB}. \tag{4.35}$$

Offset Error and Gain Error. The offset and gain errors of an AD converter are shown in Fig. 4.25. The offset error results in a horizontal displacement of the real curve compared with the dashed ideal curve of an AD converter. The gain error is expressed as the deviation from the ideal gradient of the curve.

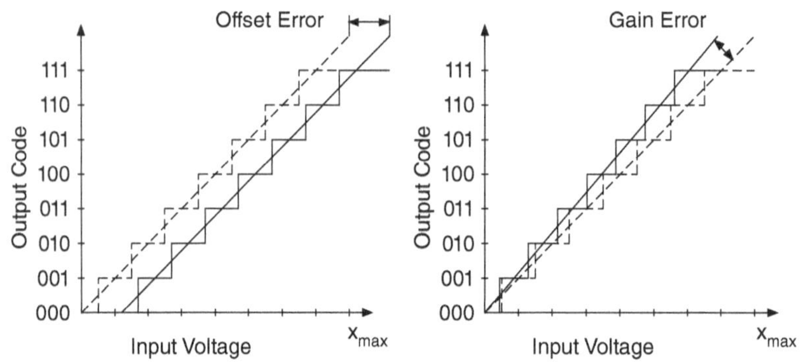

Figure 4.25 Offset error and gain error.

Differential Nonlinearity. The differential nonlinearity

$$\text{DNL} = \frac{\Delta x/Q}{\Delta x_Q} - 1 \text{ LSB} \tag{4.36}$$

describes the error of the step size of a certain code word in LSB units. For ideal quantization, the increase Δx of the input voltage up to the next output code x_Q is equal to the quantization step Q (see Fig. 4.26). The difference of two consecutive output codes is denoted as Δx_Q. When

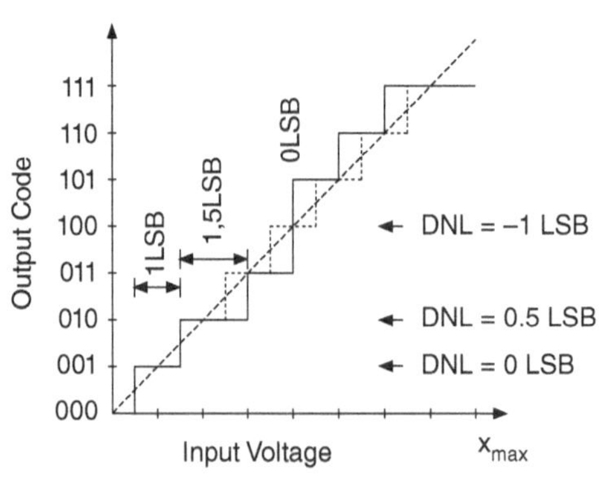

Figure 4.26 Differential nonlinearity.

the output code changes from 010 to 011, the step size is 1.5 LSB and therefore the differential nonlinearity DNL = 0.5 LSB. The step size between the codes 011 and 101 is 0 LSB and the code 200 is missing. The differential nonlinearity is DNL = −1 LSB.

Integral Nonlinearity. The integral nonlinearity (INL) describes the error between the quantized and the ideal continuous value. This error is given in LSB units. It arises owing to the accumulated error of the step size. This (see Fig. 4.27) changes itself continuously from one output code to another.

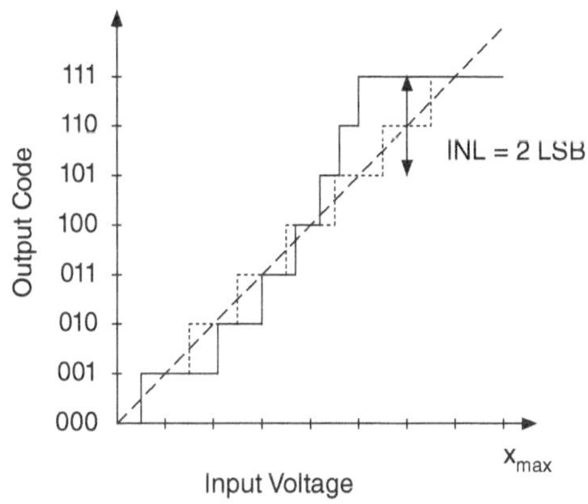

Figure 4.27 Integral nonlinearity.

Monotonicity. The progressive increase in quantizer output code for a continuously increasing input voltage and progressive decrease in quantizer output code for a continuously decreasing input voltage is called monotonicity. An example of non-monotonic behavior is shown in Fig. 4.28 where one output code does not occur.

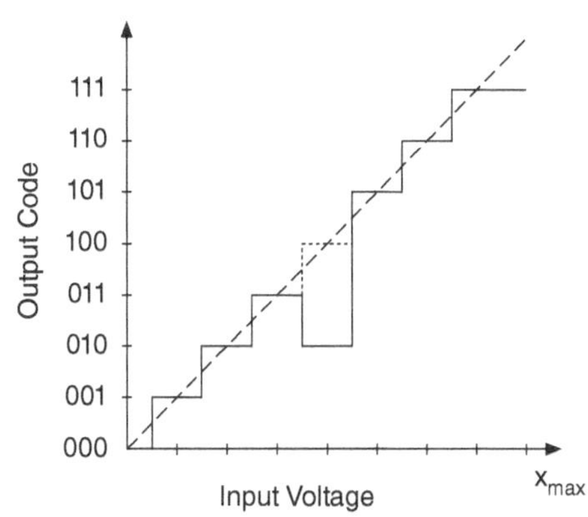

Figure 4.28 Monotonicity.

Total Harmonic Distortion. The harmonic distortion is calculated for an AD converter at full range with a sinusoid ($X_1 = 0$ dB) of given frequency. The selective measurements of harmonics of the second- to the ninth-order are used to compute

$$\mathrm{THD} = 20\ \log \sqrt{\sum_{n=2}^{\infty}[10^{(-X_n/20)}]^2}\ \mathrm{dB} \tag{4.37}$$

$$= \sqrt{\sum_{n=2}^{\infty}[10^{(-X_n/20)}]^2} \cdot 100\%, \tag{4.38}$$

where X_n are the harmonics in dB.

THD+N: Total Harmonic Distortion Plus Noise. For the calculation of harmonic distortion plus noise, the test signal is suppressed by a stopband filter. The measurement of harmonic distortion plus noise is performed by measuring the remaining broadband noise signal which consists of integral and differential nonlinearity, missing codes, aperture jitter, analog noise, and quantization error.

4.2.2 Parallel Converter

Parallel Converter. A direct method for AD conversion is called parallel conversion (that requires a flash converter). In parallel converters, the output voltage of the sample-and-hold circuit is compared with a reference voltage U_R with the help of $2^w - 1$ comparators (see Fig. 4.29). The sample-and-hold circuit is controlled with sampling rate f_S so that during the hold time t_H, a constant voltage at the output of the sample-and-hold circuit is available. The outputs of the comparators are fed at sampling clock rate into a $2^w - 1$-bit register and converted by a coding logic to a w-bit data word. This is fed at sampling clock rate to an output register. The sampling rates that can be achieved lie between 1 and 500 MHz for a resolution of up to 10 bits. Owing to the large number of comparators, the technique is not feasible for high precision.

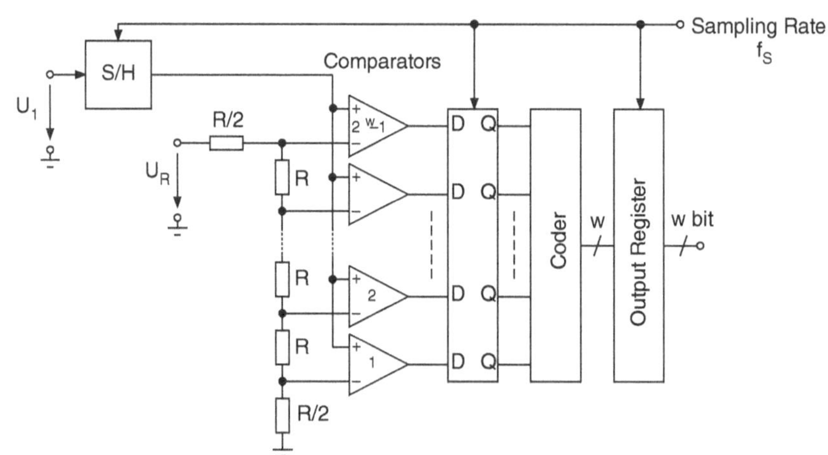

Figure 4.29 Parallel converter.

Half-flash Converter. In half-flash AD converters (Fig. 4.30), two m-bit parallel converters are used to convert two different ranges. The first m-bit AD converter gives a digital output word, which is converted into an analog voltage using an m-bit DA converter. This voltage is now subtracted from the output voltage of the sample-and-hold circuit. The difference voltage is digitized with a second m-bit AD converter. The rough and fine quantization leads to a w-bit data word with a subsequent logic.

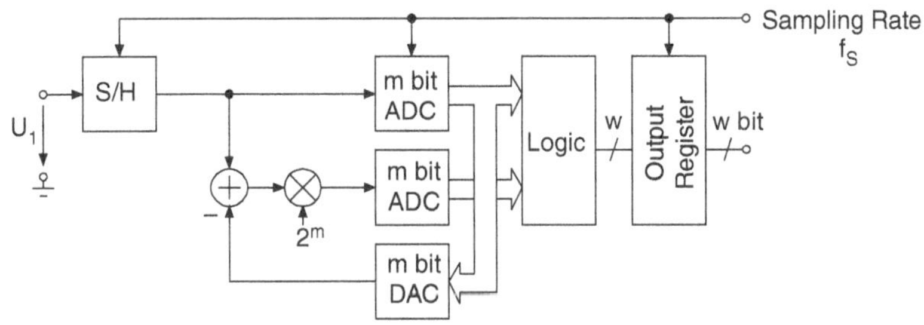

Figure 4.30 Half-flash AD converter.

Subranging Converter. A combination of the direct conversion and sequential procedure is carried out for subranging AD converters (see Fig. 4.31). In contrast to the half-flash converter, only one parallel converter is required. The switches S_1 and S_2 take the values of 0 and 1. First, the output voltage of a sample-and-hold circuit and then the difference voltage amplified by a factor 2^m are fed to an m-bit AD converter. The difference voltage is formed with the help of the output voltage of an m-bit DA converter and the output voltage of the sample-and-hold circuit. The conversion rates lie between 100 kHz and 40 MHz, where a resolution of up to 16 bits is achieved.

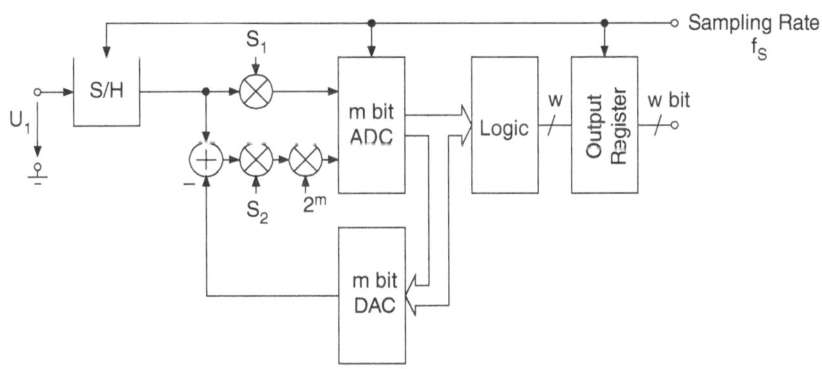

Figure 4.31 Subranging AD converter.

4.2.3 Successive Approximation

AD converters with successive approximation consist of the functional modules shown in Fig. 4.32. The analog voltage is converted into a w-bit word within w cycles. The converter consists of a comparator, a w-bit DA converter, and logic for controlling the successive approximation.

The conversion process is explained with the help of Fig. 4.33. First, whether it is a positive or negative voltage at the comparator is checked. If it is positive, the output $+0.5U_R$ is

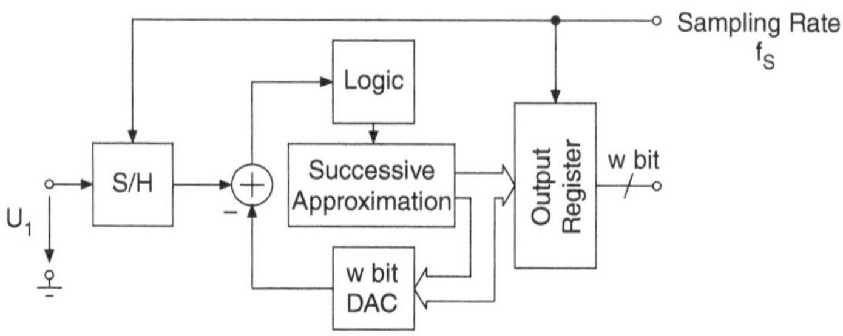

Figure 4.32 AD converter with successive approximation.

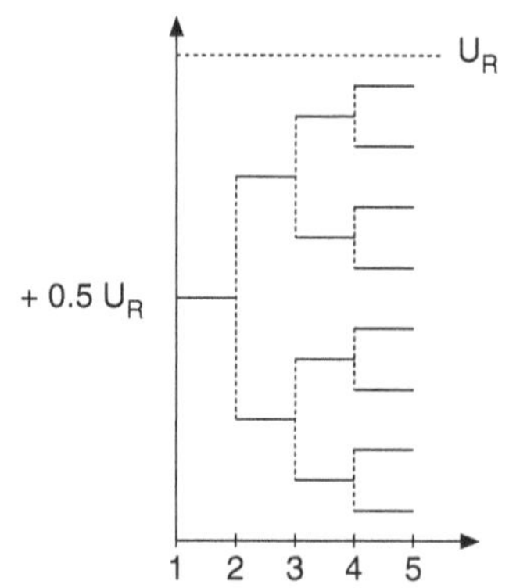

Figure 4.33 Successive approximation.

fed to a DA converter to check whether the output voltage of the comparator is greater or less than $+0.5U_R$. After that, the output of $(+0.5 \pm 0.25)U_R$ is fed to the DA comparator. The output of the comparator is then evaluated. This procedure is performed w times and leads to a w-bit word.

For a resolution of 12 bits, sampling rates of up to 1 MHz can be achieved. Higher resolutions of more than 16 bits are possible at a lower sampling rates.

4.2.4 Counter Methods

In contrast to the conversion techniques of the previous sections for high conversion rates, the following techniques are used for sampling rates smaller than 50 kHz.

Forward-backward Counter. A technique which operates like successive approximation is the forward-backward counter, shown in Fig. 4.34. A logic controls a clocked forward-backward counter whose output data word provides an analog output voltage via a w-bit DA converter. The difference signal between this voltage and the output voltage of the sample-and-hold circuit determines the direction of counting. The counter stops when the corresponding output voltage of the DA converter is equal to the output voltage of the sample-and-hold circuit.

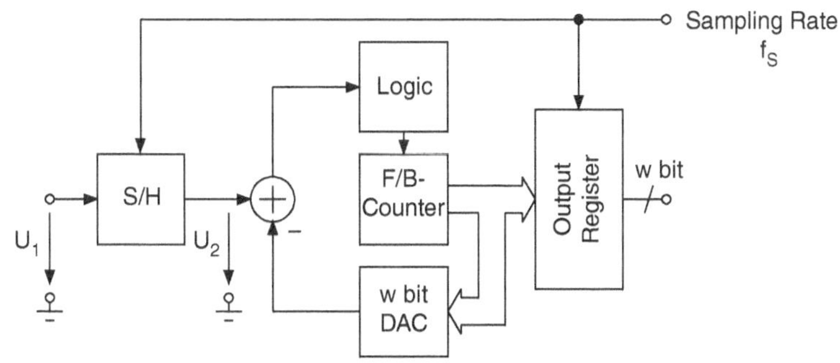

Figure 4.34 AD converter with forward-backward counter.

Single-slope Counter. The single-slope AD converter, shown in Fig. 4.35, compares the output voltage of the sample-and-hold circuit with a voltage of a sawtooth generator. The sawtooth generator is started every sampling interval. As long as the input voltage is greater than the sawtooth voltage, the clock impulses are counted. The counter value corresponds to the digital value of the input voltage.

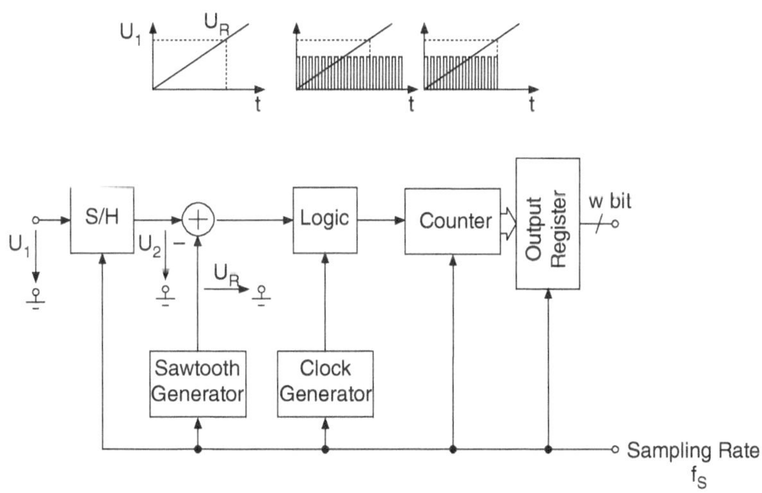

Figure 4.35 Single-slope AD converter.

Dual-slope Converter. A dual-slope AD converter is shown in Fig. 4.36. In the first phase, in which a switch S_1 is closed for a counter interval t_1, the output voltage of the sample-and-hold circuit is fed to an integrator of time constant τ. During the second phase, the switch S_2 is closed and the switch S_1 is opened. The reference voltage is switched to the integrator and the time to reach a threshold is determined by counting the clock impulses by a counter. Figure 4.36 demonstrates this for three different voltages U_2. The slope during time t_1 is proportional to the output voltage U_2 of the sample-and-hold circuit, whereas the slope is constant when the

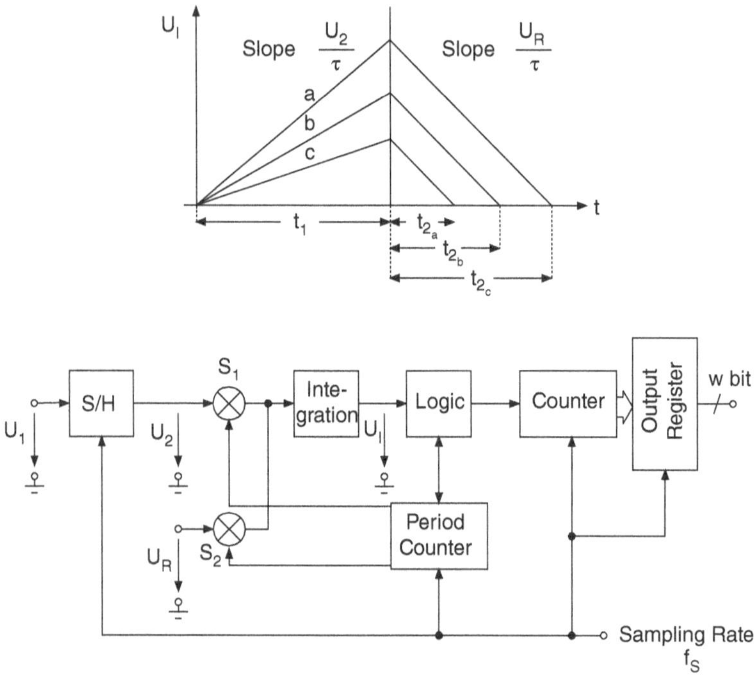

Figure 4.36 Dual-slope AD converter.

reference voltage U_R is connected to the integrator. The ratio $U_2/U_R = t_2/t_1$ leads to the digital output word.

4.2.5 Delta-sigma AD Converter

The delta-sigma AD converter, shown in Fig. 4.37, requires no sample-and-hold circuit owing to its high conversion rate. The analog band-limiting lowpass filter and the digital lowpass filter for downsampling to a sampling rate f_S are usually on the same circuit. The linear phase non-recursive digital lowpass filter shown in Fig. 4.37 has a 1-bit input signal and leads to a w-bit output signal owing to the N filter coefficients $h_0, h_1, \dots, h_{N-1}$, which are implemented with a word length of w bits. The output signal of the filter results from the summation of the filter coefficients of the non-recursive lowpass filter with either 0 or 1. The downsampling by a factor L is performed by taking every Lth sample out of the filter and writing to the output register. To reduce the number of operations, the filtering and downsampling can be performed only every Lth input sample.

The applications of delta-sigma AD converters are found at sampling rates of up to 100 kHz with a resolution of up to 24 bits.

4.3 DA Converters

Circuit principles for DA converters are mainly based on direct conversion techniques of the input code [She86]. Therefore, the achievable sampling rates are accordingly high.

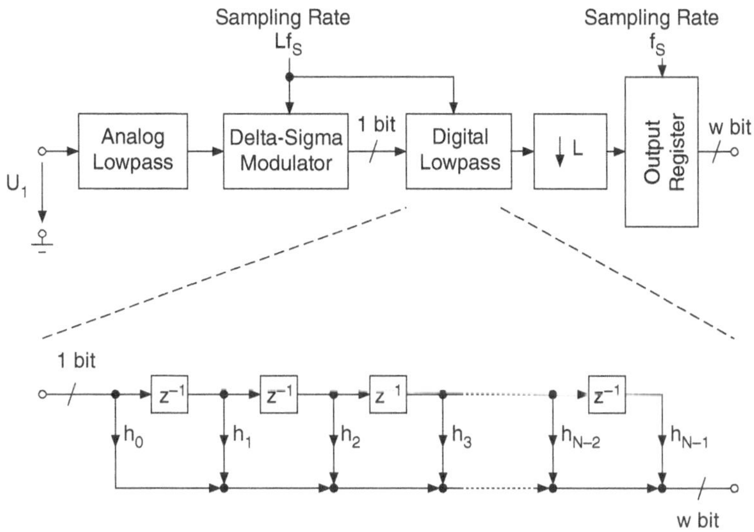

Figure 4.37 Delta-sigma AD converter.

4.3.1 Specifications

The definitions of resolution, total harmonic distortion (THD), and total harmonic distortion plus noise (THD+N) correspond to those for AD converters. Further specifications are discussed in the following.

Settling Time. The time interval between transferring a binary word and achieving the analog output value within a specific error range is called the settling time t_{SE}. The settling time determines the maximum conversion frequency $f_{S_{max}} = 1/t_{SE}$. Within this time, glitches between consecutive amplitude values can occur (see Fig. 4.38). With the help of a sample-and-hold circuit (deglitcher), the output voltage of the DA converter is sampled after the settling time and held.

Offset and Gain Error. The offset and gain errors of a DA converter are shown in Fig. 4.39.

Differential Nonlinearity. The differential nonlinearity for DA converters describes the step-size error of a code word in LSB. For ideal quantization, the increase Δx of the output voltage until the next code word corresponding to the output voltage is equal to the quantization step size Q (see Fig. 4.40). The difference of two consecutive input codes is termed Δx_Q. Differential nonlinearity is given by

$$\mathrm{DNL} = \frac{\Delta x/Q}{\Delta x_Q} - 1 \text{ [LSB]}. \tag{4.39}$$

For the code steps from 001 to 010, as shown in Fig. 4.40, the step size is 1.5 LSB, and therefore the differential nonlinearity DNL = 0.5 LSB. The step size between the codes 010 and 100 is 0.75 LSB and it follows for DNL = −0.25. The step size for the code change from 011 to 100 is 0 LSB (DNL = −1 LSB).

Integral Nonlinearity. The integral nonlinearity describes the maximum deviation of the output voltage of a real DA converter from the ideal straight line (see Fig. 4.41).

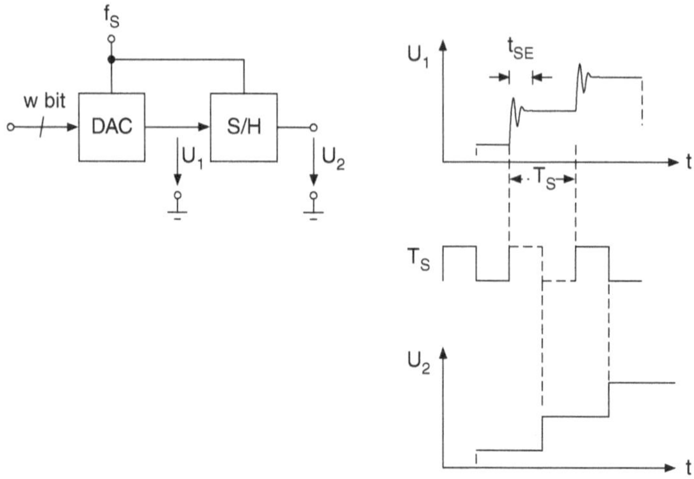

Figure 4.38 Settling time and sample-and-hold function.

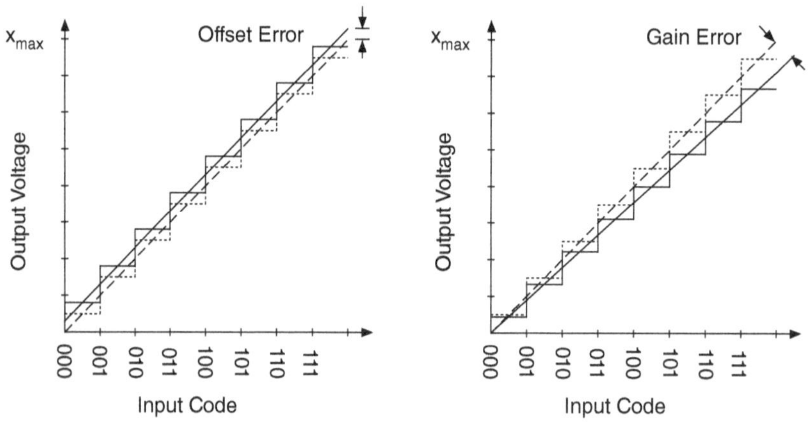

Figure 4.39 Offset and gain error.

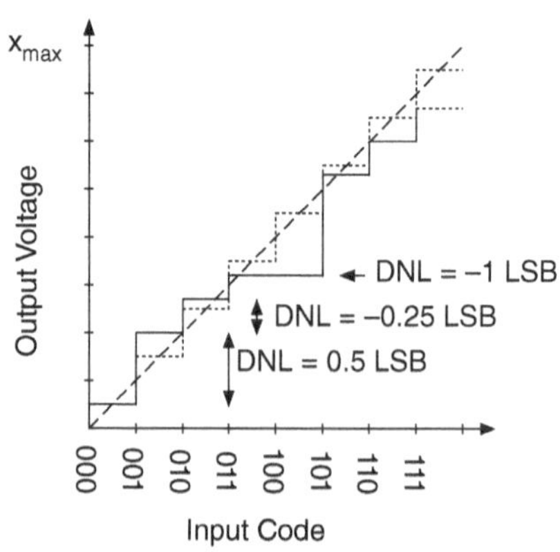

Figure 4.40 Differential nonlinearity.

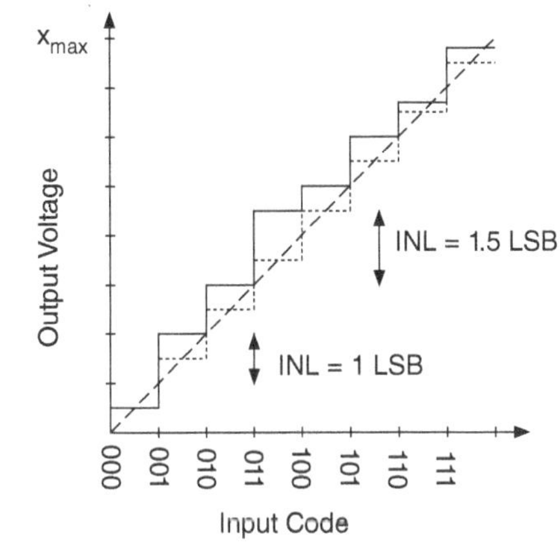

Figure 4.41 Integral nonlinearity.

Monotonicity. The continuous increase of the output voltage with increasing input code and the continuous decrease of the output voltage with decreasing input code is called monotonicity. A non-monotonic behavior is presented in Fig. 4.42.

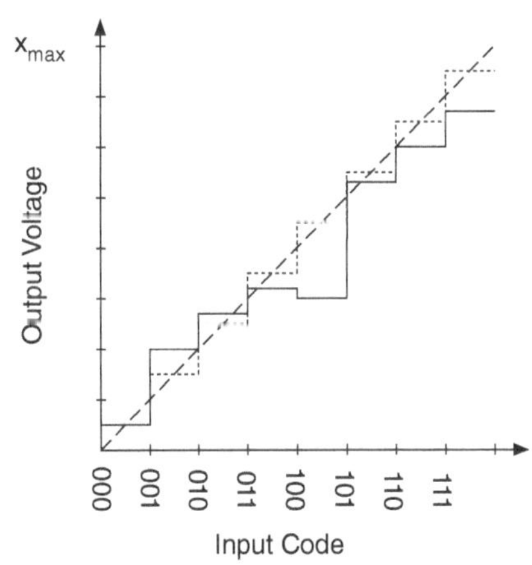

Figure 4.42 Monotonicity.

4.3.2 Switched Voltage and Current Sources

Switched Voltage Sources. The DA conversion with switched voltage sources, shown in Fig. 4.43a, is carried out with a reference voltage connected to a resistor network. The resistor network consists of 2^w resistors of equal resistance and is switched in stages to a binary-controlled decoder so that at the output, a voltage U_2 is present corresponding to the input code. Figure 4.43b shows the decoder for a 3-bit input code 101.

Switched Current Sources. DA conversion with 2^w switched current sources is shown in Fig. 4.44. The decoder switches the corresponding number of current sources onto the

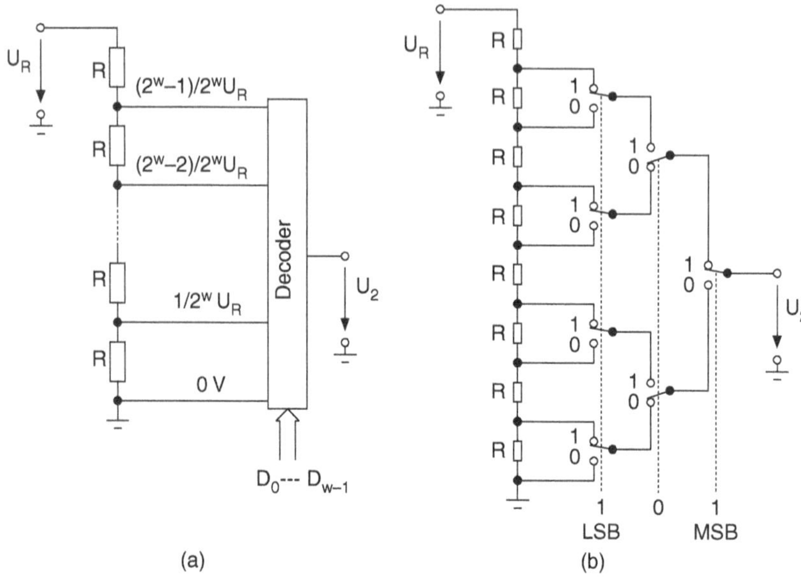

Figure 4.43 Switched voltage sources.

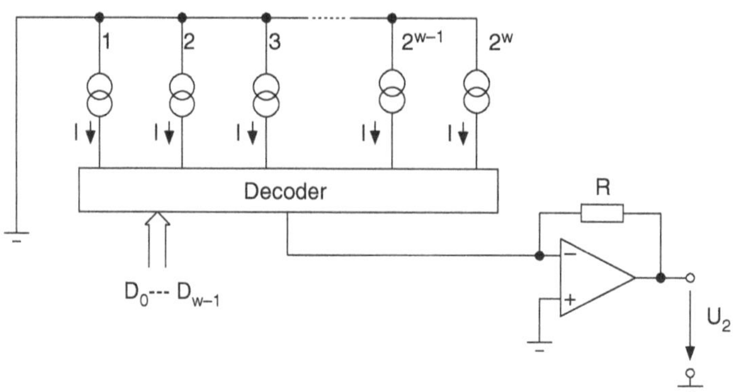

Figure 4.44 Switched current sources.

current–voltage converter. The advantage of both techniques is the monotonicity which is guaranteed for ideal switches but also for slightly deviating resistances. The large number of resistors in switched current sources or the large number of switched current sources causes problems for long word lengths. The techniques are used in combination with other methods for DA conversion of higher significant bits.

4.3.3 Weighted Resistors and Capacitors

A reduction in the number of identical resistors or current sources is achieved with the following method.

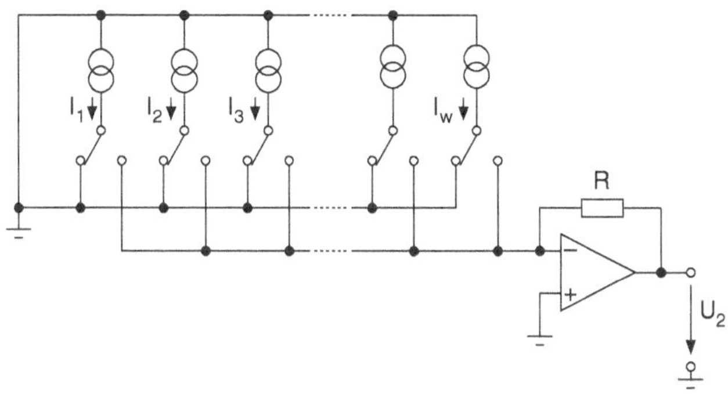

Figure 4.45 Weighted current sources.

Weighted Resistors. DA conversion with w switched current sources which are weighted according to

$$I_1 = 2I_2 = 4I_3 = \ldots = 2^{w-1}I_w \tag{4.40}$$

is shown in Fig. 4.45. The output voltage is

$$U_2 = -R \cdot I = -R \cdot (b_1 I_1 2^0 + b_2 I_2 2^1 + b_3 I_3 2^2 + \ldots + b_w I_w 2^{w-1}), \tag{4.41}$$

where b_n takes values 0 or 1. The implementation of DA conversion with switched current sources is carried out with weighted resistors, as shown in Fig. 4.46. The output voltage is

$$U_2 = R \cdot I = R\left(\frac{b_1}{2R} + \frac{b_2}{4R} + \frac{b_4}{8R} + \ldots + \frac{b_w}{2^w R}\right) U_R \tag{4.42}$$

$$= (b_1 2^{-1} + b_2 2^{-2} + b_3 2^{-3} + \ldots b_w 2^{-w}) U_R. \tag{4.43}$$

Weighted Capacitors. DA conversion with weighted capacitors is shown in Fig. 4.47. During the first phase (switch position 1 in Fig. 4.47), all capacitors are discharged. During the second phase, all capacitors that belong to 1 bits are connected to a reference voltage. Those capacitors belonging to 0 bits are connected to ground. The charge on the capacitors C_a that are connected with the reference voltage can be set equal to the total charge on all capacitors C_g, which leads to

$$U_R\, C_a = U_R\left(b_1 C + \frac{b_2 C}{2} + \frac{b_3 C}{2^2} + \ldots + \frac{b_w C}{2^{w-1}}\right) = C_g\, U_2 = 2\, C\, U_2. \tag{4.44}$$

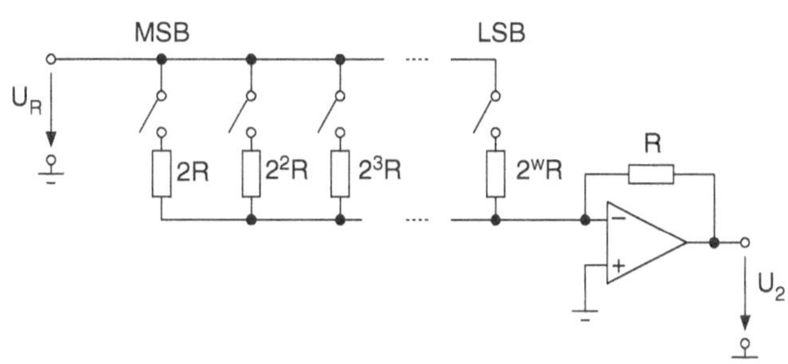

Figure 4.46 DA conversion with weighted resistors.

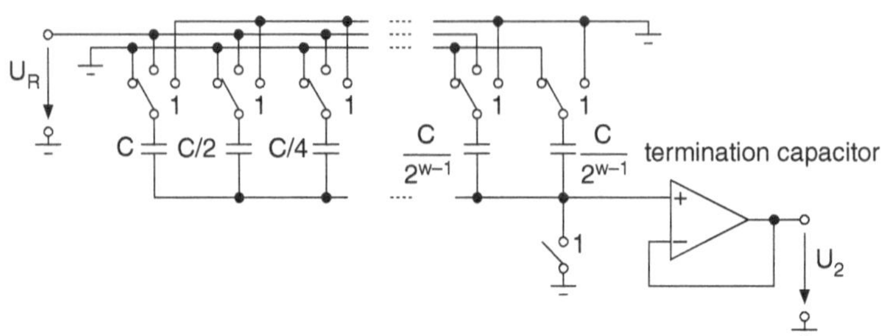

Figure 4.47 DA conversion with weighted capacitors.

Hence, the output voltage is

$$U_2 = (b_1 2^{-1} + b_2 2^{-2} + b_3 2^{-3} + \ldots + b_w 2^{-w})U_R. \tag{4.45}$$

4.3.4 R−2R Resistor Networks

The DA conversion with switched current sources can also be carried with an R−2R resistor network, as shown in Fig. 4.48. In contrast to the method with weighted resistors, the ratio of the smallest to largest resistor is reduced to 2 : 1.

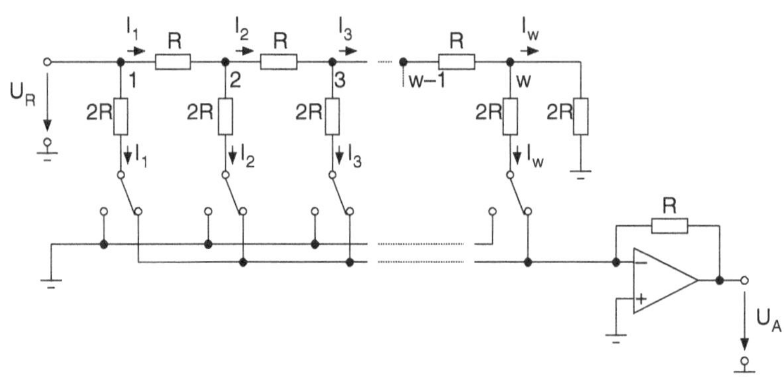

Figure 4.48 Switched current sources with R−2R resistor network.

The weighting of currents is achieved by a current division in every junction. Looking right from every junction, a resulting resistance $R + 2R \parallel 2R = 2R$ is found which is equal to the resistance in the vertical direction downwards from the junction. The current from junction 1 follows $I_1 = \frac{U_R}{2R}$, and for the current from junction 2, $I_2 = \frac{I_1}{2}$. Hence, a binary weighting of the w currents is given by

$$I_1 = 2I_2 = 4I_3 = \ldots = 2^{w-1}I_w. \tag{4.46}$$

The output voltage U_2 can be written as

$$U_2 = -RI = -R\left(\frac{b_1}{2R} + \frac{b_2}{4R} + \frac{b_3}{8R} + \ldots + \frac{b_w}{2^{w-1}R}\right)U_R \tag{4.47}$$

$$= -U_R(b_1 2^{-1} + b_2 2^{-2} + b_3 2^{-3} + \ldots + b_w 2^{-w}). \tag{4.48}$$

4.3.5 Delta-sigma DA Converter

A delta-sigma DA converter is shown in Fig. 4.49. The converter is provided with w-bit data words by an input register with the sampling rate f_S. This is followed by a sample rate conversion up to Lf_S by upsampling and a digital lowpass filter. A delta-sigma modulator converts the w-bit input signal into a 1-bit output signal. The delta-sigma modulator corresponds to the model in Section 4.1.3. Subsequently, the DA conversion of the 1-bit signal is performed followed by the reconstruction of the time-continuous signal by an analog lowpass filter.

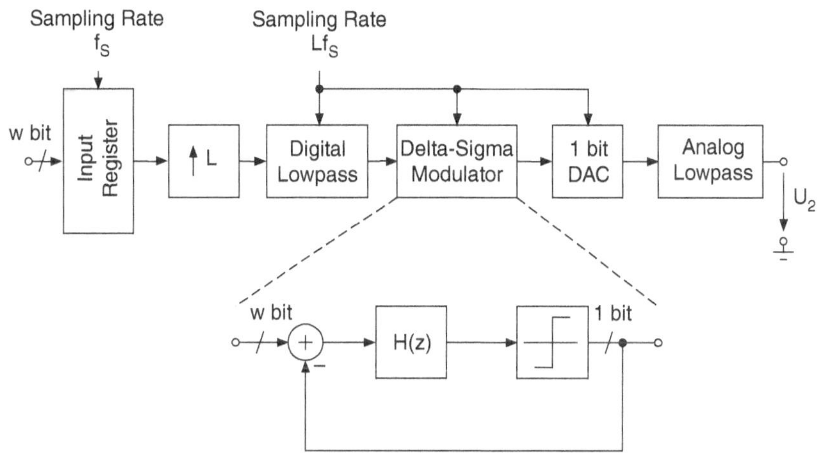

Figure 4.49 Delta-sigma DA converter.

4.4 JS Applet – Oversampling and Quantization

The applet shown in Fig. 4.50 demonstrates the influence of oversampling on the power spectral density of the quantization error. For a given quantization word-length, the noise level can be reduced by changing the oversampling factor. The graphical interface of this applet presents several quantization and oversampling values. These can be used to experiment with the noise reduction level. An additional fast Fourier transform spectral representation provides a visualization of this audio effect.

The following functions can be selected on the lower right of the graphical user interface:

- Quantizer
  - word length w leads to quantization step size $Q = 2^{-(w-1)}$;
- Dither
  - rect dither – uniform probability density function,
  - tri dither – triangular probability density function,
  - highpass dither – triangular probability density function and highpass power spectral density;

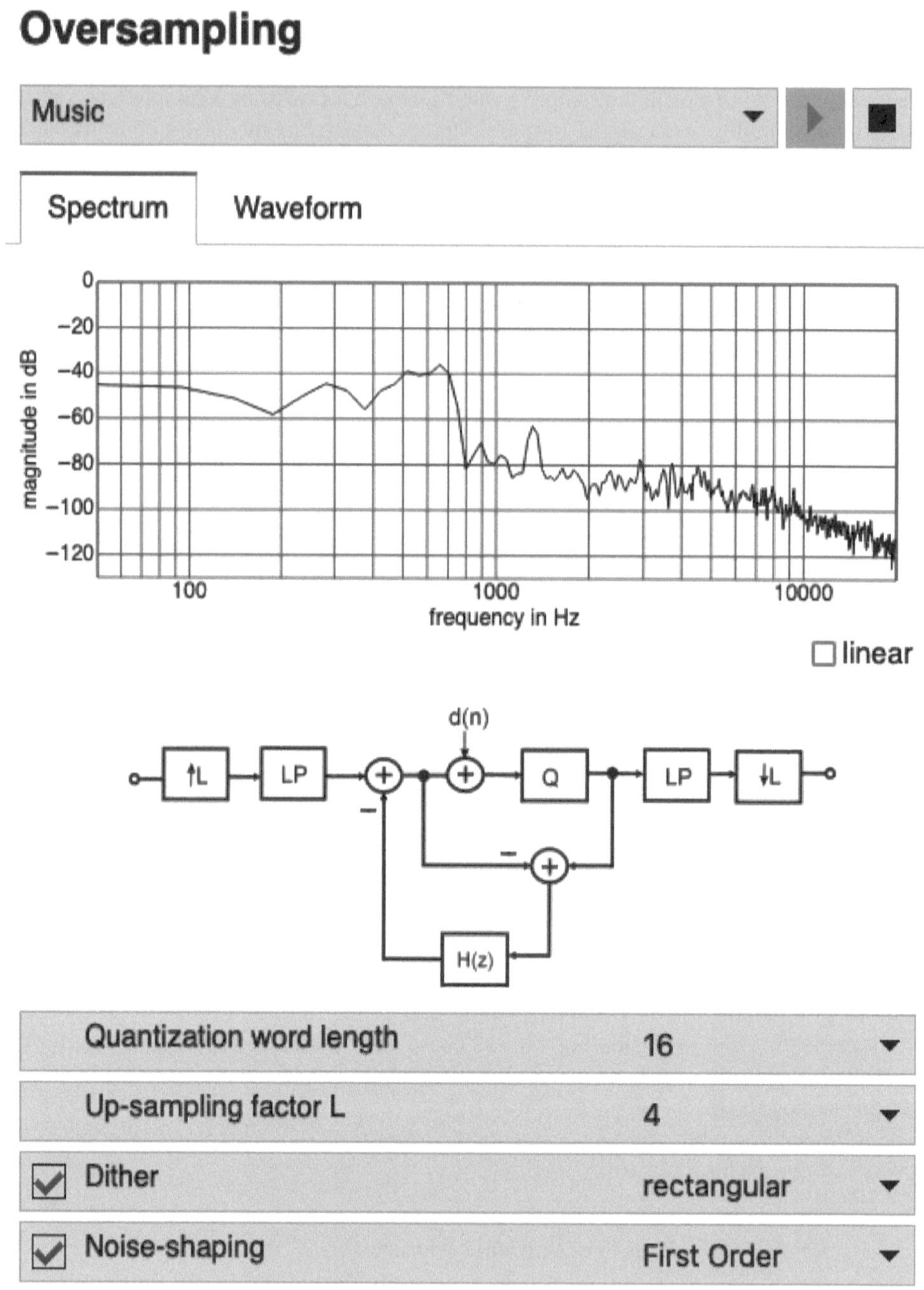

Figure 4.50 JS applet – oversampling and quantization.

- Noise shaping
  - first-order $H(z) = z^{-1}$;
- Oversampling factor
  - Factors from 4 up to 64 can be tested depending on the central processing unit's performance of your machine.

You can choose between two predefined audio files from our web server (*audio1.wav* or *audio2.wav*) or your own local WAV file to be processed [Gui05].

4.5 Exercises

1. Oversampling

1. How do we define the power spectral density (PSD) $S_{XX}(e^{j\Omega})$ of a signal $x(n)$?
2. What is the relationship between signal power σ_X^2 (variance) and power spectral density $S_{XX}(e^{j\Omega})$?
3. Why do we need to oversample a time-domain signal?
4. Explain why an oversampled pulse-code modulation (PCM) AD converter has lower quantization noise power in the baseband than a Nyquist-rate sampled PCM AD converter?
5. How do we perform oversampling by a factor of L in the time domain?
6. Explain the frequency domain interpretation of the oversampling operation.
7. What is the passband and stopband frequency of the analog anti-aliasing filter?
8. What is the passband and stopband frequency of the digital anti-aliasing filter before downsampling?
9. How is the downsampling operation performed (time domain and frequency domain explanation)?

2. Delta-sigma Conversion

1. Why can we apply noise shaping in an oversampled AD converter?
2. Show how the delta-sigma converter (DSC) has a lower quantization error power in the baseband than an oversampled PCM AD converter ?
3. How are the power spectral density and variance change related to the order of the DSC?
4. How is noise shaping achieved in an oversampled delta-sigma AD converter?

5. Show the noise-shaping effect (with Matlab plots) of a delta-sigma modulator and how the improvement of the SNR for pure oversampling and delta-sigma modulation is achieved.

6. Using the previous Matlab plots, specify which order and oversampling factor L will be needed for a 1-bit delta-sigma converter for SNR = 100 dB.

7. What is the difference between the delta-sigma modulator in the delta-sigma AD converter and the delta-sigma DA converter?

8. How do we achieve a w-bit signal representation at Nyquist sampling frequency from an oversampled 1-bit signal?

9. Why do we need to oversample a w-bit signal for a delta-sigma DA converter?

References

[Can85] J.C. Candy: *A Use of Double Integration in Sigma Delta Modulation*, IEEE Trans. Commun., vol. COM-37, pp. 249–258, March 1985.

[Can92] J.C. Candy, G.C. Temes, Ed.: *Oversampling Delta-Sigma Data Converters*, IEEE Press, Piscataway, NJ, 1992.

[Cha90] K. Chao et al: *A High Order Topology for Interpolative Modulators for Oversampling A/D Converters*, IEEE Trans. Circuits and Syst., vol. CAS-37, pp. 309–318, March 1990.

[Cro83] R.E. Crochiere, L.R. Rabiner: *Multirate Digital Signal Processing*, Prentice-Hall, Englewood Cliffs, 1983.

[Fli00] N. Fliege: *Multirate Digital Signal Processing*, J. Wiley & Sons, Chichester, 2000.

[Gui05] M. Guillemard, C. Ruwwe, U. Zölzer: *J-DAFx - Digital Audio Effects in Java*, Proc. 8th Int. Conference on Digital Audio Effects (DAFx-05), Madrid, Spain, pp.161–166, 2005.

[Ino63] H. Inose, Y. Yasuda: *A Unity Bit Coding Method by Negative Feedback*, Proc. IEEE, vol. 51, pp. 1524–1535, November 1963.

[Mat87] Y. Matsuya et al: *A 16-bit Oversampling A-to-D Conversion Technology Using Triple-Integration Noise Shaping*, IEEE J. Solid-State Circuits, vol. SC-22, pp. 921–929, Dec. 1987.

[She86] D.H. Sheingold, Ed.: *Analog-Digital Conversion Handbook*,3rd ed., Prentice-Hall, Englewood Cliffs, NJ, 1986.

[Vai93] P.P. Vaidyanathan: *Multirate Systems and Filter Banks*, Prentice-Hall, Englewood Cliffs, 1993.

Chapter 5

Audio Processing Systems

U. Zölzer and D. Ahlers

The techniques, systems, and devices for digital audio processing have substantially changed over the past decades and have now reached a status where software-based systems have pushed hardware-based systems almost off of the processing chain. Most of the processing can now be done on general purpose personal computers or tablet computers. All kinds of interfacing of microphones and loudspeakers to the computer world still need dedicated hardware solutions. Lately, microphones and loudspeakers have been improved by digital signal processing and such low-power devices need specialized hardware such as digital signal processors (DSPs) or field programmable gate arrays (FPGAs). Large-scale mixing consoles are still benefiting from hardware integration using FPGAs for interfacing and internal multiple signal processing cores. The obvious trend is toward higher integration density and smaller hardware devices integrated into audio networks.

New DSPs with extended interfaces are still used in different application areas where a computer is bulky and excessive. Newer DSPs offer a variety of interfaces to analog-to-digital convertors (ADCs)/digital-to-analog convertors (DACs), local area networks (LANs), and wireless local area networks (WLANs), which make them very attractive for innovative audio designs with low-power consumption and low latency. An overview of processing devices is shown in Table 5.1. New software-based audio systems have benefited from the progress of computer technology with different operating systems. Audio processing on central processing units (CPUs) and graphics processing units (GPUs) have impressively accelerated audio computing and recording. The range of devices from mobile phones over tablet computers to general purpose computers are all suitable for dedicated audio processing tasks but, of course, there is still a requirement for the sophisticated skills of software programmers.

Digital Audio Signal Processing, Third Edition. Udo Zölzer.
© 2022 John Wiley & Sons Ltd. Published 2022 by John Wiley & Sons Ltd.

Table 5.1 DSPs (AD/TI/NXP/STM/Cadence/Xilinx) and applications

System on a chip LAN/WLAN	Floating-point DSPs	Fixed-point DSPs	Field Program. Gate Arrays
Raspberry Pi streaming networking wireless speaker systems	PC/CPU/GPU DAWs synthesizers recording mixing mastering	Sigmastudio headphones loudspeakers synthesizers effects	Xilinx microphones mixing consoles
block latency	block latency	1–2 samples	0 samples

5.1 Digital Signal Processors

5.1.1 Fixed-point DSPs

The discrete-time and discrete-amplitude output of an AD converter is usually represented in 2's-complement format. The processing of these number sequences is carried out with fixed-point or floating-point arithmetic. The output of a processed signal is again in 2's-complement format and is fed to a DA converter. The signed fractional representation (2's complement) is the common way for algorithms in fixed-point number representation. To address the generation and modulo operations, unsigned integers are used. Figure 5.1 shows a schematic diagram of a typical fixed-point DSP. The main building blocks are program controller, arithmetic logic unit

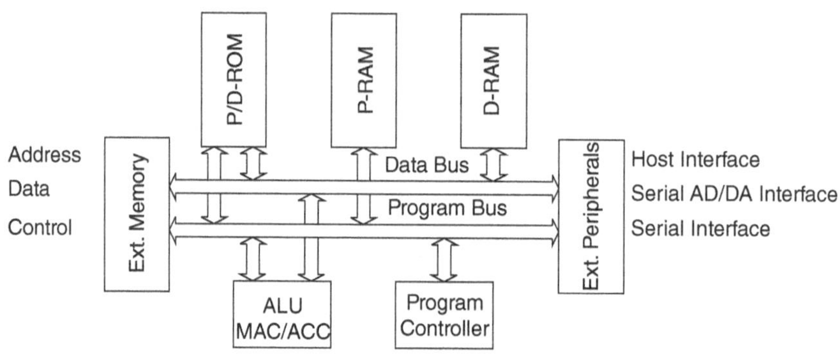

Figure 5.1 Schematic diagram of a fixed-point DSP.

(ALU) with a multiplier–accumulator (MAC), program and data memory, and interfaces to external memory and peripherals. All blocks are connected with each other by an internal bus system. The internal bus system has separate instruction and data buses. The data bus itself can consist of more than one parallel bus enabling it, for instance, to transmit both operands of a multiplication instruction to the MAC in parallel. The internal memory consists of instruction and data random access memory (RAM) and additional read-only memory (ROM). This internal memory permits a fast execution of internal instructions and data transfer. To increase the memory space, address/control and data buses are connected to external memories like erasable programmable (EP)ROM, ROM, and RAM. The connection of the external bus system to the internal bus architecture has a great influence on the efficient execution of external instructions as well as on processing external data. To connect serially operating AD/DA converters, special serial interfaces with high transmission rates are offered by several DSPs. Moreover, some processors support direct connection to an RS232 interface. The control from a microprocessor can be achieved via a host interface with a word length of 8 bits.

An overview of fixed-point DSPs with respect to word length and cycle time can be explored from the websites of the corresponding manufacturers. Basically, the precision of the arithmetic can be doubled if quantization affects the stability and numeric precision of the applied algorithm. The cycle time in connection with processing time (in processor cycles) of a combined multiplication and accumulation command gives insight into the computing power of a particular processor type. The cycle time directly results from the maximum clock frequency. The instruction processing time depends mainly on the internal instruction and data structure, as well as on the external memory connections of the processor. The fast access to external instruction and data memories is of special significance in complex algorithms and in processing huge data loads. Further attention has to be paid to the linking of serial data connections with AD/DA converters and the control by a host computer over a special host interface. Complex interface circuits could therefore be avoided. For stand-alone solutions, program loading from a simple external EPROM can also be done.

For signal processing algorithms, the following software commands are necessary:

1. MAC (multiply and accumulate)
   → combined multiplication and addition command;
2. simultaneous transfer of both operands for multiplication to the MAC (parallel move);
3. bit-reversed addressing (for fast Fourier transfer (FFT));
4. modulo addressing (for windowing and filtering).

Different signal processors have different processing times for FFT implementations. The latest signal processors with improved architecture have shorter processing times. The instruction cycles for the combined multiplication and accumulation command (application: windowing, filtering) are approximately equal for different processors, but processing cycles for external operands have to be considered.

5.1.2 Floating-point DSPs

Figure 5.2 shows the block diagram of a typical floating-point DSP. The main characteristics of the different architectures are the dual-port principle (Texas Instruments) and the external *Harvard* architecture (Analog Devices). Floating-point DSPs internally have multiple bus systems to accelerate data transfer to the processing unit. An on-chip direct memory access (DMA) controller and cache-memory support higher data transfer rates.

5.2 Digital Audio Interfaces

For transferring digital audio signals, two transmission standards have been established by the AES (Audio Engineering Society) and the EBU (European Broadcasting Union). These standards are for two-channel transmission [AES3] and for multi-channel transmission of up to 64 audio channels [AES10], known as a multichannel audio digital interface (MADI).

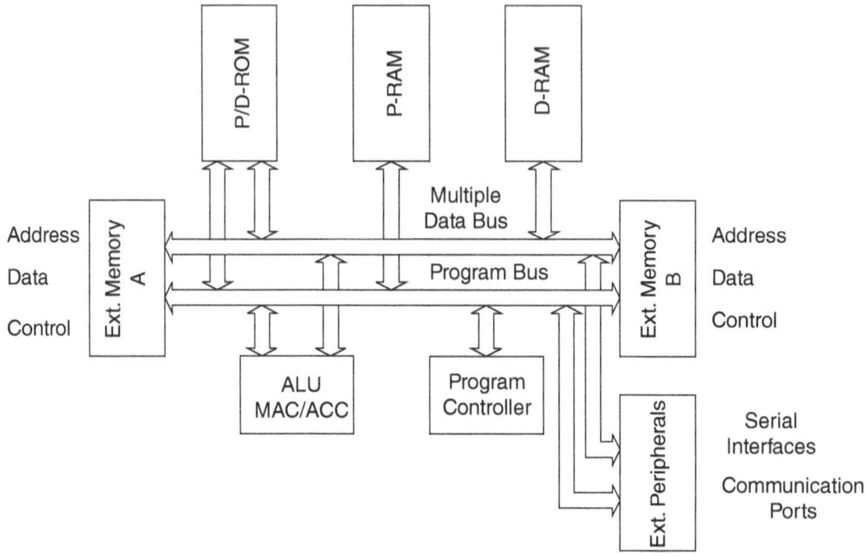

Figure 5.2 Block diagram of a floating-point digital signal processor.

5.2.1 Two-channel AES/EBU Interface

For the two-channel AES/EBU (also known as AES3) interface, professional and consumer modes are defined. The outer frame is identical for both modes and is shown in Fig. 5.3. For a sampling period, a frame is defined so that it consists of two subframes, for channel 1 with preamble X and for channel 2 with preamble Y. A total of 192 frames form a block. The block start is characterized by a special preamble Z. The bit allocation of a subframe consists of 32 bits, as shown in Fig. 5.4. The preamble consists of 4 bits (bits 0–3) and the audio data of up to 24 bits (bits 4–27). The last four bits of the subframe characterize Validity (validity of data word or error), User Status (usable bit), Channel Status (from 192bits/block = 24 bytes coded status information for the channel), and Parity (even parity). The transmission of the serial data bits is carried out with a bi-phase mark code. This is done with the help of an XOR relationship between clock (of double bit rate) and the serial data bits (Fig. 5.5). At the receiver, clock retrieval is

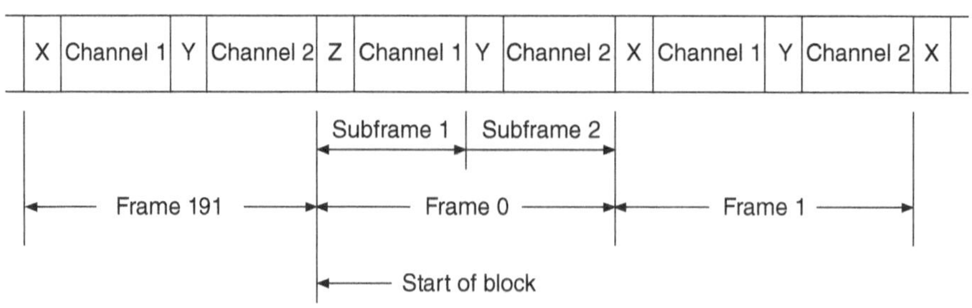

Figure 5.3 Two-channel format.

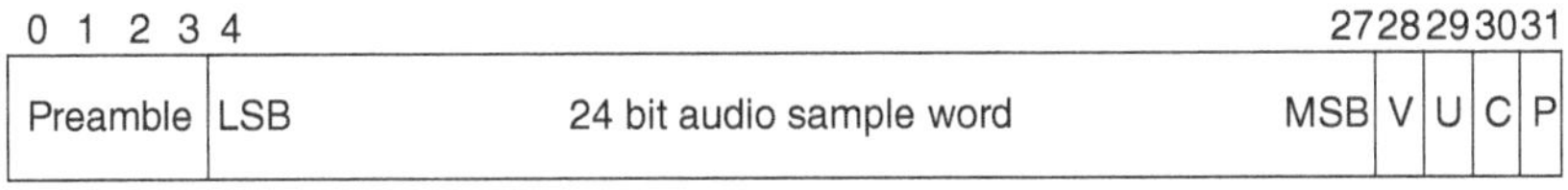

Figure 5.4 Two-channel format (subframe).

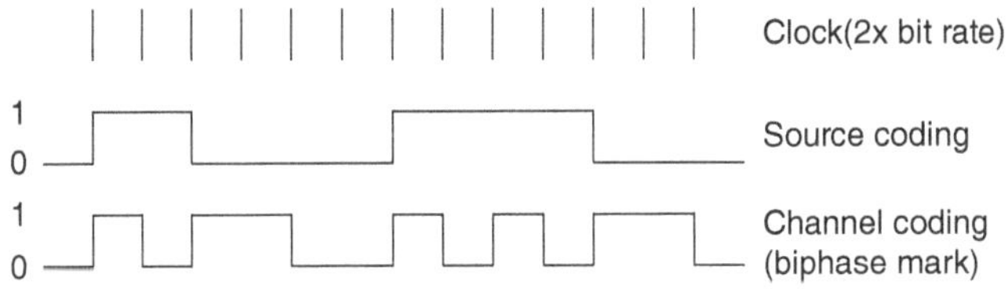

Figure 5.5 Channel coding.

achieved by detecting the preamble (X=11100010, Y=11100100, Z=11101000) as it violates the coding rule (see Fig. 5.6). The meaning of the 24 bytes for channel status information is summarized in Table 5.2. An exact bit allocation of the first three important bytes of this channel status information is presented in Fig. 5.7. In the individual fields of byte 0, pre-emphasis and sampling rate are specified in addition to professional/consumer modes and the characterization of data/audio (see Tables 5.3 and 5.4). Byte 1 determines the channel mode (Table 5.5). The consumer format (often labeled as SPDIF, Sony/Philips Digital Interface Format) differs from the professional format in the definition of the channel status information and the technical specifications for inputs and outputs. The bit allocation for the first four bits of the channel information is shown in Fig. 5.8. For consumer applications, a coaxial cable with RCA connectors or a fiber optic cable with TOSLINK connectors are used. For professional use, shielded two-wired leads with XLR connectors and symmetrical inputs and outputs (professional format) are used. Table 5.6 shows the electrical specifications for professional AES/EBU interfaces.

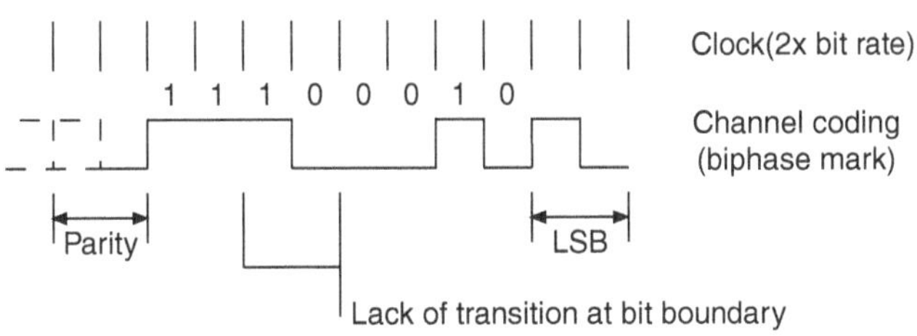

Figure 5.6 Preamble X.

5.2.2 MADI Interface

To connect an audio processing system at different locations, a MADI interface is used [AES10]. A system link by MADI is presented in Fig. 5.9. Analog/digital I/O systems consisting of AD/DA

Table 5.2 Channel status bytes

Byte	Description
0	emphasis, sampling rate
1	channel use
2	sample length
3	vector for byte 1
4	reference bits
5	reserved
6–9	4 bytes of ASCII origin
10–13	4 bytes of ASCII destination
14–17	4 bytes of local address
18–21	time code
22	flags
23	CRC

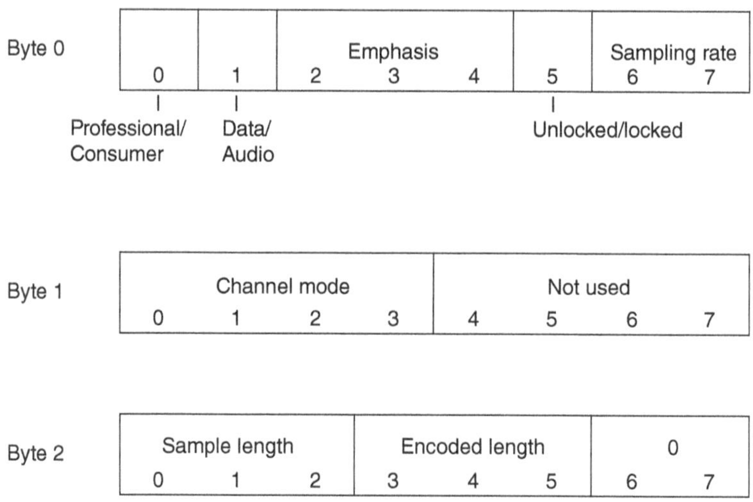

Figure 5.7 Bytes 0–2 of channel status information.

Table 5.3 Emphasis field

0	none indicated, override enabled
4	none indicated, override disabled
6	50/15 μ s emphasis
7	CCITT J.17 emphasis

Table 5.4 Sampling rate field

0	none indicated (48 kHz default)
1	48 kHz
2	44.1 kHz
3	32 kHz

Table 5.5 Channel mode

0	none indicated (2 channel default)
1	two channel
2	monaural
3	primary/secondary (A=primary, B=secondary)
4	stereo (A=left, B=right)
7	vector to byte 3

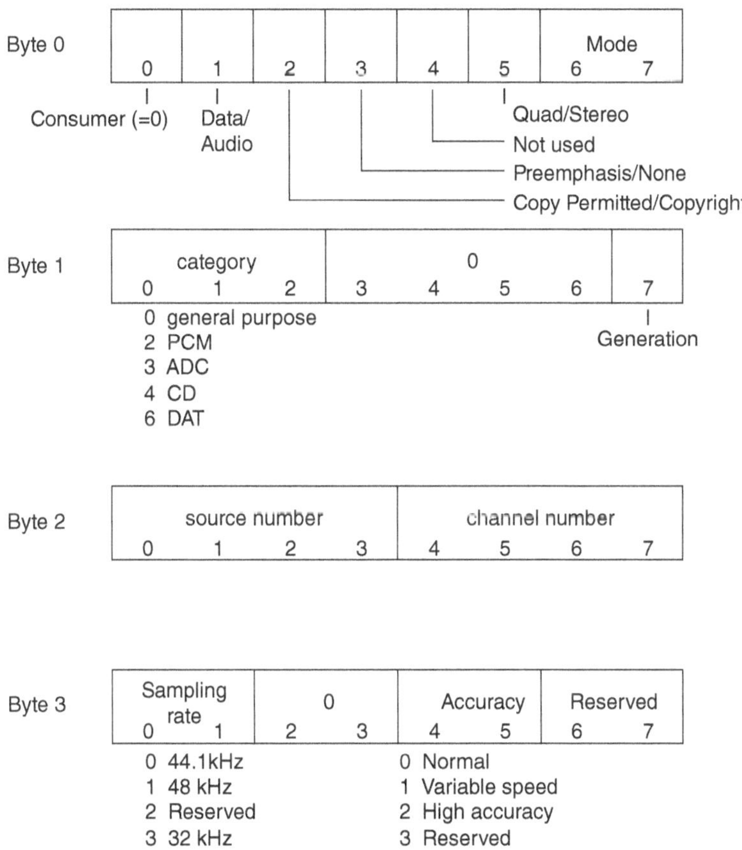

Figure 5.8 Bytes 0–3 (consumer format).

Table 5.6 Electrical specifications of professional interfaces

Output impedance	Signal amplitude	Jitter
110 Ω	2–7 V	max. 20 ns
Input impedance	Signal amplitude	Connect.
110 Ω	min. 200 mV	XLR

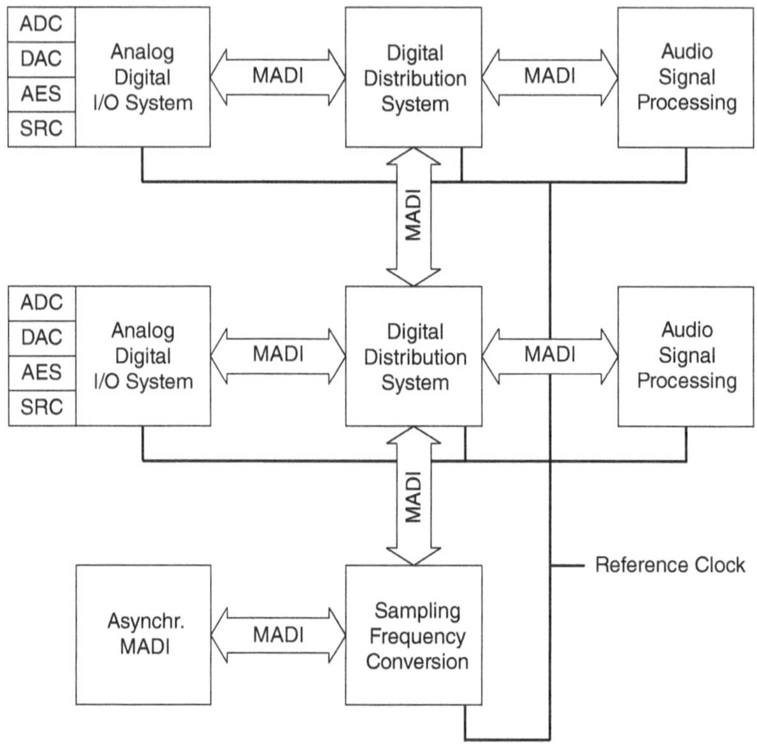

Figure 5.9 A system link by MADI.

converters, AES/EBU interfaces (AES), and sampling rate converters (SRC) are connected to digital distribution systems with bi-directional MADI links. The actual audio signal processing is performed in special DSP systems which are connected to the digital distribution systems by MADI links. The MADI format is standardized as AES10 and is derived from the two-channel AES/EBU format and allows the transmission of 64 digital mono channels (see Fig. 5.10) within

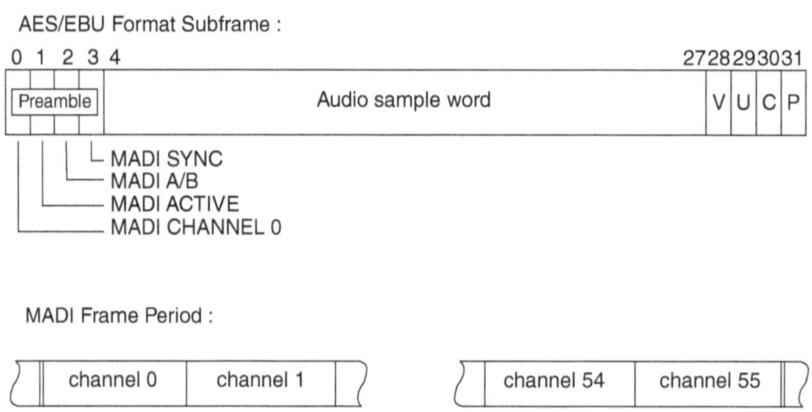

Figure 5.10 MADI frame format.

a sampling period. The MADI frame consists of 64 AES/EBU subframes. Each channel has a preamble containing the information shown in Fig. 5.10. The bit 0 is responsible for identifying the first MADI channel (MADI Channel 0). Table 5.7 shows the sampling rates and the corresponding data transfer rates. The maximum data rate of 98.304 Mbit/s is required at a sampling rate of 48 kHz. The transmission is done via 75-Ohm coaxial cables with BNC connectors (50 m, signal amplitude 0.3–0.6 V) or via fiber optic lines with SC connectors (up to 2 km).

Table 5.7 MADI specifications

Sampling rate	32–48 kHz ± 12.5%
Transmission rate	125 Mbit/s
Data transfer rate	100 Mbit/s
Max. data transfer rate	98.304 Mbit/s (64 channels at 48 kHz)

5.2.3 Audio in HDMI

High-definition multimedia interface (HDMI) is a proprietary audio/video interface standard. The digital audio data are sent, combined together with video and auxiliary data, within the tree TMDS (transition minimized differential signaling) data channels, as depicted in Fig. 5.11. Figure 5.12 shows the physical pinout of the channels, where one can see that each TMDS channel is separately shielded and transmitted as a differential signal over a twisted pair. The audio data can consist of any compressed, uncompressed, pulse code modulation (PCM), single or multichannel format of up to 8 channels (since HDMI 2.0, 32 channels) and is being transmitted during the video blanking intervals encapsulated inside the so called data island period [1]. To make sure the transmitter only sends audio or video data, the receiver is capable of receiving, there is an

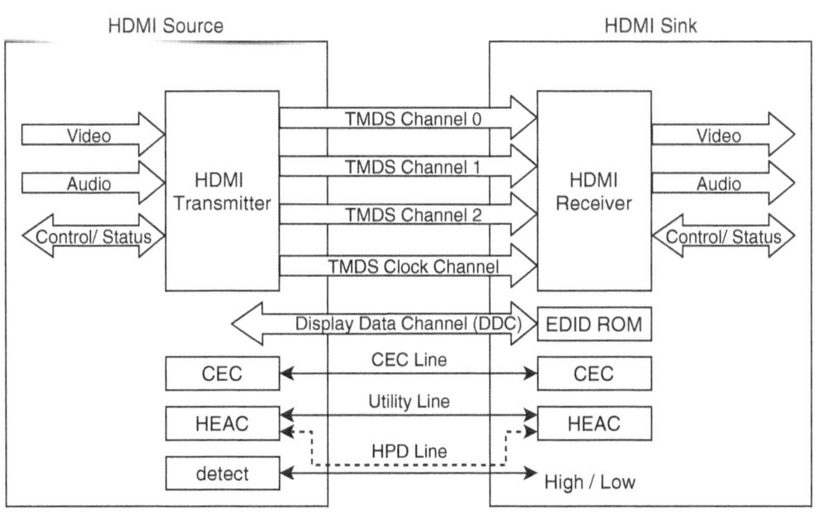

Figure 5.11 HDMI block diagram.

[1] HDMI Specification Version 1.3a, 2006.

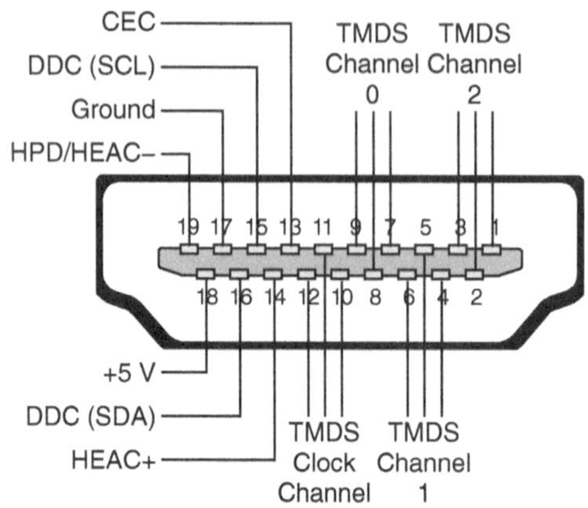

Figure 5.12 HDMI pinout since HDMI 1.4.

information exchange in the display data channel (DDC) that is I^2C-based during the connection establishment. The standard for that is called extended display identification data (EDID) and is published by the Video Electronics Standards Association (VESA). It contains the manufacturer ID and model name as well as information of the capable video formats like resolution, aspect ratio, bit depth, and refresh rate. The audio descriptor of the EDID contains information like audio format, number of channels, sampling rate, and bit depth.

In addition to the audio transmission from the transmitter to the receiver, HDMI specifies an audio transmission from the receiver back to the transmitter, called audio return channel (ARC), a feature of HDMI 1.4a and an updated version called eARC (enhanced ARC) since HDMI 2.1. This is useful if an AV receiver is connected to a TV as display output but the TV is the source of the signal, for example when receiving a television program. Then the audio of the television program can be sent back to the AV receiver within the same HDMI cable. ARC is carried together with an Ethernet connection on the HDMI Ethernet and audio return channel (HEAC), a separate connection in an HDMI cable. HEAC shares the pin with the HPD (hot plug detection), as depicted in Fig. 5.12. The remaining unexplained connection in Fig. 5.11 is the consumer electronics control (CEC). A bidirectional connection that makes it possible to control all connected HDMI devices by using only on remote control.

Currently, the maximum data rate that can be transmitted over HDMI is $42.\overline{6}$ Gbit/s defined by the HDMI 2.1 standard[2].

5.2.4 Audio Computer Interfaces

There are several computer interfaces for audio, the most common being:

- Universal Serial Bus (USB) (device class definition for audio data formats[3]);
- Thunderbolt;
- Bluetooth audio.

[2] HDMI Specification Version 2.1, 2017.
[3] Audio Data Formats 1.0, 1998.

In particular, USB and Thunderbolt allow a high data rate (see Table 5.8) and therefore a high number of audio channels for both input and output signals. For comparison, a common sampling rate of 48 kHz @ 24 bits needs 1.152 Mbit/s. Even professional audio recordings with a sampling rate of 192 kHz @ 24 bits only create 4.608 Mbit/s of pure data. The audio data will be transmitted uncompressed as PCM values, which are divided into blocks.

Table 5.8 Computer interfaces

Type	Bitrate
USB 2.0	480 Mbit/s
USB 3.0	5 Gbit/s
USB 3.1	10 Gbit/s
Thunderbolt 1.0	10 Gbit/s
Thunderbolt 2.0	20 Gbit/s
Thunderbolt 3.0	40 Gbit/s

While Bluetooth offers a wireless communication widely used in the consumer market, USB and Thunderbolt are still in the lead for professional equipment. Both offer short-range connection between external audio devices and the computer using specific interface cables. For longer ranges, network interfaces will become the main solution.

On the software side, a device driver is required that creates a software interface and makes the hardware accessible to the operation system (OS) and other programs. For audio devices, normally this driver talks to the OS audio engine. The audio engine then provides standard APIs for the applications. Over the past few years, multiple APIs have been created. A list of the most common ones is provided in Table 5.9 and there are a few more.

Table 5.9 Computer interfaces

OS	Audio Engine
Linux	ALSA (Advanced Linux Sound Architecture)
	OSS (Open Sound System) (outdated)
Windows	ASIO (Audio Stream Input/Output)
	WASAPI (Windows Audio Session API) to connect to
	WDM (Windows Driver Model) Audio drivers
macOS and iOS	CoreAudio

5.2.5 Audio Network Interfaces

Network interfaces for audio can be divided into two main use cases. For the first use case, network-based audio protocols are described that are mainly used in LAN for audio transmission with specialized hardware interfaces. They are commonly used for live events or in recordings scenarios and they are increasingly replacing the analog audio transmission. Because low latency is very important in such scenarios and the data rate on LAN is not so small, audio is sent mostly uncompressed in packets as PCM values.

For the second use case, the more internet-based audio protocols are described that are often only software based and normally used for telecommunication. It is a more general term for Voice-over-Internet Protocol (VoIP) and often combined with video in conferencing software.

However, almost every live audio transmission like music streaming or online radio is done in a similar way. Because internet bandwidth is very costly, normally the audio gets compressed by encoding algorithms.

Network-based audio

There is a very large number of network-based audio protocols that are incompatible with each other. Some are open standards and some are proprietary, but they all share the same physical connection: Ethernet cables. These cables are mostly made from copper but fiber connections are also possible. The fundamental difference between the protocols is the used network layer on which they operate. This also defines the topology of the network, the interoperability to existing Ethernet networks, and can influence the reliability and latency. The protocols are grouped into layer-1, layer-2, and layer-3 protocols. This categorization is based on the ISO OSI model.

Layer-1 network protocol. This protocol only uses a physical connection of Ethernet cables but defines their own cable pin assignment and does not use the Ethernet frame structure. Therefore only point-to-point connections between special hardware that supports the protocol are used. This means every connection requires an additional wire. Because there is a dedicated network, it is possible to create a reliable connection with guaranteed latency. As an example, AES50 [AES50] is described in more detail here and a typical network for a live show recording is shown in Fig. 5.13. It is mostly used by the Music Tribe Group (Midas, Klark Teknik, Behringer).

- AES50 [AES50]
  - Uses CAT5 copper cables up to 100 m (100 Mbit/s),
  - 48 input /48 output channels @ 48 kHz/24 bit or
  - 24 input /24 output channels channels @ 96 kHz/24 bits,
  - 5 Mbit/s auxiliary data connection, for example, used for remote gain control of ADCs,
  - latency $< 100\ \mu$ s,

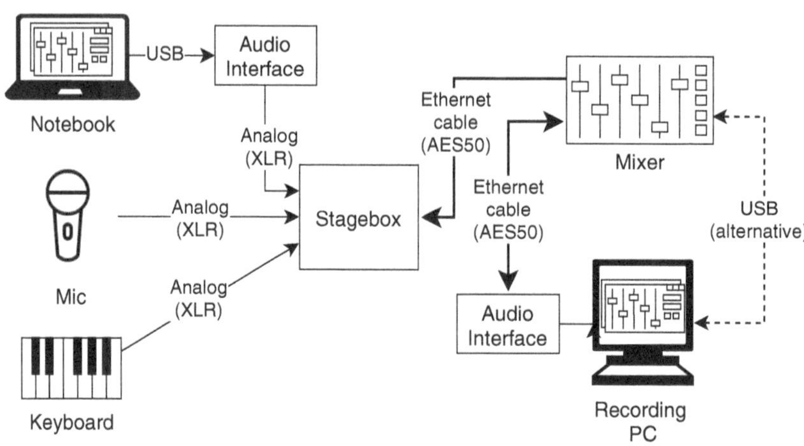

Figure 5.13 Example layer-1 network (AES50 replaces the traditional multicore wires in this live scenario).

- uses AES/EBU streams (enhancement of AES10 / MADI),
- separate signal pairs for audio and sync connection.

Layer-2 network protocol. Audio data get encapsulated in Ethernet packages and can then be transmitted on an Ethernet network. However, because these packages are below the Internet Protocol (IP)-level, special Ethernet hubs and switches are required that can handle the audio protocol and IP packets at the same time. As an example, AVB, the most prominent protocol that is layer-2 based, is described here and a typical network for a live show recording is shown in Fig. 5.14.

- AVB (Audio Video Bridging)
  - Key aspects: timing and synchronization, stream reservation protocol. Reliability and latency can be guaranteed.
  - Requires special hardware to create a network (AVB compatible switches).
  - Can coexist with an IP connection on the same link, if hardware supports it.
  - Supports audio up to 196 kHz/32 bit.
  - Multiple channels are encapsulated in AVB streams. One stream can contain up to 64 channels. With multiple streams between two devices, the number of channels is only limited by hardware limitations and link bandwidth.
  - Typical interface: 128 input / 128 output channels @ 48 kHz/24 bit (14% utilization of 1 Gigabit Ethernet).
  - 100 Mbit/s, 1 Gbit/s, or fiber connection possible.

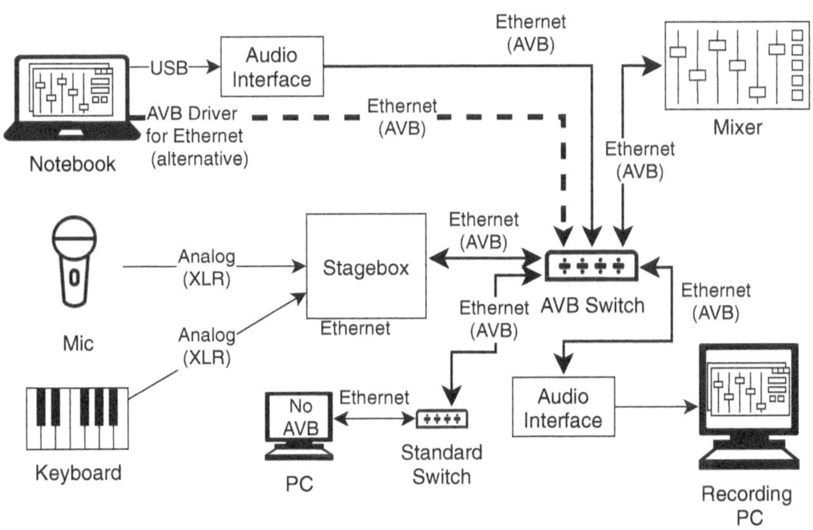

Figure 5.14 Example layer-2 network (the PC cannot receive the AVB stream because it is connected via a standard router with no AVB support).

- Synchronization done via precision time protocol (PTP) with nanosecond precision.
- Network contains one Master clock that must be configured (primary clock) and the other devices as slaves (secondary clock).
- Network contains talker, listener, and AVB capable switches.
- Multiple stream reservation protocol (MSRP) used to reserve bandwidth throughout the entire network path.
- Bandwidth gets reserved before transmission and blocked, and is therefore guaranteed.
- Latency is, by default, bounded to 2 ms, which is comparable to 7 hops in a network. The playback will be delayed if the packet arrives earlier at the receiver. The advantage is that the delay is constant.
- AVB is defined in IEEE 802.1 (open standard, no license fee).
- Remote gain control can be done via IP link but is not defined in the standard and therefore may likely not be compatible among vendors.

Layer-3 network protocol. Unlike the previous protocol standards, layer-3 protocols can be easily integrated in existing local area networks with standard and therefore cheap hardware. One thing to keep in mind is that this protocol relays on the IP that only provides best-effort delivery and is, by design, unreliable. To compensate for this, larger buffers need to be added that result in higher delays and if the network gets overloaded by other protocols, some packages could get lost, which will result in audio glitches. However, the latency is still small enough to use it in live applications. Figure 5.15 shows such a live application with a PC recording station. The following examples show three different layer-3 protocols.

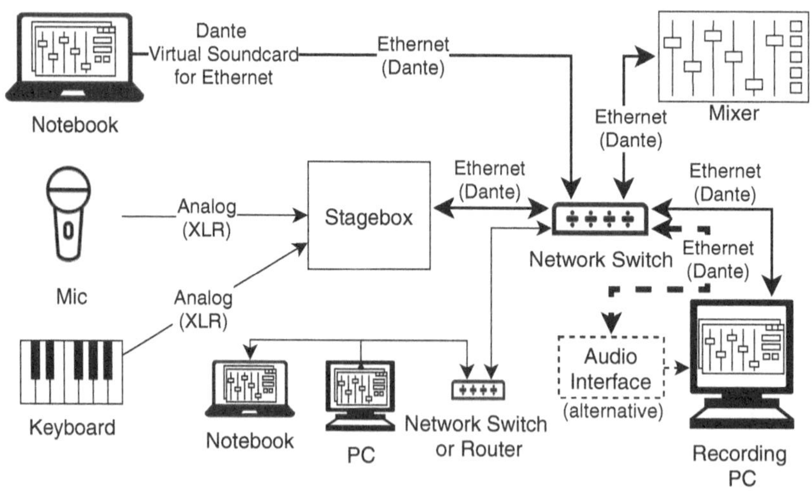

Figure 5.15 Example layer-3 network (all devices in the network can connect to any Dante device independent of the used network hardware).

- Dante – Proprietary
  - Uses UDP (user datagram protocol) packets.
  - Supports audio up to 196 kHz/32 bit.
  - Easy to set up with the Dante Controller software.
  - PTP used for clock synchronization.
  - Unicast and multicast connection possible.
  - Specific Dante devices are capable of sending and receiving AES67 streams.
- Ravenna
  - Based on RTP (real-time transport protocol).
  - Uses linear PCM with 16 or 24 bit (L16/L24).
  - Uses PTP for synchronization.
  - For QoS (quality of service), DiffServ is used.
  - Ravenna is compatible with AES67.
- AES67 [AES67]
  - Designed to create interoperability between various layer-3 networking protocols like RAVENNA, Dante and co.
  - Supports sampling rates of 44.1 kHz, 48 kHz, and 96 kHz.
  - Uses linear PCM with 16 or 24 bit (L16/L24).

After introducing the various types of network audio protocols, it must be said that all of these protocols only specify the audio delivery. Control messages like ADC/DAC gain-control are not part of the protocol standards. To control the devices, proprietary apps or software are most often used. Most likely, they are TCP-/IP-based and incompatible across vendors. However, there are of course other standards to achieve interoperability like OSC, MIDI, or OCA (open control architecture/AES70) but normally they require some sort of configuration to control everything inside by software, and often do not provide the full feature set.

Internet-based audio

While in local networks, the data are normally sent as uncompressed PCM values, in communication over the internet, most audio data get encoded. The encoding and decoding decrease the required data rate per stream but takes time (<40 ms) and is computationally expensive. On local networks, the data rate normally is not the limiting factor and to reduce latency, it will be omitted. On audio transmission over the internet, a small data rate is beneficial and latency is not that critical. There are special codecs specifically designed for transmission that use VoIP and telecommunications software.

There are a lot of VoIP and telecommunications software available. The fundamental structure of computer-based audio over IP is given in Fig. 5.16, which shows an example of a unidirectional transmission beginning with an analog audio source, like a microphone, that is being sent to a speaker. After the ADC converter, the audio gets processed by the OS audio engine to the VoIP application. In the application, depending on the use case, a lot of audio processing can be performed, such as noise cancellation, and filters like an EQ or a noise gate can be applied. Afterwards, the audio will be split into packets and encoded by an encoding algorithm and sent through the network. At the receiver application, the encoded packages get ordered and decoded and then given back to the OS audio engine to be then played back on the speaker which is connected to the DAC.

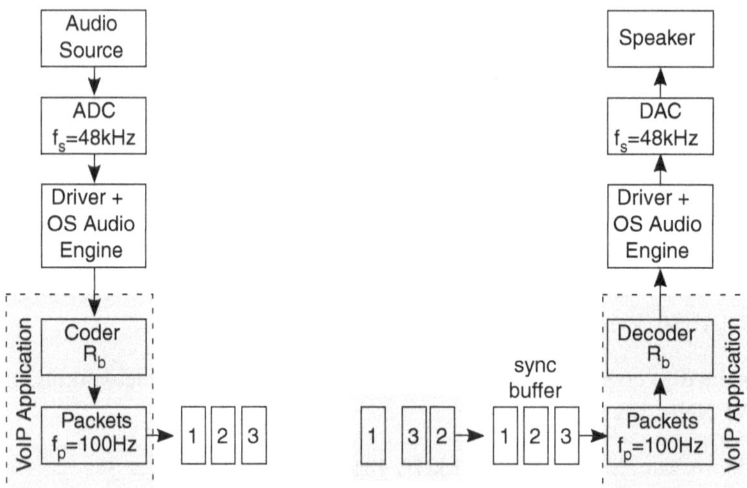

Figure 5.16 Audio over IP.

5.3 Two-channel Systems

Most two-channel systems for stereo processing can be realized with one DSP connected via serial interfacing to ADCs and DACs, and some control logic for parameter adjustment. Figure 5.17 gives an overview. For further functionalities (such as USB, HDMI, LAN, WLAN), system on a chip (SoC) devices offer DSP computation and extended communication interfaces.

5.4 Multi-channel Systems

Multi-channel systems can be realized by parallel DSP processing of multiple audio channels using multicore CPUs, GPUs, and FPGAs (field programmable gate arrays). A mixing console with N input and K output channels, such as depicted in Fig. 5.18, can be realized by a single FPGA and adapted audio interface to ADCs/DACs/AES-EBU/MADI devices and of course standard computer interfaces such as X Gbit/s USB or Ethernet.

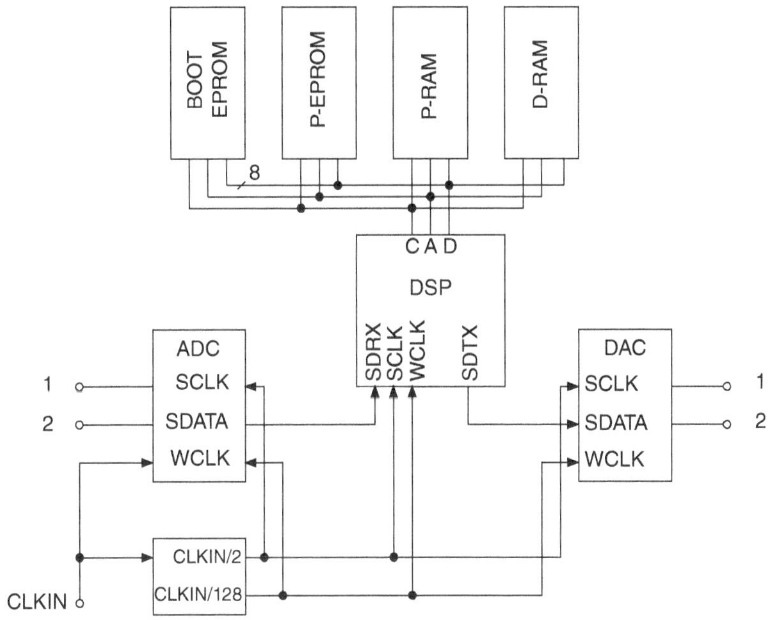

Figure 5.17 Two-channel DSP system with two-channel AD/DA converters (C = control, A = address, D = data, SDATA = serial data, SCLK = bit clock, WCLK = word clock, SDRX = serial input, SDTX = serial output).

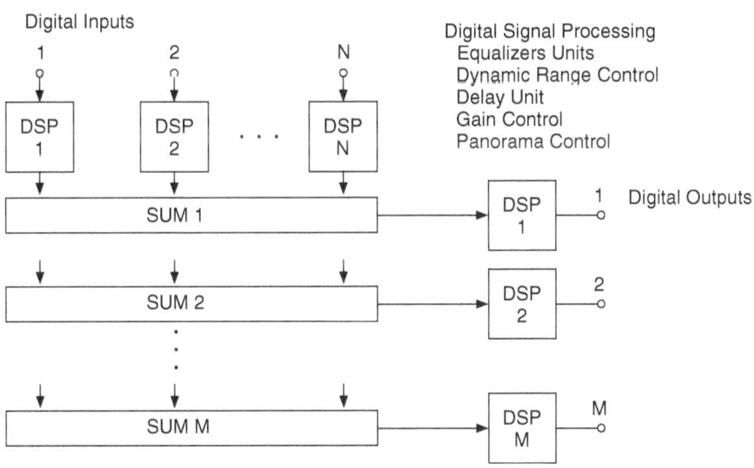

Figure 5.18 Mixing console application of a multichannel system.

References

[AES3] AES3-2009: AES Recommended Practice for Digital Audio Engineering – Serial Transmission Format for Two-Channel Linearly Represented Digital Audio.

[AES10] AES10-2020: AES Recommended Practice for Digital Audio Engineering – Serial Multichannel Audio Digital Interface (MADI).

[AES50] AES50-2011: AES standard for digital audio engineering – High-resolution multi-channel audio interconnection (HRMAI).

[AES67] AES67-2018: AES standard for audio applications of networks – High-performance streaming audio-over-IP interoperability.

Chapter 6

Equalizers

U. Zölzer

Spectral sound equalization is one of the most important methods for processing audio signals. Equalizers are found in various forms in the transmission of audio signals from a sound studio to the listener. The more complex filter functions are used in sound studios. However, in almost every consumer product like car radios, hi-fi amplifiers etc., simple filter functions are used for sound equalization. We first discuss basic filter types followed by the design and implementation of recursive audio filters. In the third and fourth sections, linear-phase non-recursive filter structures and their implementation are introduced.

6.1 Basics

For filtering of audio signals, the following filter types are used.

- **Lowpass and highpass filters** with cutoff frequency f_c (3 dB cutoff frequency) are shown with their magnitude response in Fig. 6.1. They have a passband in the lower and higher frequency range, respectively.

- **Bandpass and bandstop filters** (magnitude responses in Fig. 6.1) have a center frequency f_c and a lower and upper f_l and f_u cutoff frequency. They have a pass- and stopband in the middle of the frequency range. For the bandwidth of a bandpass or a bandstop filter, we have

$$f_b = f_u - f_l. \tag{6.1}$$

  Bandpass filters with a constant relative bandwidth f_b/f_c are very important for audio applications [Cre03]. The bandwidth is proportional to the center frequency, which is given by $f_c = \sqrt{f_l \cdot f_u}$ (see Fig. 6.2).

Digital Audio Signal Processing, Third Edition. Udo Zölzer.
© 2022 John Wiley & Sons Ltd. Published 2022 by John Wiley & Sons Ltd.

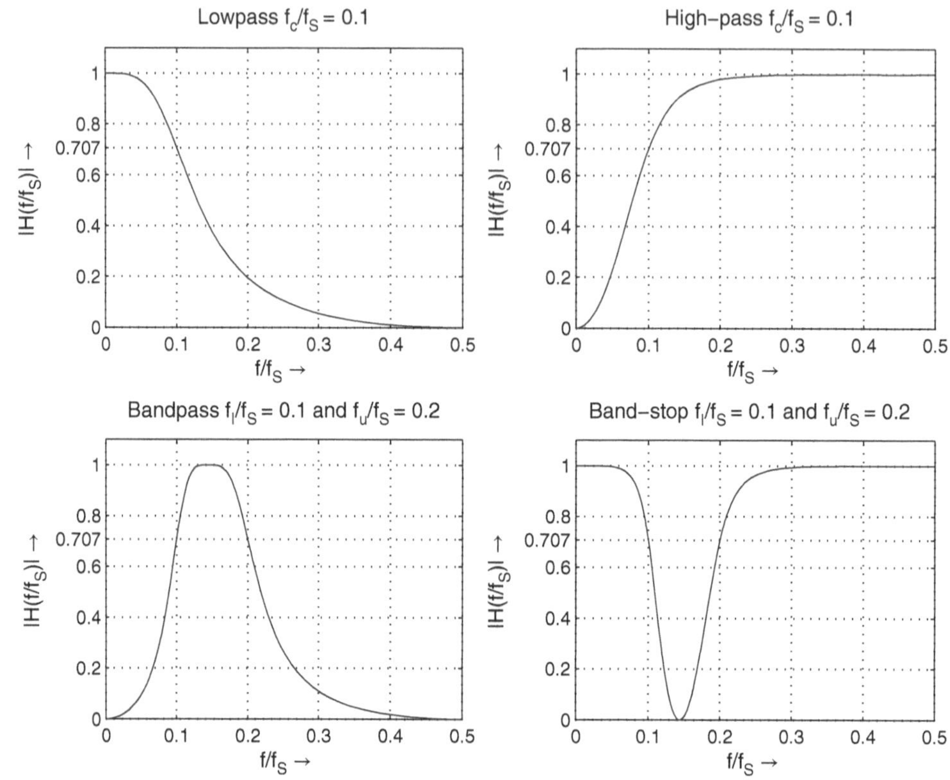

Figure 6.1 Linear magnitude responses of lowpass, highpass, bandpass, and bandstop filters.

- **Octave filters** are bandpass filters with special cutoff frequencies given by

$$f_u = 2 \cdot f_l, \tag{6.2}$$

$$f_c = \sqrt{f_l \cdot f_u} = \sqrt{2} \cdot f_l. \tag{6.3}$$

A spectral decomposition of the audio frequency range with octave filters is shown in Fig. 6.3. At the lower and upper cutoff frequency, an attenuation of −3 dB occurs. The upper octave band is represented as a highpass. A parallel connection of octave filters can be used for a spectral analysis of the audio signal in octave frequency bands. This decomposition is used for the signal power distribution across the octave bands. For the center frequencies of octave bands, we get $f_{c_i} = 2 \cdot f_{c_{i-1}}$. The weighting of octave bands with gain factors A_i and summation of the weighted octave bands represents an octave equalizer for sound processing (see Fig. 6.4). For this application, the lower and upper cutoff frequencies need an attenuation of −6 dB, such that a sinusoid at the crossover frequency has gain of 0 dB. The attenuation of −6 dB is achieved through a series connection of two octave filters with −3 dB attenuation.

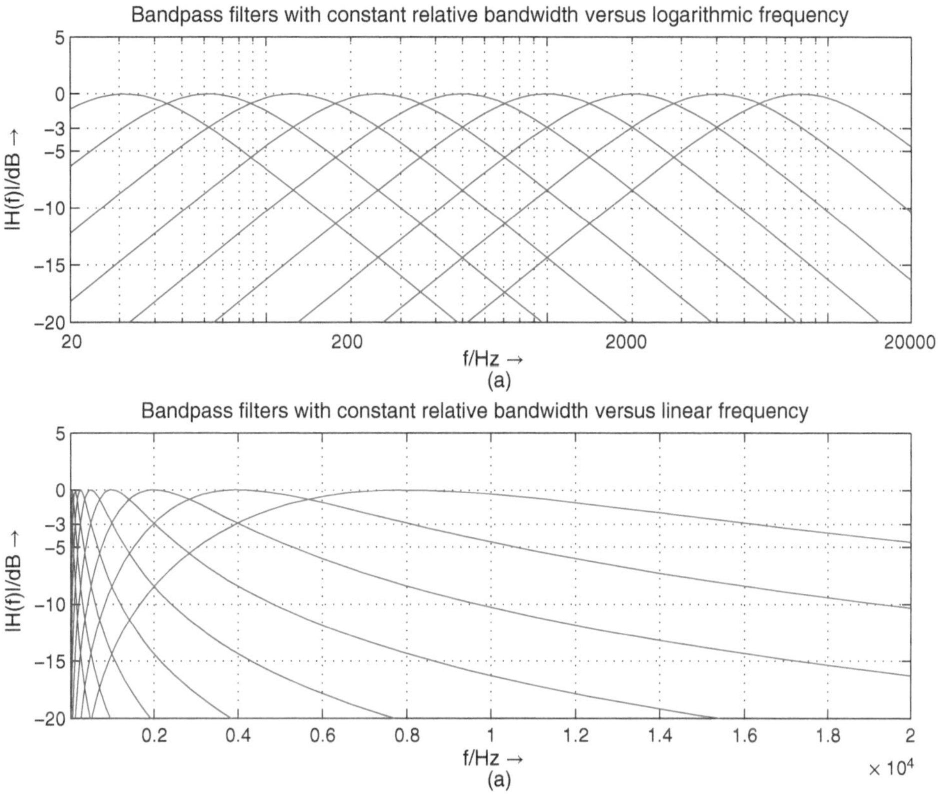

Figure 6.2 Logarithmic magnitude responses of bandpass filters with constant relative bandwidth.

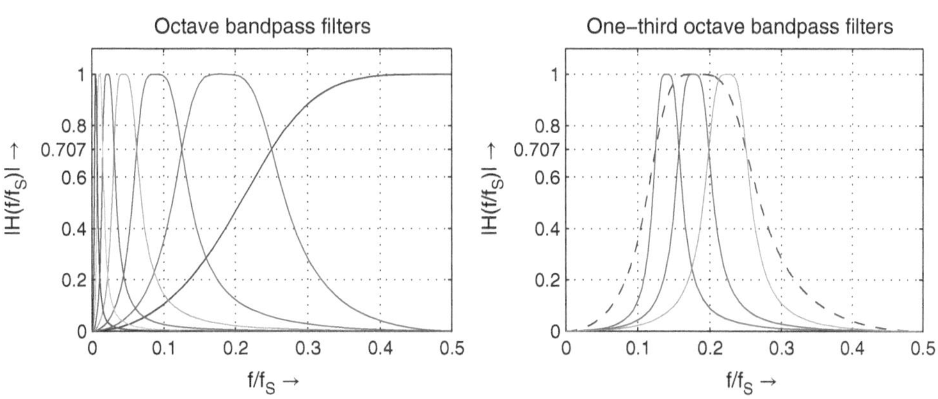

Figure 6.3 Linear magnitude responses of octave filters and decomposition of an octave band by three one-third octave filters.

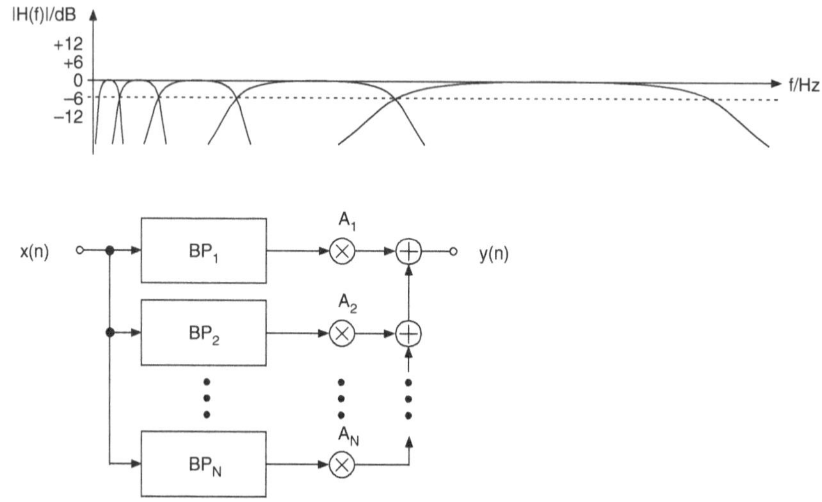

Figure 6.4 Parallel connection of bandpass filters (BP) for octave/one-third octave equalizers with gain factors (A_i for octave or one-third octave band).

- **One-third octave filters** are bandpass filters (see Fig. 6.3) with cutoff frequencies given by

$$f_u = \sqrt[3]{2} \cdot f_l, \tag{6.4}$$

$$f_c = \sqrt[6]{2} \cdot f_l. \tag{6.5}$$

The attenuation at the lower and upper cutoff frequency is −3 dB. One-third octave filters split an octave into three frequency bands (see Fig. 6.3).

- **Shelving filters** and **peak filters** are special weighting filters, which are based on lowpass/highpass/bandpass filters and a direct path (see Section 6.2.2). They have no stopband compared with lowpass/highpass/bandpass filters. They are used in a series connection of shelving and peak filters, as shown in Fig. 6.5. The lower frequency range is equalized by lowpass shelving filters and the higher frequencies are modified by highpass shelving

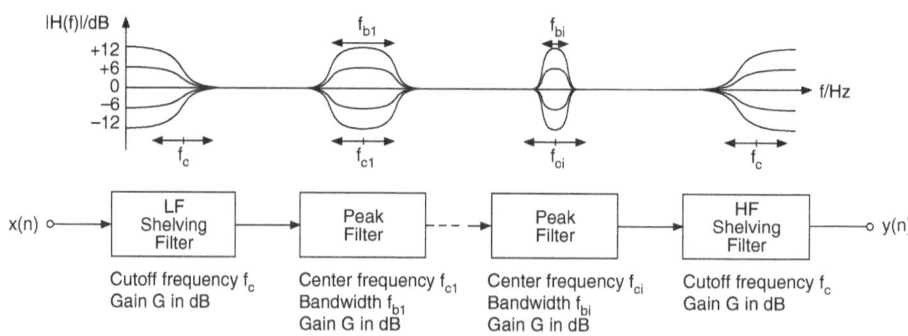

Figure 6.5 Series connection of shelving and peak filters (low-frequency LF, high-frequency HF).

filters. Both filter types allow the adjustment of the cutoff frequency and gain factor. For the mid-frequency range, a series connection of peak filters with variable center frequency, bandwidth, and gain factor are used. These shelving and peak filters can also be applied for octave and one-third octave equalizers in a series connection.

- **Weighting filters** are used for signal level and noise measurement applications. The signal from a device under test is first passed through the weighting filter and then a root-mean-square or peak value measurement is performed. The two most commonly used filters are the A-weighting filter and the CCIR-468 weighting filter (see Fig. 6.6). Both weighting filters take the increased sensitivity of human perception in the 1–6 kHz frequency range into account. The 0 dB of the magnitude response of both filters is crossed at 1 kHz. The CCIR-468 weighting filter has a gain of 12 dB at 6 kHz. A variant of the CCIR 468 filter is the ITU-ARM 2-kHz weighting filter, which is a 5.6-dB down-tilted version of the CCIR-468 filters and passes 0 dB at 2 kHz.

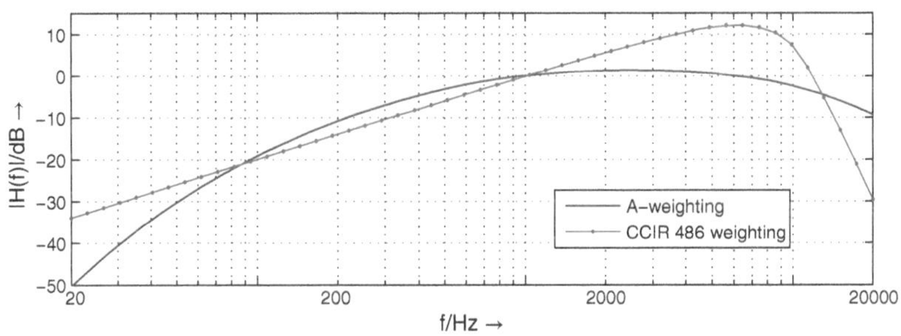

Figure 6.6 Magnitude responses of weighting filters for root-mean-square and peak value measurements.

6.2 Recursive Audio Filters

6.2.1 Design

A certain filter response can be approximated by two kinds of transfer function. On the one hand, the combination of poles and zeros leads to a very low-order transfer function $H(z)$ in fractional form, which solves the given approximation problem. The digital implementation of this transfer function needs recursive procedures owing to its poles. On the other hand, the approximation problem can be solved by placing only zeros in the z-plane. This transfer function $H(z)$ has, in addition to its zeros, a corresponding number of poles at the origin of the z-plane. The order of this transfer function, for the same approximation conditions, is substantially higher than for transfer functions consisting of poles and zeros. In view of an economical implementation of a filter algorithm in terms of complexity, recursive filters achieve shorter computing time owing to their lower order. For a sampling rate of 48 kHz, the algorithm has 20.83 μs of processing time available. With the digital signal processors (DSPs) presently available, it is easily possible to implement recursive digital filters for audio applications within this sampling period using only one DSP. To design the typical audio equalizers, we will start with filter designs in the S-domain. These filters will then be mapped to the Z-domain by the bilinear transformation.

Lowpass/Highpass Filters. To limit the audio spectrum, lowpass and highpass filters with Butterworth response are used in analog mixers. They offer a monotonic passband and a monotonically decreasing stopband attenuation per octave ($n \cdot 6$ dB/oct.) that is determined by the filter order. Lowpass filters of the second and fourth order are commonly used. The normalized and denormalized second-order lowpass transfer functions are given by

$$H_{LP}(s) = \frac{1}{s^2 + \frac{1}{Q_\infty}s + 1} \quad \text{and} \quad H_{LP}(s) = \frac{\omega_c^2}{s^2 + \frac{\omega_c}{Q_\infty}s + \omega_c^2}, \tag{6.6}$$

where ω_c is the cutoff frequency and Q_∞ is the pole quality factor. The Q-factor Q_∞ of a Butterworth approximation is equal to $1/\sqrt{2}$. The denormalization of a transfer function is obtained by replacing the Laplace variable s by $\frac{s}{\omega_g}$ in the normalized transfer function.

The corresponding second-order highpass transfer functions

$$H_{HP}(s) = \frac{s^2}{s^2 + \frac{1}{Q_\infty}s + 1} \quad \text{and} \quad H_{HP}(s) = \frac{s^2}{s^2 + \frac{\omega_c}{Q_\infty}s + \omega_c^2} \tag{6.7}$$

are obtained by a lowpass to highpass transformation. Figure 6.7 shows the pole-zero locations in the s-plane. The amplitude frequency response of a highpass filter with a 3-dB cutoff frequency of 50 Hz and a lowpass filter with a 3-dB cutoff frequency of 5000 Hz are shown in Fig. 6.8. Second- and fourth-order filters are shown.

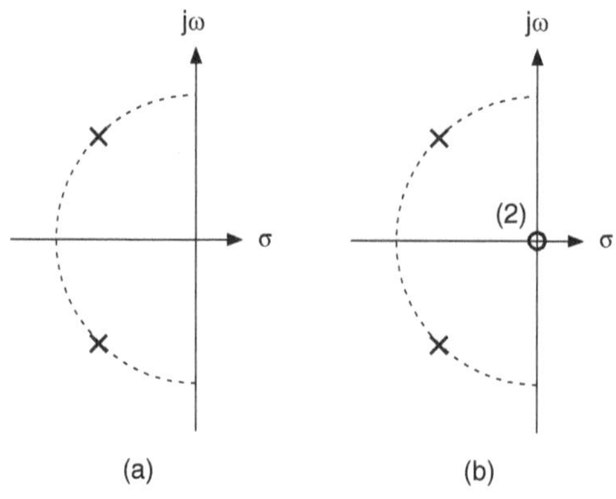

Figure 6.7 Pole-zero location for (a) second-order lowpass and (b) second-order highpass.

Table 6.1 summarizes the transfer functions of lowpass and highpass filters with Butterworth response.

Bandpass and bandstop filters. The normalized and denormalized bandpass transfer functions of second order are

$$H_{BP}(s) = \frac{\frac{1}{Q_\infty}s}{s^2 + \frac{1}{Q_\infty}s + 1} \quad \text{and} \quad H_{BP}(s) = \frac{\frac{\omega_c}{Q_\infty}s}{s^2 + \frac{\omega_c}{Q_\infty}s + \omega_c^2}, \tag{6.8}$$

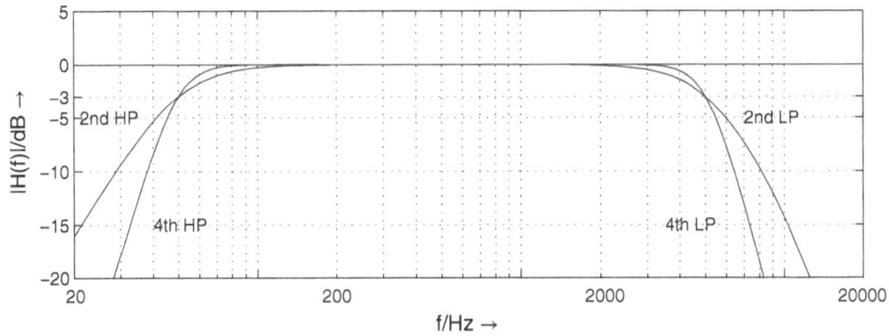

Figure 6.8 Frequency response of lowpass and highpass filters – highpass $f_c = 50$ Hz (second/fourth order), lowpass $f_c = 5000$ Hz (second/fourth order).

Table 6.1 Transfer functions of lowpass and highpass filters.

Lowpass	$H(s) = \dfrac{1}{s^2 + \sqrt{2}s + 1}$	second order
	$H(s) = \dfrac{1}{(s^2 + 1.848s + 1)(s^2 + 0.765s + 1)}$	fourth order
Highpass	$H(s) = \dfrac{s^2}{s^2 + \sqrt{2}s + 1}$	second order
	$H(s) = \dfrac{s^4}{(s^2 + 1.848s + 1)(s^2 + 0.765s + 1)}$	fourth order

and the bandstop transfer functions are given by

$$H_{BS}(s) = \frac{s^2 - 1}{s^2 + \frac{1}{Q_\infty}s + 1} \quad \text{and} \quad H_{BS}(s) = \frac{s^2 - \omega_c^2}{s^2 + \frac{\omega_c}{Q_\infty}s + \omega_c^2}. \tag{6.9}$$

The relative bandwidth can be expressed by the Q-factor

$$Q_\infty = \frac{f_c}{f_b}, \tag{6.10}$$

which is the ratio of center frequency f_c and the 3-dB bandwidth given by f_b. The magnitude responses of bandpass filters with constant relative bandwidth are shown in Fig. 6.2. Such kinds of filters are also called *constant-Q filters*. The geometric symmetric behavior of the frequency response regarding the center frequency f_c is clearly noticeable (symmetry regarding the center frequency using a logarithmic frequency axis).

Shelving Filters. In addition to the purely band-limiting filters like lowpass and highpass filters, shelving filters are used to perform weighting of certain frequencies. A simple approach for a first-order lowpass shelving filter is given by

$$H(s) = 1 + H_{LP}(s) = 1 + \frac{H_0}{s+1}. \tag{6.11}$$

It consists of a first-order lowpass filter with dc amplification of H_0 connected in parallel with an allpass system of transfer function equal to 1. Equation (6.11) can be written as

$$H(s) = \frac{s + (1 + H_0)}{s + 1} = \frac{s + V_0}{s + 1}, \tag{6.12}$$

where V_0 determines the amplification at $\omega = 0$. By changing the parameter V_0, any desired boost ($V_0 > 1$) and cut ($V_0 < 1$) level can be adjusted. Figure 6.9 shows the frequency responses for $f_c = 100$ Hz. For $V_0 < 1$, the cutoff frequency is dependent on V_0 and is moved toward lower frequencies.

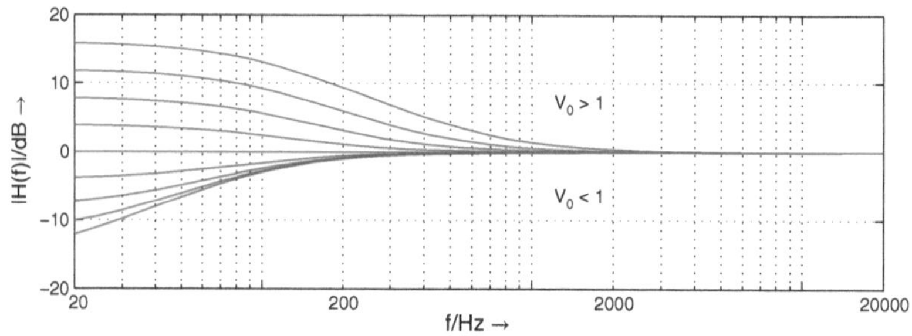

Figure 6.9 Frequency response of transfer function (6.12) with varying V_0 and cutoff frequency $f_c = 100$ Hz.

To obtain a symmetrical frequency response with respect to the zero-decibel line without changing the cutoff frequency, it is necessary to invert the transfer function (6.12) in the case of cut ($V_0 < 1$). This has the effect of swapping poles with zeros and leads to the transfer function

$$H(s) = \frac{s + 1}{s + V_0} \tag{6.13}$$

for the cut case. Figure 6.10 shows the corresponding frequency responses for varying V_0.

Finally, Figure 6.11 shows the locations of poles and zeros for both the boost and the cut cases. By moving zeros and poles on the negative σ-axis, boost and cut can be adjusted.

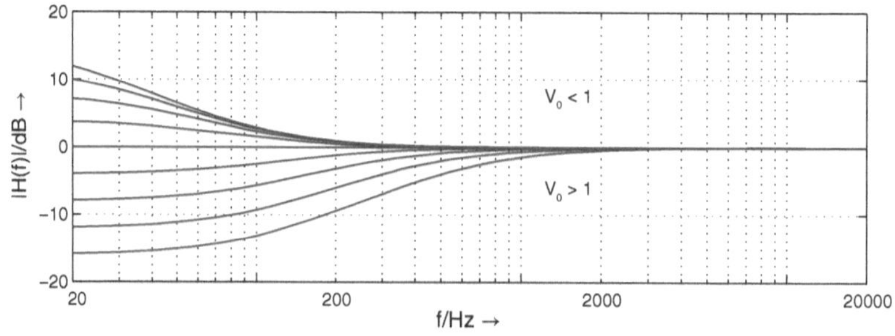

Figure 6.10 Frequency responses of transfer function (6.13) with varying V_0 and cutoff frequency $f_c = 100$ Hz.

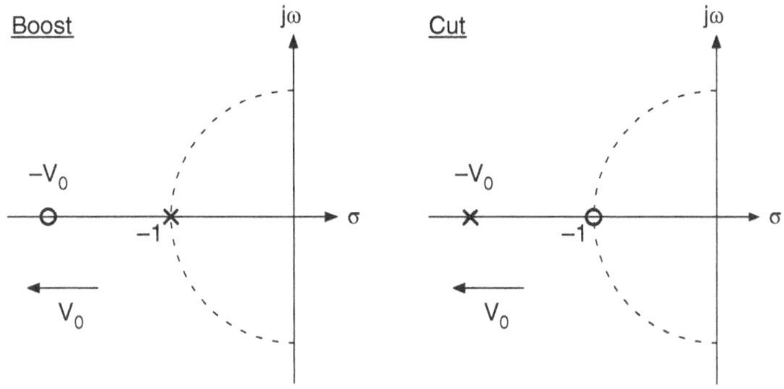

Figure 6.11 Pole-zero locations of a first-order low-frequency shelving filter.

The equivalent shelving filter for high frequencies can be obtained by

$$\boxed{H(s) = 1 + H_{HP}(s) = 1 + \frac{H_0 s}{s+1},} \tag{6.14}$$

which is a parallel connection of a first-order highpass with gain H_0 and a system with transfer function equal to 1. In the case of boost, the transfer function can be written with $V_0 = H_0 + 1$ as

$$H(s) = \frac{sV_0 + 1}{s+1}, \quad V_0 > 1, \tag{6.15}$$

and for cut, we get

$$H(s) = \frac{s+1}{sV_0 + 1}, \quad V_0 > 1. \tag{6.16}$$

The parameter V_0 determines the value of the transfer function $H(s)$ at $\omega = \infty$ for high-frequency shelving filters.

To increase the slope of the filter response in the transition band, a general second-order transfer function

$$H(s) = \frac{a_2 s^2 + a_1 s + a_0}{s^2 + \sqrt{2}s + 1} \tag{6.17}$$

is considered, in which complex zeros are added to the complex poles. The calculation of poles leads to

$$s_{\infty\ 1/2} = \sqrt{\frac{1}{2}}(-1 \pm j). \tag{6.18}$$

If the complex zeros

$$s_{\circ\ 1/2} = \sqrt{\frac{V_0}{2}}(-1 \pm j) \tag{6.19}$$

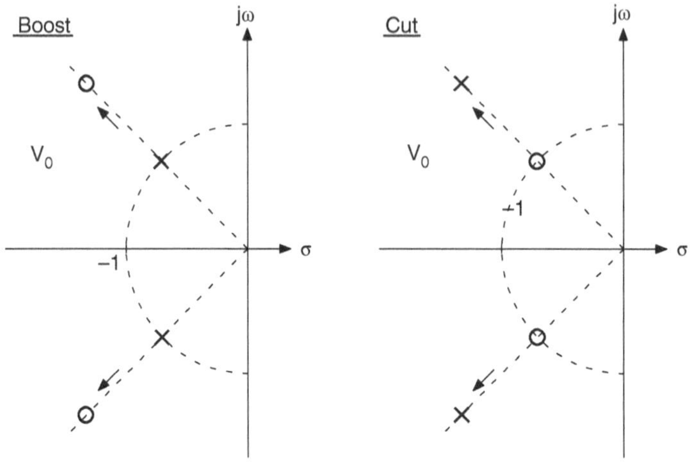

Figure 6.12 Pole-zero locations of a second-order low-frequency shelving filter.

are moved on a straight line with the help of the parameter V_0 (see Fig. 6.12), the transfer function

$$H(s) = \frac{s^2 + \sqrt{2V_0}s + V_0}{s^2 + \sqrt{2}s + 1} \tag{6.20}$$

of a second-order low-frequency shelving filter is obtained. The parameter V_0 determines the boost for low frequencies. The cut case can be achieved by inversion of Eq. (6.20).

A lowpass to highpass transformation of Eq. (6.20) provides the transfer function

$$H(s) = \frac{V_0 s^2 + \sqrt{2V_0}s + 1}{s^2 + \sqrt{2}s + 1} \tag{6.21}$$

of a second-order high-frequency shelving filter. The zeros

$$s_{\circ\ 1/2} = \sqrt{\frac{1}{2V_0}}(-1 \pm j) \tag{6.22}$$

are moved on a straight line toward the origin with increasing V_0 (see Fig. 6.13). The cut case is obtained by inverting the transfer function (6.21). Figure 6.14 shows the amplitude frequency response of a second-order low-frequency shelving filter with cutoff frequency 100 Hz and a second-order high-frequency shelving filter with cutoff frequency 5000 Hz (parameter V_0).

Peak Filter. Another equalizer used for boosting or cutting any desired frequency is the peak filter. A peak filter can be obtained by a parallel connection of a direct path and a bandpass according to

$$\boxed{H(s) = 1 + H_{BP}(s).} \tag{6.23}$$

With the help of a second-order bandpass transfer function

$$H_{BP}(s) = \frac{(H_0/Q_\infty)s}{s^2 + \frac{1}{Q_\infty}s + 1}, \tag{6.24}$$

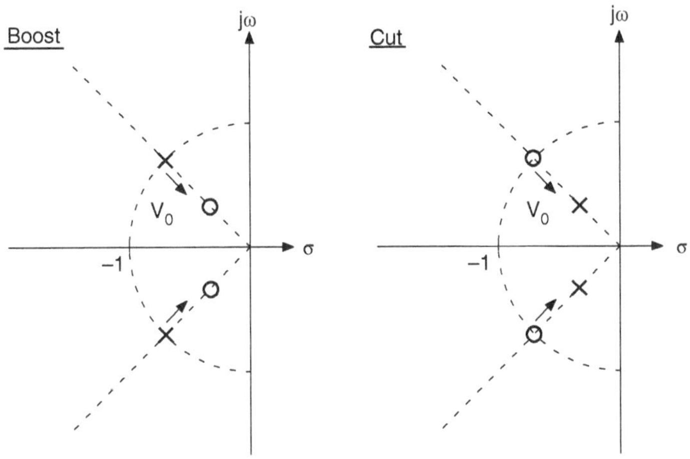

Figure 6.13 Pole-zero locations of second-order high-frequency shelving filter.

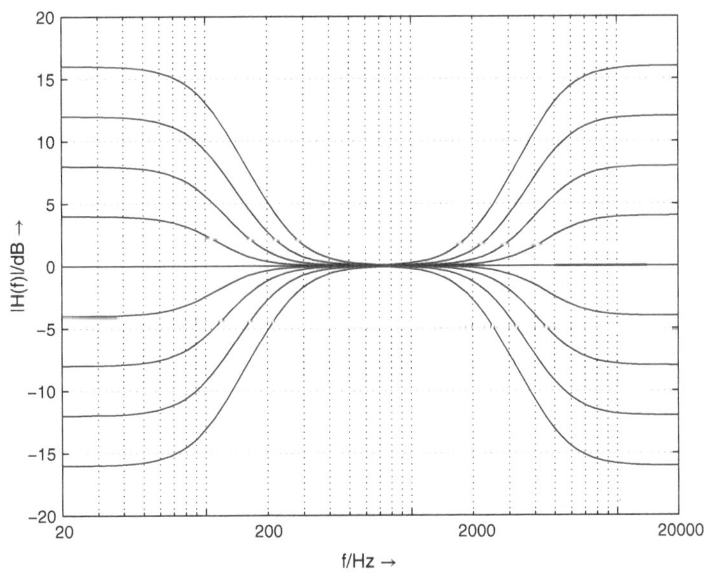

Figure 6.14 Frequency responses of second-order low-/high-frequency shelving filters – low-frequency shelving filter $f_c = 100\,\text{Hz}$ (parameter V_0), high-frequency shelving filter $f_c = 5000\,\text{Hz}$ (parameter V_0).

the transfer function

$$H(s) = 1 + H_{BP}(s)$$
$$= \frac{s^2 + \frac{1+H_0}{Q_\infty}s + 1}{s^2 + \frac{1}{Q_\infty}s + 1}$$

$$= \frac{s^2 + \frac{V_0}{Q_\infty}s + 1}{s^2 + \frac{1}{Q_\infty}s + 1} \tag{6.25}$$

of a peak filter can be derived. It can be shown that the maximum of the amplitude frequency response at the center frequency is determined by the parameter V_0. The relative bandwidth is fixed by the Q-factor. The geometrical symmetry of the frequency response relative to the center frequency remains constant for the transfer function of a peak filter given by Eq. (6.25). The poles and zeros lie on the unit circle. By adjusting the parameter V_0, the complex zeros are moved with respect to the complex poles. Figure 6.15 shows this for the boost and cut cases. With increasing Q-factor, the complex poles move toward the $j\omega$-axis on the unit circle.

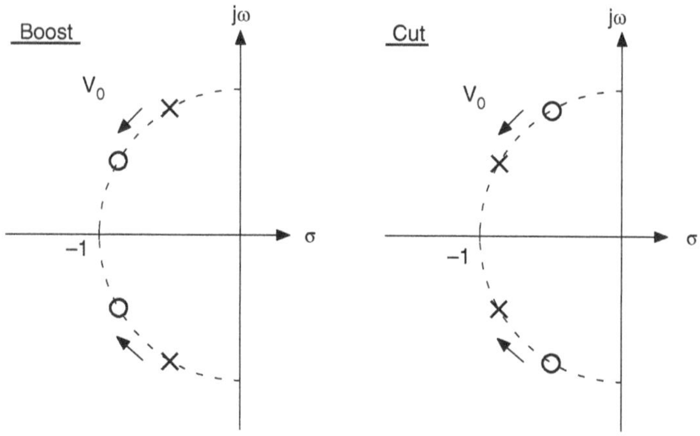

Figure 6.15 Pole-zero locations of a second-order peak filter.

Figure 6.16 shows the amplitude frequency response of a peak filter by changing the parameter V_0 at a center frequency of 500 Hz and a Q-factor of 1.25. Figure 6.17 shows the variation of the Q-factor Q_∞ at a center frequency of 500 Hz, a boost/cut of ± 16 dB, and Q-factor of 1.25. Finally, the variation of the center frequency with boost and cut of ± 16 dB and a Q-factor 1.25 is shown in Fig. 6.18.

Mapping to Z-domain. To implement a digital filter, the filter designed in the S-domain with transfer function $H(s)$ is converted to the Z-domain with the help of a suitable transformation to obtain the transfer function $H(z)$. The impulse-invariant transformation is not suitable as it leads to overlapping effects if the transfer function $H(s)$ is not band limited to half the sampling rate. An independent mapping of poles and zeros from the S-domain into poles and zeros in the Z-domain is possible with help of the bilinear transformation given by

$$s = \frac{2}{T}\frac{z-1}{z+1}. \tag{6.26}$$

Tables 6.2, 6.3, 6.4, and 6.5 contain the coefficients of the second-order transfer function

$$H(z) = \frac{a_0 + a_1 z^{-1} + a_2 z^{-2}}{1 + b_1 z^{-1} + b_2 z^{-2}}, \tag{6.27}$$

which are determined by the bilinear transformation and the auxiliary variable $K = \tan(\omega_c T/2)$ for all discussed audio filter types. Further filter designs of peak and shelving filters are discussed in

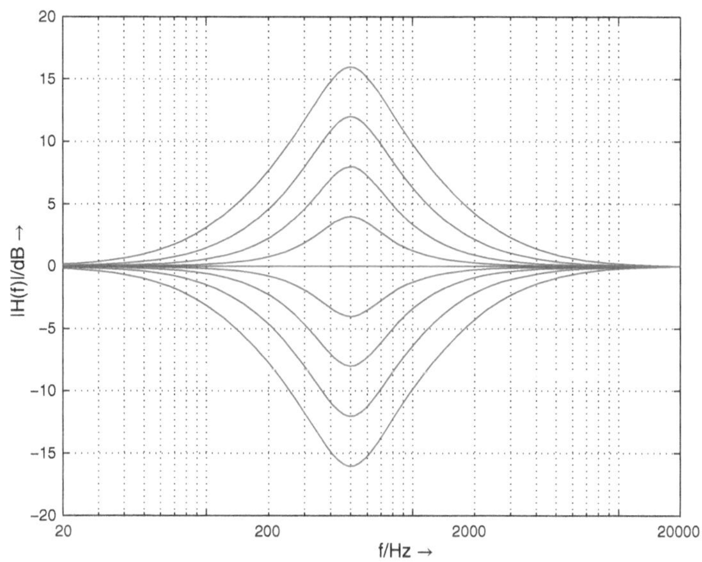

Figure 6.16 Frequency response of a peak filter – $f_c = 500\,\text{Hz}$, $Q_\infty = 1.25$, cut parameter V_0.

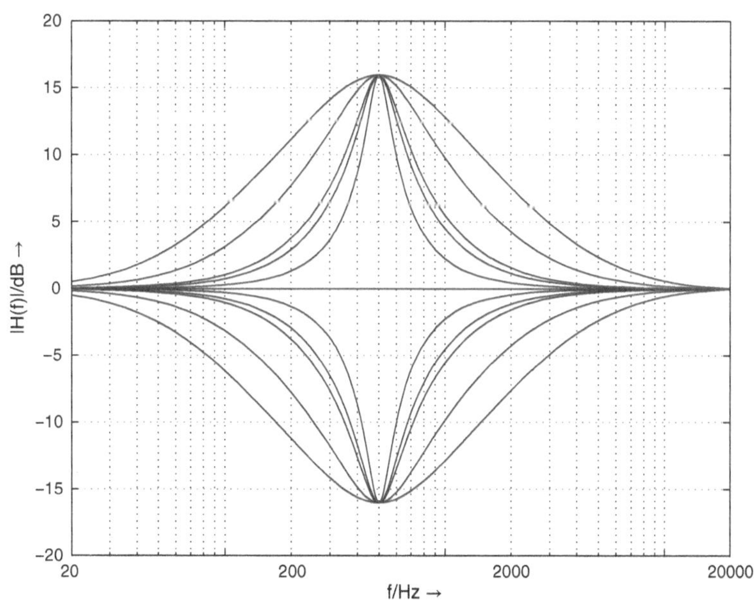

Figure 6.17 Frequency responses of peak filters – $f_c = 500\,\text{Hz}$, boost/cut $\pm 16\,\text{dB}$, $Q_\infty = 0.707, 1.25, 2.5, 3, 5$.

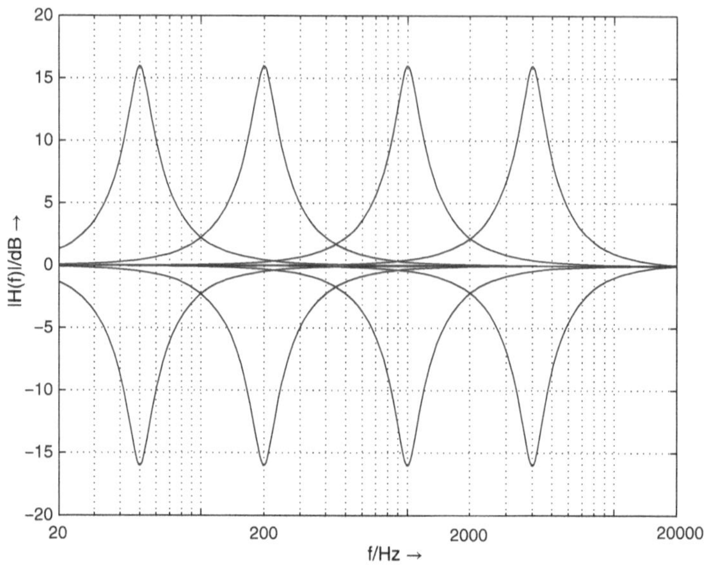

Figure 6.18 Frequency responses of peak filters – boost/cut ±16 dB, $Q_\infty = 1.25, f_c = 50, 200, 1000, 4000$ Hz.

Table 6.2 Lowpass/highpass/bandpass filter design.

		Lowpass (second-order)		
a_0	a_1	a_2	b_1	b_2
$\frac{K^2}{1+\sqrt{2}K+K^2}$	$\frac{2K^2}{1+\sqrt{2}K+K^2}$	$\frac{K^2}{1+\sqrt{2}K+K^2}$	$\frac{2(K^2-1)}{1+\sqrt{2}K+K^2}$	$\frac{1-\sqrt{2}K+K^2}{1+\sqrt{2}K+K^2}$
		Highpass (second-order)		
a_0	a_1	a_2	b_1	b_2
$\frac{1}{1+\sqrt{2}K+K^2}$	$\frac{-2}{1+\sqrt{2}K+K^2}$	$\frac{1}{1+\sqrt{2}K+K^2}$	$\frac{2(K^2-1)}{1+\sqrt{2}K+K^2}$	$\frac{1-\sqrt{2}K+K^2}{1+\sqrt{2}K+K^2}$
		Bandpass (second-order)		
a_0	a_1	a_2	b_1	b_2
$\frac{\frac{1}{Q}K}{1+\frac{1}{Q}K+K^2}$	0	$-\frac{\frac{1}{Q}K}{1+\frac{1}{Q}K+K^2}$	$\frac{2(K^2-1)}{1+\frac{1}{Q}K+K^2}$	$\frac{1-\frac{1}{Q}K+K^2}{1+\frac{1}{Q}K+K^2}$

[Moo83, Whi86, Sha92, Bri94, Orf96a, Dat97, Cla00, Väl16]. A method for reducing the warping effect of the bilinear transform is proposed in [Orf96b]. Strategies for time-variant switching of audio filters can be found in [Rab88, Mou90, Zöl93, Din95, Väl98].

6.2.2 Parametric Filter Structures

Parametric filter structures allow direct access to the parameters of the transfer function, like center/cutoff frequency, bandwidth, and gain, through the control of associated coefficients. To

Table 6.3 Peak filter design with gain G in dB.

Peak (boost $V_0 = 10^{G/20}$)				
a_0	a_1	a_2	b_1	b_2
$\frac{1+\frac{V_0}{Q_\infty}K+K^2}{1+\frac{1}{Q_\infty}K+K^2}$	$\frac{2(K^2-1)}{1+\frac{1}{Q_\infty}K+K^2}$	$\frac{1-\frac{V_0}{Q_\infty}K+K^2}{1+\frac{1}{Q_\infty}K+K^2}$	$\frac{2(K^2-1)}{1+\frac{1}{Q_\infty}K+K^2}$	$\frac{1-\frac{1}{Q_\infty}K+K^2}{1+\frac{1}{Q_\infty}K+K^2}$
Peak (cut $V_0 = 10^{-G/20}$)				
a_0	a_1	a_2	b_1	b_2
$\frac{1+\frac{1}{Q_\infty}K+K^2}{1+\frac{V_0}{Q_\infty}K+K^2}$	$\frac{2(K^2-1)}{1+\frac{V_0}{Q_\infty}K+K^2}$	$\frac{1-\frac{1}{Q_\infty}K+K^2}{1+\frac{V_0}{Q_\infty}K+K^2}$	$\frac{2(K^2-1)}{1+\frac{V_0}{Q_\infty}K+K^2}$	$\frac{1-\frac{V_0}{Q_\infty}K+K^2}{1+\frac{V_0}{Q_\infty}K+K^2}$

Table 6.4 Low-frequency shelving filter design with gain G in dB.

Low-frequency shelving (Boost $V_0 = 10^{G/20}$)				
a_0	a_1	a_2	b_1	b_2
$\frac{1+\sqrt{2V_0}K+V_0K^2}{1+\sqrt{2}K+K^2}$	$\frac{2(V_0K^2-1)}{1+\sqrt{2}K+K^2}$	$\frac{1-\sqrt{2V_0}K+V_0K^2}{1+\sqrt{2}K+K^2}$	$\frac{2(K^2-1)}{1+\sqrt{2}K+K^2}$	$\frac{1-\sqrt{2}K+K^2}{1+\sqrt{2}K+K^2}$
Low-frequency shelving (Cut $V_0 = 10^{-G/20}$)				
a_0	a_1	a_2	b_1	b_2
$\frac{1+\sqrt{2}K+K^2}{1+\sqrt{2V_0}K+V_0K^2}$	$\frac{2(K^2-1)}{1+\sqrt{2V_0}K+V_0K^2}$	$\frac{1-\sqrt{2}K+K^2}{1+\sqrt{2V_0}K+V_0K^2}$	$\frac{2(V_0K^2-1)}{1+\sqrt{2V_0}K+V_0K^2}$	$\frac{1-\sqrt{2V_0}K+V_0K^2}{1+\sqrt{2V_0}K+V_0K^2}$

Table 6.5 High-frequency shelving filter design with gain G in dB.

High-frequency shelving (Boost $V_0 = 10^{G/20}$)				
a_0	a_1	a_2	b_1	b_2
$\frac{V_0+\sqrt{2V_0}K+K^2}{1+\sqrt{2}K+K^2}$	$\frac{2(K^2-V_0)}{1+\sqrt{2}K+K^2}$	$\frac{V_0-\sqrt{2V_0}K+K^2}{1+\sqrt{2}K+K^2}$	$\frac{2(K^2-1)}{1+\sqrt{2}K+K^2}$	$\frac{1-\sqrt{2}K+K^2}{1+\sqrt{2}K+K^2}$
High-frequency shelving (Cut $V_0 = 10^{-G/20}$)				
a_0	a_1	a_2	b_1	b_2
$\frac{1+\sqrt{2}K+K^2}{V_0+\sqrt{2V_0}K+K^2}$	$\frac{2(K^2-1)}{V_0+\sqrt{2V_0}K+K^2}$	$\frac{1-\sqrt{2}K+K^2}{V_0+\sqrt{2V_0}K+K^2}$	$\frac{2(K^2/V_0-1)}{1+\sqrt{2/V_0}K+K^2/V_0}$	$\frac{1-\sqrt{2/V_0}K+K^2/V_0}{1+\sqrt{2/V_0}K+K^2/V_0}$

modify one of these parameters, it is therefore not necessary to compute a complete set of coefficients for a second-order transfer function, but instead only one coefficient in the filter structure is calculated.

An independent control of gain, cutoff/center frequency, and bandwidth for shelving and peak filters is achieved by a feed-forward (FF) structure for *boost* and a feed-backward (FB) structure for *cut*, as shown in Fig. 6.19. The corresponding transfer functions are

$$G_{FW}(z) = 1 + H_0 H(z), \tag{6.28}$$

$$G_{FB}(z) = \frac{1}{1 + H_0 H(z)}. \tag{6.29}$$

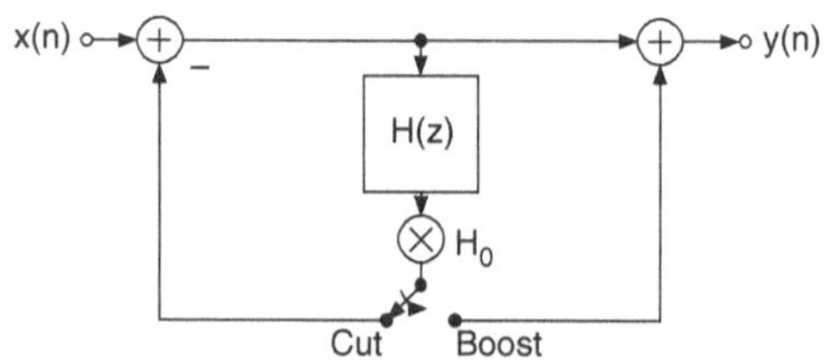

Figure 6.19 Filter structure for implementing boost and cut filters.

The boost/cut factor is $V_0 = 1 + H_0$. For digital filter implementations, it is necessary for the FB case that the inner transfer function be of the form $H(z) = z^{-1}H_1(z)$ to ensure causality. A parametric filter structure proposed by Harris [Har93] is based on the FF/FB technique, but the frequency response shows slight deviations near $z = 1$ and $z = -1$ from the desired one. This arises from the z^{-1} in the FF/FB branch. Delay-free loops inside filter computations can be solved by the methods presented in [Här98, Fon01, Fon03]. Higher-order parametric filter designs have been introduced in [Kei04, Orf05, Hol06a, Hol06b, Hol06c, Hol06d]. It is possible to implement typical audio filters with only an FF structure. The complete decoupling of the control parameters is possible for the boost case, but there remains a coupling between bandwidth and gain factor for the cut case. In the following, two approaches for parametric audio filter structures based on an allpass decomposition of the transfer function will be discussed.

Regalia filter [Reg87]. The denormalized transfer function of a first-order shelving filter is given by

$$H(s) = \frac{s + V_0 \omega_c}{s + \omega_c} \tag{6.30}$$

with

$$H(0) = V_0,$$
$$H(\infty) = 1.$$

A decomposition of Eq. (6.30) leads to

$$H(s) = \frac{s}{s + \omega_c} + V_0 \frac{\omega_c}{s + \omega_c}. \tag{6.31}$$

The lowpass and highpass transfer functions in Eq. (6.31) can be expressed by an allpass decomposition of the form

$$\frac{s}{s+\omega_c} = \frac{1}{2}\left[1 + \frac{s-\omega_c}{s+\omega_c}\right], \tag{6.32}$$

$$\frac{V_0\omega_c}{s+\omega_c} = \frac{V_0}{2}\left[1 - \frac{s-\omega_c}{s+\omega_c}\right]. \tag{6.33}$$

With the allpass transfer function

$$A_B(s) = \frac{s-\omega_c}{s+\omega_c} \tag{6.34}$$

for *boost*, Eq. (6.30) can be rewritten as

$$H(s) = \frac{1}{2}[1 + A_B(s)] + \frac{1}{2}V_0[1 - A_B(s)]. \tag{6.35}$$

The bilinear transformation $s = \frac{2}{T}\frac{z-1}{z+1}$ leads to

$$H(z) = \frac{1}{2}[1 + A_B(z)] + \frac{1}{2}V_0[1 - A_B(z)] \tag{6.36}$$

with

$$A_B(z) = -\frac{z^{-1} + a_B}{1 + a_B z^{-1}} \tag{6.37}$$

and the frequency parameter

$$a_B = \frac{\tan(\omega_c T/2) - 1}{\tan(\omega_c T/2) + 1}. \tag{6.38}$$

A filter structure for direct implementation of Eq. (6.36) is presented in Fig. 6.20a. Other possible structures can be seen in Fig. 6.20b,c. For the cut case $V_0 < 1$, the cutoff frequency of the filter moves towards lower frequencies [Reg87].

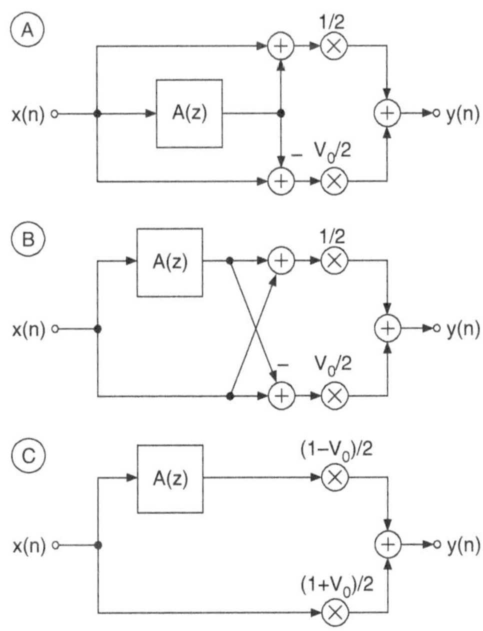

Figure 6.20 Filter structures by Regalia.

To retain the cutoff frequency for the cut case [Zöl95], the denormalized transfer function of a first-order shelving filter (cut)

$$H(s) = \frac{s + \omega_c}{s + \omega_c/V_0} \tag{6.39}$$

with the boundary conditions

$$H(0) = V_0,$$
$$H(\infty) = 1$$

can be decomposed as follows:

$$H(s) = \frac{s}{s + \omega_c/V_0} + \frac{\omega_c}{s + \omega_c/V_0}. \tag{6.40}$$

With the allpass decompositions

$$\frac{s}{s + \omega_c/V_0} = \frac{1}{2}\left[1 + \frac{s - \omega_c/V_0}{s + \omega_c/V_0}\right], \tag{6.41}$$

$$\frac{\omega_c}{s + \omega_c/V_0} = \frac{V_0}{2}\left[1 - \frac{s - \omega_c/V_0}{s + \omega_c/V_0}\right], \tag{6.42}$$

and the allpass transfer function

$$A_C(s) = \frac{s - \omega_c/V_0}{s + \omega_c/V_0} \tag{6.43}$$

for *cut*, Eq. (6.39) can be rewritten as

$$H(s) = \frac{1}{2}[1 + A_C(s)] + \frac{V_0}{2}[1 - A_C(s)]. \tag{6.44}$$

The bilinear transformation leads to

$$H(z) = \frac{1}{2}[1 + A_C(z)] + \frac{V_0}{2}[1 - A_C(z)] \tag{6.45}$$

with

$$A_C(z) = -\frac{z^{-1} + a_C}{1 + a_C z^{-1}} \tag{6.46}$$

and the frequency parameter

$$a_C = \frac{\tan(\omega_c T/2) - V_0}{\tan(\omega_c T/2) + V_0}. \tag{6.47}$$

Owing to Eqs. (6.45) and (6.36), boost and cut can be implemented with the same filter structure (see Fig. 6.20). However, it has to be noted that the frequency parameter a_C, as in Eq. (6.47) for cut, depends on the cutoff frequency and gain.

A second-order peak filter is obtained by a lowpass to bandpass transformation according to

$$z^{-1} \to -z^{-1}\frac{z^{-1} + d}{1 + dz^{-1}}. \tag{6.48}$$

For an allpass, as given in Eqs. (6.37) and (6.46), the second-order allpass is given by

$$A_{BC}(z) = \frac{z^{-2} + d(1 + a_{BC})z^{-1} + a_{BC}}{1 + d(1 + a_{BC})z^{-1} + a_{BC}z^{-2}} \tag{6.49}$$

with parameters (cut as in [Zöl95])

$$d = -cos(\Omega_c), \tag{6.50}$$

$$V_0 = H(e^{j\Omega_c}), \tag{6.51}$$

$$a_B = \frac{1 - \tan(\omega_b T/2)}{1 + \tan(\omega_b T/2)}, \tag{6.52}$$

$$a_C = \frac{V_0 - \tan(\omega_b T/2)}{V_0 + \tan(\omega_b T/2)}. \tag{6.53}$$

The center frequency f_c is fixed by the parameter d, the bandwidth f_b by the parameters a_B and a_C, and gain $G = 20 \log_{10} V_0$ by the parameter V_0.

Simplified Allpass Decomposition [Zöl95]. The transfer function of a first-order low-frequency shelving filter can be a decomposed as

$$\begin{aligned} H(s) &= \frac{s + V_0\omega_c}{s + \omega_c} \\ &= 1 + H_0 \frac{\omega_c}{s + \omega_c} \qquad (6.54) \\ &= 1 + \frac{H_0}{2}\left[1 - \frac{s - \omega_c}{s + \omega_c}\right] \qquad (6.55) \end{aligned}$$

with

$$V_0 = H(s = 0), \tag{6.56}$$

$$H_0 = V_0 - 1, \tag{6.57}$$

$$V_0 = 10^{\frac{G}{20}} \qquad \text{(G in dB)}. \tag{6.58}$$

The transfer function (6.55) is composed of a direct branch and a lowpass filter. The first-order lowpass filter is again implemented by an allpass decomposition. Applying the bilinear transformation to Eq. (6.55) leads to

$$H(z) = 1 + \frac{H_0}{2}[1 - A(z)] \tag{6.59}$$

with

$$A(z) = -\frac{z^{-1} + a_B}{1 + a_B z^{-1}}. \tag{6.60}$$

For cut, the following decomposition can be derived:

$$\begin{aligned} H(s) &= \frac{s + \omega_c}{s + \omega_c/V_0} \qquad (6.61) \\ &= 1 + \underbrace{(V_0 - 1)}_{H_0} \frac{\omega_c/V_0}{s + \omega_c/V_0} \qquad (6.62) \\ &= 1 + \frac{H_0}{2}\left[1 - \frac{s - \omega_c/V_0}{s + \omega_c/V_0}\right]. \qquad (6.63) \end{aligned}$$

The bilinear transformation applied to Eq. (6.63) again gives Eq. (6.59). The filter structure is identical for boost and cut. The frequency parameter a_B for boost and a_C for cut can be calculated as

$$a_B = \frac{\tan(\omega_c T/2) - 1}{\tan(\omega_c T/2) + 1}, \tag{6.64}$$

$$a_C = \frac{\tan(\omega_c T/2) - V_0}{\tan(\omega_c T/2) + V_0}. \tag{6.65}$$

The transfer function of a first-order low-frequency shelving filter can be calculated as

$$H(z) = \frac{1 + (1 + a_{BC})\frac{H_0}{2} + (a_{BC} + (1 + a_{BC})\frac{H_0}{2})z^{-1}}{1 + a_{BC}z^{-1}}. \tag{6.66}$$

With $A_1(z) = -A(z)$, the signal flow chart in Fig. 6.21 shows a first-order lowpass filter and a first-order low-frequency shelving filter.

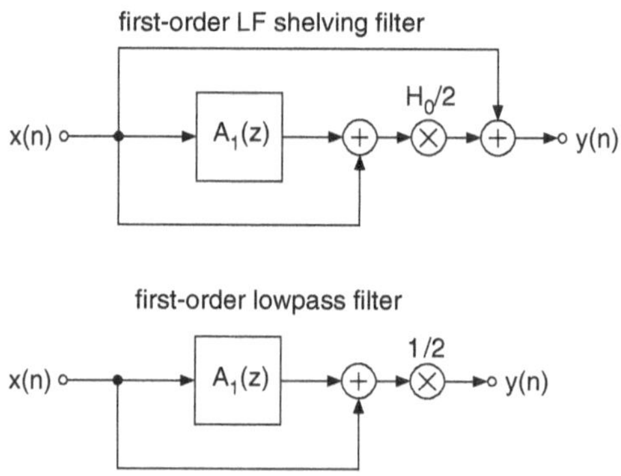

Figure 6.21 Low-frequency shelving filter and first-order lowpass filter.

The decomposition of a denormalized transfer function of a first-order high-frequency shelving filter can be given in the form of

$$\begin{aligned} H(s) &= \frac{sV_0 + \omega_c}{s + \omega_c} \\ &= 1 + H_0 \frac{s}{s + \omega_c} && (6.67) \\ &= 1 + \frac{H_0}{2}\left[1 + \frac{s - \omega_c}{s + \omega_c}\right], && (6.68) \end{aligned}$$

where

$$V_0 = H(s = \infty), \tag{6.69}$$

$$H_0 = V_0 - 1. \tag{6.70}$$

The transfer function results by adding a highpass filter to a constant. Applying the bilinear transformation to Eq. (6.68) gives

$$H(z) = 1 + \frac{H_0}{2}[1 + A(z)] \tag{6.71}$$

with

$$A(z) = -\frac{z^{-1} + a_B}{1 + a_B z^{-1}}. \tag{6.72}$$

For cut, the decomposition can be given by

$$H(s) = \frac{s + \omega_c}{s/V_0 + \omega_c} \tag{6.73}$$

$$= 1 + \underbrace{(V_0 - 1)}_{H_0}\frac{s}{s + V_0\omega_c} \tag{6.74}$$

$$= 1 + \frac{H_0}{2}\left[1 + \frac{s - V_0\omega_c}{s + V_0\omega_c}\right], \tag{6.75}$$

which in return results in Eq. (6.71) after a bilinear transformation. The boost and cut parameters can be calculated as

$$a_B = \frac{\tan(\omega_c T/2) - 1}{\tan(\omega_c T/2) + 1}, \tag{6.76}$$

$$a_C = \frac{V_0\ \tan(\omega_c T/2) - 1}{V_0\ \tan(\omega_c T/2) + 1}. \tag{6.77}$$

The transfer function of a first-order high-frequency shelving filter can then be written as

$$H(z) = \frac{1 + (1 - a_{BC})\frac{H_0}{2} + (a_{BC} + (a_{BC} - 1)\frac{H_0}{2})z^{-1}}{1 + a_{BC}z^{-1}}. \tag{6.78}$$

With $A_1(z) = -A(z)$, the signal flow chart in Fig. 6.22 shows a first-order highpass filter and a high-frequency shelving filter.

The implementation of a second-order peak filter can be carried out with a lowpass to bandpass transformation of a first-order shelving filter. However, the addition of a second-order bandpass

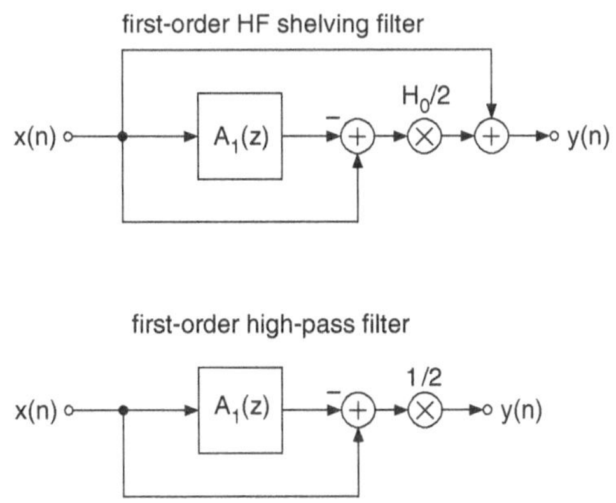

Figure 6.22 First-order high-frequency shelving and highpass filters.

filter to a constant branch also results in a peak filter. With the help of an allpass implementation of a bandpass filter, as given by

$$H(z) = \frac{1}{2}\left[1 - A_2(z)\right] \tag{6.79}$$

and

$$A_2(z) = \frac{-a_B + (d - da_B)z^{-1} + z^{-2}}{1 + (d - da_B)z^{-1} - a_B z^{-2}}, \tag{6.80}$$

a second-order peak filter can be expressed as

$$H(z) = 1 + \frac{H_0}{2}\left[1 - A_2(z)\right]. \tag{6.81}$$

The bandwidth parameters a_B and a_C for boost and cut are given by

$$a_B = \frac{\tan(\omega_b T/2) - 1}{\tan(\omega_b T/2) + 1}, \tag{6.82}$$

$$a_C = \frac{\tan(\omega_b T/2) - V_0}{\tan(\omega_b T/2) + V_0}. \tag{6.83}$$

The center frequency parameter d and the coefficient H_0 are given by

$$d = -cos(\Omega_c), \tag{6.84}$$

$$V_0 = H(e^{j\Omega_c}), \tag{6.85}$$

$$H_0 = V_0 - 1. \tag{6.86}$$

The transfer function of a second-order peak filter results in

$$H(z) = \frac{1 + (1 + a_{BC})\frac{H_0}{2} + d(1 - a_{BC})z^{-1} + (-a_{BC} - (1 + a_{BC})\frac{H_0}{2})z^{-2}}{1 + d(1 - a_{BC})z^{-1} - a_{BC}z^{-2}}. \tag{6.87}$$

The signal flow charts for a second-order peak filter and a second-order bandpass filter are shown in Fig. 6.23.

The frequency responses for high-frequency shelving, low-frequency shelving, and peak filters are shown in Figs. 6.24, 6.25, and 6.26.

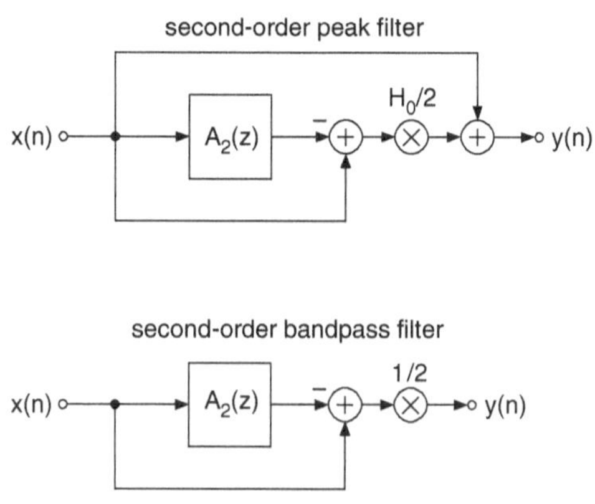

Figure 6.23 Second-order peak filter and bandpass filter.

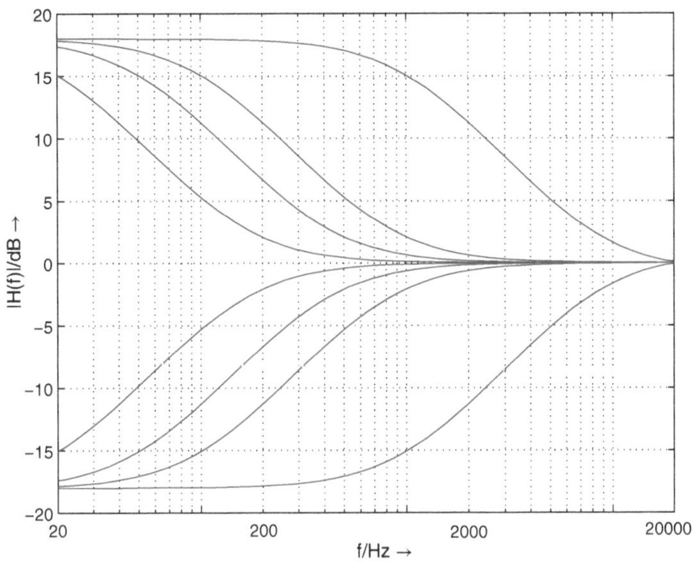

Figure 6.24 Low-frequency first-order shelving filter ($G = \pm 18$ dB; $f_c = 20, 50, 100, 1000$ Hz).

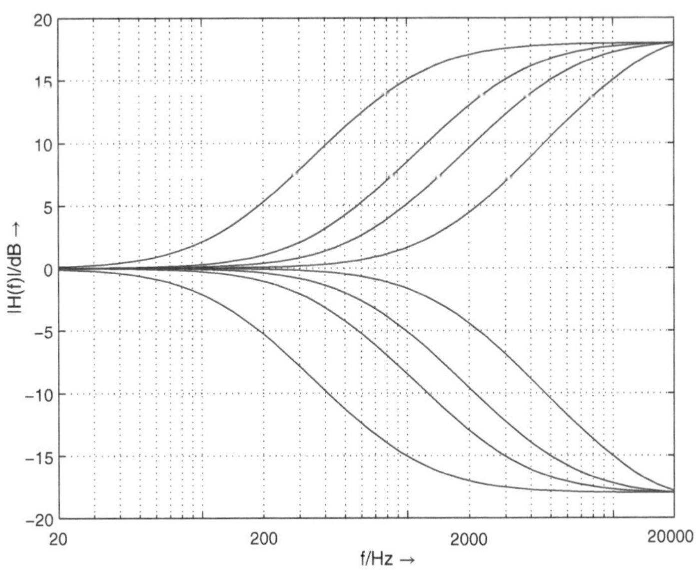

Figure 6.25 First-order high-frequency shelving filter ($G = \pm 18$ dB; $f_c = 1, 3, 5, 10, 16$ kHz).

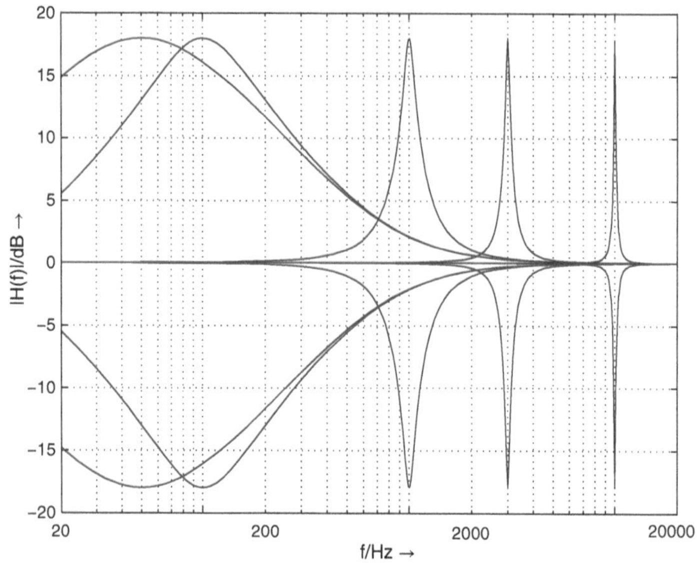

Figure 6.26 Second-order peak filter ($G = \pm 18$ dB; $f_c = 50, 100, 1000, 3000, 10000$ Hz; $f_b = 100$ Hz).

6.2.3 Quantization Effects

The limited word length for digital recursive filters leads to two different types of quantization error. The quantization of the coefficients of a digital filter results in linear distortion which can be noticed as a deviation from the ideal frequency response. The quantization of the signal inside a filter structure is responsible for the maximum dynamic range and determines the noise behavior of the filter. Owing to rounding operations in a filter structure, round-off noise is produced. Another effect of the signal quantization is limit cycles. They can be classified as overflow limit cycles, small-scale limit cycles, and limit cycles correlated with the input signal. Limit cycles are very disturbing owing to their small-band (sinusoidal) nature. The overflow limit cycles can be avoided by suitable scaling of the input signal. The effects of other errors mentioned above can be reduced by increasing the word lengths of the coefficient and the state variables of the filter structure.

The noise behavior and coefficient sensitivity of a filter structure depend on the topology and the cutoff frequency (position of the poles in the Z-domain) of the filter. Because common audio filters operate between 20 Hz and 20 kHz at a sampling rate of 48 kHz, the filter structures are subjected to specially strict criteria with respect to error behavior. The frequency range for equalizers is between 20 Hz and 4–6 kHz because the human voice and many musical instruments have their formants in that frequency region. For given coefficient and signal word-lengths (like in a digital signal processor), a filter structure with low round-off noise for audio application can lead to a suitable solution. For this, the following second-order filter structures are compared.

The basis of the following considerations is the relationship between the coefficient sensitivity and round-off noise. This was first stated by Fettweis [Fet72]. By increasing the pole density in a certain region of the z-plane, the coefficient sensitivity and the round-off noise of the filter structure are reduced. Owing to these improvements, the coefficient word-length as well as signal word-length can be reduced. Work in designing digital filters with minimum word-length for coefficients and state variables was first carried out by Avenhaus [Ave71].

Typical audio filters like highpass/lowpass, peak/shelving filters can be described by the second-order transfer function

$$H(z) = \frac{a_0 + a_1 z^{-1} + a_2 z^{-2}}{1 + b_1 z^{-1} + b_2 z^{-2}}. \tag{6.88}$$

The recursive part of the difference equation, which can be derived from the transfer function (6.88), is considered more closely, because it plays a major role in affecting the error behavior. Owing to the quantization of the coefficients in the denominator in Eq. (6.88), the distribution of poles in the z-plane is restricted (see Fig. 6.27 for 6-bit quantization of coefficients). The pole distribution in the second quadrant of the z-plane is the mirror image of the first quadrant. Figure 6.28 shows a block diagram of the recursive part. Another equivalent representation of the denominator is given by

$$H(z) = \frac{N(z)}{1 - 2r\cos\varphi z^{-1} + r^2 z^{-2}}. \tag{6.89}$$

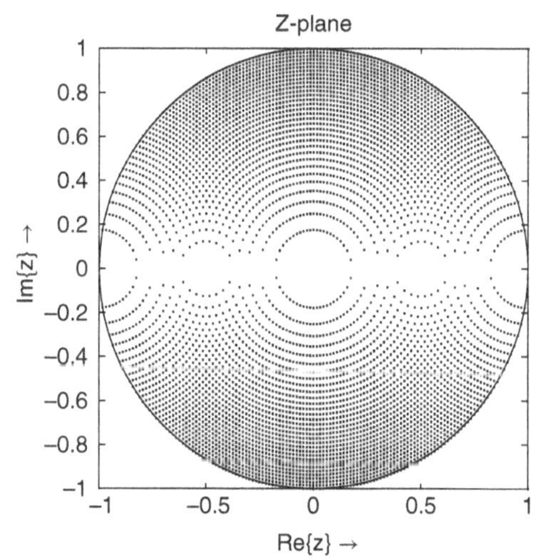

Figure 6.27 Direct-form structure – pole distribution (6-bit quantization).

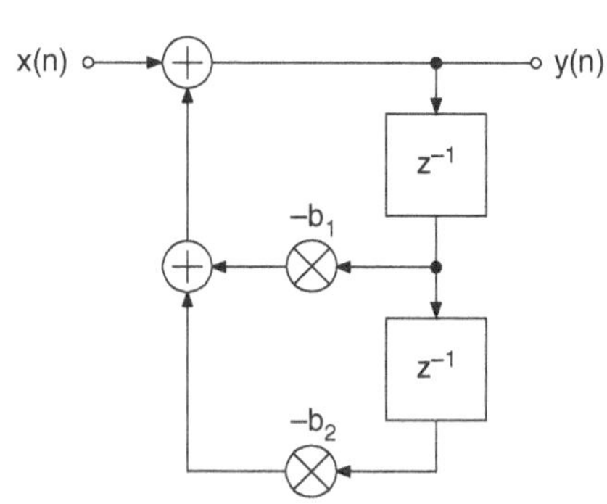

Figure 6.28 Direct-form structure – block diagram of recursive part.

Here r is the radius and φ the corresponding phase of the complex poles. By quantizing these parameters, the pole distribution is altered in contrast to the case where b_1 and b_2 are quantized, as in Eq. (6.88).

The state variable structure [Mul76, Bom85] is based on the approach by Gold and Rader [Gol67], which is given by

$$H(z) = \frac{N(z)}{1 - 2\mathrm{Re}\{z_\infty\}z^{-1} + (\mathrm{Re}\{z_\infty\}^2 + \mathrm{Im}\{z_\infty\}^2)z^{-2}}. \tag{6.90}$$

The possible pole locations are shown in Fig. 6.29 for 6-bit quantization (block diagram of recursive part is shown in Fig. 6.30). Owing to the quantization of real and imaginary parts, a uniform grid of different pole locations results. In contrast to direct quantization of the coefficients b_1 and b_2 in the denominator, the quantization of the real and imaginary parts leads to an increase in the pole density at $z = 1$. The possible pole locations in the second quadrant in the z-plane are the mirror images of those in the first quadrant.

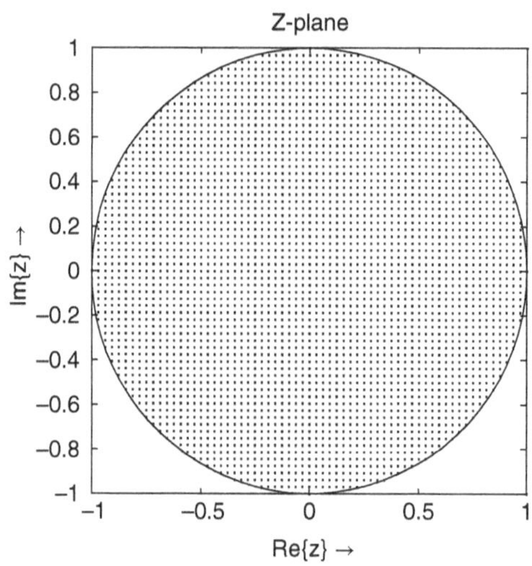

Figure 6.29 Gold and Rader – pole distribution (6-bit quantization).

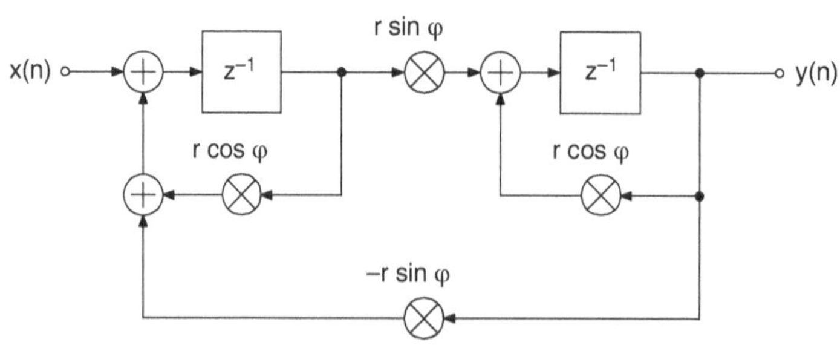

Figure 6.30 Gold and Rader – block diagram of recursive part.

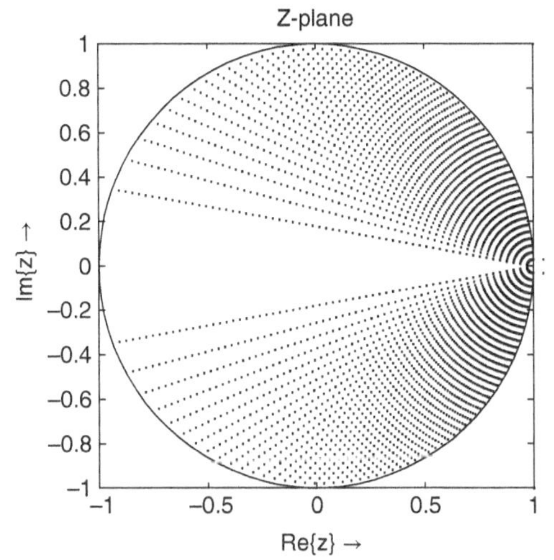

Figure 6.31 Kingsbury – pole distribution (6-bit quantization).

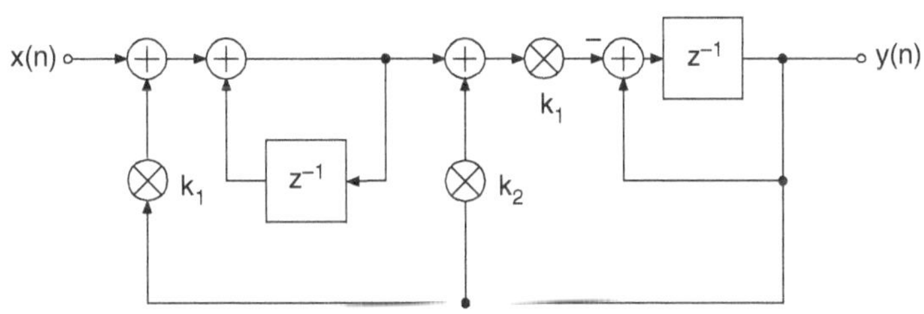

Figure 6.32 Kingsbury – block diagram of recursive part.

In [Kin72], a filter structure is suggested which has a pole distribution as shown in Fig. 6.31 (for the block diagram of recursive part, see Fig. 6.32).

The corresponding transfer function

$$H(z) = \frac{N(z)}{1 - (2 - k_1 k_2 - k_1^2)z^{-1} + (1 - k_1 k_2)z^{-2}} \tag{6.91}$$

shows that in this case, the coefficients b_1 and b_2 can be obtained by a linear combination of the quantized coefficients k_1 and k_2. The distance d of the pole from the point $z = 1$ determines the coefficients

$$k_1 = d = \sqrt{1 - 2r\cos\varphi + r^2}, \tag{6.92}$$

$$k_2 = \frac{1 - r^2}{k_1}, \tag{6.93}$$

as illustrated in Fig. 6.33.

The filter structures under consideration show that by a suitable linear combination of quantized coefficients, any desired pole distribution can be obtained. An increase of the pole density

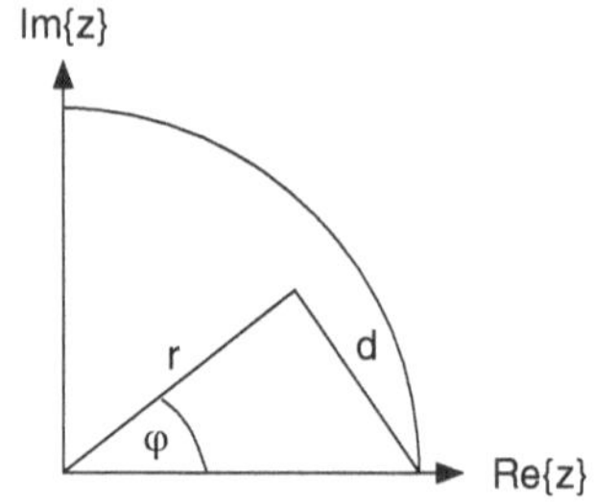

Figure 6.33 Geometric interpretation.

at $z = 1$ can be achieved by influencing the linear relationship between the coefficient k_1 and the distance d from $z = 1$ [Zöl89, Zöl90]. The nonlinear relationship of the new coefficients gives the following structure with the transfer function

$$H(z) = \frac{N(z)}{1 - (2 - z_1 z_2 - z_1^3)z^{-1} + (1 - z_1 z_2)z^{-2}} \tag{6.94}$$

and coefficients

$$z_1 = \sqrt[3]{1 + b_1 + b_2}, \tag{6.95}$$

$$z_2 = \frac{1 - b_2}{z_1}, \tag{6.96}$$

with

$$z_1 = \sqrt[3]{d^2}. \tag{6.97}$$

The pole distribution of this structure is shown in Fig. 6.34. The block diagram of the recursive part is illustrated in Fig. 6.35. An increase in the pole density at $z = 1$ in contrast to previous

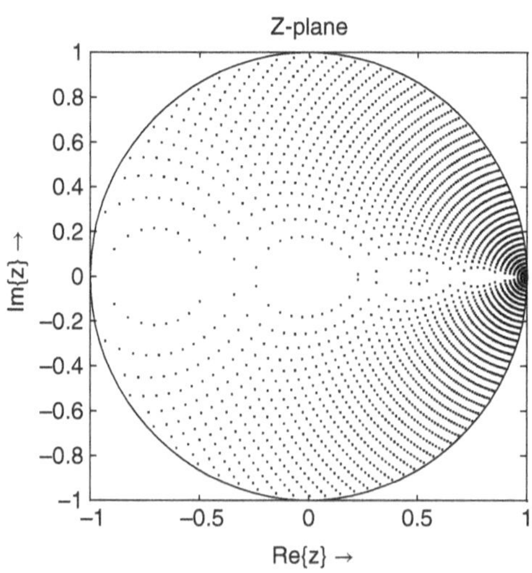

Figure 6.34 Zölzer – pole distribution (6-bit quantization).

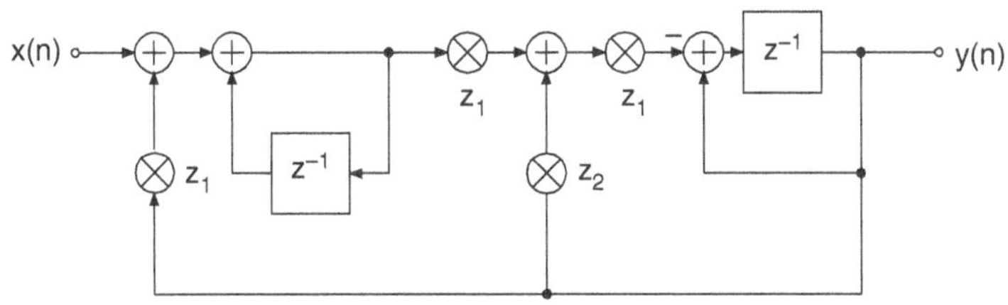

Figure 6.35 Zölzer – block diagram of recursive part.

pole distributions is noticed. The pole distributions of the Kingsbury and Zölzer structures show a decrease in the pole density for higher frequencies. For the pole density, a symmetry with respect to the imaginary axis, as in the case of the direct-form structure and the Gold and Rader structure, is not possible. However, changing the sign in the recursive part of the difference equation results in a mirror image of the pole density. The mirror image can be achieved through a change of sign in the denominator polynomial. The denominator polynomial

$$D(z) = 1 \overbrace{\quad \pm \quad}^{!} (2 - z_1 z_2 - z_1^3)z^{-1} + (1 - z_1 z_2)z^{-2} \tag{6.98}$$

shows that the real part depends on the coefficient of z^{-1}.

Analytical Comparison of Noise Behavior of Different Filter Structures

In this section, recursive filter structures are analyzed in terms of their noise behavior in fixed-point arithmetic [Zöl89, Zöl90, Zol94]. The block diagrams provide the basis of an analytical calculation of noise power owing to the quantization of state variables. First of all, the general case is considered in which quantization is performed after multiplication. For this purpose, the transfer function $G_i(z)$ of every multiplier output to the output of the filter structure is determined.

For this error analysis, it is assumed that the signal within the filter structure covers the whole dynamic range so that the quantization error $e_i(n)$ is not correlated with the signal. Consecutive quantization error samples are not correlated with each other so that a uniform power density spectrum results [Sri77]. It can also be assumed that different quantization errors $e_i(n)$ are uncorrelated within the filter structure. Owing to the uniform distribution of the quantization error, the variance can be given by

$$\sigma_E^2 = \frac{Q^2}{12}. \tag{6.99}$$

The quantization error is added at every point of quantization and is filtered by the corresponding transfer function $G(z)$ to the output of the filter. The variance of the output quantization noise (owing to the noise source $e(n)$) is given by

$$\sigma_{ye}^2 = \sigma_E^2 \frac{1}{2\pi j} \oint_{z=e^{j\Omega}} G(z)G(z^{-1})z^{-1}dz. \tag{6.100}$$

Exact solutions for the ring integral (6.100) can be found in [Jur64] for transfer functions up to the fourth order. With the L_2 norm of a periodic function

$$\| G \|_2 = \left[\frac{1}{2\pi} \int_{-\pi}^{\pi} |G(e^{j\Omega})|^2 d\Omega \right]^{\frac{1}{2}}, \tag{6.101}$$

the superposition of the noise variances leads with Eq. (6.101) to the total output noise variance

$$\sigma_{ye}^2 = \sigma_E^2 \sum_i \| G_i \|_2^2 . \tag{6.102}$$

The signal-to-noise ratio (SNR) for a full-range sinusoid can be written as

$$\text{SNR} = 10 \log_{10} \frac{0.5}{\sigma_{ye}^2} \quad [\text{dB}]. \tag{6.103}$$

The ring integral

$$I_n = \frac{1}{2\pi j} \oint_{z=e^{j\Omega}} \frac{A(z)A(z^{-1})}{B(z)B(z^{-1})} z^{-1}\, dz \tag{6.104}$$

is given in [Jur64] for first-order systems by

$$G(z) = \frac{a_0 z + a_1}{b_0 z + b_1}, \tag{6.105}$$

$$I_1 = \frac{(a_0^2 + a_1^2) b_0 - 2 a_0 a_1 b_1}{b_0 (b_0^2 - b_1^2)}, \tag{6.106}$$

and for second-order systems by

$$G(z) = \frac{a_0 z^2 + a_1 z + a_2}{b_0 z^2 + b_1 z + b_2}, \tag{6.107}$$

$$I_2 = \frac{A_0 b_0 c_1 - A_1 b_0 b_1 + A_2 (b_1^2 - b_2 c_1)}{b_0 [(b_0^2 - b_2^2) c_1 - (b_0 b_1 - b_1 b_2) b_1]}, \tag{6.108}$$

$$A_0 = a_0^2 + a_1^2 + a_2^2, \tag{6.109}$$

$$A_1 = 2(a_0 a_1 + a_1 a_2), \tag{6.110}$$

$$A_2 = 2 a_0 a_2, \tag{6.111}$$

$$c_1 = b_0 + b_2. \tag{6.112}$$

In the following, an analysis of the noise behavior for different recursive filter structures will be made. The noise transfer functions of individual recursive parts are responsible for noise shaping.

The error transfer function of a second-order direct-form structure (see Fig. 6.36) has only complex poles (see Table 6.6).

The implementation of poles near the unit circle leads to high amplification of the quantization error. The effect of the pole radius on the noise variance can be observed in the equation for output noise variance. The coefficient $b_2 = r^2$ approaches 1 which leads to a huge increase in the output noise variance.

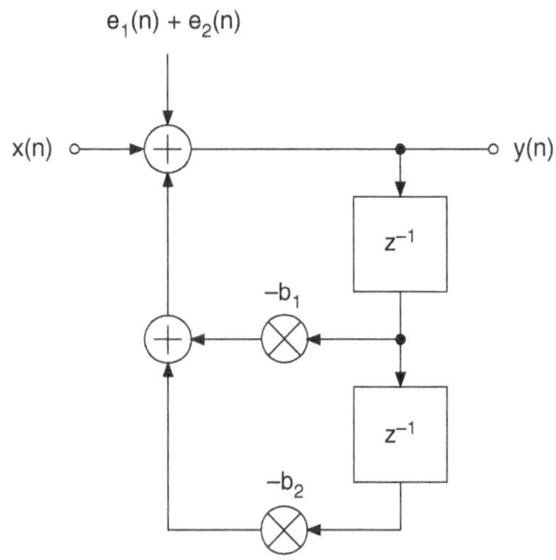

Figure 6.36 Direct form with additive error signal.

Table 6.6 Direct-form – a) noise transfer function, b) quadratic L_2 norm, and c) output noise variance in the case of quantization after every multiplication.

a)	$G_1(z) = G_2(z) = \dfrac{z^2}{z^2 + b_1 z + b_2}$
b)	$\Vert G_1 \Vert_2^2 = \Vert G_2 \Vert_2^2 = \dfrac{1 + b_2}{1 - b_2} \dfrac{1}{(1 + b_2)^2 - b_1^2}$
c)	$\sigma_{ye}^2 = \sigma_E^2 2 \dfrac{1 + b_2}{1 - b_2} \dfrac{1}{(1 + b_2)^2 - b_1^2}$

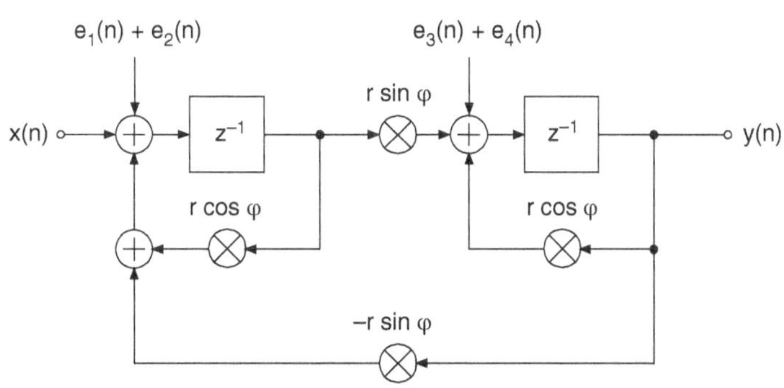

Figure 6.37 Gold and Rader structure with additive error signals.

The Gold and Rader filter structure (Fig. 6.37) has an output noise variance that depends on the pole radius (see Table 6.7) and is independent of the pole phase. The latter fact is because of the uniform grid of the pole distribution. An additional zero on the real axis ($z = r\cos\varphi$) directly beneath the poles reduces the effect of the complex poles.

Table 6.7 Gold and Rader – a) noise transfer function, b) quadratic L_2 norm, and c) output noise variance in the case of quantization after every multiplication.

a)	$G_1(z) = G_2(z)$	$= \dfrac{r \sin \varphi}{z^2 - 2r \cos \varphi z + r^2}$
	$G_3(z) = G_4(z)$	$= \dfrac{z - r \cos \varphi}{z^2 - 2r \cos \varphi z + r^2}$
b)	$\parallel G_1 \parallel_2^2 = \parallel G_2 \parallel_2^2$	$= \dfrac{1 + b_2}{1 - b_2} \dfrac{(r \sin \varphi)^2}{(1 + b_2)^2 - b_1^2}$
	$\parallel G_3 \parallel_2^2 = \parallel G_4 \parallel_2^2$	$= \dfrac{1}{1 - b_2} \dfrac{[1 + (r \sin \varphi)^2](1 + b_2)^2 - b_1^2}{(1 + b_2)^2 - b_1^2}$
c)	$\sigma_{ye}^2 = \sigma_E^2 2 \dfrac{1}{1 - b_2}$	

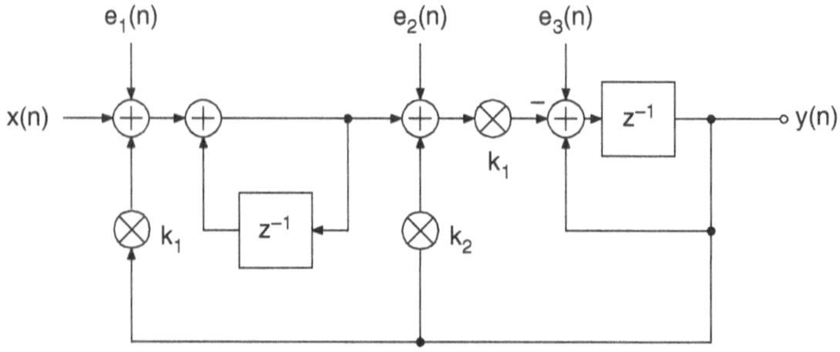

Figure 6.38 Kingsbury structure with additive error signals.

The Kingsbury filter (Fig. 6.38 and Table 6.8) and the Zölzer filter (Fig. 6.39 and Table 6.9), which is derived from it, show that the noise variance depends on the pole radius. The noise transfer functions have a zero at $z = 1$ in addition to the complex poles. This zero reduces the amplifying effect of the pole near the unit circle at $z = 1$.

Figure 6.40 shows the SNR versus the cutoff frequency for the four filter structures presented above. The signals are quantized to 16 bit. Here, the poles move with increasing cutoff frequency on the curve characterized by the Q-factor $Q_\infty = 0.7071$ in the z-plane. For very small cutoff frequencies, the Zölzer filter shows an improvement of 3 dB in terms of SNR compared with the Kingsbury filter and an improvement of 6 dB compared with the Gold and Rader filter. Up to 5 kHz, the Zölzer filter yields better results (see Fig. 6.41). From 6 kHz onwards, the reduction of pole density in this filter leads to a decrease in the SNR (see Fig. 6.41).

With regard to the implementation of the these filters with digital signal processors, a quantization after every multiplication is not necessary. Quantization takes place when the accumulator has to be stored to memory. This can be seen in Figs. 6.42, 6.43, 6.44, and 6.45 by introducing quantizers where they really occur. The resulting output noise variances are also shown. The SNR is plotted versus the cutoff frequency in Figs. 6.46 and 6.47. In the case of direct-form and Gold and Rader filters, the SNR increases by 3 dB whereas the output noise variance for the Kingsbury filter remains unchanged. The Kingsbury filter and the Gold and Rader filters exhibit similar results up to a frequency of 200 kHz (see Fig. 6.46). The Zölzer filter demonstrates an improvement of 3 dB compared with these structures. For frequencies of up to 2 kHz (see Fig. 6.47), it is seen that the increased pole density leads to an improvement of the SNR as well as a reduced effect owing to coefficient quantization.

Table 6.8 Kingsbury – a) noise transfer function, b) quadratic L_2 norm, and c) output noise variance in the case of quantization after every multiplication.

a)	$G_1(z)$	$= \frac{-k_1 z}{z^2-(2-k_1k_2-k_1^2)z+(1-k_1k_2)}$
	$G_2(z)$	$= \frac{-k_1(z-1)}{z^2-(2-k_1k_2-k_1^2)z+(1-k_1k_2)}$
	$G_3(z)$	$= \frac{z-1}{z^2-(2-k_1k_2-k_1^2)z+(1-k_1k_2)}$
b)	$\lVert G_1 \rVert_2^2$	$= \frac{1}{k_1k_2}\frac{2-k_1k_2}{2(2-k_1k_2)-k_1^2}$
	$\lVert G_2 \rVert_2^2$	$= \frac{k_1}{k_2}\frac{2}{2(2-k_1k_2)-k_1^2}$
	$\lVert G_3 \rVert_2^2$	$= \frac{1}{k_1k_2}\frac{2}{2(2-k_1k_2)-k_1^2}$
c)		$\sigma_{ye}^2 = \sigma_E^2 2\frac{5+2b_1+3b_2}{(1-b_2)(1+b_2-b_1)}$

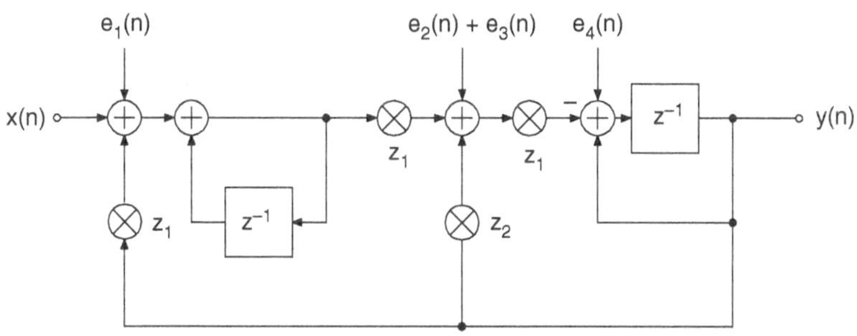

Figure 6.39 Zölzer structure with additive error signals.

Table 6.9 Zölzer – a) noise transfer function, b) quadratic L_2 norm, and c) output noise variance in the case of quantization after every multiplication.

a)	$G_1(z)$	$= \frac{-z_1^2 z}{z^2-(2-z_1z_2-z_1^3)z+(1-z_1z_2)}$
	$G_2(z) = G_3(z)$	$= \frac{-z_1(z-1)}{z^2-(2-z_1z_2-z_1^3)z+(1-z_1z_2)}$
	$G_4(z)$	$= \frac{z-1}{z^2-(2-z_1z_2-z_1^3)z+(1-z_1z_2)}$
b)	$\lVert G_1 \rVert_2^2$	$= \frac{z_1^4}{z_1z_2}\frac{2-z_1z_2}{2z_1^3(2-z_1z_2)-z_1^6}$
	$\lVert G_2 \rVert_2^2 = \lVert G_3 \rVert_2^2$	$= \frac{z_1^6}{z_1z_2}\frac{2}{2z_1^3(2-z_1z_2)-z_1^6}$
	$\lVert G_4 \rVert_2^2$	$= \frac{z_1^3}{z_1z_2}\frac{2}{2z_1^3(2-z_1z_2)-z_1^6}$
c)		$\sigma_{ye}^2 = \sigma_E^2 2\frac{6+4(b_1+b_2)+(1+b_2)(1+b_1+b_2)^{1/3}}{(1-b_2)(1+b_2-b_1)}$

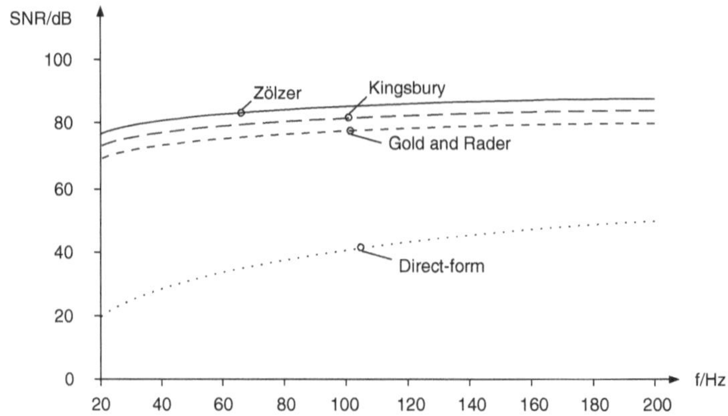

Figure 6.40 SNR versus cutoff frequency – quantization of products (f_c < 200 Hz).

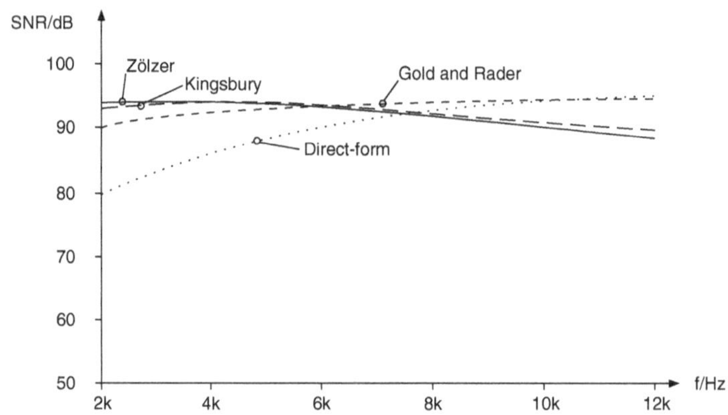

Figure 6.41 SNR versus cutoff frequency – quantization of products (f_c > 2 kHz).

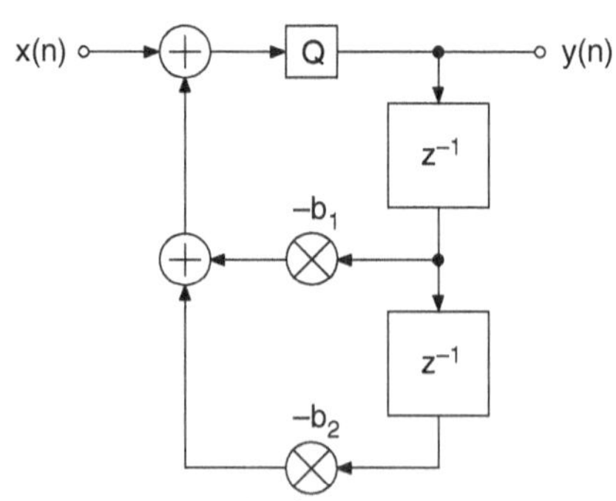

$$\sigma_{ye}^2 = \sigma_E^2 \frac{1+b_2}{1-b_2} \frac{1}{(1+b_2)^2 - b_1^2}$$

Figure 6.42 Direct-form filter – quantization after accumulator.

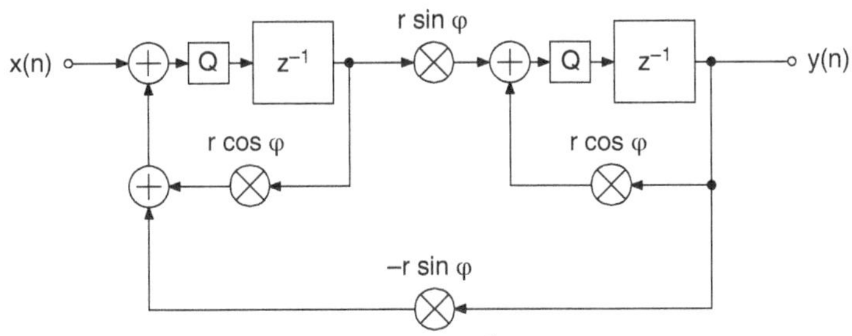

$$\sigma_{ye}^2 = \sigma_E^2 \frac{1}{1-b_2}$$

Figure 6.43 Gold and Rader filter – quantization after accumulator.

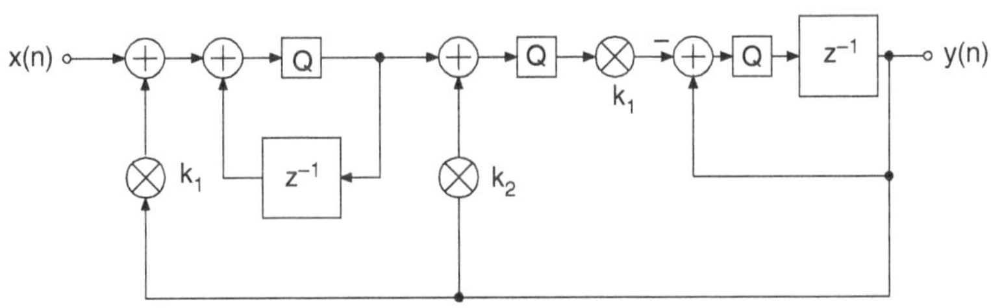

$$\sigma_{ye}^2 = \sigma_E^2 \frac{5+2b_1+3b_2}{(1-b_2)(1+b_2-b_1)}$$

Figure 6.44 Kingsbury filter – quantization after accumulator.

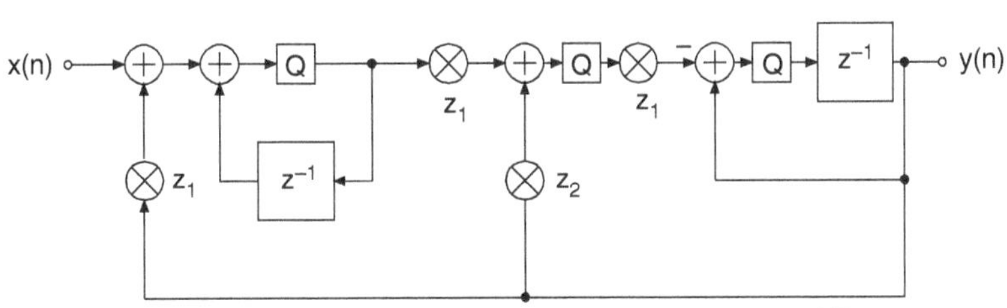

$$\sigma_{ye}^2 = \sigma_E^2 \frac{2(2+b_1+b_2)+(1+b_2)(1+b_1+b_2)^{1/3}}{(1-b_2)(1+b_2-b_1)}$$

Figure 6.45 Zölzer filter – quantization after accumulator.

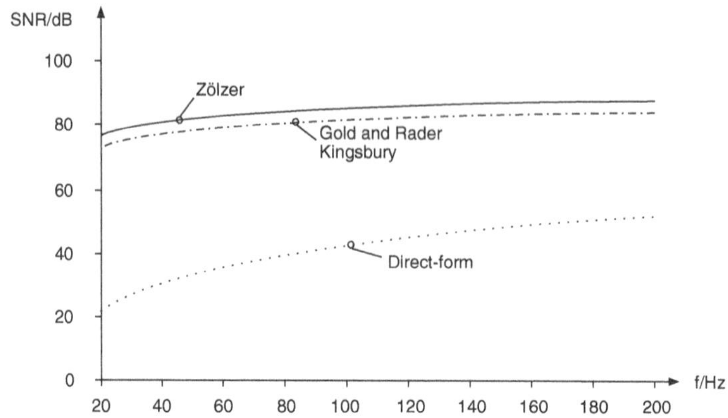

Figure 6.46 SNR versus cutoff frequency – quantization after accumulator (f_c < 200 Hz).

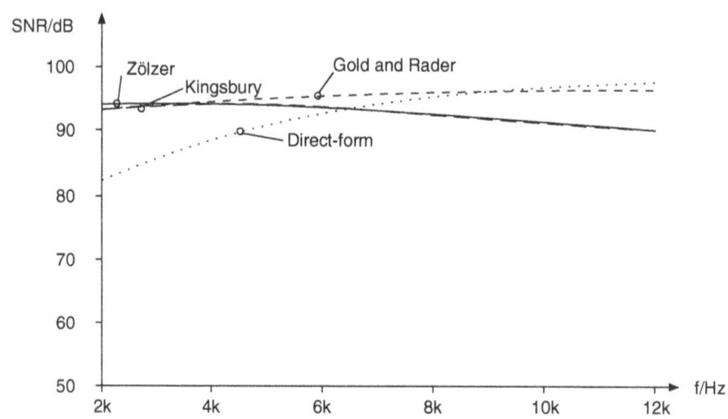

Figure 6.47 SNR versus cutoff frequency – quantization after accumulator (f_c > 2 kHz).

Noise Shaping in Recursive Filters

The analysis of the noise transfer function of different structures shows that for three structures with low round-off noise, a zero at $z = 1$ occurs in the transfer functions $G(z)$ of the error signals in addition to the complex poles. This zero near the poles reduces the amplifying effect of the pole. If it is now possible to introduce another zero into the noise transfer function, then the effect of the poles is compensated for to a larger extent. The procedure of feeding back the quantization error, as shown in Chapter 2, produces an additional zero in the noise transfer function [Tra77, Cha78, Abu79, Bar82, Zöl89]. The feedback of the quantization error is first demonstrated with the help of the direct-form structure, as shown in Fig. 6.48. This generates a zero at $z = 1$ in the noise transfer function given by

$$G_{1.O}(z) = \frac{1 - z^{-1}}{1 + b_1 z^{-1} + b_2 z^{-2}}. \tag{6.113}$$

The resulting variance σ^2 of the quantization error at the output of the filter is presented in Fig. 6.48. To produce two zeros at $z = 1$, the quantization error is fed back over two delays

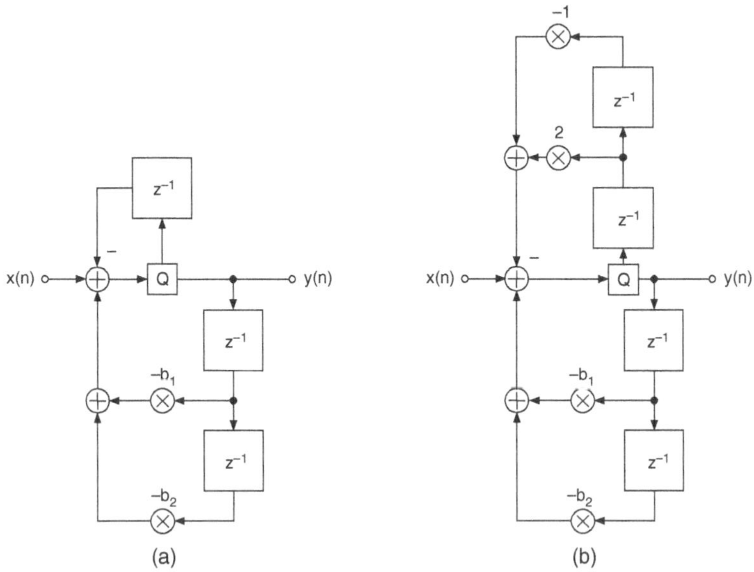

$$\sigma^2_{DF1} = \sigma^2_E \frac{2}{(1-b_2)(1+b_2-b_1)}$$

$$\sigma^2_{DF2} = \sigma^2_E \frac{6+2b_1-2b_2}{(1-b_2)(1+b_2-b_1)}$$

Figure 6.48 Direct-form with noise shaping.

weighted with 2 and −1 (see Fig. 6.48b). The noise transfer function is, hence, given by

$$G_{2.O}(z) = \frac{1-2z^{-1}+z^{-2}}{1+b_1z^{-1}+b_2z^{-2}}. \tag{6.114}$$

The SNR of the direct form is plotted versus the cutoff frequency in Fig. 6.49. Even a single zero significantly improves the SNR in the direct form. The coefficients b_1 and b_2 approach −2 and 1, respectively, with the decrease of the cutoff frequency. With this, the error is filtered with a second-order highpass. The introduction of the additional zeros in the noise transfer function only affects the noise signal of the filter. The input signal is only affected by the transfer function $H(z)$. If the feedback coefficients are chosen equal to the coefficients b_1 and b_2 in the denominator polynomial, complex zeros are produced that are identical with the complex poles. The noise transfer function $G(z)$ is then reduced to unity. The choice of complex zeros directly at the location of the complex poles corresponds to double-precision arithmetic.

In [Abu79], an improvement of the noise behavior for the direct form in any desired location of the z-plane is achieved by placing additional simple-to-implement complex zeros near the poles. For implementing filter algorithms with digital signal processors, these kinds of suboptimal zero are easily realized. Because the Gold and Rader, Kingsbury, and Zölzer filter structures already have zeros in their respective noise transfer functions, it is sufficient to use a simple feedback for the quantization error. By virtue of this extension, the block diagrams in Figs. 6.50, 6.51, and 6.52 are obtained.

The effect of noise shaping on SNR is shown in Figs. 6.53 and 6.54. The almost ideal noise behavior of all filter structures for 16-bit quantization and very small cutoff frequencies can be

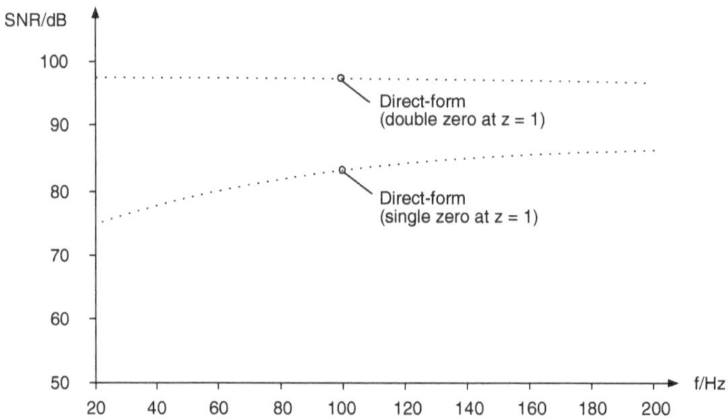

Figure 6.49 SNR – Noise shaping in direct-form filter structures.

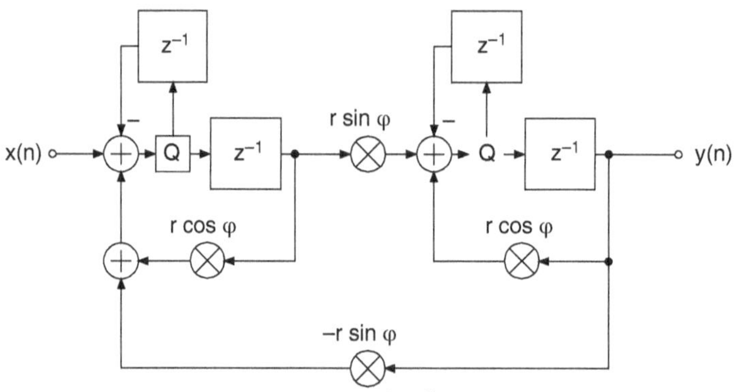

$$\sigma_{ye}^2 = \sigma_E^2 \frac{2 + b_1}{1 - b_2}$$

Figure 6.50 Gold and Rader filter with noise shaping.

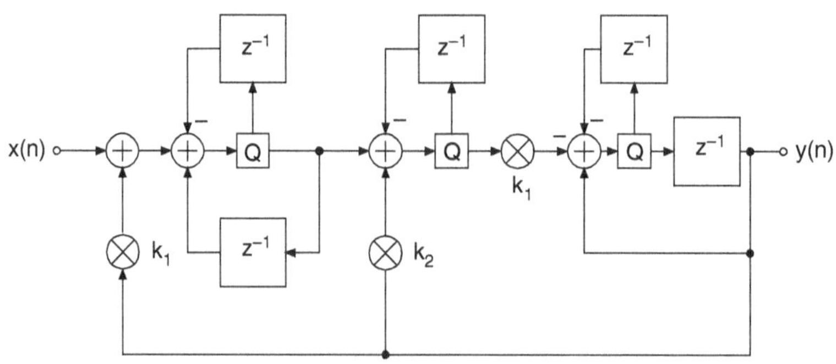

$$\sigma_{ye}^2 = \sigma_E^2 \frac{(1 + k_1^2)((1 + b_2)(6 - 2b_2) + 2b_1^2 + 8b_1) + 2k_1^2(1 + b_1 + b_2)}{(1 - b_2)(1 + b_2 - b_1)(1 + b_2 - b_1)}$$

Figure 6.51 Kingsbury filter with noise shaping.

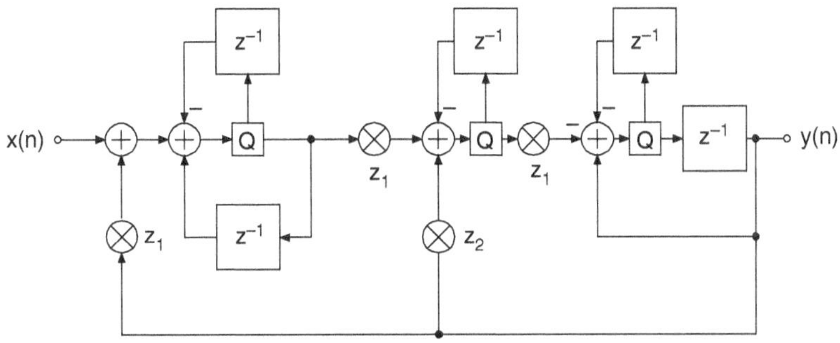

$$\sigma_{ye}^2 = \sigma_E^2 \frac{(1+z_1^2)((1+b_2)(6-2b_2)+2b_1^2+8b_1)+2z_1^4(1+b_1+b_2)}{(1-b_2)(1+b_2-b_1)(1+b_2-b_1)}$$

Figure 6.52 Zölzer filter with noise shaping.

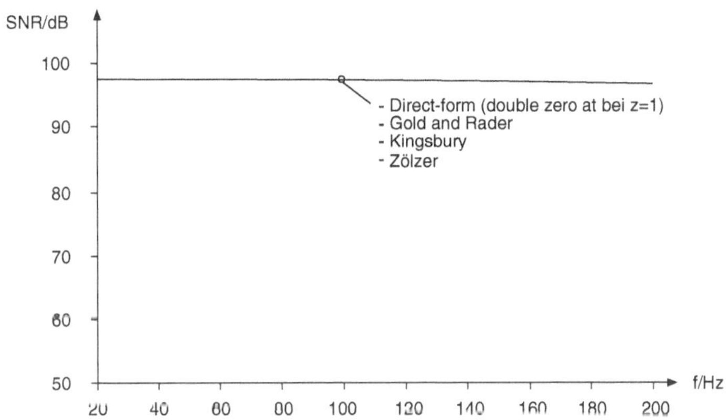

Figure 6.53 SNR – noise shaping (20–200 Hz).

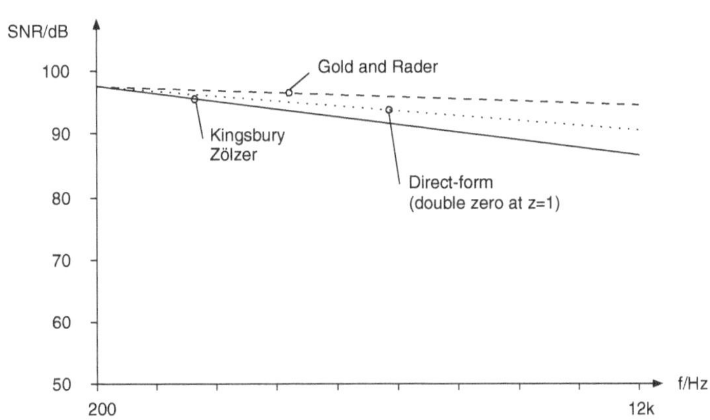

Figure 6.54 SNR – noise shaping (200 Hz–12 kHz).

observed. The effect of this noise shaping for increasing cutoff frequencies is shown in Fig. 6.54. The compensating effect of the two zeros at $z = 1$ is reduced.

Scaling

In a fixed-point implementation of a digital filter, a transfer function from the input of the filter to a junction within the filter has to be determined, as well as the transfer function from the input to the output. By scaling the input signal, it has to be guaranteed that the signals remain within the number range at each junction and at the output.

To calculate scaling coefficients, different criteria can be used. The L_p norm is defined as

$$L_p = \| H\|_p = \left[\frac{1}{2\pi}\int_{-\pi}^{\pi} |H(e^{j\Omega})|^p d\Omega\right]^{1/p}, \tag{6.115}$$

and an expression for the L_∞ norm follows for $p = \infty$:

$$L_\infty = \| H(e^{j\Omega})\|_\infty = \max_{0\leq\Omega\leq\pi} |H(e^{j\Omega})|. \tag{6.116}$$

The L_∞ norm represents the maximum of the amplitude frequency response. In general, the modulus of the output is

$$|y(n)| \leq \| H\|_p \ \| X\|_q \tag{6.117}$$

with

$$\frac{1}{p} + \frac{1}{q} = 1 \qquad p, q \geq 1. \tag{6.118}$$

For the L_1, L_2, and L_∞ norms, the explanations in Table 6.10 can be used.
With

$$|y_i(n)| \leq \| H_i(e^{j\Omega})\|_\infty \ \| X(e^{j\Omega})\|_1, \tag{6.119}$$

the L_∞ norm is given by

$$L_\infty = \| h_i\|_\infty = \max_{k=0}^{\infty}|h_i(k)|. \tag{6.120}$$

For a sinusoidal input signal of amplitude one, we get $\| X(e^{j\Omega})\|_1 = 1$. For $|y_i(n)| \leq 1$ to be valid, the scaling factor must be chosen to be

$$S_i = \frac{1}{\| H_i(e^{j\Omega})\|_\infty}. \tag{6.121}$$

Table 6.10 Commonly used scaling.

p	q	
1	∞	given max. value of input spectrum scaling w.r.t. the L_1 norm of $H(e^{j\Omega})$
∞	1	given L_1 norm of input spectrum $X(e^{j\Omega})$ scaling w.r.t. the L_∞ norm of $H(e^{j\Omega})$
2	2	given L_2 norm of input spectrum $X(e^{j\Omega})$ scaling w.r.t. the L_2 norm of $H(e^{j\Omega})$

The scaling of the input signal is carried out with the maximum of the amplitude frequency response with the goal that for $|x(n)| \leq 1$, $|y_i(n)| \leq 1$. As a scaling coefficient for the input signal, the highest scaling factor S_i is chosen. For determining the maximum of the transfer function

$$\| H(e^{j\Omega})\|_\infty = \max_{0\leq\Omega\leq\pi} |H(e^{j\Omega})| \tag{6.122}$$

of a second-order system

$$\begin{aligned} H(z) &= \frac{a_0 + a_1 z^{-1} + a_2 z^{-1}}{1 + b_1 z^{-1} + b_2 z^{-1}} \\ &= \frac{a_0 z^2 + a_1 z + a_2}{z^2 + b_1 z + b_2}, \end{aligned}$$

the maximum value can be calculated as

$$|H(e^{j\Omega})|^2 = \frac{\overbrace{\frac{a_0 a_2}{b_2}}^{\alpha_0}\cos^2(\Omega) + \overbrace{\frac{a_1(a_0+a_2)}{2b_2}}^{\alpha_1}\cos(\Omega) + \overbrace{\frac{(a_0-a_2)^2+a_1^2}{4b_2}}^{\alpha_2}}{\cos^2(\Omega) + \underbrace{\frac{b_1(1+b_2)}{2b_2}}_{\beta_1}\cos(\Omega) + \underbrace{\frac{(1-b_2)^2+b_1^2}{4b_2}}_{\beta_2}} = S^2. \tag{6.123}$$

With $x = \cos(\Omega)$, it follows that

$$(S^2 - \alpha_0)x^2 + (\beta_1 S^2 - \alpha_1)x + (\beta_2 S^2 - \alpha_2) = 0. \tag{6.124}$$

The solution of Eq. (6.124) leads to $x = \cos(\Omega_{max/min})$, which must be real ($-1 \leq x \leq 1$) for the maximum/minimum to occur at a real frequency. For a single solution (repeated roots) of the above quadratic equation, the discriminant must be $D = (p/2)^2 - q = 0\,(x^2 + px + q = 0)$. It follows that

$$D = \frac{(\beta_1 S^2 - \alpha_1)^2}{4(S^2 - \alpha_0)^2} - \frac{\beta_2 S^2 - \alpha_2}{S^2 - \alpha_0} = 0 \tag{6.125}$$

and

$$S^4(\beta_1^2 - 4\beta_2) + S^2(4\alpha_2 + 4\alpha_0\beta_2 - 2\alpha_1\beta_1) + (\alpha_1^2 - 4\alpha_0\alpha_2) = 0. \tag{6.126}$$

The solution of Eq. (6.126) gives two solutions for S^2. The solution with the larger value is chosen. If the discriminant D is not greater than zero, the maximum lies at $x = 1$ ($z = 1$) or $x = -1$ ($z = -1$), as given by

$$S^2 = \frac{\alpha_0 + \alpha_1 + \alpha_2}{1 + \beta_1 + \beta_2} \tag{6.127}$$

or

$$S^2 = \frac{\alpha_0 - \alpha_1 + \alpha_2}{1 - \beta_1 + \beta_2}. \tag{6.128}$$

Limit Cycles

Limit cycles are periodic processes in a filter which can be measured as sinusoidal signals. They arise owing to the quantization of state variables. The different types of limit cycle and the methods necessary to prevent them are briefly listed below:

- overflow limit cycles
  → saturation curve
  → scaling

- limit cycles for vanishing input
  → noise shaping
  → dithering

- limit cycles correlated with the input signal
  → noise shaping
  → dithering

6.3 Non-recursive Audio Filters

For implementing linear phase audio filters, non-recursive filters are used. The basis of an efficient implementation is the *fast convolution*

$$y(n) = x(n) * h(n) \circ\!\!-\!\!\bullet\; Y(k) = X(k) \cdot H(k), \tag{6.129}$$

where the convolution in the time domain is performed by transforming the signal and the impulse response into the frequency domain, multiplication of the corresponding Fourier transforms, and inverse Fourier transform of the product into the time domain signal (see Fig. 6.55). The transform is carried out by a discrete Fourier transform of length N, such that $N = N_1 + N_2 - 1$ is valid and time-domain aliasing is avoided.

First we will discuss the basics. We then introduce the convolution of long sequences followed by a filter design for linear phase filters.

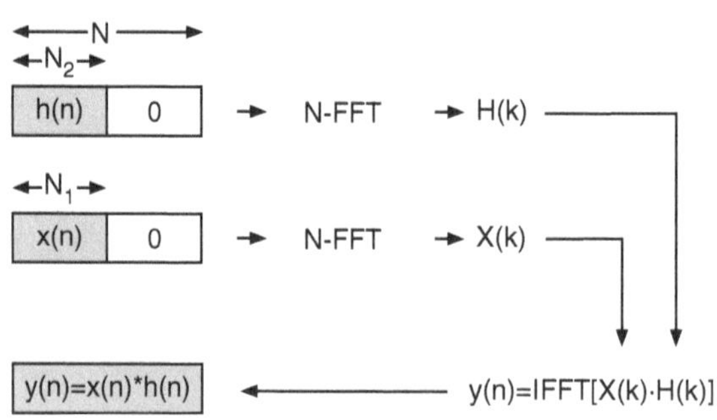

Figure 6.55 Fast convolution of signal $x(n)$ of length N_1 and impulse response $h(n)$ of length N_2 delivers the convolution result $y(n) = x(n) * h(n)$ of length $N_1 + N_2 - 1$.

6.3.1 Basics of Fast Convolution

IDFT Implementation with DFT Algorithm. The discrete Fourier transformation (DFT) is described by

$$X(k) = \sum_{n=0}^{N-1} x(n)W_N^{nk} = \mathrm{DFT}_k[x(n)], \tag{6.130}$$

$$W_N = e^{-j2\pi/N}, \tag{6.131}$$

and the inverse discrete Fourier transformation (IDFT) by

$$x(n) = \frac{1}{N}\sum_{k=0}^{N-1} X(k)W_N^{-nk}. \tag{6.132}$$

Without scaling factor $1/N$, we write

$$x'(n) = \sum_{k=0}^{N-1} X(k)W_N^{-nk} = \mathrm{IDFT}_n[X(k)], \tag{6.133}$$

so that the following symmetrical transformation algorithms hold:

$$X'(k) = \frac{1}{\sqrt{N}}\sum_{n=0}^{N-1} x(n)W_N^{nk}, \tag{6.134}$$

$$x(n) = \frac{1}{\sqrt{N}}\sum_{k=0}^{N-1} X'(k)W_N^{-nk}. \tag{6.135}$$

The IDFT differs from the DFT only by its sign in the exponential term.

An alternative approach for calculating the IDFT with the help of a DFT is described as follows [Cad87, Duh88]. We will make use of the relationships

$$x(n) = a(n) + j \cdot b(n), \tag{6.136}$$

$$j \cdot x^*(n) = b(n) + j \cdot a(n). \tag{6.137}$$

Conjugating Eq. (6.133) gives

$$x'^*(n) = \sum_{k=0}^{N-1} X^*(k)W_N^{nk}. \tag{6.138}$$

The multiplication of Eq. (6.138) by j leads to

$$j \cdot x'^*(n) = \sum_{k=0}^{N-1} j \cdot X^*(k)W_N^{nk}. \tag{6.139}$$

Conjugating and multiplying Eq. (6.139) by j results in

$$x'(n) = j \cdot \left[\sum_{k=0}^{N-1} (j \cdot X^*(k)W_N^{nk}\right]^*. \tag{6.140}$$

An interpretation of Eqs. (6.137) and (6.140) suggests the following way of performing the IDFT with the DFT algorithm:

1. exchange the real with the imaginary part of the spectral sequence

$$Y(k) = Y_I(k) + jY_R(k);$$

2. transformation with DFT algorithm

$$\mathrm{DFT}[Y(k)] = y_I(n) + jy_R(n);$$

3. exchange the real with the imaginary part of the time sequence

$$y(n) = y_R(n) + jy_I(n).$$

For implementation on a digital signal processor, the use of DFT saves memory for IDFT.

Discrete Fourier Transformation of Two Real Sequences. In many applications, stereo signals that consist of a left and right channel are processed. With the help of the DFT, both channels can be transformed simultaneously into the frequency domain [Sor87, Ell82].

For a real sequence $x(n)$,

$$X(k) = X^*(-k) \qquad k = 0, 1, \ldots, N-1 \tag{6.141}$$

$$= X^*(N-k). \tag{6.142}$$

For a discrete Fourier transformation of two real sequences $x(n)$ and $y(n)$, a complex sequence is first formed according to

$$z(n) = x(n) + jy(n). \tag{6.143}$$

The Fourier transformation gives

$$\mathrm{DFT}[z(n)] = \mathrm{DFT}[x(n) + jy(n)]$$

$$= Z_R(k) + jZ_I(k) \tag{6.144}$$

$$= Z(k), \tag{6.145}$$

where

$$Z(k) = Z_R(k) + jZ_I(k) \tag{6.146}$$

$$= X_R(k) + jX_I(k) + j[Y_R(k) + jY_I(k)] \tag{6.147}$$

$$= X_R(k) - Y_I(k) + j[X_I(k) + Y_R(k)]. \tag{6.148}$$

Because $x(n)$ and $y(n)$ are real sequences, it follows from Eq. (6.142) that

$$Z(N-k) = Z_R(N-k) + jZ_I(N-k) = Z^*(k) \tag{6.149}$$

$$= X_R(k) - jX_I(k) + j[Y_R(k) - jY_I(k)] \tag{6.150}$$

$$= X_R(k) + Y_I(k) - j[X_I(k) - Y_R(k)]. \tag{6.151}$$

Considering the real part of $Z(k)$, adding Eqs. (6.148) and (6.151) gives

$$2X_R(k) = Z_R(k) + Z_R(N-k) \tag{6.152}$$

$$\rightarrow X_R(k) = \frac{1}{2}[Z_R(k) + Z_R(N-k)] \tag{6.153}$$

and subtraction of Eq. (6.151) from Eq. (6.148) results in

$$2Y_I(k) = Z_R(N-k) - Z_R(k) \tag{6.154}$$

$$\rightarrow Y_I(k) = \frac{1}{2}[Z_R(N-k) - Z_R(k)]. \tag{6.155}$$

Considering the imaginary part of $Z(k)$, adding Eqs. (6.148) and (6.151) gives

$$2Y_R(k) = Z_I(k) + Z_I(N-k) \tag{6.156}$$

$$\rightarrow Y_R(k) = \frac{1}{2}[Z_I(k) + Z_I(N-k)] \tag{6.157}$$

and subtraction of Eq. (6.151) from Eq. (6.148) results in

$$2X_I(k) = Z_I(k) - Z_I(N-k) \tag{6.158}$$

$$\rightarrow X_I(k) = \frac{1}{2}[Z_I(k) - Z_I(N-k)]. \tag{6.159}$$

Hence the spectral functions are given by

$$X(k) = \text{DFT}[x(n)] = X_R(k) + jX_I(k) \tag{6.160}$$

$$= \frac{1}{2}[Z_R(k) + Z_R(N-k)]$$
$$+j\frac{1}{2}[Z_I(k) - Z_I(N-k)]$$

$$k = 0, 1, \ldots, \frac{N}{2}, \tag{6.161}$$

$$Y(k) = \text{DFT}[y(n)] = Y_R(k) + jY_R(k) \tag{6.162}$$

$$= \frac{1}{2}[Z_I(k) + Z_I(N-k)]$$
$$+j\frac{1}{2}[Z_R(N-k) - Z_R(k)]$$

$$k = 0, 1, \ldots, \frac{N}{2}, \tag{6.163}$$

and

$$X_R(k) + jX_I(k) = X_R(N-k) - jX_I(N-k), \tag{6.164}$$

$$Y_R(k) + jY_I(k) = Y_R(N-k) - jY_I(N-k),$$

$$k = \frac{N}{2} + 1, \ldots, N-1. \tag{6.165}$$

Fast Convolution if Spectral Functions are Known. The spectral functions $X(k)$, $Y(k)$, and $H(k)$ are known. With the help of Eq. (6.148), the spectral sequence can be formed by

$$Z(k) = Z_R(k) + jZ_I(k) \tag{6.166}$$

$$= X_R(k) - Y_I(k) + j[X_I(k) + Y_R(k)]$$

$$k = 0, 1, \ldots, N-1. \tag{6.167}$$

Filtering is done by multiplication in the frequency domain:

$$Z'(k) = [Z_R(k) + jZ_I(k)][H_R(k) + jH_I(k)]$$

$$= Z_R(k)H_R(k) - Z_I(k)H_I(k)$$

$$+ j[Z_R(k)H_I(k) + Z_I(k)H_R(k)]. \tag{6.168}$$

The inverse transformation gives

$$z'(n) = [x(n) + jy(n)] * h(n) = x(n) * h(n) + jy(n) * h(n) \tag{6.169}$$

$$= \mathrm{IDFT}[Z'(k)]$$

$$= z'_R(n) + jz'_I(n), \tag{6.170}$$

so that the filtered output sequence is given by

$$x'(n) = z'_R(n), \tag{6.171}$$

$$y'(n) = z'_I(n). \tag{6.172}$$

The filtering of a stereo signal can hence be done by transformation into the frequency domain, multiplication of the spectral functions, and inverse transformation of left and right channels.

6.3.2 Fast Convolution of Long Sequences

The fast convolution of two real input sequences $x_l(n)$ and $x_{l+1}(n)$ of length N_1 with the impulse response $h(n)$ of length N_2 leads to the output sequences

$$y_l(n) = x_l(n) * h(n), \tag{6.173}$$

$$y_{l+1}(n) = x_{l+1}(n) * h(n), \tag{6.174}$$

of length $N_1 + N_2 - 1$. The implementation of a non-recursive filter with fast convolution becomes more efficient than the direct implementation of an FIR filter for filter lengths $N > 30$. Therefore, the following procedure will be performed:

- formation of a complex sequence

$$z(n) = x_l(n) + jx_{l+1}(n); \tag{6.175}$$

- Fourier transformation of the impulse response $h(n)$ that is padded with zeros to a length $N \geq N_1 + N_2 - 1$

$$H(k) = \mathrm{DFT}[h(n)] \quad \text{(FFT-length } N\text{)}; \tag{6.176}$$

- Fourier transformation of the sequence $z(n)$ that is padded with zeros to a length $N \geq N_1 + N_2 - 1$

$$Z(k) = \mathrm{DFT}[z(n)] \quad \text{(FFT-length } N\text{)}; \tag{6.177}$$

- formation of a complex output sequence

$$e(n) = \mathrm{IDFT}[Z(k)H(k)] \tag{6.178}$$

$$= z(n) * h(n) \tag{6.179}$$

$$= x_l(n) * h(n) + jx_{l+1}(n) * h(n); \tag{6.180}$$

- formation of a real output sequence

$$y_l(n) = \mathrm{Re}\{e(n)\}, \tag{6.181}$$

$$y_{l+1}(n) = \mathrm{Im}\{e(n)\}. \tag{6.182}$$

For the convolution of an infinite-length input sequence (see Fig. 6.56) with an impulse response $h(n)$, the input sequence is partitioned into sequences $x_m(n)$ of length L:

$$x_m(n) = \begin{cases} x(n) & (m-1)L \leq n \leq mL - 1, \\ 0 & \text{otherwise.} \end{cases} \tag{6.183}$$

The input sequence is given by superposition of finite-length sequences according to

$$x(n) = \sum_{m=1}^{\infty} x_m(n). \tag{6.184}$$

The convolution of the input sequence with the impulse response $h(n)$ of length M gives

$$y(n) = \sum_{k=0}^{M-1} h(k)x(n-k) \tag{6.185}$$

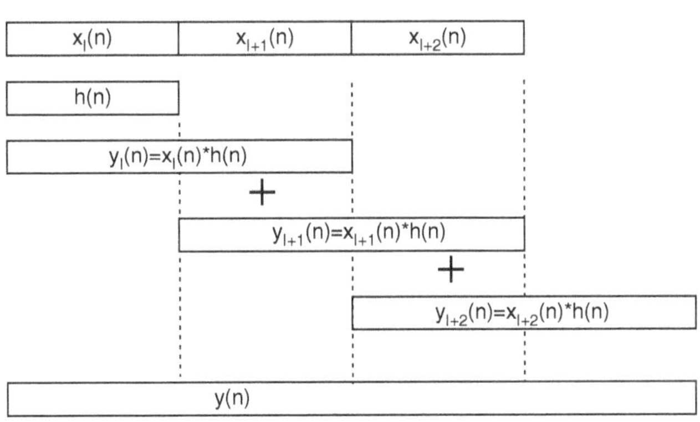

Figure 6.56 Fast convolution with partitioning of the input signal $x(n)$ into blocks of length L.

$$= \sum_{k=0}^{M-1} h(k) \sum_{m=1}^{\infty} x_m(n-k) \tag{6.186}$$

$$= \sum_{m=1}^{\infty} \left[\sum_{k=0}^{M-1} h(k) x_m(n-k) \right]. \tag{6.187}$$

The term in brackets corresponds to the convolution of a finite-length sequence $x_m(n)$ of length L with the impulse response of length M. The output signal can be given as a superposition of convolution products of length $L + M - 1$. With these partial convolution products,

$$y_m(n) = \begin{cases} \sum_{k=0}^{M-1} h(k) x_m(n-k) & (m-1)L \le n \le mL + M - 2, \\ 0 & \text{otherwise}, \end{cases} \tag{6.188}$$

the output signal can be written as

$$y(n) = \sum_{m=1}^{\infty} y_m(n). \tag{6.189}$$

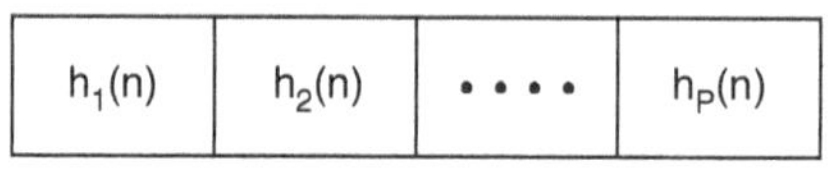

Figure 6.57 Partitioning of the impulse response $h(n)$.

If the length M of the impulse response is very long, it can be similarly partitioned into P parts each of length M/P (see Fig. 6.57). With

$$h_p\left[n - (p-1)\frac{M}{P}\right] = \begin{cases} h(n) & (p-1)\frac{M}{P} \le n \le p\frac{M}{P} - 1, \\ 0 & \text{otherwise}, \end{cases} \tag{6.190}$$

it follows that

$$h(n) = \sum_{p=1}^{P} h_p\left(n - (p-1)\frac{M}{P}\right). \tag{6.191}$$

With $M_p = pM/P$ and Eq. (6.189), the following partitioning can be done:

$$y(n) = \sum_{m=1}^{\infty} \left[\underbrace{\sum_{k=0}^{M-1} h(k) x_m(n-k)}_{y_m(n)} \right] \tag{6.192}$$

$$= \sum_{m=1}^{\infty} \left[\sum_{k=0}^{M_1-1} h(k) x_m(n-k) + \sum_{k=M_1}^{M_2-1} h(k) x_m(n-k) + \ldots + \sum_{k=M_{P-1}}^{M-1} h(k) x_m(n-k) \right]. \tag{6.193}$$

This can be rewritten as

$$y(n) = \sum_{m=1}^{\infty} \Bigg[\underbrace{\sum_{k=0}^{M_1-1} h_1(k)x_m(n-k)}_{y_{m1}} + \underbrace{\sum_{k=0}^{M_1-1} h_2(k)x_m(n-M_1-k)}_{y_{m2}}$$

$$+ \underbrace{\sum_{k=0}^{M_1-1} h_3(k)x_m(n-2M_1-k)}_{y_{m3}}$$

$$\ldots + \underbrace{\sum_{k=0}^{M_1-1} h_P(k)x_m(n-(P-1)M_1-k)}_{y_{mP}} \Bigg]$$

$$= \sum_{m=1}^{\infty} \underbrace{[y_{m1}(n) + y_{m2}(n-M_1) + \ldots + y_{mP}(n-(P-1)M_1)]}_{y_m(n)}. \tag{6.194}$$

An example of partitioning the impulse response into $P = 4$ parts is graphically shown in Fig. 6.58. This leads to

$$y(n) = \sum_{m=1}^{\infty} \Bigg[\underbrace{\sum_{k=0}^{M_1-1} h_1(k)x_m(n-k)}_{y_{m1}} + \underbrace{\sum_{k=0}^{M_1-1} h_2(k)x_m(n-M_1-k)}_{y_{m2}}$$

$$+ \underbrace{\sum_{k=0}^{M_1-1} h_3(k)x_m(n-2M_1-k)}_{y_{m3}} + \underbrace{\sum_{k=0}^{M_1-1} h_4(k)x_m(n-3M_1-k)}_{y_{m4}} \Bigg]$$

$$= \sum_{m=1}^{\infty} \underbrace{[y_{m1}(n) + y_{m2}(n-M_1) + y_{m3}(n-2M_1) + y_{m4}(n-3M_1)]}_{y_m(n)}. \tag{6.195}$$

The procedure of a fast convolution by partitioning the input sequence $x(n)$, as well as the impulse response $h(n)$, is given in the following for the example in Fig. 6.58.

1. Decomposition of the impulse response $h(n)$ of length $4M$:

$$h_1(n) = h(n) \qquad 0 \le n \le M-1; \tag{6.196}$$

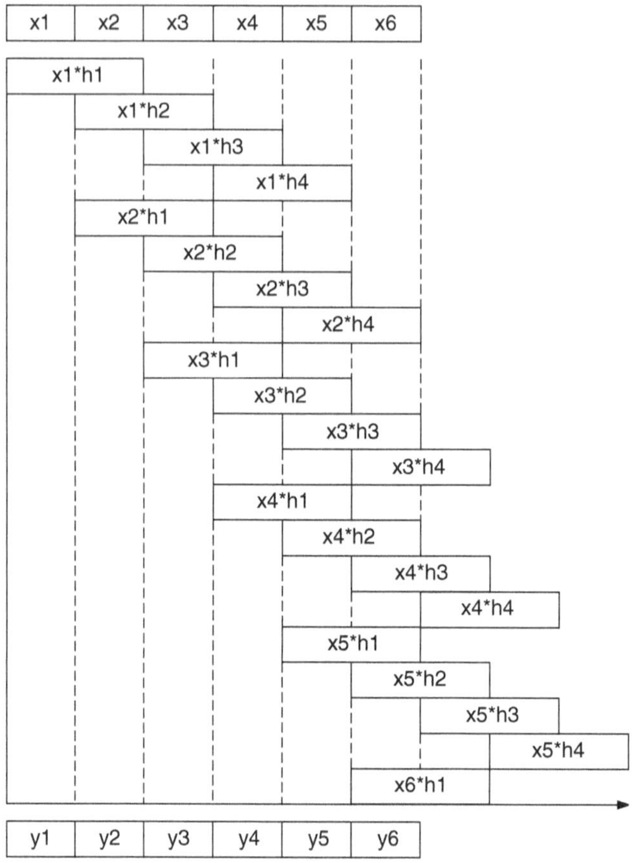

Figure 6.58 Scheme for a fast convolution with $P = 4$.

$$h_2(n - M) = h(n) \qquad M \leq n \leq 2M - 1; \tag{6.197}$$

$$h_3(n - 2M) = h(n) \qquad 2M \leq n \leq 3M - 1; \tag{6.198}$$

$$h_4(n - 3M) = h(n) \qquad 3M \leq n \leq 4M - 1. \tag{6.199}$$

2. Zero-padding of partial impulse responses up to a length $2M$:

$$h_1(n) = \begin{cases} h_1(n) & 0 \leq n \leq M - 1, \\ 0 & M \leq n \leq 2M - 1; \end{cases} \tag{6.200}$$

$$h_2(n) = \begin{cases} h_2(n) & 0 \leq n \leq M - 1, \\ 0 & M \leq n \leq 2M - 1; \end{cases} \tag{6.201}$$

$$h_3(n) = \begin{cases} h_3(n) & 0 \leq n \leq M - 1, \\ 0 & M \leq n \leq 2M - 1; \end{cases} \tag{6.202}$$

$$h_4(n) = \begin{cases} h_4(n) & 0 \leq n \leq M - 1, \\ 0 & M \leq n \leq 2M - 1. \end{cases} \tag{6.203}$$

3. Calculating and storing

$$H_i(k) = \text{DFT}[h_i(n)] \quad i = 1, \dots, 4 \quad \text{(FFT-length } 2M\text{)}. \tag{6.204}$$

4. Decomposition of the input sequence $x(n)$ into partial sequences $x_l(n)$ of length M:

$$x_l(n) = x(n) \qquad (l-1)M \leq n \leq lM - 1 \quad l = 1, \dots \infty. \tag{6.205}$$

5. Nesting partial sequences:

$$z_m(n) = x_l(n) + jx_{l+1}(n) \quad m = 1, \dots, \infty. \tag{6.206}$$

6. Zero-padding of complex sequence $z_m(n)$ up to a length $2M$:

$$z_m(n) = \begin{cases} z_m(n) & (l-1)M \leq n \leq lM - 1, \\ 0 & lM \leq n \leq (l+1)M - 1. \end{cases} \tag{6.207}$$

7. Fourier transformation of the complex sequences $z_m(n)$:

$$Z_m(k) = \text{DFT}[z_m(n)] = Z_{mR}(k) + jZ_{mI}(k) \quad \text{(FFT-length } 2M\text{)}. \tag{6.208}$$

8. Multiplication in the frequency domain:

$$\begin{aligned}[Z_R(k) + jZ_I(k)][H_R(k) + jH_I(k)] = \\ Z_R(k)H_R(k) - Z_I(k)H_I(k) \\ +j[Z_R(k)H_I(k) + Z_I(k)H_R(k)]\end{aligned} \tag{6.209}$$

$$E_{m1}(k) = Z_m(k)H_1(k) \quad k = 0, 1, \dots, 2M-1; \tag{6.210}$$

$$E_{m2}(k) = Z_m(k)H_2(k) \quad k = 0, 1, \dots, 2M-1; \tag{6.211}$$

$$E_{m3}(k) = Z_m(k)H_3(k) \quad k = 0, 1, \dots, 2M-1; \tag{6.212}$$

$$E_{m4}(k) = Z_m(k)H_4(k) \quad k = 0, 1, \dots, 2M-1. \tag{6.213}$$

9. Inverse transformation:

$$e_{m1}(n) = \text{IDFT}[Z_m(k)H_1(k)] \quad n = 0, 1, \dots, 2M-1; \tag{6.214}$$

$$e_{m2}(n) = \text{IDFT}[Z_m(k)H_2(k)] \quad n = 0, 1, \dots, 2M-1; \tag{6.215}$$

$$e_{m3}(n) = \text{IDFT}[Z_m(k)H_3(k)] \quad n = 0, 1, \dots, 2M-1; \tag{6.216}$$

$$e_{m4}(n) = \text{IDFT}[Z_m(k)H_4(k)] \quad n = 0, 1, \dots, 2M-1. \tag{6.217}$$

10. Determination of partial convolutions:

$$\text{Re}\{e_{m1}(n)\} = x_l * h_1; \tag{6.218}$$

$$\text{Im}\{e_{m1}(n)\} = x_{l+1} * h_1; \tag{6.219}$$

$$\text{Re}\{e_{m2}(n)\} = x_l * h_2; \tag{6.220}$$

$$\text{Im}\{e_{m2}(n)\} = x_{l+1} * h_2; \tag{6.221}$$

$$\text{Re}\{e_{m3}(n)\} = x_l * h_3; \tag{6.222}$$

$$\mathrm{Im}\{e_{m3}(n)\} = x_{l+1} * h_3; \quad (6.223)$$

$$\mathrm{Re}\{e_{m4}(n)\} = x_l * h_4; \quad (6.224)$$

$$\mathrm{Im}\{e_{m4}(n)\} = x_{l+1} * h_4. \quad (6.225)$$

11. *Overlap-Add* of partial sequences, increments $l = l + 2$ and $m = m + 1$, and back to step 5.

Based on the partitioning of the input signal and the impulse response and the following Fourier transform, the result of each single convolution is only available after a delay of one block of samples. Different methods to reduce computational complexity or overcome the block delay have been proposed [Soo90, Gar95, Ege96, Mül99, Mül01, Garc02]. These methods make use of a hybrid approach where the first part of the impulse response is used for time-domain convolution and the other parts are used for fast convolution in the frequency domain. Figure 6.59a,b demonstrates a simple derivation of the hybrid convolution scheme, which can be described by the decomposition of the transfer function according to

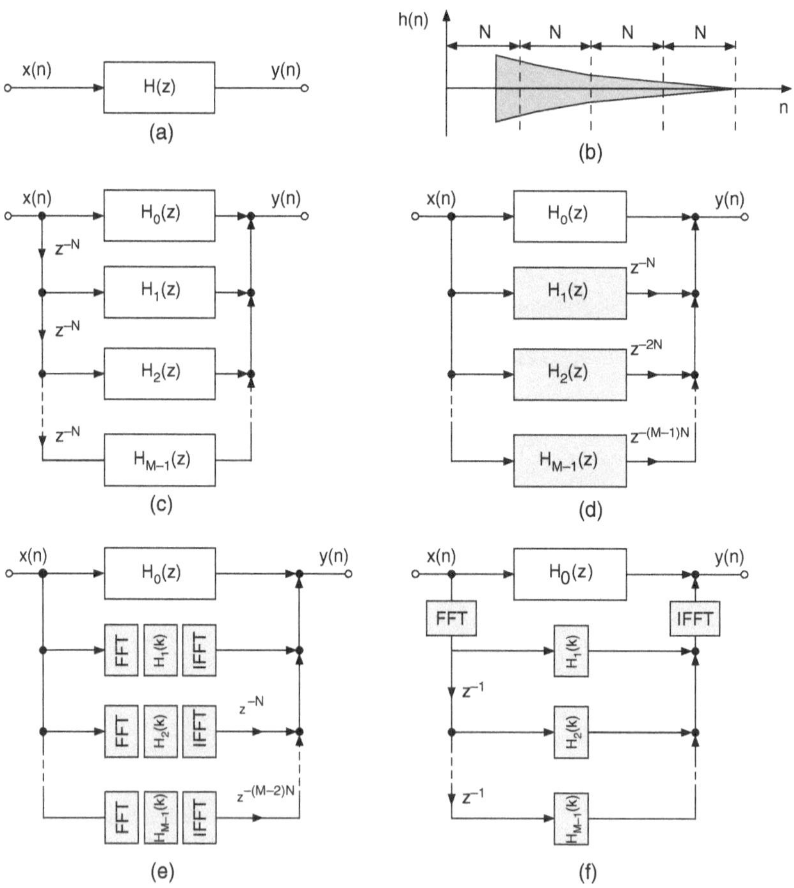

Figure 6.59 Hybrid fast convolution.

$$H(z) = \sum_{i=0}^{M-1} z^{-iN} H_i(z), \tag{6.226}$$

where the impulse response has length $M \cdot N$ and M is the number of smaller partitions of length N. Figure 6.59c,d shows two different signal flow graphs for the decomposition given by Eq. (6.226) of the entire transfer function. Especially, Fig. 6.59d highlights (with gray background) that in each branch $i = 1, \dots, M-1$, a delay of $i \cdot N$ occurs and each filter $H_i(z)$ has the same length and makes use of the same state variables. This means that they can be computed in parallel in the frequency domain with $2N$-FFTs/IFFTs and the outputs have to be delayed according to $(i-1) \cdot N$, as shown in Fig. 6.59e. A further simplification shown in Fig. 6.59f leads to one input $2N$-FFT and block delays z^{-1} for the frequency vectors. Then, parallel multiplications with $H_i(k)$ of length $2N$ and the summation of all intermediate products are performed before one output $2N$-IFFT for the overlap and add operation in the time domain is used. The first part of the impulse response, represented by $H_0(z)$, is performed by direct convolution in the time domain. The frequency- and time-domain parts are then overlapped and added. An alternative realization for fast convolution is based on the overlap and save operation.

6.3.3 Filter Design by Frequency Sampling

Audio filter design for non-recursive filter realizations by fast convolution can be carried out by the frequency sampling method. For linear phase systems,

$$H(e^{j\Omega}) = A(e^{j\Omega}) e^{-j\frac{N_F-1}{2}\Omega}, \tag{6.227}$$

where $A(e^{j\Omega})$ is a real valued amplitude response and N_F is the length of the impulse response. The magnitude $|H(e^{j\Omega})|$ is calculated by sampling in the frequency domain at equidistant places

$$\frac{f}{f_S} = \frac{k}{N_F} \quad \text{with} \quad k = 0, 1, \cdots, N_F - 1, \tag{6.228}$$

according to

$$|H(e^{j\Omega})| = A(e^{j2\pi k/N_F}) \qquad k = 0, 1, \cdots, \frac{N_F}{2} - 1. \tag{6.229}$$

Hence, a filter can be designed by fulfilling conditions in the frequency domain. The linear phase is determined as

$$e^{-j\frac{N_F-1}{2}\Omega} = e^{-j2\pi\frac{N_F-1}{2}\frac{k}{N_F}} \tag{6.230}$$

$$= \cos(2\pi\frac{N_F-1}{2}\frac{k}{N_F}) - j\sin(2\pi\frac{N_F-1}{2}\frac{k}{N_F})$$

$$k = 0, 1, \cdots, \frac{N_F}{2} - 1. \tag{6.231}$$

Owing to the real transfer function $H(z)$ for an even filter length, we have to fulfill

$$H(k = \frac{N_F}{2}) = 0 \wedge H(k) = H^*(N_F - k) \qquad k = 0, 1, \cdots \frac{N_F}{2} - 1. \tag{6.232}$$

This has to be taken into consideration when designing filters of even length N_F. The impulse response $h(n)$ is obtained through an N_F-point IDFT of the spectral sequence $H(k)$. This impulse

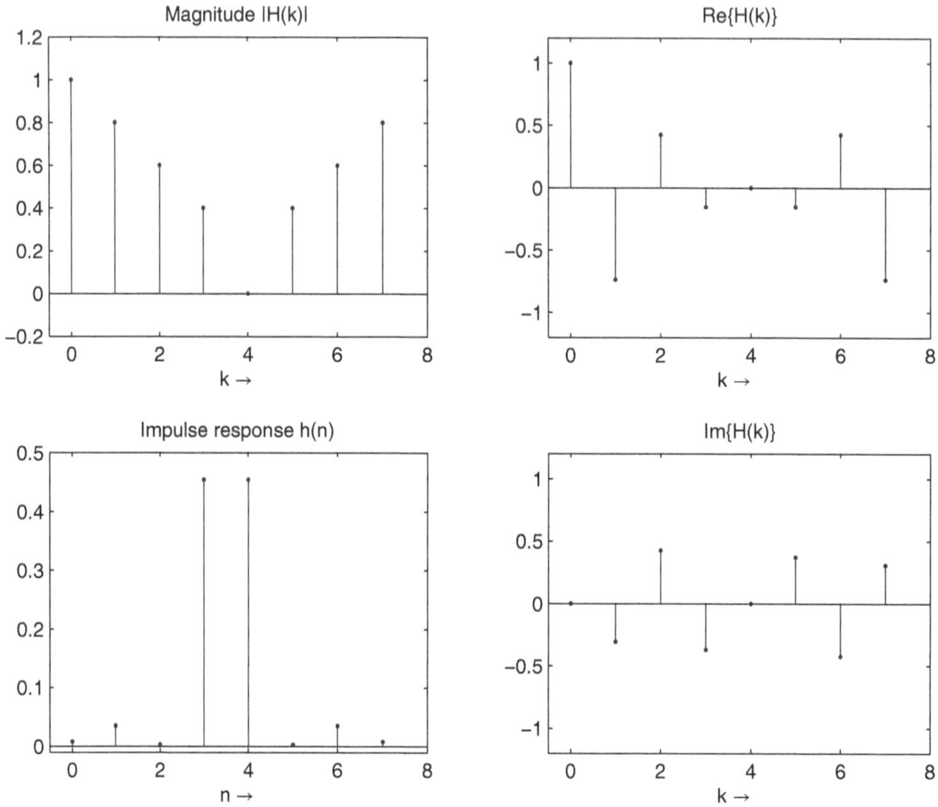

Figure 6.60 Filter design by frequency sampling (N_F even).

response is extended with *zero-padding* to the length N and then transformed by an N-point DFT resulting in the spectral sequence $H(k)$ of the filter.

Example: For $N_F = 8$, $|H(k)| = 1$ $(k = 0, 1, 2, 3, 5, 6, 7)$ and $|H(4)| = 0$, the group delay is $t_G = 3.5$. Figure 6.60 shows the amplitude, real part, and imaginary part of the transfer function and the impulse response $h(n)$.

6.4 Multi-complementary Filter Bank

The sub-band processing of audio signals is mainly used in source coding applications for efficient transmission and storing. The basis for the sub-band decomposition is critically sampled filter banks [Vai93, Fli00]. These filter banks allow a perfect reconstruction of the input provided there is no processing within the sub-bands. They consist of an analysis filter bank for decomposing the signal in critically sampled sub-bands and a synthesis filter bank for reconstructing the broadband output. The aliasing in the sub-bands is eliminated by the synthesis filter bank. Nonlinear methods are used for coding the sub-band signals. The reconstruction error of the filter bank is negligible compared with the errors arising from the coding/decoding process. Using a critically sampled filter bank as a multiband equalizer, multiband dynamic range control, or multiband room simulation, the processing in the sub-bands leads to aliasing at the output. To avoid aliasing, a multi-complementary filter bank [Fli92, Zöl92, Fli00] is presented which enables an aliasing-free processing in the sub-bands and leads to a perfect reconstruction of the output. It allows a decomposition into octave frequency bands which are matched to the human ear.

6.4.1 Principles

Figure 6.61 shows an octave-band filter bank with *critical sampling*. It performs a successive lowpass/highpass decomposition into half-bands followed by downsampling by a factor of two. The decomposition leads to the sub-bands Y_1 to Y_N (see Fig. 6.62). The transition frequencies of this decomposition are given by

$$\Omega_{Ck} = \frac{\pi}{2} 2^{-k+1} \quad \text{with} \quad k = 1, 2, \cdots, N-1. \tag{6.233}$$

To avoid aliasing in sub-bands, a modified octave-band filter bank is considered, which is shown in Fig. 6.63 for a two-band decomposition. The cutoff frequency of the modified filter bank is moved from $\frac{\pi}{2}$ to a lower frequency. This means that in downsampling the lowpass branch, no aliasing occurs in the transition band (e.g. cutoff frequency $\frac{\pi}{3}$). The broader highpass branch cannot be downsampled. A continuation of the described two-band decomposition leads to the modified octave-band filter bank shown in Fig. 6.64. The frequency bands are depicted in Fig. 6.65 showing that besides the cutoff frequencies

$$\Omega_{Ck} = \frac{\pi}{3} 2^{-k+1} \quad \text{with} \quad k = 1, 2, \cdots, N-1, \tag{6.234}$$

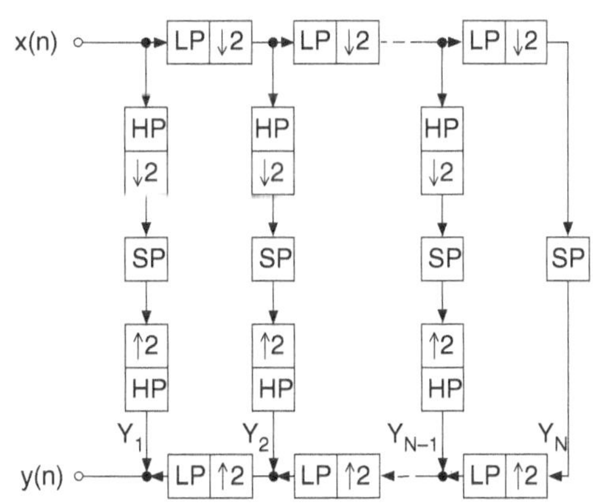

Figure 6.61 Octave-band QMF bank (SP = signal processing, LP = lowpass, HP = highpass).

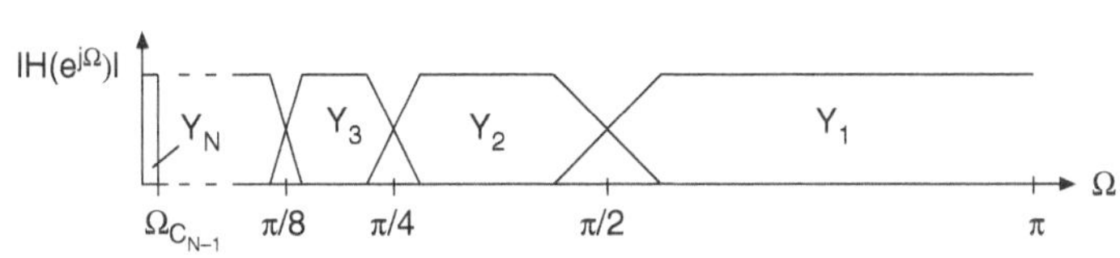

Figure 6.62 Octave-frequency bands.

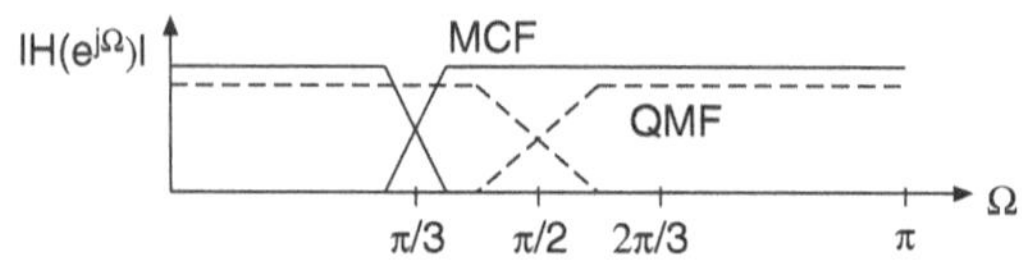

Figure 6.63 Two-band decomposition.

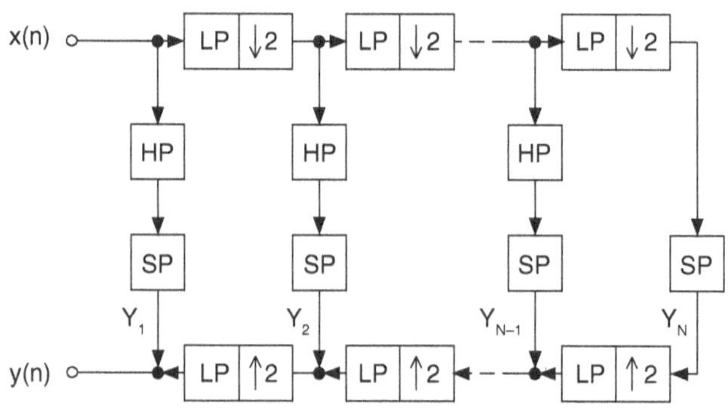

Figure 6.64 Modified octave-band filter bank.

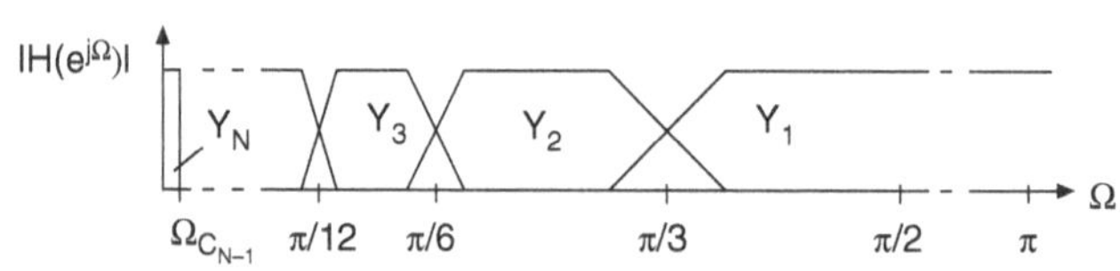

Figure 6.65 Modified octave decomposition.

the bandwidth of the sub-bands is reduced by a factor of two. The highpass sub-band Y_1 is an exception.

The special lowpass/highpass decomposition is carried out by a two-band complementary filter bank, as shown in Fig. 6.66. The frequency responses of a decimation filter $H_D(z)$, interpolation filter $H_I(z)$, and kernel filter $H_K(z)$ are shown in Fig. 6.67.

The lowpass filtering of a signal $x_1(n)$ is done with the help of a decimation filter $H_D(z)$, the downsampler of factor two, and the kernel filter $H_K(z)$, and leads to $y_2(2n)$. The Z-transform of $y_2(2n)$ is given by

$$Y_2(z) = \frac{1}{2}[H_D(z^{\frac{1}{2}})X_1(z^{\frac{1}{2}})H_K(z) \\ +H_D(-z^{\frac{1}{2}})X_1(-z^{\frac{1}{2}})H_K(z)]. \tag{6.235}$$

The interpolated lowpass signal $y_{1L}(n)$ is generated by upsampling by a factor of two and filtering with the interpolation filter $H_I(z)$. The Z-transform of $y_{1L}(n)$ is given by

$$Y_{1L}(z) = Y_2(z^2)H_I(z) \tag{6.236}$$

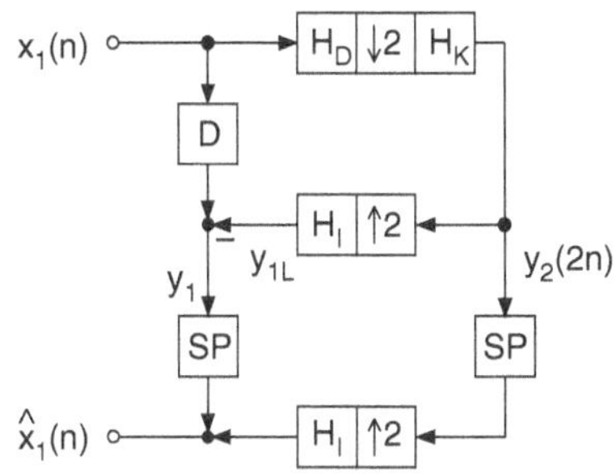

Figure 6.66 Two-band complementary filter bank.

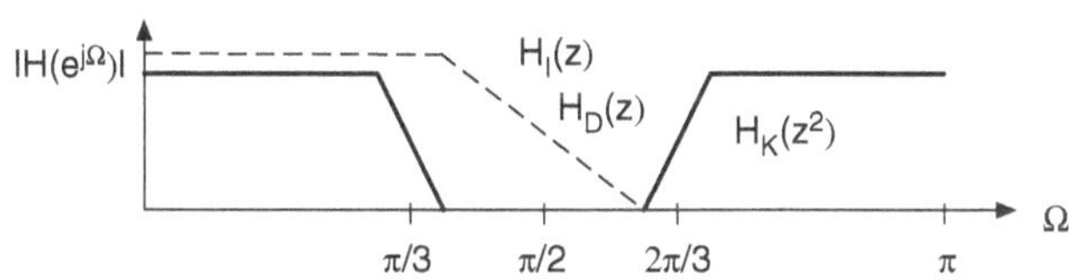

Figure 6.67 Design of $H_D(z)$, $H_I(z)$, and $H_K(z)$.

$$= \underbrace{\frac{1}{2}H_D(z)H_I(z)H_K(z^2)X_1(z)}_{G_1(z)}$$

$$+ \underbrace{\frac{1}{2}H_D(-z)H_I(z)H_K(z^2)X_1(-z)}_{G_2(z)}. \tag{6.237}$$

The highpass signal $y_1(n)$ is obtained by subtracting the interpolated lowpass signal $y_{1L}(n)$ from the delayed input signal $x_1(n-D)$. The Z-transform of the highpass signal is given by

$$Y_1(z) = z^{-D}X_1(z) - Y_{1L}(z) \tag{6.238}$$

$$= [z^{-D} - G_1(z)]X_1(z) - G_2(z)X_1(-z). \tag{6.239}$$

The lowpass and highpass signals are processed individually. The output signal $\hat{x}_1(n)$ is formed by adding the highpass signal to the upsampled and filtered lowpass signal. With Eqs. (6.237) and (6.239), the Z-transform of $\hat{x}_1(n)$ can be written as

$$\hat{X}_1(z) = Y_{1L}(z) + Y_1(z) = z^{-D}X_1(z). \tag{6.240}$$

Equation (6.240) shows the perfect reconstruction of the input signal which is delayed by D sampling units.

The extension to N sub-bands and implementing the kernel filter using complementary techniques [Ram88, Ram90] leads to the multi-complementary filter bank, as shown in Fig. 6.68. Delays are integrated in the highpass (Y_1) and bandpass sub-bands (Y_2 to Y_{N-2}) to compensate for the group delay. The filter structure consists of N horizontal stages. The kernel filter is implemented as a complementary filter in S vertical stages. The design of the latter shall be discussed

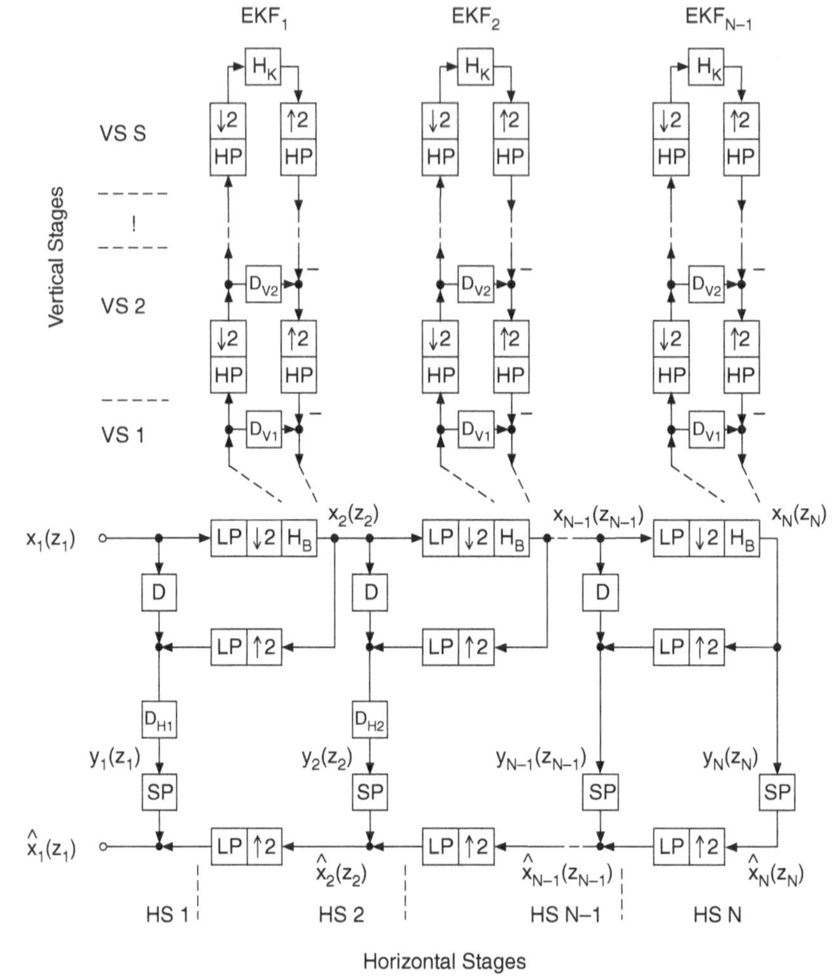

Figure 6.68 Multi-complementary filter bank.

later. The vertical delays in the extended kernel filters (EKF_1 to EKF_{N-1}) compensate group delays caused by forming the complementary component. At the end of each of these vertical stages is the kernel filter H_K. With

$$z_k = z^{2^{-(k-1)}} \quad \text{and} \quad k = 1, \cdots, N, \tag{6.241}$$

the signals $\hat{X}_k(z_k)$ can be written as a function of the signals $X_k(z_k)$ as

$$\hat{\mathbf{X}} = \mathrm{diag}\, [z_1^{-D_1} \quad z_2^{-D_2} \cdots \quad z_N^{-D_N}]\mathbf{X}, \tag{6.242}$$

with

$$\begin{aligned} \hat{\mathbf{X}} &= [\hat{X}_1(z_1) \;\; \hat{X}_2(z_2) \;\; \cdots \;\; \hat{X}_N(z_N)]^T, \\ \mathbf{X} &= [X_1(z_1) \;\; X_2(z_2) \;\; \cdots \;\; X_N(z_N)]^T, \end{aligned}$$

and with $k = N - l$, the delays are given by

$$D_{k=N} = 0, \tag{6.243}$$

$$D_{k=N-l} = 2D_{N-l+1} + D \qquad l = 1, \cdots, N-1. \tag{6.244}$$

Perfect reconstruction of the input signal can be achieved if the horizontal delays D_{Hk} are given by

$$D_{H_{k=N}} = 0,$$
$$D_{H_{k=N-1}} = 0,$$
$$D_{H_{k=N-l}} = 2D_{N-l+1} \qquad l = 2, \cdots, N-1.$$

The implementation of the extended vertical kernel filters is done by calculating complementary components, as shown in Fig. 6.69. After upsampling, interpolating with a highpass (Fig. 6.69b) and forming the complementary component, the kernel filter H_K with frequency response as in Fig. 6.69a becomes lowpass with frequency response as illustrated in Fig. 6.69c. The slope of the filter characteristic remains constant whereas the cutoff frequency is doubled. A subsequent upsampling with an interpolation highpass (Fig. 6.69d) and complement filtering leads to the frequency response in Fig. 6.69e. With the help of this technique, the kernel filter

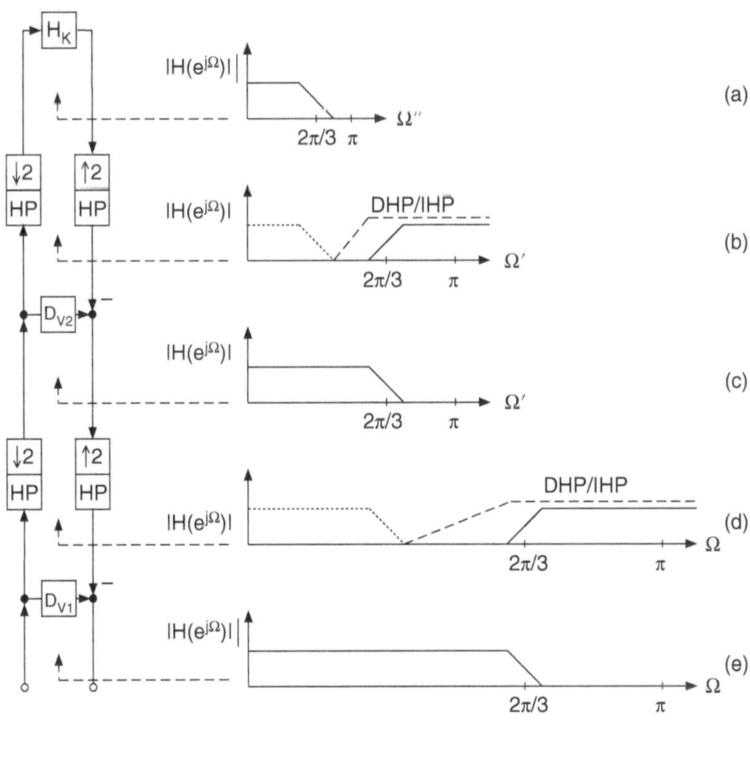

Figure 6.69 Multirate complementary filter.

is implemented at a reduced sampling rate. The cutoff frequency is moved to a desired cutoff frequency by using decimation/interpolation stages with complement filtering.

Computational Complexity. For an N-band multi-complementary filter bank with $N-1$ decomposition filters, where each is implemented by a kernel filter with S stages, the horizontal complexity is given by

$$\mathrm{HC} = \mathrm{HC}_1 + \mathrm{HC}_2 \left(\frac{1}{2} + \frac{1}{4} + \cdots + \frac{1}{2^N} \right). \tag{6.245}$$

Here HC_1 denotes the number of operations that are carried out at the input sampling rate. These operations occur in the horizontal stage HS_1 (see Fig. 6.68). Furthermore, HC_2 denotes the number of operations (horizontal stage HS_2) that are performed at half of the sampling rate. The number of operations in the stages from HS_2 to HS_N are approximately identical but are calculated at sampling rates that are successively halved.

The complexities VC_1 to VC_{N-1} of the vertical kernel filters EKF_1 to EKF_{N-1} are calculated as

$$\mathrm{VC}_1 = \frac{1}{2}\mathrm{V}_1 + \mathrm{V}_2 \left(\frac{1}{4} + \frac{1}{8} + \cdots + \frac{1}{2^{S+1}} \right),$$

$$\mathrm{VC}_2 = \frac{1}{4}\mathrm{V}_1 + \mathrm{V}_2 \left(\frac{1}{8} + \frac{1}{16} + \cdots + \frac{1}{2^{S+2}} \right) = \frac{1}{2}\mathrm{VC}_1,$$

$$\mathrm{VC}_3 = \frac{1}{8}\mathrm{V}_1 + \mathrm{V}_2 \left(\frac{1}{16} + \frac{1}{32} + \cdots + \frac{1}{2^{S+3}} \right) = \frac{1}{4}\mathrm{VC}_1,$$

$$\mathrm{VC}_{N-1} = \frac{1}{2^{N-1}}\mathrm{V}_1 + \mathrm{V}_2 \left(\frac{1}{2^N} + \cdots + \frac{1}{2^{S+N-1}} \right) = \frac{1}{2^{N-1}}\mathrm{VC}_1,$$

where V_1 depicts the complexity of the first stage VS_1 and V_2 is the complexity of the second stage VS_2 (see Fig. 6.68). It can be seen that the total vertical complexity is given by

$$\mathrm{VC} = \mathrm{VC}_1 \left(1 + \frac{1}{2} + \frac{1}{4} + \cdots + \frac{1}{2^{N-1}} \right). \tag{6.246}$$

The upper bound of the total complexity results is the sum of horizontal and vertical complexities and can be written as

$$C_{tot} = \mathrm{HC}_1 + \mathrm{HC}_2 + 2\mathrm{VC}_1. \tag{6.247}$$

The total complexity C_{tot} is independent of the number of frequency bands N and vertical stages S. This means that for real-time implementation with finite computation power, any desired number of sub-bands with however narrow transition bands can be implemented.

6.4.2 Example: Eight-band Multi-complementary Filter Bank

To implement the frequency decomposition into the eight bands shown in Fig. 6.70, the multi-rate filter structure of Fig. 6.71 is employed. The individual parts of the system provide means of downsampling (D=decimation), upsampling (I=interpolation), kernel filtering (K), signal processing (SP), delays (N_1=Delay 1, N_2=Delay 2), and group delay compensation M_i in the ith band. The frequency decomposition is carried out successively from the highest to the lowest frequency band. In the two lowest frequency bands, a compensation for group delay is not required.

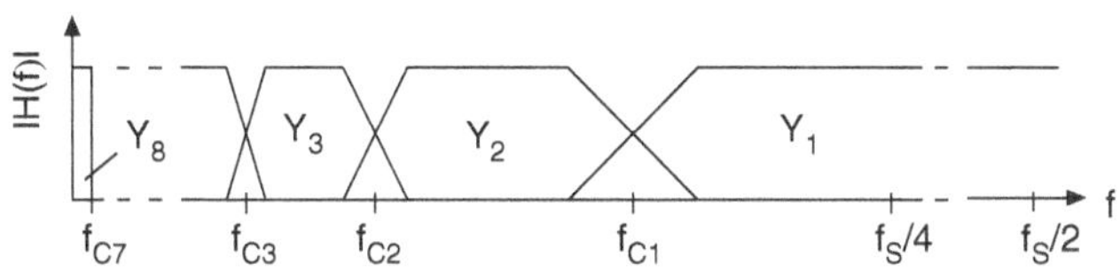

Figure 6.70 Modified octave decomposition of the frequency band.

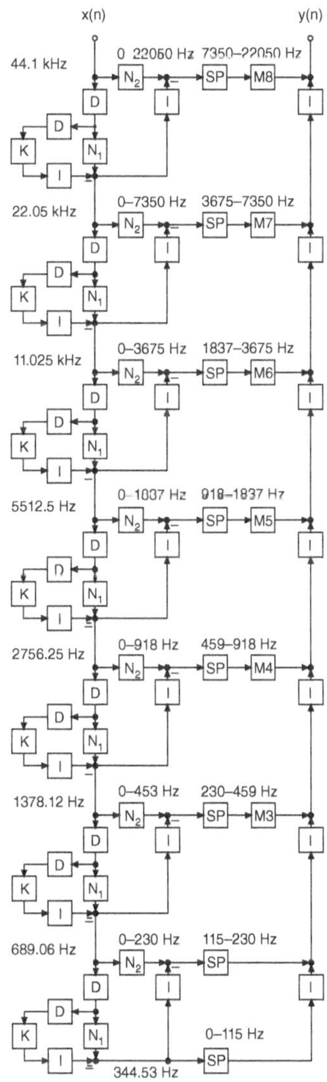

Figure 6.71 Linear phase 8-band equalizer.

Table 6.11 Transition frequencies f_{Ci} and transition bandwidths TB in an 8-band equalizer.

$f_S[kHz]$	$f_{C1}[Hz]$	$f_{C2}[Hz]$	$f_{C3}[Hz]$	$f_{C4}[Hz]$	$f_{C5}[Hz]$	$f_{C6}[Hz]$	$f_{C7}[Hz]$
44.1	7350	3675	1837.5	918.75	459.375	≈ 230	≈ 115
TB$[Hz]$	1280	640	320	160	80	40	20

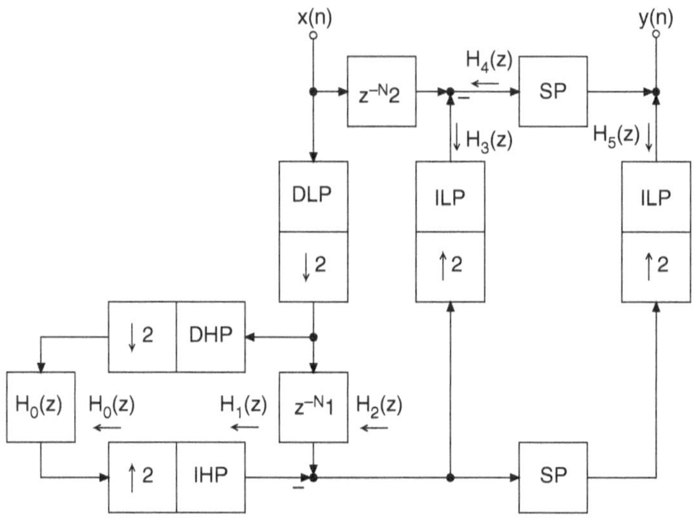

Figure 6.72 Kernel complementary filter structure.

The slope of the filter response can be adjusted with the kernel complementary filter structure shown in Fig. 6.72 which consists of one stage. The specifications of an 8-band equalizer are listed in Table 6.11. The stopband attenuation of the sub-band filters is chosen to be 100 dB.

Filter Design

The design of different decimation and interpolation filters is mainly determined by the transition bandwidth and the stopband attenuation for the lower frequency band. As an example, a design is made for an 8-band equalizer. The kernel complementary filter structure for both lower frequency bands is illustrated in Fig. 6.72. The design specifications for the kernel lowpass, decimation, and interpolation filters are presented in Fig. 6.73.

Kernel Filter Design. The transition bandwidth of the kernel filter is known if the transition bandwidth is given for the lower frequency band. This kernel filter must be designed for a sampling rate of $f_S'' = 44100/(2^8)$. For a given transition bandwidth f_{TB} at a frequency $f'' = f_S''/3$, the normalized passband frequency is

$$\frac{\Omega_{Pb}''}{2\pi} = \frac{f'' - f_{TB}/2}{f_S''} \tag{6.248}$$

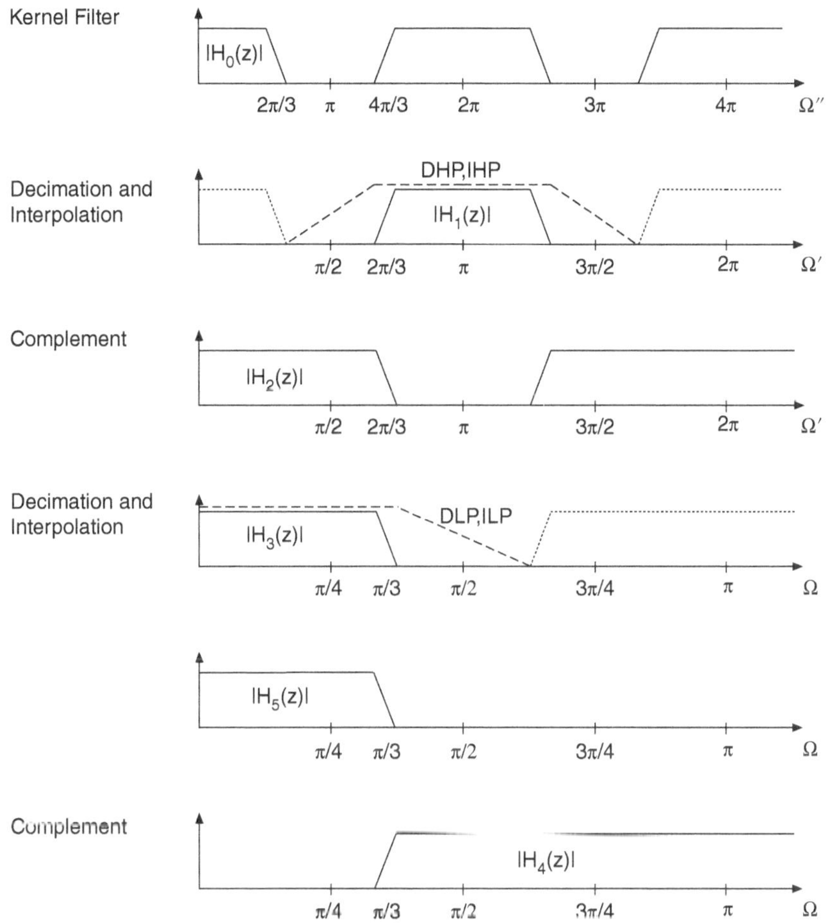

Figure 6.73 Decimation and interpolation filters.

and the normalized stopband frequency

$$\frac{\Omega''_{Sb}}{2\pi} = \frac{f'' + f_{TB}/2}{f''_S}. \tag{6.249}$$

With the help of these parameters, the filter can be designed. Making use of the Parks–McClellan program, the frequency response shown in Fig. 6.74 is obtained for a transition bandwidth of $f_{TB} = 20\,\text{Hz}$. The necessary filter length for a stopband attenuation of 100 dB is 53 taps.

Decimation and Interpolation Highpass Filter. These filters are designed for a sampling rate of $f'_S = 44100/(2^7)$ and are half-band filters, as illustrated in Fig. 6.73. First a lowpass filter is designed followed by a highpass to lowpass transformation. For a given transition bandwidth f_{TB}, the normalized passband frequency is

$$\frac{\Omega'_{Pb}}{2\pi} = \frac{f'' + f_{TB}/2}{f'_S} \tag{6.250}$$

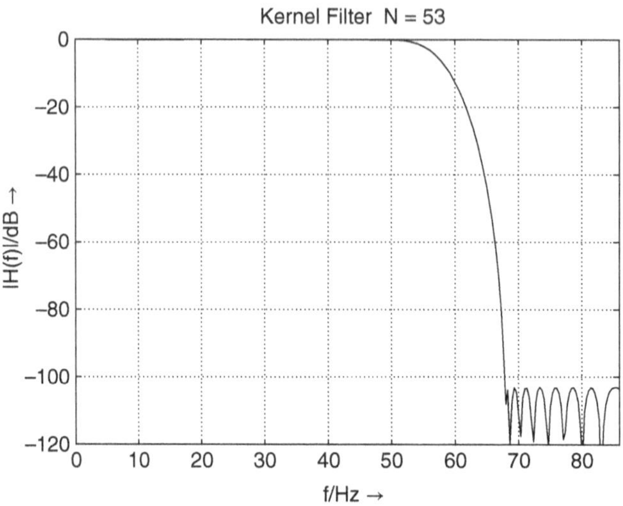

Figure 6.74 Kernel lowpass filter with a transition bandwidth of 20 Hz.

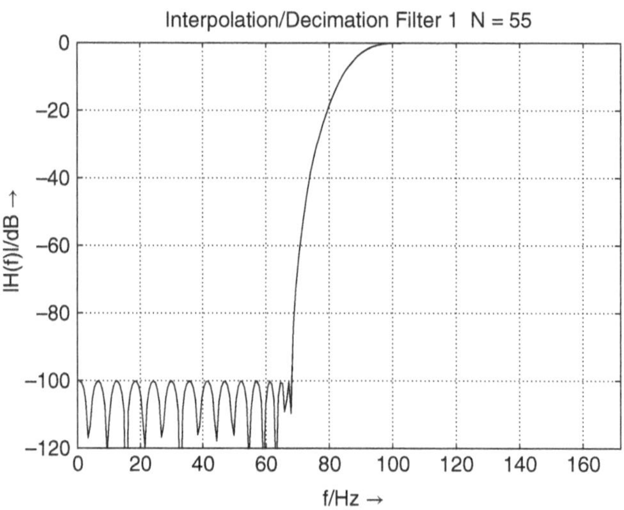

Figure 6.75 Decimation and interpolation highpass filter.

and the normalized stopband frequency is given by

$$\frac{\Omega'_{Sb}}{2\pi} = \frac{2f'' - f_{TB}/2}{f'_S}. \tag{6.251}$$

With these parameters, the design of a half-band filter is carried out. Figure 6.75 shows the frequency response. The necessary filter length for a stopband attenuation of 100 dB is 55 taps.

Decimation and Interpolation Lowpass Filter. These filters are designed for a sampling rate of $f_S = 44100/(2^6)$ and are also half-band filters. For a given transition bandwidth f_{TB}, the normalized passband frequency is

$$\frac{\Omega_{Pb}}{2\pi} = \frac{2f'' + f_{TB}/2}{f_S} \tag{6.252}$$

and the normalized stopband frequency is given by

$$\frac{\Omega_{Sb}}{2\pi} = \frac{4f'' - f_{TB}/2}{f_S}. \tag{6.253}$$

With these parameters, the design of a half-band filter is carried out. Figure 6.76 shows the frequency response. The necessary filter length for a stopband attenuation of 100 dB is 43 taps. These filter designs are used in every decomposition stage so that the transition frequencies and bandwidths are obtained as listed in Table 6.11.

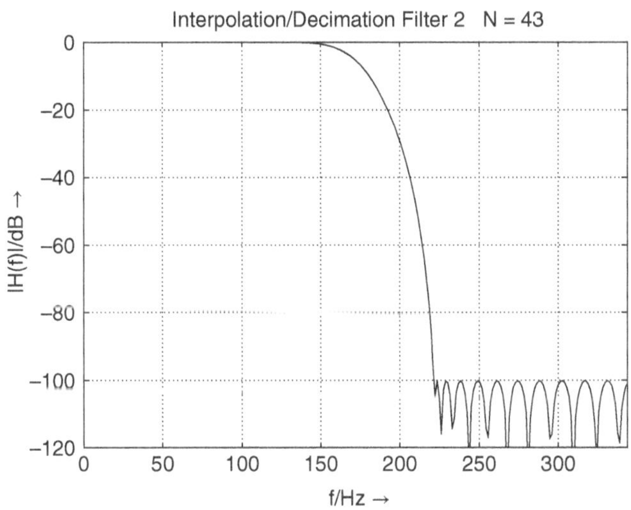

Figure 6.76 Decimation and interpolation lowpass filter.

Memory Requirements and Latency Time. The memory requirements depend directly on the transition bandwidth and the stopband attenuation. Here, the memory operations for the actual kernel, decimation, and interpolation filters have to be differentiated from the group delay compensations in the frequency bands. The compensating group delay N_1 for decimation and interpolation highpass filters of order $O_{\mathrm{DHP/IHP}}$ is calculated with the help of the kernel filter order O_{KF} according to

$$N_1 = O_{\mathrm{KF}} + O_{\mathrm{DHP/IHP}}. \tag{6.254}$$

The group delay compensation N_2 for the decimation and interpolation lowpass filters of order $O_{\mathrm{DLP/ILP}}$ is given by

$$N_2 = 2N_1 + O_{\mathrm{DLP/ILP}}. \tag{6.255}$$

The delays $M_3 \ldots M_8$ in the individual frequency bands are calculated recursively starting from the two lowest frequency bands:

$$M_3 = 2N_2;$$
$$M_4 = 6N_2;$$
$$M_5 = 14N_2;$$
$$M_6 = 30N_2;$$
$$M_7 = 62N_2;$$
$$M_8 = 126N_2.$$

The memory requirements per decomposition stage are listed in Table 6.12. The memory for the delays can be computed by $\sum_i M_i = 240N_2$. The latency time (delay) is given by $t_D = \frac{M_8}{44100} 10^3$ ms ($t_D = 725$ ms).

Table 6.12 Memory requirements.

Kernel filter	O_{KF}
DHP/IHP	$2 \cdot O_{DHP/IHP}$
DLP/ILP	$3 \cdot O_{DLP/ILP}$
N_1	$O_{KF} + O_{DHP/IHP}$
N_2	$2 \cdot N_1 + O_{DLP/ILP}$

6.5 Delay-based Audio Effects

So far, we have discussed linear time-invariant filtering using FIR and IIR systems with fixed coefficients over time. We will now introduce a time-varying system which consists of a delay line where the length of that delay line is varied by a control signal $m(n) = M(n) + \text{frac}(n)$, which is shown in the block diagram of Fig. 6.77. Such a delay line modification [Dis99] performs a time-varying delay of the input given by the relations:

$$y(n) = x(n) * \delta(n - m(n)); \tag{6.256}$$

$$Y(e^{j\Omega}) = X(e^{j\Omega}) \cdot e^{-j\Omega m(n)}; \tag{6.257}$$

$$y(n) = (1\text{-frac}) \cdot x(n - M) + \text{frac} \cdot x(n - (M + 1)). \tag{6.258}$$

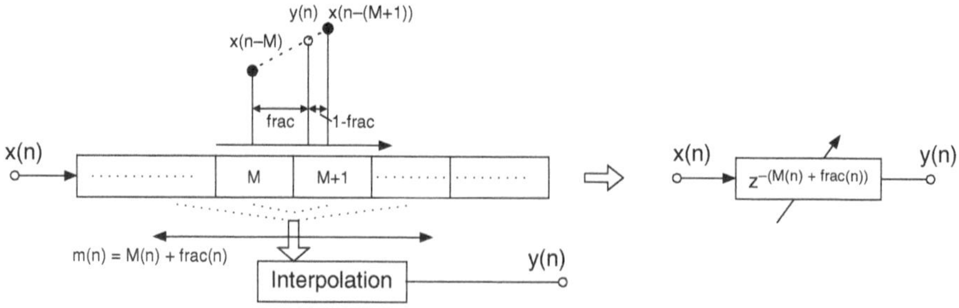

Figure 6.77 Phase modulation with delay line.

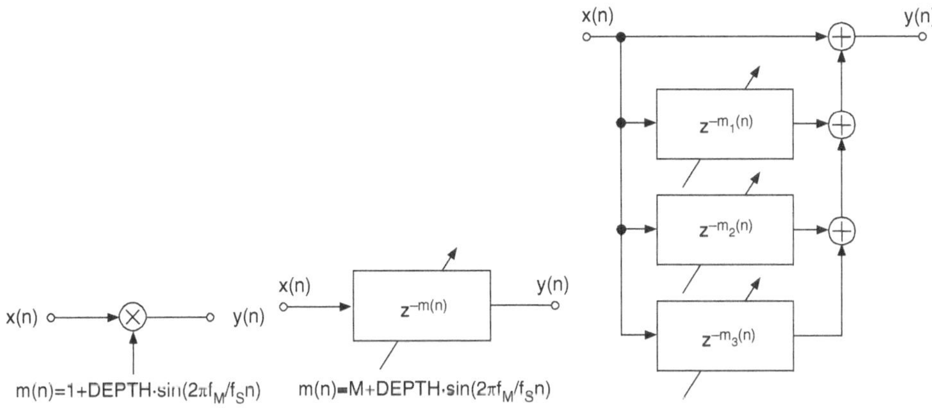

Figure 6.78 Tremolo, vibrato, and chorus effect.

The time-varying delay of Eq. (6.256) leads to a phase modulation of the input $x(n)$, as shown by the Fourier transform given in Eq. (6.257). This delay line modulation can be achieved by linear interpolation between consecutive samples as shown by Eq. (6.258) performing fractional sample computation of the output $y(n)$. Amplitude and phase modulation are used in the following four audio effects.

- Tremolo is the rapid repetition of one note in music terminology or a rapid alternation between two or more notes (string instruments, mandolin, flute, piano, harpsichord, spinet, guitar). Tremolo is a rapid repetitive increase and decrease in volume. The effect is based on modifying the time-domain envelope (amplitude) of a signal, as shown in the left part of Fig. 6.78.

- Vibrato is the rapid and repetitive variation in pitch for the duration of a note (string instruments, guitar, singing voice, yodeling). As shown above, the effect is based on modifying the pitch of a signal by delay line modulation (phase modulation) shown in the middle part of Fig. 6.78.

- Chorus simulates the result of an ensemble of musicians (singers, violins) playing together with slight pitch and amplitude differences. The chorus effect performs a loudness increase. These amplitude and pitch differences are simulated by amplitude modulation and delay line modulation (phase modulation) and all signals are summed together to form the output signal shown in the right part of Fig. 6.78.

- A Flanger simulates a special filter effect which occurs if a slowly time-varying single reflection is added to the direct signal (jet airplane effect). This effect can be achieved by playing two copies of a single signal from two tape machines while slowing down the replay speed of one tape machine by placing a finger to the flange of one tape reel.

6.6 JS Applet – Audio Filters

The applet shown in Fig. 6.79 demonstrates audio filters. It is designed for a first insight into the perceptual effect of filtering an audio signal. In addition to the different filter types and their

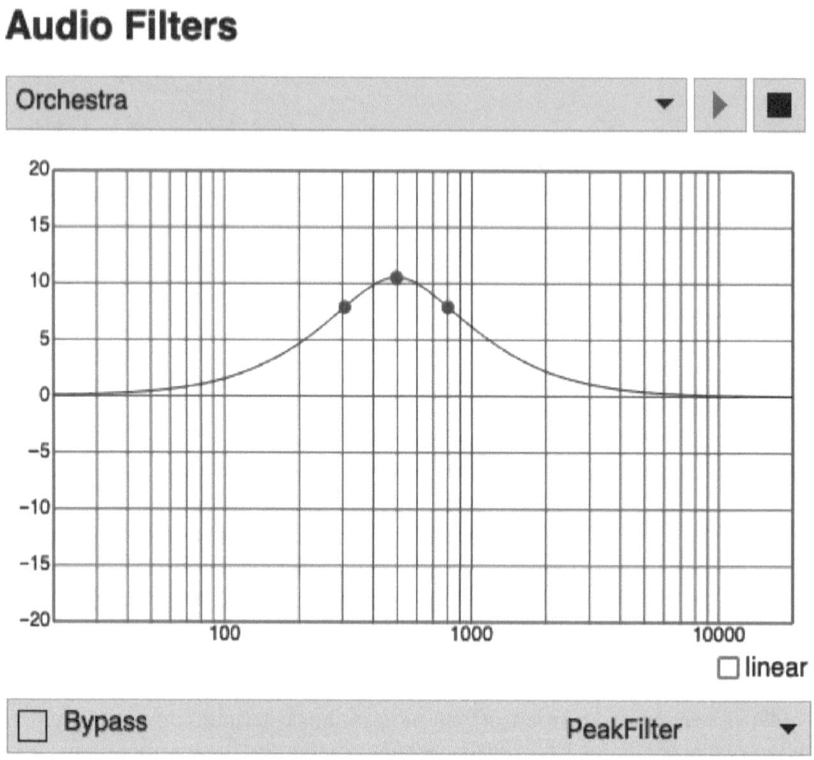

Figure 6.79 JS applet – audio filters.

acoustical effect, the applet offers a first insight into the logarithmic behavior of loudness and frequency resolution of our human acoustical perception.

The following filter functions can be selected on the lower right of the graphical user interface:

- Lowpass/highpass filter (LP/HP) with control parameter
  - cutoff frequency f_c in Hertz (lower horizontal slider),
  - all frequencies above (LP) or below (HP the cutoff frequency are attenuated according to the shown frequency response);
- Low-/high-frequency shelving filter (LFS/HFS) with control parameters
  - cutoff frequency f_c in Hertz (lower horizontal slider),
  - boost/cut in dB (left vertical slider with + for boost or - for cut),
  - all frequencies below (LFS) or above (HFS) the cutoff frequency are boosted/cut according to the selected boost/cut;
- Peak filter with control parameters

- center frequency f_c in Hertz (lower horizontal slider),
- boost/cut in dB (left vertical slider with + for boost or - for cut),
- Q-factor $Q = \frac{f_c}{f_b}$ (right vertical slider), which controls the bandwidth f_b of the boost/cut around the adjusted center frequency f_c. Lower Q-factor means wider bandwidth,
- the peak filter boosts/cuts the center frequency with a bandwidth adjusted by the Q-factor.

The center window shows the frequency response (filter gain versus frequency) of the selected filter functions. You can choose between a linear and a logarithmic frequency axis.
You can choose between two predefined audio files from our web server (*audio1.wav* or *audio2.wav*) or your own local WAV file to be processed [Gui05].

6.7 Exercises

1. Design of Recursive Audio Filters

1. How can we design a low-frequency shelving filter? Which parameters define the filter? Explain the control parameters.
2. How can we derive a high-frequency shelving filter? Which parameters define the filter?
3. What is the difference between first- and second order shelving filters?
4. How can we design a peak filter? Which parameters define the filter? What is the filter order? Describe the control parameters. Describe the Q-factor.
5. How do we derive the digital transfer function?
6. Derive the digital transfer functions for the first-order shelving filters.

2. Parametric Audio Filters

1. What is the basic idea for parametric filters?
2. What is the difference between the Regalia and the Zölzer filter structures? Count the number of multiplications and additions for both filter structures.
3. Derive a signal flow graph for first- and second-order parametric Zölzer filters with a direct-form implementation of the allpass filters.
4. Is there a complete decoupling of all control parameters for boost and cut? Which parameters are decoupled?

3. Shelving Filter: Direct Form

Derive a first-order low-shelving filter from a purely band-limiting first-order lowpass filter. Use bilinear transform and give the transfer function of the low-shelving filter.

1. State the filter coefficients and calculate the poles/zeros as functions of V_0 and T. Which gain factor do you have if $z = \pm 1$?
2. What is the difference between purely band-limiting filters and shelving filter?
3. How can you describe the boost and cut effect related to poles/zeros of the filter?
4. How do we get a transfer function for the boost case from the cut case?
5. How do we go from a low-shelving filter to a high-shelving filter?

4. Shelving Filter: Allpass Form

Implement a first-order high-shelving filter for the boost and cut case with the sampling rate $f_S = 44.1\,\text{kHz}$, the cutoff frequency $f_c = 10\,\text{kHz}$, and the gain $G = 12\,\text{dB}$.

1. Define the allpass parameters and coefficients for the boost and cut case.
2. Derive from the allpass decomposition the complete transfer function of the shelving filter.
3. Using Matlab, give the magnitude frequency response for boost and cut. Show the result for the case where the boost and cut filters are in a series connection.
4. If the input signal to the system is a unit impulse, give the spectrum of the input and output signal for the boost and cut case. What will you expect as result in this case when boost and cut are again cascaded?

5. Quantization of Filter Coefficients

For the quantization of the filter coefficients, different methods have been proposed: direct form, Gold and Rader, Kingsbury, and Zölzer.

1. What is the motivation behind this?
2. Plot a pole distribution using the quantized polar representation of a second-order IIR filter

$$H(z) = \frac{N(z)}{1 - 2r\cos\varphi z^{-1} + r^2 z^{-2}}.$$

6. Signal Quantization Inside the Audio Filter

Now we combine coefficient and signal quantization.

1. Design a digital highpass filter (second-order IIR), with a cutoff frequency $f_c = 50\,\text{Hz}$. (Use the Butterworth, Chebyshev, or elliptic design methods implemented in Matlab.)

2. Quantize the signal only when it leaves the accumulator (i.e. before it is saved in any state variable).

3. Now quantize the coefficients (direct form) too.

4. Extend your quantization to every arithmetic operation (i.e. after each addition/multiplication).

7. Quantization Effects in Recursive Audio Filters

1. Why is the quantization of signals inside a recursive filter of special interest?

2. Derive the noise transfer function of the second-order direct-form filter. Apply a first- and second-order noise shaping to the quantizer inside the direct-form structure and discuss the influence. What is the difference between second-order noise shaping and double precision arithmetic?

3. Write a Matlab implementation of a second-order filter structure for quantization and noise shaping.

8. Fast Convolution

For an input sequence $x(n)$ of length $N_1 = 500$ and the impulse response $h(n)$ of length $N_2 = 31$, perform the discrete time convolution.

1. Give the discrete time convolution sum formula.

2. Define with Matlab $x(n)$ as the sum of two sinusoids and derive $h(n)$ with Matlab function *firpm(..)*.

3. Realize the filter operation with Matlab using:

   - the function *conv*(x, h);
   - the sample by the sample convolution sum method;
   - the FFT method;
   - the FFT with the overlap and add method.

4. Describe FIR filtering with the fast convolution technique. Which conditions have to be fulfilled for the input signal and the impulse response if convolution is performed by equivalent frequency-domain processing?

5. What happens if the input signal and impulse response are as long as the FFT transform length?

6. How can we perform IFFT with the FFT algorithm?

7. Explain the processing steps:

- for a segmentation of the input signal into blocks and fast convolution;
- for a stereo signal by the fast convolution technique;
- for the segmentation of the impulse response.

8. What is the processing delay of the fast convolution technique?
9. Write a Matlab program for fast convolution.
10. How does quantization of the signal influence the roundoff noise behavior of an FIR filter?

9. FIR Filter Design by Frequency Sampling

1. Why is frequency sampling an important design method for audio equalizers? How do we sample magnitude and phase response?
2. What is a linear phase frequency response of a system? What is the effect on an input signal passing through such a system?
3. Explain the derivation of the magnitude and phase response for a linear phase FIR filter.
4. What is the condition for a real-valued impulse response of even length N? What is the group delay?
5. Write a Matlab program for the design of an FIR filter and verify the example in the chapter.
6. Now the desired frequency response is an ideal lowpass filter of length $N_F = 31$ with cutoff frequency $\Omega_c = \pi/2$, derive the impulse response of this system. What will be the result for $N_F = 32$ and $\Omega_c = \pi$?

10. Multi-complementary Filter Bank

1. What is octave-spaced frequency splitting and how can we design a filter bank for that task?
2. How can we perform aliasing-free sub-band processing? How can we achieve narrow transition bands for a filter bank? What is the computational complexity of an octave-spaced filter bank?

References

[Abu79] A.I. Abu-El-Haija, A.M. Peterson: *An Approach to Eliminate Roundoff Errors in Digital Filters*, IEEE Trans. ASSP, pp. 195–198, April 1979.

[Ave71] E. Avenhaus: *Zum Entwurf digitaler Filter mit minimaler Speicherwortlänge für Koeffizienten und Zustandsgrössen*, Ausgewählte Arbeiten über Nachrichtensysteme, Nr. 13, herausgegeben von Prof. Dr.-Ing. W. Schüssler, Erlangen 1971.

[Bar82] C.W. Barnes: *Error Feedback in Normal Realizations of Recursive Digital Filters*, IEEE Trans. Circuits and Systems, pp. 72–75, Jan. 1982.

[Bom85] B.W. Bomar: *New Second-Order State-Space Structures for Realizing Low Roundoff Noise Digital Filters*, IEEE Trans. ASSP, pp. 106–110, Feb. 1985.

[Bri94] R. Bristow-Johnson: *The Equivalence of Various Methods of Computing Biquad Coefficients for Audio Parametric Equalizers*, Proc. 97th AES Convention, Preprint No. 3906, November 1994.

[Cad87] J.A. Cadzow: *Foundations of Digital Signal Processing and Data Analysis*, New York: Macmillan Publishing Company, 1987.

[Cha78] T.L. Chang: *A Low Roundoff Noise Digital Filter Structure*, Proc. Int. Symp. on Circuits and Systems, pp. 1004–1008, May 1978.

[Cla00] R.J. Clark, E.C. Ifeachor, G.M. Rogers, and P.W.J. Van Eetvelt: *Techniques for Generating Digital Equalizer Coefficients*, J. Audio Eng. Soc., Vol. 48, pp. 281–298, April 2000.

[Cre03] L. Cremer, M. Möser: *Technische Akustik*, Springer-Verlag, Berlin, 2003.

[Dat97] J. Dattorro: *Effect design - Part 1: Reverberator and other Filters*, J. Audio Eng. Soc., 45(19):660–684, September 1997.

[Din95] Yinong Ding, D. Rossum: *Filter Morphing of Parametric Equalizers and Shelving Filters for Audio Signal Processing*, J. Audio Eng. Soc., Vol. 43, No.10, pp. 821–826, October 1995.

[Dis99] S. Disch, U. Zölzer: *Modulation and delay line based digital audio effects*, In Proceedings of the COST G-6 Conference on Digital Audio Effects (DAFX–99), Trondheim, Norway, December 9–11, 1999.

[Duh88] P. Duhamel, B. Piron, J. Etcheto: *On Computing the Inverse DFT*, IEEE Trans. Acoust., Speech, Signal Processing, Vol. 36, No. 2, pp. 285–286, February 1988.

[Ege96] G. P. M. Egelmeers and P. C. W. Sommen: *A new method for efficient convolution in frequency domain by nonuniform partitioning for adaptive filtering*, IEEE Trans. Signal Processing, Vol. 44, pp. 3123–3192, Dec. 1996.

[Ell82] D.F. Elliott, K.R. Rao: *Fast Transforms: Algorithms, Analyses, Applications*, New York: Academic Press, 1982.

[Fet72] A. Fettweis: *On the Connection Between Multiplier Wordlength Limitation and Roundoff Noise in Digital Filters*, IEEE Trans. Circuit Theory, pp. 486–491, Sept. 1972.

[Fli92] N.J. Fliege, U. Zölzer: *Multi-Complementary Filter Bank: A New Concept with Aliasing-Free Subband Signal Processing and Perfect Reconstruction*, Proc. EUSIPCO-92, Br′′ussel, pp. 207–210, August 1992.

[Fli93] N.J. Fliege, U. Zölzer: *Multi-Complementary Filter Bank*, Proc. ICASSP-93, Minneapolis, pp. 193–196, April 1993.

[Fli00] N. Fliege: *Multirate Digital Signal Processing*, J. Wiley & Sons, Chichester, 2000.

[Fon01] F. Fontana, M. Karjalainen: *Magnitude-complementary Filters for Dynamic Equalization*, In Proc. of the DAFX-01, Limerick, Ireland, pp. 97–101, Dec. 2001.

[Fon03] F. Fontana, M. Karjalainen: *A Digital Bandpass/Bandstop Complementary Equalization Filter with Independent Tuning Characteristics*, IEEE Signal Processing Letters, vol. 10, n. 4, pp.119–122, Apr. 2003.

[Gar95] W. G. Gardner: *Efficient Convolution Without Input-output Delay*, J. Audio Eng. Soc., 43(3), pp. 127–136,1995.

[Garc02] G. Garcia: *Optimal Filter Partition for Efficient Convolution with Short Input/Output Delay*, Proc. 113th AES Convention, Los Angeles, Preprint No. 5660, October 2002.

[Gol67] B. Gold, C.M. Rader: *Effects of Parameter Quantization on the Poles of a Digital Filter*, Proc. IEEE, pp. 688–689, May 1967.

[Gui05] M. Guillemard, C. Ruwwe, U. Zölzer: *J-DAFx - Digital Audio Effects in Java*, Proc. 8th Int. Conference on Digital Audio Effects (DAFx-05), Madrid, Spain, pp.161–166, 2005.

[Har93] F.J. Harris, E. Brooking: *A Versatile Parametric Filter Using an Imbedded All-Pass Sub-Filter to Independently Adjust Bandwidth, Center Frequency and Boost or Cut*, Proc. 95th AES Convention, San Francisco, Preprint No. 3757, 1993.

[Här98] A. Härmä: *Implementation of Recursive Filters Having Delay Free Loops*, Proc. IEEE Int. Conf. Acoustics, Speech, and Signal Processing (ICASSP'98), Vol. 3, pp. 1261–1264, Seattle, Washington, 1998.

[Hol06a] M. Holters, U. Zölzer: *Parametric Recursive Higher-Order Shelving Filters*, Jahrestagung Akustik DAGA, 20.-23. März 2006, Braunschweig, 2006.

[Hol06b] M. Holters, U. Zölzer: *Parametric recursive higher-order shelving filters*, Proc. 120th Conv. Audio Eng. Soc., Preprint No. 6722, Paris, 2006.

[Hol06c] M. Holters, U. Zölzer: *Parametric Higher-Order Shelving Filters*, Proc. EUSIPCO-06, Florence, 2006.

[Hol06d] M. Holters, U. Zölzer: *Graphic Equalizer Design Using Higher-Order Recursive Filters*, Proc. of the 9th Int. Conference on Digital Audio Effects (DAFx-06), Montreal, Canada, September 18-20, pp. 37–40, 2006.

[Jur64] E.I. Jury: *Theory and Application of the z-Transform Method*, Wiley, 1964.

[Kei04] F. Keiler, U. Zölzer, *Parametric Second- and Fourth-Order Shelving Filters for Audio Applications*, Proc. of IEEE 6th International Workshop on Multimedia Signal Processing, Siena, Italy, September 29 - October 1, 2004.

[Kin72] N.G. Kingsbury: *Second-Order Recursive Digital Filter Element for Poles Near the Unit Circle and the Real z-Axis*, Electronic Letters, pp. 155–156, March 1972.

[Moo83] J.A. Moorer: *The Manifold Joys of Conformal Mapping*, J. Audio Eng. Soc., Vol. 31, pp. 826–841, 1983.

[Mou90] J.N. Mourjopoulos, E.D. Kyriakis-Bitzaros, C.E. Goutis: *Theory and Real-time Implementation of Time-varying Digital Audio Filters*, J. Audio Eng. Soc., Vol. 38, pp. 523–536, July/August 1990.

[Mul76] C.T. Mullis, R.A. Roberts: *Synthesis of Minimum Roundoff Noise Fixed Point Digital Filters*, IEEE Trans. Circuits and Systems, pp. 551–562, Sept. 1976.

[Mül99] C. Müller-Tomfelde: *Low Latency Convolution for Real-time Application*, In Proceedings of the AES 16th International Conference: Spatial Sound Reproduction, Rovaniemi, Finland, April 10-12, pp. 454–460, 1999.

[Mül01] C. Müller-Tomfelde: *Time-Varying Filter in Non-Uniform Block Convolution*, In Proceedings of the COST G-6 Conference on Digital Audio Effects (DAFX-01), Limerick, Ireland, December 6–8, 2001.

[Orf96a] S.J. Orfanidis: *Introduction to Signal Processing*. Prentice-Hall, 1996.

[Orf96b] S.J. Orfanidis: *Digital Parametric Equalizer Design with Prescribed Nyquist-Frequency Gain*, Proc. 101st Convention Audio Engineering Society, Preprint No. 4361, November 1996.

[Orf05] S.J. Orfanidis: *High-order digital parametric equalizer design*, J. Audio Eng. Soc., Vol. 53, No. 11, pp. 1026–1046, 2005.

[Rab88] R. Rabenstein: *Minimization of Transient Signals in Recursive Time-varying Digital Filters*, Circuits, Systems, and Signal Processing, Vol. 7, No. 3, pp. 345–359, 1988.

[Ram88] T.A. Ramstad, T. Saramäki: *Efficient Multirate Realization for Narrow Transition-Band FIR Filters*, Proc. IEEE Int. Symp. on Circuits and Syst. (Espoo, Finland), pp. 2019–2022, June 1988.

[Ram90] T.A. Ramstad, T. Saramäki: *Multistage, Multirate FIR Filter Structures for Narrow Transition-Band Filters*, Proc. IEEE Int. Symp. on Circuits and Syst. (New Orleans, USA), pp. 2017–2021, May 1990.

[Reg87] P.A. Regalia, S.K. Mitra: *Tunable Digital Frequency Response Equalization Filters*, IEEE Transactions on Acoustics, Speech, and Signal Processing, Vol. 35, No. 1, pp. 118–120, January 1987.

[Sha92] D.J. Shpak: *Analytical Design of Biquadradic Filter Sections for Parametric Filters*, J. Audio Eng. Soc., Vol. 40, pp. 876–885, November 1992.

[Soo90] J.S. Soo, K.K. Pang: *Multidelay Block Frequency Domain Adaptive Filter*, IIEEE Transactions on Acoustics, Speech, and Signal Processing, Vol. 38, No. 2, pp. 373–376, Feb. 1990.

[Sor87] H.V. Sorensen, D.J. Jones, M.T. Heideman, C.S. Burrus: *Real-Valued Fast Fourier Transform Algorithms*, IEEE Trans. Acoust., Speech, Signal Processing, Vol. 35, No. 6, pp. 849–863, June 1987.

[Sri77] A.B. Sripad, D.L. Snyder: *A Necessary and Sufficient Condition for Quantization Errors to be Uniform and White*, IEEE Trans. ASSP, Vol. 25, pp. 442–448, Oct. 1977.

[Tra77] Tran-Thong, B. Liu: *Error Spectrum Shaping in Narrow Band Recursive Filters*, IEEE Trans. ASSP, pp. 200–203, April 1977.

[Vai93] P.P. Vaidyanathan: *Multirate Systems and Filter Banks*, Prentice-Hall, Englewood Cliffs, 1993.

[Väl98] V. Välimäki, T.I. Laakso: *Suppression of Transients in Time-varying Recursive Filters for Audio Signals*, Proc. IEEE Int. Conf. Acoustics, Speech, and Signal Processing (ICASSP'98), Vol. 6, pp. 3569–3572, Seattle, Washington, 1998.

[Väl16] V. Välimäki and J.D. Reiss: *All about audio equalization: Solutions and frontiers*. Applied Sciences, 6(5), 2016.

[Whi86] S.A. White: *Design of a Digital Biquadratic Peaking or Notch Filter for Digital Audio Equalization*, J. Audio Eng. Soc., Vol. 34, pp. 479–483, 1986.

[Zöl89] U. Zölzer: *Entwurf digitaler Filter für die Anwendung im Tonstudiobereich*, Wissenschaftliche Beiträge zur Nachrichtentechnik und Signalverarbeitung, TU Hamburg-Harburg, Juni 1989.

[Zöl90] U. Zölzer: *A Low Roundoff Noise Digital Audio Filter*, Proc. EUSIPCO-90, Barcelona, pp. 529–532, 1990.

[Zöl92] U. Zölzer, N. Fliege: *Logarithmic Spaced Analysis Filter Bank for Multiple Loudspeaker Channels*, Proc. 93rd AES Convention, Preprint No. 3453, San Francisco 1992.

[Zöl93] U. Zölzer, B. Redmer, J. Bucholtz: *Strategies for Switching Digital Audio Filters*, Proc. 95th AES Convention, New York, Preprint No. 3714, October 1993.

[Zöl94] U. Zölzer: *Roundoff Error Analysis of Digital Filters*, J. Audio Eng. Soc., Vol. 42, No. 4, pp. 232–244, April 1994.

[Zöl95] U. Zölzer, T. Boltze: *Parametric Digital Filter Structures*, Proc. 99th AES Convention, New York, Preprint No. 4099, October 1995.

Chapter 7

Room Simulation

U. Zölzer, P. Nowak, and P. Bhattacharya

Room simulation artificially reproduces the acoustics of a room. The foundations of room acoustics are found in [Cre78, Kut91]. Room simulation is mainly used for post-processing signals in which a microphone is located in the vicinity of an instrument or a voice. The direct signal, without additional room impression, is mapped to a certain acoustical room, for example, a concert hall or a church. In terms of signal processing, the post-processing of an audio signal with room simulation corresponds to the convolution of the audio signal with a room impulse response.

7.1 Basics

7.1.1 Room Acoustics

The room impulse response between two points in a room can be classified as shown in Fig. 7.1. The impulse response consists of the direct signal, early reflections (from walls), and subsequent reverberation. The number of early reflections continuously increases with time and leads to a random signal with exponential decay called subsequent reverberation. The *reverberation time* (decrease of sound pressure level by 60 dB) can be calculated using the geometry of the room and the partial areas that absorb sound in the room according to

$$T_{60} = 0.163\frac{V}{\alpha S} = \frac{0.163}{[\mathrm{m/s}]}\frac{V}{\sum_n \alpha_n S_n}, \tag{7.1}$$

T_{60} = reverberation time in s;

V = volume of the room m^3;

Digital Audio Signal Processing, Third Edition. Udo Zölzer.
© 2022 John Wiley & Sons Ltd. Published 2022 by John Wiley & Sons Ltd.

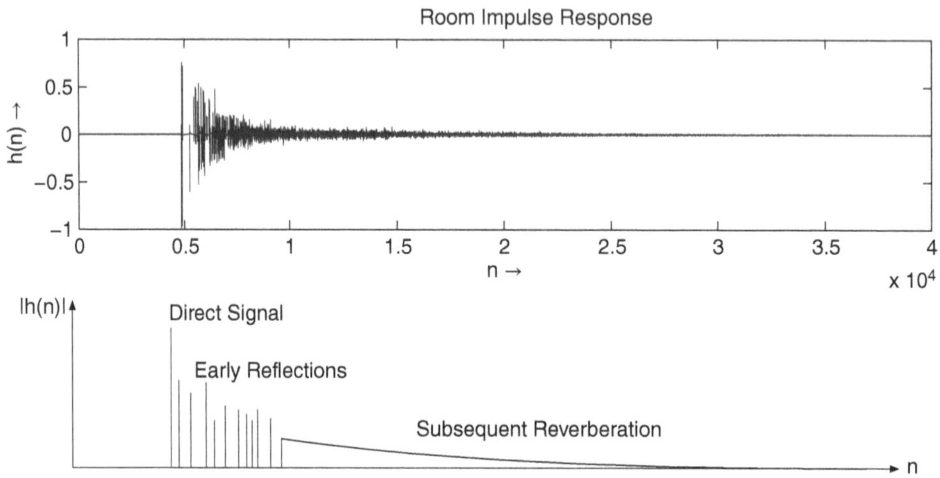

Figure 7.1 Room impulse response $h(n)$ and simplified decomposition into direct signal, early reflections, and subsequent reverberation (with $|h(n)|$).

$$S_n = \text{partial areas m}^2;$$

$$\alpha_n = \text{absorption coefficient of partial area } S_n.$$

The geometry of the room also determines the eigenfrequencies of a three-dimensional rectangular room:

$$f_e = \frac{c}{2}\sqrt{\left(\frac{n_x}{l_x}\right)^2 + \left(\frac{n_y}{l_y}\right)^2 + \left(\frac{n_z}{l_z}\right)^2}, \tag{7.2}$$

with

n_x, n_y, n_z integer number of half-waves (0,1,2, ...);

l_x, l_y, l_z dimensions of a rectangular room;

c sound velocity.

For larger rooms, the eigenfrequencies start from very low frequencies. In contrast, the lowest eigenfrequencies of smaller rooms are shifted toward higher frequencies. The mean frequency between two extrema of the frequency response of a large room is approximately inversely proportional to the reverberation time [Schr87]:

$$\Delta f \sim 1/T_{60}. \tag{7.3}$$

The distance between two eigenfrequencies decreases with increasing number of half-waves. Above a *critical frequency*

$$f_c > 4000\sqrt{T_{60}/V}, \tag{7.4}$$

the density of eigenfrequencies becomes so large that they overlap each other [Schr87].

7.1.2 Model-based Room Impulse Responses

The methods for analytically determining a room impulse response are based on the ray-tracing model [Schr70] or image model [All79]. In the case of the ray-tracing model, a point source with radial emission is assumed. The path length of rays and the absorption coefficients of walls, roofs, and floors are used to determine the room impulse response (see Fig. 7.2). For the image model, image rooms with secondary image sources are formed, which in turn have further image rooms and image sources. The summation of all image sources with corresponding delays and attenuations provides the estimated room impulse response. Both methods are applied in room acoustics to get insight into the acoustical properties when planning concert halls, theaters, etc.

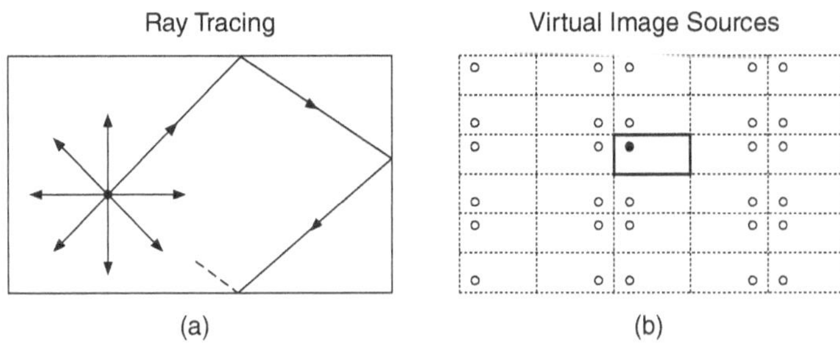

Figure 7.2 Model-based methods for calculating room impulse responses.

To simulate the room impulse response between two points inside a rectangular room, Allen and Berkley proposed the image source model in 1979 [All79]. The aim of this simulation method was to be simple, easy to use, and fast. The underlying principle is the mirroring of the original room, including the source at the walls of the room, infinite times in all room dimensions. In this way, the image rooms are created containing the image sources. Figure 7.3 illustrates this principle in a two-dimensional representation while focusing on first- and second-order image rooms.

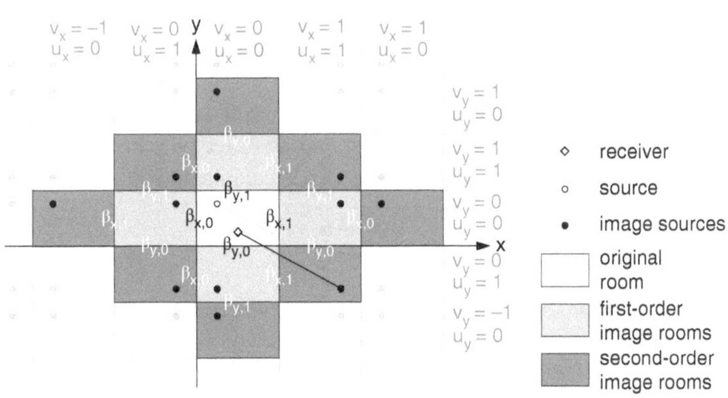

Figure 7.3 Two-dimensional representation of the image source model highlighting first- and second-order image rooms including the parameterization of the individual image sources. Additionally, an exemplary second-order reflection is drawn.

Afterward, the room impulse response is estimated by the summation of the attenuation of all image sources $A(\mathbf{u}, \mathbf{v})$ at the corresponding delays $\tau(\mathbf{u}, \mathbf{v})$ as

$$h(t) = \sum_{\mathbf{u}=0}^{1} \sum_{\mathbf{v}=-\infty}^{\infty} A(\mathbf{u}, \mathbf{v}) \cdot \delta(t - \tau(\mathbf{u}, \mathbf{v})), \tag{7.5}$$

where

$$\mathbf{u} = (u_x, u_y, u_z)^{\mathrm{T}} \quad \text{with} \quad u_x, u_y, u_z \in \{0,1\} \text{ and} \tag{7.6}$$

$$\mathbf{v} = (v_x, v_y, v_z)^{\mathrm{T}} \quad \text{with} \quad v_x, v_y, v_z \in \mathbb{N} \tag{7.7}$$

denote the different image rooms, as shown in Fig. 7.3. Here, the original room is characterized by $\mathbf{u} = (0, 0, 0)^{\mathrm{T}}$ and $\mathbf{v} = (0, 0, 0)^{\mathrm{T}}$.

In the following, the calculation of the attenuation $A(\mathbf{u}, \mathbf{v})$ and the delays $\tau(\mathbf{u}, \mathbf{v})$ is explained in detail. First, the room dimensions are defined as

$$\mathbf{l}_{\mathbf{room}} = (l_{\mathrm{x}}, l_{\mathrm{y}}, l_{\mathrm{z}})^{\mathrm{T}} \tag{7.8}$$

with the origin of the coordinate system being in one of the corners of the room, as shown in Fig. 7.3. Afterward, the positions of the receiver and the source inside the original room are defined as

$$\mathbf{p}_{\mathrm{r}} = (x_{\mathrm{r}}, y_{\mathrm{r}}, z_{\mathrm{r}})^{\mathrm{T}} \text{ and} \tag{7.9}$$

$$\mathbf{p}_{\mathrm{s}} = (x_{\mathrm{s}}, y_{\mathrm{s}}, z_{\mathrm{s}})^{\mathrm{T}}, \tag{7.10}$$

respectively. From this, the positions of the image sources can be calculated by

$$\begin{aligned} \mathbf{p_{is}}(\mathbf{u}, \mathbf{v}) = &- \operatorname{diag}(2u_x - 1, 2u_y - 1, 2u_z - 1) \cdot \mathbf{p}_{\mathrm{s}} \\ &+ \operatorname{diag}(v_x, v_y, v_z) \cdot 2\mathbf{l}_{\mathbf{room}}, \end{aligned} \tag{7.11}$$

where diag(·) denotes a diagonal matrix with the arguments as diagonal elements [Leh08]. The distances from these image sources to the receiver are given as

$$d(\mathbf{u}, \mathbf{v}) = \left\| \mathbf{p}_{\mathrm{r}} - \mathbf{p_{is}}(\mathbf{u}, \mathbf{v}) \right\| \tag{7.12}$$

with $\| \cdot \|$ being the Euclidean norm. Based on this distance $d(\mathbf{u}, \mathbf{v})$ between an image source and the receiver, which equals the length of the corresponding reflected path from the original source to the receiver, the time of arrival of the reflected path is calculated by

$$\tau(\mathbf{u}, \mathbf{v}) = \frac{d(\mathbf{u}, \mathbf{v})}{c}, \tag{7.13}$$

where c denotes the speed of sound. To stay simple during the calculation of the attenuation of the reflected paths, two assumptions are made [All79]. First, the point image model that is only exact for rigid walls is also used for non-rigid walls. Second, the reflection coefficient β is assumed to be frequency- and direction-independent. In this way, the attenuation of a reflected path can be determined as

$$A(\mathbf{u}, \mathbf{v}) = \frac{\beta_{x,0}^{|v_x - u_x|} \beta_{x,1}^{|v_x|} \beta_{y,0}^{|v_y - u_y|} \beta_{y,1}^{|v_y|} \beta_{z,0}^{|v_z - u_z|} \beta_{z,1}^{|v_z|}}{4\pi d(\mathbf{u}, \mathbf{v})}, \tag{7.14}$$

where the reflection coefficients β are assigned to the different walls, as shown in Fig. 7.3. As can be seen, the attenuation of a reflected path depends on two factors: the propagation loss arising from the length of the reflected path $d(\mathbf{u}, \mathbf{v})$ and the energy absorption of the walls hit by the reflected path. The relation of the absorption coefficient α and the reflection coefficient β of a wall is given as

$$\beta = -\sqrt{(1-\alpha)}\,, \tag{7.15}$$

where the negative sign ensures simulated reverberation tails similar to those of real acoustic measurements [Ant02]. To implement the image source model in the time domain according to Eq. (7.5), the continuous-time room impulse response $h(t)$ has to be converted into the discrete-time room impulse response $h(n)$ by the sampling operation. However, before the sampling operation, the unit pulses $\delta(t)$ from Eq. (7.5) have to be band limited to $f_S/2$, which results in sinc-like pulses. In this way, the discrete-time room impulse responses are given as

$$h(n) = \sum_{\mathbf{u}=0}^{1} \sum_{\mathbf{v}=-\infty}^{\infty} A(\mathbf{u}, \mathbf{v}) \cdot \mathrm{sinc}(n - \tau(\mathbf{u}, \mathbf{v}) \cdot f_S) \tag{7.16}$$

with

$$\mathrm{sinc}(n) = \frac{\sin(\pi n)}{\pi n}\,. \tag{7.17}$$

In Fig. 7.4, an exemplary room impulse response calculated via the image source model is shown. Here, the room dimensions are given as $\mathbf{l}_{\mathbf{room}} = (6\text{ m}, 4\text{ m}, 2.5\text{ m})^{\mathrm{T}}$, and the receiver and source

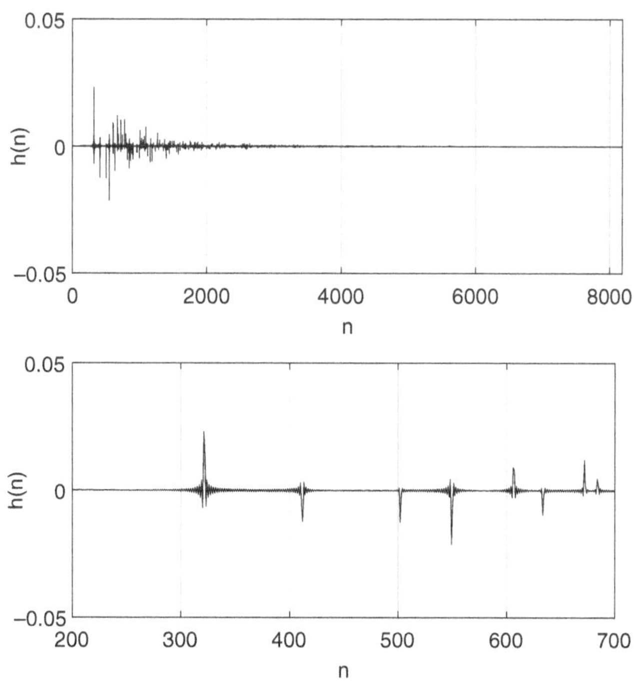

Figure 7.4 Simulated room impulse response via image source model with a length of $L_h = 8192$ (top) and cutout of the direct path and the first reflections (bottom). The parameters are given as $\mathbf{l}_{\mathbf{room}} = (6\text{ m}, 4\text{ m}, 2.5\text{ m})^{\mathrm{T}}$, $\mathbf{p}_{\mathrm{r}} = (1.5\text{ m}, 3\text{ m}, 1.5\text{ m})^{\mathrm{T}}$, $\mathbf{p}_{\mathrm{s}} = (3\text{ m}, 1\text{ m}, 1.5\text{ m})^{\mathrm{T}}$, and $\alpha = 0.6$ for all walls.

are positioned at $\mathbf{p}_r = (1.5\text{ m}, 3\text{ m}, 1.5\text{ m})^T$ and $\mathbf{p}_s = (3\text{ m}, 1\text{ m}, 1.5\text{ m})^T$, respectively. Additionally, the absorption coefficient is set to $\alpha = 0.6$ for all walls. In Fig. 7.4 (top), the room impulse response is plotted for a length of $L_h = 8192$. Contrarily, Fig. 7.4 (bottom) focuses on the direct path and the first reflections, which illustrate also the sinc-like characteristic of the individual reflections.

Finally, the order k of a reflected path can be determined by

$$k(\mathbf{u}, \mathbf{v}) = |v_x - u_x| + |v_x| + |v_y - u_y| + |v_y| + |v_z - u_z| + |v_z| \,. \tag{7.18}$$

Although the image source model is able to calculate reflections up to an arbitrary order, practically, only low-order reflections are calculated because of the strong increase of the number of image sources with rising order of reflections [Väl12].

7.1.3 Measurement of Room Impulse Responses

The direct measurement of a room impulse response is carried out by impulse excitation. Better measurement results are obtained by correlation measurement of the room impulse responses by using pseudo-random sequences as the excitation signal. Pseudo-random sequences can be generated by feedback shift registers [Mac76]. The pseudo-random sequence is periodic with period $L = 2^N - 1$, where N is the number of states of the shift register. The autocorrelation function (ACF) of such a random sequence is given by

$$r_{XX}(n) = \begin{cases} a^2 & n = 0, L, 2L, \ldots, \\ \dfrac{-a^2}{L} & \text{elsewhere,} \end{cases} \tag{7.19}$$

where a is the maximum value of the pseudo-random sequence. The ACF also has a period L. After going through a DA converter, the pseudo-random signal is fed through a loudspeaker into a room (see Fig. 7.5).

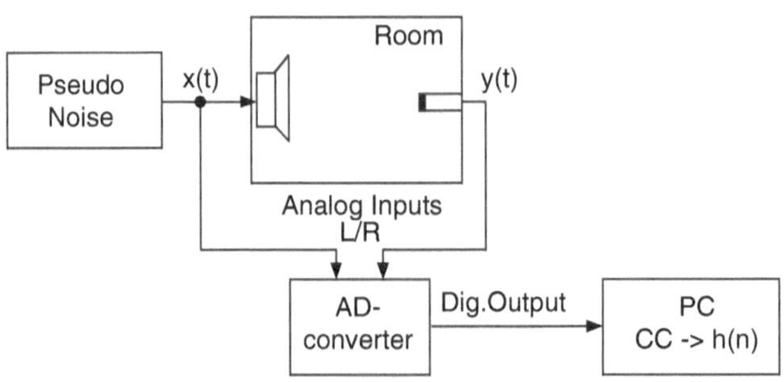

Figure 7.5 Measurement of room impulse response with pseudo-random signal $x(t)$.

At the same time, the pseudo-random signal and the room signal captured by a microphone are recorded on a personal computer. The impulse response is obtained with the cyclic cross-correlation:

$$r_{XY}(n) = r_{XX}(n) * h(n) \approx \tilde{h}(n). \tag{7.20}$$

For the measurement of room impulse responses, it has to be considered that the periodic length of the pseudo-random sequence must be longer than the length of the room impulse response.

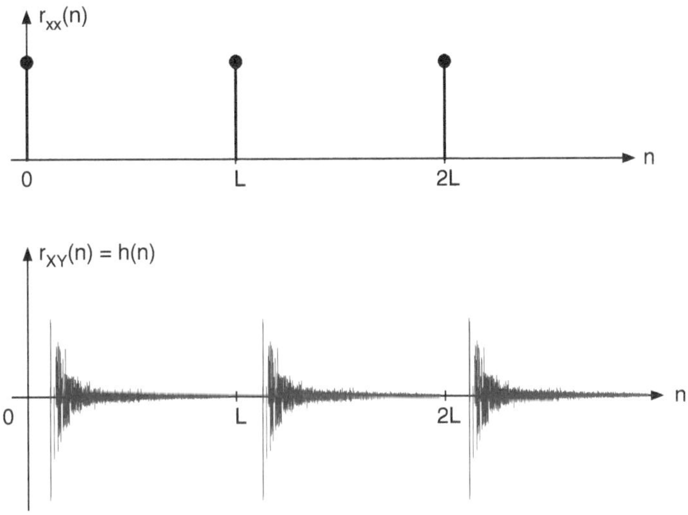

Figure 7.6 Periodic auto-correlation of pseudo-random sequence and periodic cross-correlation.

Otherwise, aliasing in the periodic cross-correlation $r_{XY}(n)$ (see Fig. 7.6) occurs. To improve the SNR of the measurement, the average of several periods of the cross-correlation is calculated.

In [Far00, Far07], Farina proposed to use exponential sine sweeps as the excitation signal for measuring impulse responses. In comparison with other signals, the use of exponential sine sweeps has two major advantages [Hol09, Mül01, Sta02]. First, owing to the exponential increase in frequency, more energy is present at low frequencies, resulting in a higher SNR at low frequencies, which is particularly desirable in audio applications. Second, nonlinearities of the system under test are separated from the linear impulse response by the exponential sine sweep method.

A continuous-time exponential sine sweep is defined as

$$x(t) = \sin\left[\frac{\omega_1 \cdot T}{\ln\left(\frac{\omega_2}{\omega_1}\right)} \cdot \left(e^{\frac{t}{T}\cdot\ln\left(\frac{\omega_2}{\omega_1}\right)} - 1\right)\right], \tag{7.21}$$

where T is the duration of the sweep in seconds, and $\omega_1 = 2\pi f_1$ and $\omega_2 = 2\pi f_2$ define the instantaneous angular frequencies at the beginning ($t = 0$ s) and the end of the sweep ($t = T$), respectively. Figure 7.7 illustrates the first half-second of an exponential sine sweep $x(t)$ with $T = 3$ s, $f_1 = 20$ Hz, and $f_2 = 20$ kHz. The evaluation of the period across time clearly indicates the increase in the frequency of the sine wave.

By calculating the derivative of the argument of the sine sweep $\arg[x(t)]$ with respect to time t, the instantaneous angular frequency is determined as

$$\omega(t) = \frac{d\arg[x(t)]}{dt} = \omega_1 \cdot e^{\frac{t}{T}\cdot\ln\left(\frac{\omega_2}{\omega_1}\right)}. \tag{7.22}$$

From this, the definition of the instantaneous angular frequencies at the beginning ($\omega(0) = \omega_1$) and the end ($\omega(T) = \omega_2$) of the sweep can be confirmed. Additionally, the exponential increase in frequency with time is explicitly shown. This increase is also visible in the spectrogram of the sine sweep shown in Fig. 7.8.

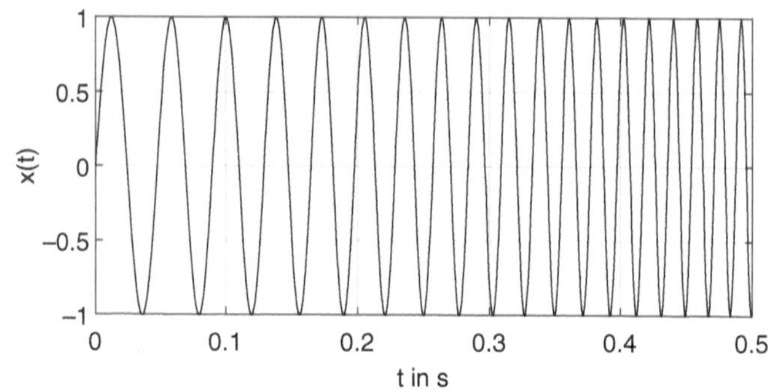

Figure 7.7 Beginning of an exponential sine sweep $x(t)$ with $T = 3$ s, $f_1 = 20$ Hz, and $f_2 = 20$ kHz.

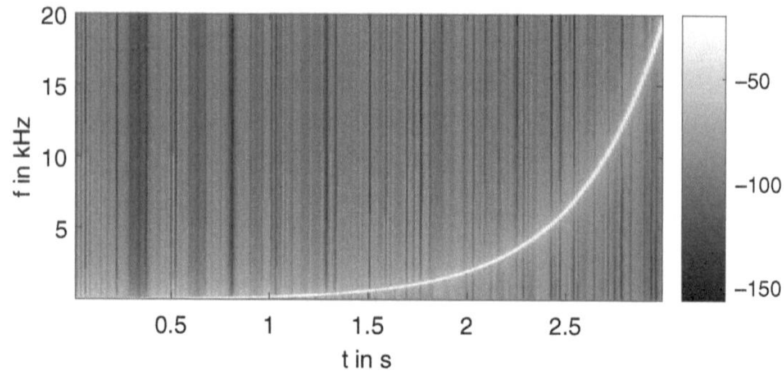

Figure 7.8 Spectrogram of an exponential sine sweep with $T = 3$ s, $f_1 = 20$ Hz, and $f_2 = 20$ kHz.

For a digital implementation with a sampling rate of f_S, the discrete-time exponential sine sweep

$$x(n) = \sin\left[\frac{\Omega_1 \cdot (L-1)}{\ln\left(\frac{\Omega_2}{\Omega_1}\right)} \cdot \left(e^{\frac{n}{L-1}\cdot\ln\left(\frac{\Omega_2}{\Omega_1}\right)} - 1\right)\right] \tag{7.23}$$

is used, where $L = Tf_S + 1$ specifies the length of the sweep in samples, and $\Omega_1 = 2\pi f_1/f_S$ and $\Omega_2 = 2\pi f_2/f_S$ define the instantaneous normalized angular frequencies at the beginning ($n = 0$) and the end ($n = L - 1$) of the sweep [Hol09]. Additionally, an inverse sine sweep $x_{\text{inv}}(n)$ can be defined as

$$x_{\text{inv}}(n) = x(L-1-n) \cdot \left(\frac{\Omega_2}{\Omega_1}\right)^{\frac{-n}{L-1}}, \tag{7.24}$$

where the first factor flips the sine sweep in time and the second factor is a correction factor that accounts for the equalization of the magnitude of the sine sweep. As can be seen in Fig. 7.9, the exponential increase of the frequency in the sine sweep $x(n)$ results in a higher magnitude for low frequencies. Because a simple flipping of the sine sweep in time will not change the magnitude response, a correction factor is included to change the magnitude of the inverse sweep, as shown in Fig. 7.9.

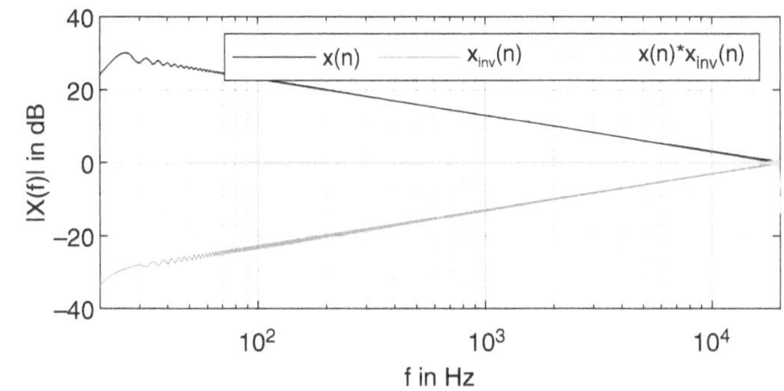

Figure 7.9 Magnitude responses of the exponential sine sweep $x(n)$, the inverse sine sweep $x_{\text{inv}}(n)$, and the convolution result $x(n) * x_{\text{inv}}(n)$. Note that the magnitude responses are scaled by $1/\sqrt{C}$.

The convolution of the exponential sine sweep $x(n)$ and the inverse sine sweep $x_{\text{inv}}(n)$ results in a scaled and time-shifted unit impulse

$$x(n) * x_{\text{inv}}(n) \approx C \cdot \delta(n - n_0), \tag{7.25}$$

where $n_0 = L - 1$ depends on the length of the inverse sweep and the correlation factor C is given in [Hol09] as

$$C = \frac{\pi L \cdot \left(\frac{\Omega_1}{\Omega_2} - 1\right)}{2(\Omega_2 - \Omega_1)\ln\left(\frac{\Omega_1}{\Omega_2}\right)}. \tag{7.26}$$

Here, the approximation of the unit impulse $\delta(n)$ results from the band limitation of the exponential sine sweep $x(n)$ in the range of Ω_1 to Ω_2. Correcting the time shift and scaling using Eq. (7.25) yields the band-limited unit impulse shown in Fig. 7.10. Furthermore, the magnitude response of the band-limited unit impulse is shown in Fig. 7.9. In addition, the band limitation also overshoots and passband ripples can be seen in the magnitude responses, which can be reduced by applying fade in and fade out on the sine sweep in the time domain [Hol09].

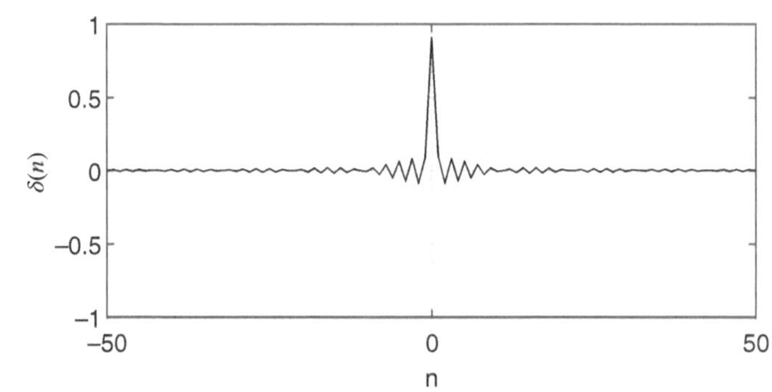

Figure 7.10 Result of the convolution of the exponential sine sweep $x(n)$ and the inverse sine sweep $x_{\text{inv}}(n)$.

When using an exponential sine sweep as an excitation signal during impulse response measurements, the recorded signal $y(n)$ is determined as

$$y(n) = x(n) * h(n), \tag{7.27}$$

where $h(n)$ defines the impulse response of the system under test. Finally, the convolution of the recorded signal $y(n)$ and the inverse sine sweep $x_{\text{inv}}(n)$ determines the measured impulse response $\hat{h}(n)$ as

$$y(n) * x_{\text{inv}}(n) = C \cdot \hat{h}(n - L + 1) \tag{7.28}$$

with C being the correlation factor defined in Eq. (7.26). Owing to the characteristics of the exponential sine sweep, this convolution separates the linear impulse response and the harmonic impulse responses of a nonlinear system in time [Hol09]. Here, a given frequency inside the kth harmonic is reached

$$\Delta n_k = (L - 1) \cdot \frac{\ln(k+1)}{\ln\left(\frac{\omega_2}{\omega_1}\right)} \tag{7.29}$$

samples before the excitation signal reaches this frequency. Thus, the kth harmonic impulse response will be visible at $n = -\Delta n_k$ in the anti-causal part of the measured impulse response. In Fig. 7.11, an exemplary measured room impulse response $\hat{h}(n)$ is shown. Here, the parameters of the sweep are $T = 3$ s, $f_S = 44.1$ kHz, $f_1 = 55$ Hz, and $f_2 = f_S/2$. For a better representation, the maximum absolute amplitude of the impulse response is set to 1 and the linear impulse response is moved to $n = 0$ by reversing the time shift $n_0 = L - 1$ introduced by the convolution. Furthermore, the impulse response is plotted in decibels rather than in linear scale. In addition to the linear impulse response at $n = 0$, the first and second harmonic impulse responses arise at $n_1 = -15300$ and $n_2 = -24250$, respectively.

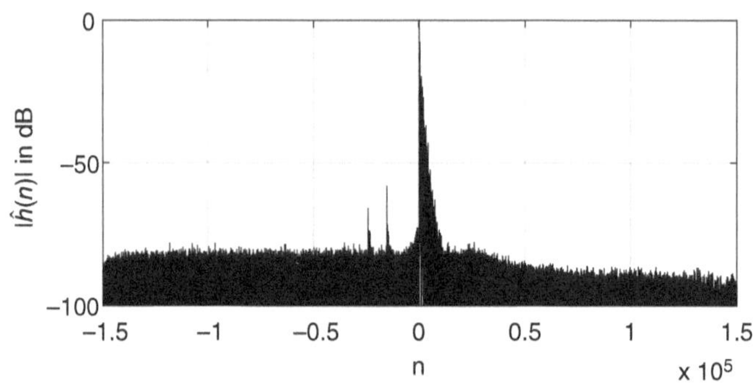

Figure 7.11 Exemplary measured room impulse response including the first and second harmonic impulse responses at $n = -\Delta n_k$ in the anti-causal part. The sweep parameters are $T = 3$ s, $f_S = 44.1$ kHz, $f_1 = 55$ Hz, and $f_2 = f_S/2$.

7.1.4 Simulation of Room Impulse Responses

The just described methods provide a means for calculating the impulse response out of the geometry of a room and for measuring the impulse response of a real room. The reproduction of such an impulse response is basically possible with the help of the *fast convolution* method, as described in Chapter 6. The ear signals at a listening position inside the room are computed by

$$y_L(n) = \sum_{k=0}^{N-1} x(k) \cdot h_L(n-k), \tag{7.30}$$

$$y_R(n) = \sum_{k=0}^{N-1} x(k) \cdot h_R(n-k), \tag{7.31}$$

where $h_L(n)$ and $h_R(n)$ are the measured impulse responses between the source inside the room, which generates the signal $x(n)$, and a dummy head with two ear microphones. Special implementations of fast convolution with low latency are described in [Soo90, Gar95, Rei95, Ege96, Joh00] and a hybrid approach based on convolution and recursive filters can be found in [Bro01]. Investigations regarding fast convolution with sparse psychoacoustic based room impulse responses are discussed in [Iid95, Lee03a, Lee03b].

In Sections 7.2 and 7.3, we will consider special approaches for early reflections and subsequent reverberation, respectively, which allow a parametric adjustment of all relevant parameters of a room impulse response. With this approach, an accurate room impulse response is not possible, but with a moderate computational complexity, a satisfying solution from an acoustical point of view can be achieved, as shown in Section 7.4. In Section 7.5, an efficient implementation of the convolutions (7.30) and (7.31) with a multi-rate signal processing approach [Zöl90, Sch92, Sch93, Sch94] will be discussed.

7.2 Early Reflections

Early reflections decisively affect room perception. *Spatial impression* is produced by early reflections which reach the listener laterally. The significance of lateral reflections in creating *spatial impression* was investigated by Barron [Bar71, Bar81]. Fundamental investigations of concert halls and their different acoustics have been described by Ando [And90].

7.2.1 Ando's Investigations

The results of the investigations by Ando are summarized in the following:

- Preferred *delay time of a single reflection*: with the ACF of the signal, the delay is determined by $|r_{xx}(\Delta t_1)| = 0.1 \cdot r_{xx}(0)$.
- Preferred *direction of a single reflection*: $\pm(55^\circ \pm 20^\circ)$.
- Preferred *amplitude of a single reflection*: $A_1 = \pm 5$ dB.
- Preferred *spectrum of a single reflection*: no spectral shaping.
- Preferred *delay time of a second reflection*: $\Delta t_2 = 1.8 \cdot \Delta t_1$.
- Preferred *reverberation time*: $T_{60} = 23 \cdot \Delta t_1$.

These results show that in terms of perception, a preferred pattern of reflections as well as the reverberation time depend decisively on the audio signal. Hence, for different audio signals like classical music, pop music, speech, or musical instruments, entirely different requirements for early reflections and reverberation time have to be considered.

7.2.2 Gerzon Algorithm

The commonly used method of simulating early reflections is shown in Figs. 7.12 and 7.13. The signal is weighted and fed into a system generating early reflections, followed by an addition to the input signal. The first M reflections are implemented by reading samples from a delay line and weighting these samples with a corresponding factor g_i (see Fig. 7.13). The design of a system for simulating early reflections will now be described, as proposed by Gerzon [Ger92].

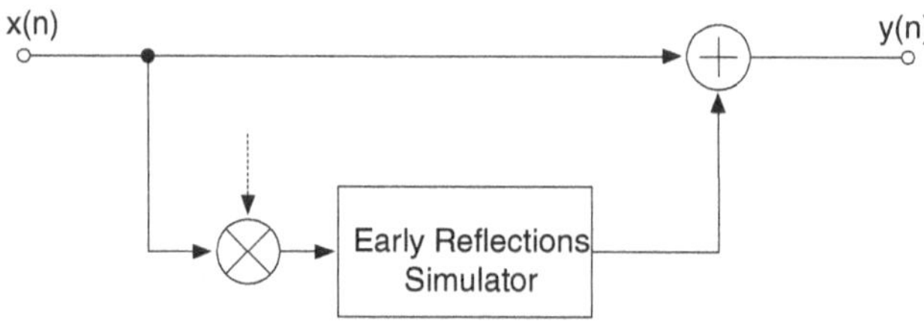

Figure 7.12 Simulation of early reflections.

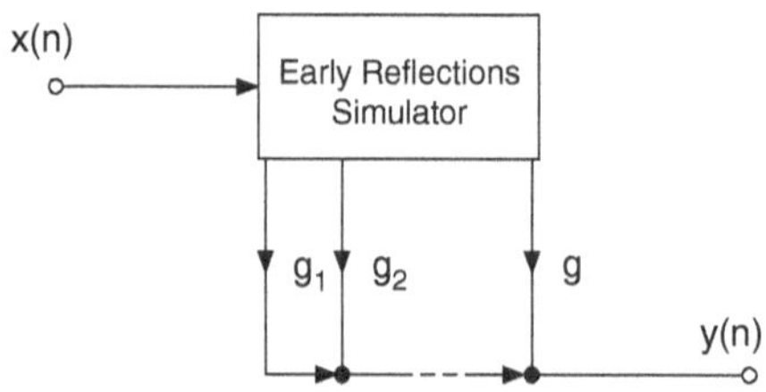

Figure 7.13 Early reflections.

Craven Hypothesis. The Craven hypothesis [Ger92] states that a human's perception of the distance to a sound source is evaluated with the help of the amplitude and delay time ratios of the direct signal and early reflections, as given by

$$g = \frac{d}{d'}, \tag{7.32}$$

$$T_D = \frac{d' - d}{c}, \tag{7.33}$$

$$\Rightarrow d = \frac{cT_D}{g^{-1} - 1}, \tag{7.34}$$

with

d distance of source;

d' distance of image source of the first reflection;

g relative amplitude of direct signal to first reflection;

c sound velocity;

T_D relative delay time of first reflection to direct signal.

Without a reflection, human beings are unable to determine the distance d to a sound source. The extended Craven hypothesis includes the absorption coefficient r for determining

$$g = \frac{d}{d'} \exp(-rT_D), \tag{7.35}$$

$$T_D = \frac{d' - d}{c}, \tag{7.36}$$

$$\rightarrow d = \frac{cT_D}{g^{-1} \exp(-rT_D) - 1}, \tag{7.37}$$

$$\rightarrow g = \frac{\exp(-rT_D)}{1 + cT_D/d}. \tag{7.38}$$

For a given reverberation time T_{60}, the absorption coefficient can be calculated by using $\exp(-rT_{60}) = 1/1000$ according to

$$r = (\ln 1000)/T_{60}. \tag{7.39}$$

With the relationships (7.36) and (7.38), the parameters for an early reflections simulator, as shown in Fig. 7.12, can be determined.

Gerzon's Distance Algorithm. For a system simulating early reflections produced by more than one sound source, Gerzon's distance algorithm can be used [Ger92], where several sound sources are placed with different distances as well as in the stereo position into a stereophonic sound field. An application of this technique is mainly used in multichannel mixing consoles.

By shifting a sound source by $-\delta$ (decrease of relative delay time), it follows that from the relative delay time of the first reflection $T_D - \delta/c = \frac{d'-(d+\delta)}{c}$, and the relative amplitude according to (7.38),

$$g_\delta = \frac{1}{1 + \frac{c(T_D - \delta/c)}{d + \delta}} \exp(-r(T_D - \delta/c)) = \left[\frac{d+\delta}{d} \exp(r\delta/c)\right] \frac{\exp(-rT_D)}{1 + cT_D/d}. \tag{7.40}$$

This results in a delay and a gain factor for the direct signal (see Fig. 7.14) as given by

$$d_2 = d + \delta, \tag{7.41}$$

$$t_D = \delta/c, \tag{7.42}$$

$$g_D = \frac{d}{d+\delta} \exp(-r\delta/c). \tag{7.43}$$

By shifting a sound source by $+\delta$ (increase of relative delay time), the relative delay time of the first reflection is $T_D - \delta/c = \frac{d'-(d-\delta)}{c}$. As a consequence, a delay and a gain factor for the effect signal (see Fig. 7.15) are given by

$$d_2 = d - \delta, \tag{7.44}$$

$$t_E = \delta/c, \tag{7.45}$$

$$g_E = \frac{d}{d+\delta} \exp(-r\delta/c). \tag{7.46}$$

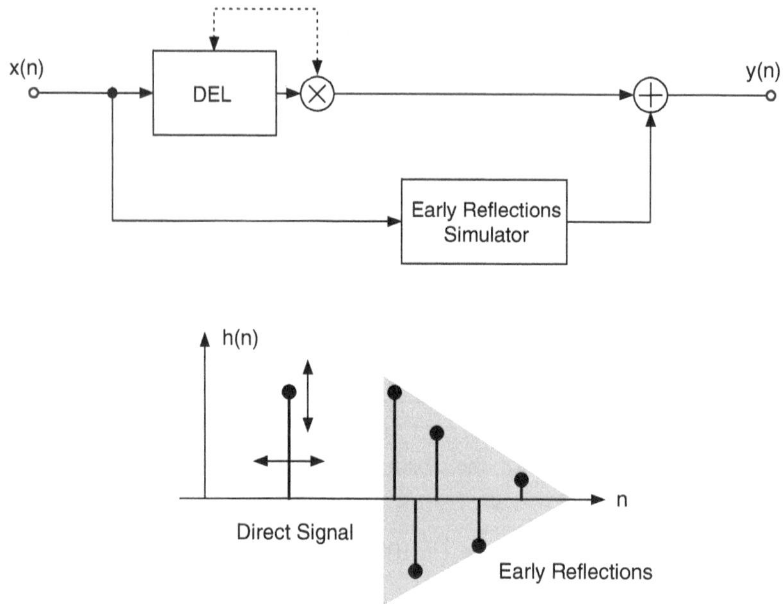

Figure 7.14 Delay and weighting of the direct signal.

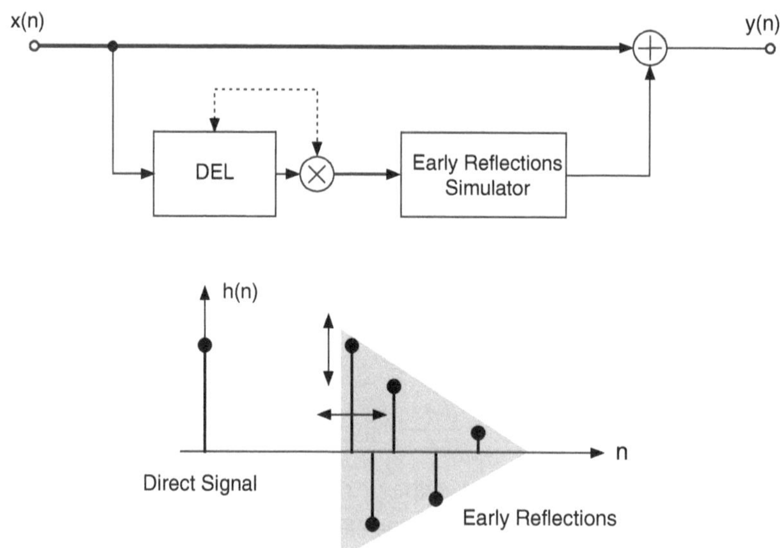

Figure 7.15 Delay and weighting of effect signal.

Using two delay systems in the direct signal as well as in the reflection path, two coupled weighting factors and delay lengths (see Fig. 7.16) can be obtained. For multichannel applications like digital mixing consoles, the scheme in Fig. 7.17 is suggested by Gerzon [Ger92]. Only one system for implementing early reflections is necessary.

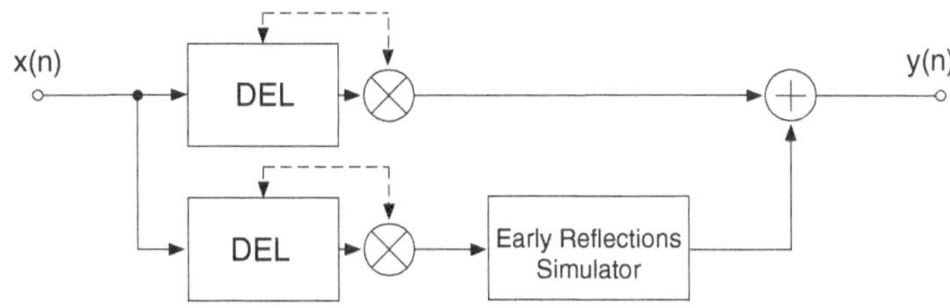

Figure 7.16 Coupled factors and delays.

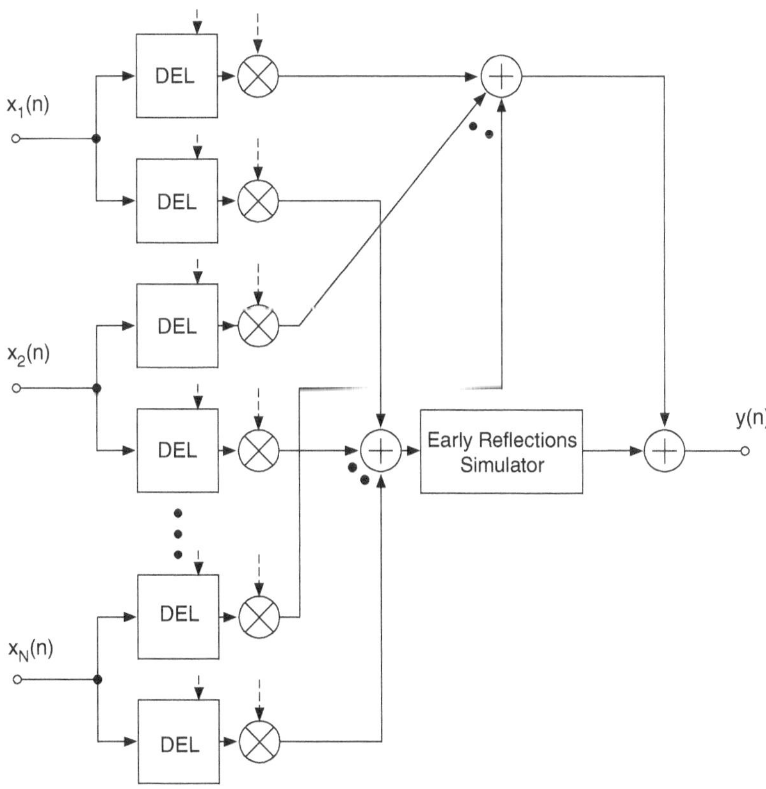

Figure 7.17 Multichannel application.

Stereo Implementation. In many applications, stereo signals have to be processed (see Fig. 7.18). For this, reflections from both sides with positive and negative angles are implemented to avoid stereo displacements. The weighting is done with

$$g_i = \frac{\exp(-rT_i)}{1 + cT_i/d},$$

$$\mathbf{G}_i = g_i \begin{pmatrix} \cos\Theta_i & -\sin\Theta_i \\ \sin\Theta_i & \cos\Theta_i \end{pmatrix}. \tag{7.47}$$

For each reflection, a weighting factor and an angle have to be considered.

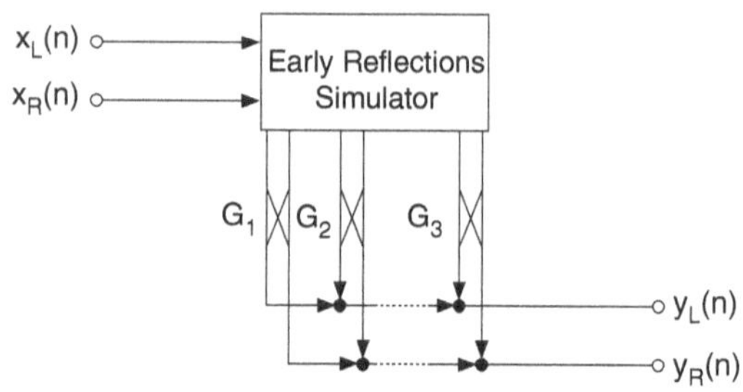

Figure 7.18 Stereo reflections.

Generation of early reflection with increasing time density. In [Schr61], it is stated that the time density of reflections increases proportional to the square of time:

$$\text{Number of reflections per second} = (4\pi c^3/V) \cdot t^2. \tag{7.48}$$

After time t_C, the reflections have a statistical decay behavior. For a pulse width of Δt, individual reflections overlap after

$$t_C = 5 \cdot 10^{-5}\sqrt{V/\Delta t}. \tag{7.49}$$

To avoid an overlap of reflections, Gerzon [Ger92] suggests the increase of the density of reflections with t^p (for example, $p = 1, 0.5$ leads to t or $t^{0.5}$). In the interval (0,1], with initial value x_0 and a number k between 0.5 and 1, the following procedure is performed:

$$y_i = x_0 + ik(\text{mod } 1) \qquad i = 0,1, \ldots, M-1. \tag{7.50}$$

The numbers y_i in the interval (0,1] are now transformed to time delays T_i in the interval $[T_{min}, T_{min} + T_{max}]$ by

$$b = T_{min}^{1+p}, \tag{7.51}$$

$$a = (T_{max} + T_{min})^{1+p} - b, \tag{7.52}$$

$$T_i = (ay_i + b)^{1/(1+p)}. \tag{7.53}$$

The increase of the density of reflections is shown by the example in Fig. 7.19.

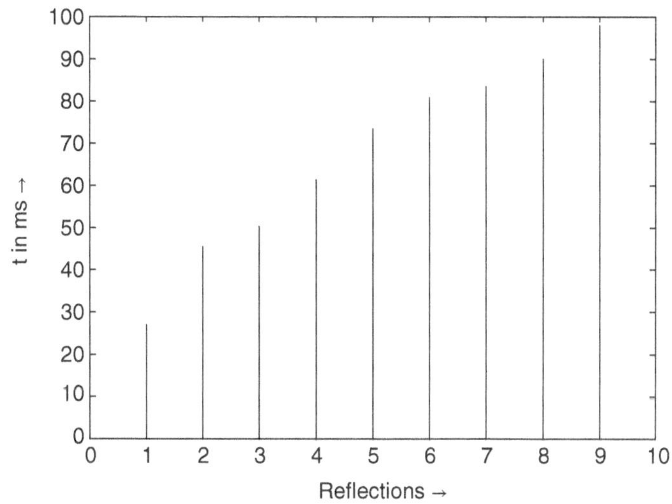

Figure 7.19 Increase of density for nine reflections.

7.3 Subsequent Reverberation

This section deals with techniques for reproducing subsequent reverberation. The first approaches by Schroeder [Schr61, Schr62] and their extension by Moorer [Moo78] will be described. Further developments by Stautner/Puckette [Sta82], Smith [Smi85], Dattarro [Dat97], and Gardner [Gar98] led to general feedback networks [Ger71, Ger76, Jot91, Jot92, Roc95, Roc97a, RS97b, Roc02], which have a random impulse response with exponential decay. An extensive discussion on the analysis and synthesis parameters of subsequent reverberation can be found in [Ble01]. An important parameter of subsequent reverberation [Cre03] is, in addition to the echo density, the quadratic increase of

$$\text{Frequency density} = \frac{4\pi V}{c^3} \cdot f^2 \tag{7.54}$$

with frequency. The following systems perform the quadratic increase in echo density and frequency density.

7.3.1 Schroeder Algorithm

The first software implementations of room simulation algorithms were carried out in 1961 by Schroeder. The basis for simulating an impulse response with exponential decay is a recursive comb filter, shown in Fig. 7.20.

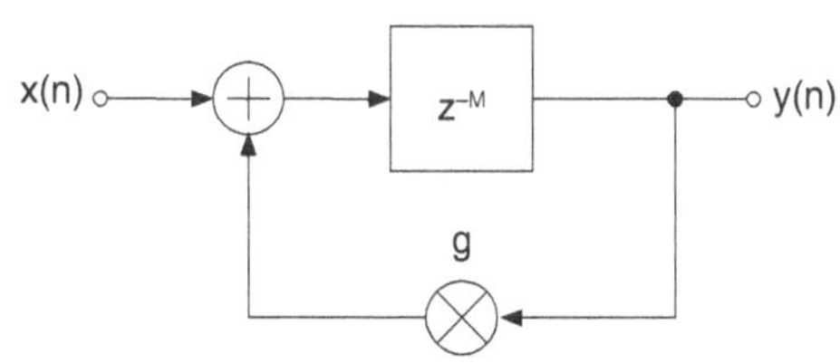

Figure 7.20 Recursive comb filter (g = feedback factor, M = delay length).

The transfer function is given by

$$H(z) = \frac{z^{-M}}{1 - gz^{-M}} \tag{7.55}$$

$$= \sum_{k=0}^{M-1} \frac{A_k}{z - z_k} \tag{7.56}$$

with

$$A_k = \frac{z_k}{Mg} \qquad \text{residues;} \tag{7.57}$$

$$z_k = re^{j2\pi k/M} \qquad \text{poles;} \tag{7.58}$$

$$r = g^{1/M} \qquad \text{pole radius.} \tag{7.59}$$

With the correspondence of the Z-transform $a/(z-a) \circ\!\!-\!\!\bullet\ \epsilon(n-1)a^n$, the impulse response is given by

$$H(z) \circ\!\!-\!\!\bullet\ h(n) = \frac{\epsilon(n-1)}{Mg} \sum_{k=0}^{M-1} z_k^n,$$

$$h(n) = \frac{\epsilon(n-1)}{Mg} r^n \sum_{k=0}^{M-1} e^{j\Omega_k n}. \tag{7.60}$$

The complex poles are combined as pairs so that the impulse response can be written as

$$h(n) = \frac{\epsilon(n-1)}{Mg} r^n \sum_{k=1}^{\frac{M}{2}-1} \cos \Omega_k n \qquad M \text{ even} \tag{7.61}$$

$$= \frac{\epsilon(n-1)}{Mg} r^n \left[1 + \sum_{k=1}^{\frac{M+1}{2}-1} \cos \Omega_k n \right] \qquad M \text{ uneven.} \tag{7.62}$$

The impulse response is expressed as a summation of cosine oscillations with frequencies Ω_k. These frequencies correspond to the eigenfrequencies of a room. They decay with an exponential envelope r^n, where r is the damping constant (see Fig. 7.22a). The overall impulse response is weighted by $\frac{1}{Mg}$. The frequency response of the comb filter is shown in Fig. 7.22c and is given by

$$|H(e^{j\Omega})| = \sqrt{\frac{1}{1 - 2g\cos(\Omega M) + g^2}}. \tag{7.63}$$

For positive g, it shows maxima at $\Omega = 2\pi k/M \quad (k = 0,1, \ldots, M-1)$ of magnitude

$$|H(e^{j\Omega})|_{\max} = \frac{1}{1-g} \tag{7.64}$$

and minima at $\Omega = (2k+1)\pi/M \quad (k = 0,1, \ldots, M-1)$ of magnitude

$$|H(e^{j\Omega})|_{\min} = \frac{1}{1+g}. \tag{7.65}$$

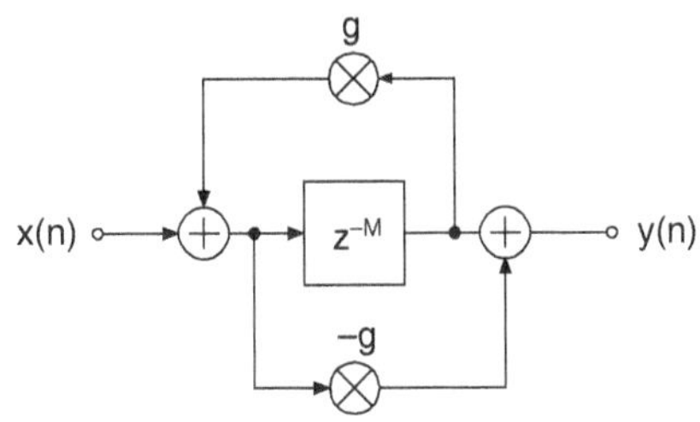

Figure 7.21 Allpass filter (M = delay length).

Another basis of the Schroeder algorithm is the allpass filter, shown in Fig. 7.21, with transfer function

$$H(z) = \frac{z^{-M} - g}{1 - gz^{-M}} \tag{7.66}$$

$$= \frac{z^{-M}}{1 - gz^{-M}} - \frac{g}{1 - gz^{-M}}. \tag{7.67}$$

From Eq. (7.67), it can be seen that the impulse response can also be expressed as a summation of cosine oscillations.

The impulse responses and the frequency responses of a comb filter and an allpass filter are presented in Fig. 7.22 with a negative g. Both impulse responses show an exponential decay.

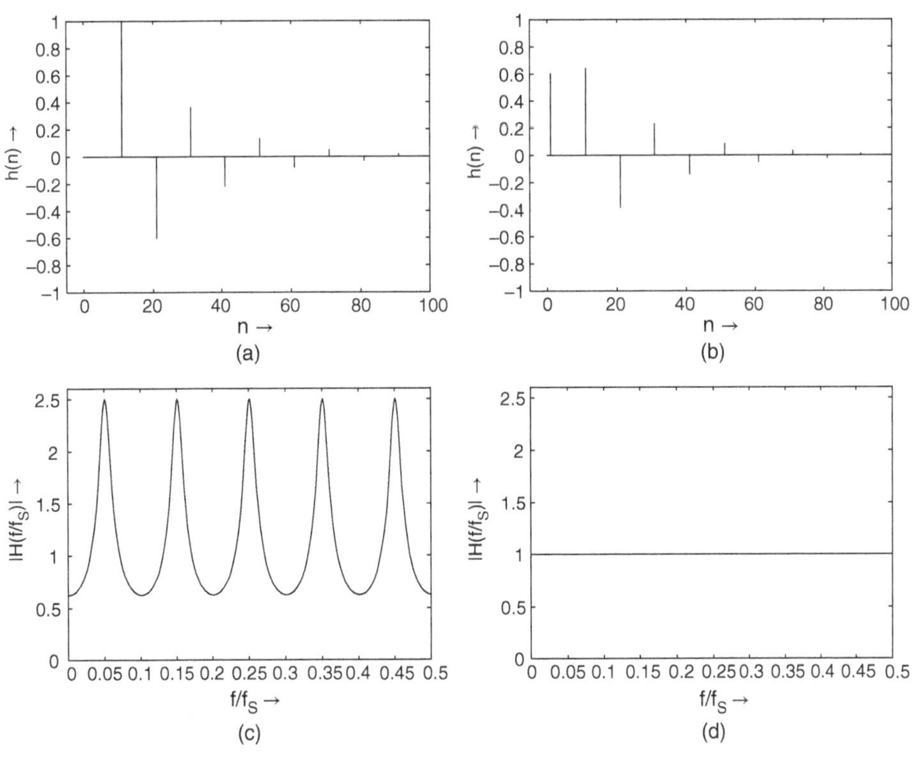

Figure 7.22 (a) Impulse response of a comb filter ($M = 10$, $g = -0.6$). (b) Impulse response of an allpass filter ($M = 10$, $g = -0.6$). (c) Frequency response of a comb filter. (d) Frequency response of an allpass filter.

A sample in the impulse response occurs every M sampling periods. The density of samples in the impulse responses does not increase with time. For the recursive comb filter, spectral shaping, owing to the maxima at the corresponding poles of the transfer function, is observed.

Frequency Density

The frequency density describes the number of eigenfrequencies per Hertz and is defined for a comb filter [Jot91] as

$$D_f = M \cdot T_S \quad [1/\text{Hz}]. \tag{7.68}$$

A single comb filter gives M resonances in the interval $[0, 2\pi]$, which are separated by a frequency distance of $\Delta f = \frac{f_S}{M}$. To increase the frequency density, a parallel circuit (see Fig. 7.23) of P comb filters is used, which leads to

$$H(z) = \sum_{p=1}^{P} \frac{z^{-M_p}}{1 - g_p z^{-M_p}} = \left[\frac{z^{-M_1}}{1 - g_1 z^{-M_1}} + \frac{z^{-M_2}}{1 - g_2 z^{-M_2}} + \cdots\right]. \tag{7.69}$$

The choice of the delay systems [Schr62] is suggested as

$$M_1 : M_P = 1 : 1.5 \tag{7.70}$$

and leads to a frequency density

$$D_f = \sum_{p=1}^{P} M_p \cdot T_S = P \cdot \overline{M} \cdot T_S. \tag{7.71}$$

In [Schr62], a necessary frequency density of $D_f = 0.15$ eigenfrequencies per Hertz is proposed.

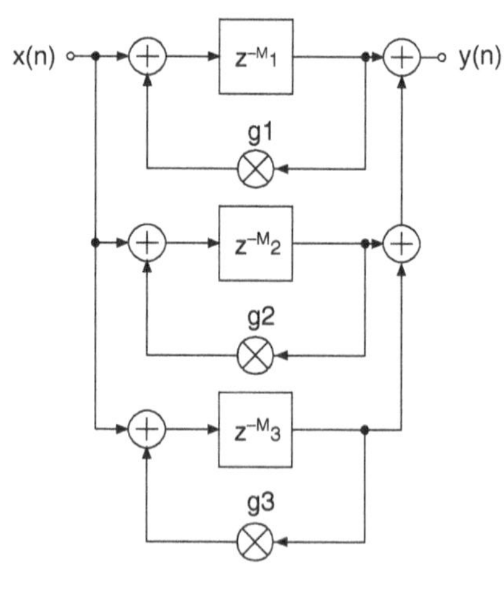

Figure 7.23 Parallel circuit of comb filters.

Echo Density

The echo density is the number of reflections per second and is defined for a comb filter [Jot91] as

$$D_t = \frac{1}{M \cdot T_S} \quad [1/\mathrm{s}]. \tag{7.72}$$

For a parallel circuit of comb filters, the echo density is given by

$$D_t = \sum_{p=1}^{P} \frac{1}{M_p \cdot T_S} = P \frac{1}{\overline{M} \cdot T_S}. \tag{7.73}$$

With Eqs. (7.71) and (7.73), the number P of parallel comb filters and the mean delay length M,

$$P = \sqrt{D_f \cdot D_t}, \tag{7.74}$$

$$\overline{M}T_S = \sqrt{D_f / D_t}, \tag{7.75}$$

are obtained. For a frequency density $D_f = 0.15$ and an echo density $D_t = 1000$, it can be concluded that the number of parallel comb filters is $P = 12$ and the mean delay length is $\overline{M}T_S =$ 12 ms. Because the frequency density is proportional to the reverberation time, the number of parallel comb filters has to be increased accordingly.

A further increase of the echo density is achieved by a cascade circuit of P_A allpass filters (see Fig. 7.24) with transfer function

$$H(z) = \prod_{p=1}^{P_A} \frac{z^{-M_p} - g_p}{1 - g_p z^{-M_p}}. \tag{7.76}$$

These allpass sections are connected in series with the parallel circuit of comb filters. For a sufficient echo density, 10000 reflections per second are necessary [Gri89].

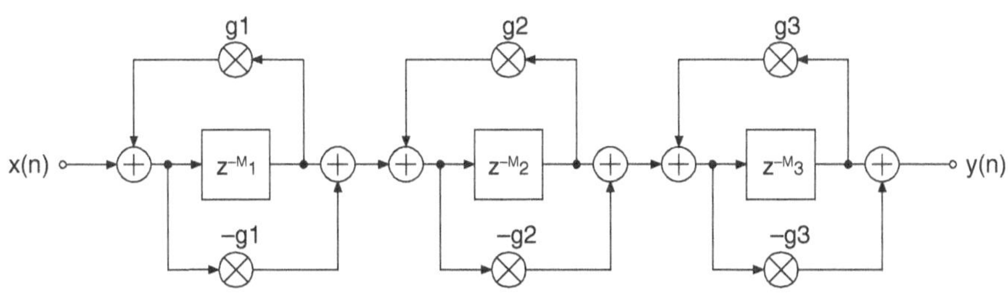

Figure 7.24 Cascade circuit of allpass filters.

Avoiding Unnatural Resonances

Because the impulse response of a single comb filter can be described as a sum of M (delay length) decaying sinusoidal oscillations, the short-time FFT of consecutive parts from this impulse

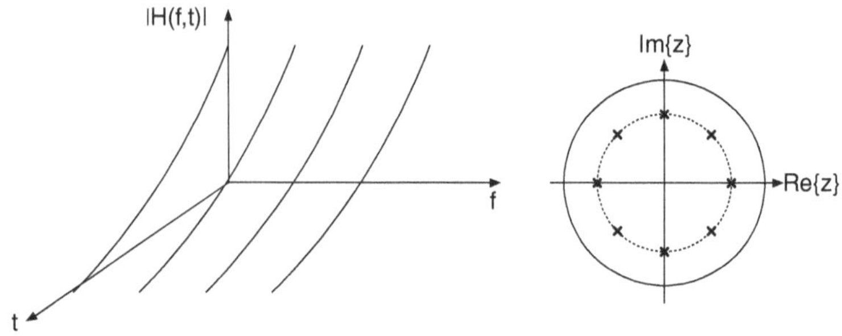

Figure 7.25 Short-time spectra of a comb filter ($M = 8$).

response gives the frequency response shown in Fig. 7.25 in the time-frequency domain. Only the maxima are presented. The parallel circuit of comb filters with the condition (7.70) leads to radii of the pole distribution, as given by $r_p = g_p^{1/M_p}$ ($p = 1, 2, \ldots, P$). To avoid unnatural resonances, the radii of the pole distribution of a parallel circuit of comb filters must satisfy the condition:

$$r_p = \text{const.} = g_p^{1/M_p} \quad \text{for} \quad p = 1, 2, \ldots, P. \tag{7.77}$$

This leads to the short-time spectra and the pole distribution, as shown in Fig. 7.26. Figure 7.27 shows the impulse response and the echogram (logarithmic presentation of the amplitude of the impulse response) of a parallel circuit of comb filters with equal and unequal pole radii. For unequal pole radius, the different decay times of the eigenfrequencies can be seen.

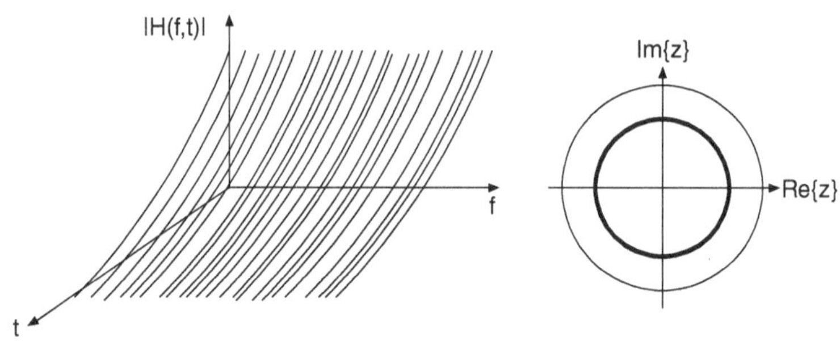

Figure 7.26 Short-time spectra of a parallel circuit of comb filters.

Reverberation Time

The reverberation time of a recursive comb filter can be adjusted with the feedback factor g, which describes the ratio

$$g = \frac{h(n)}{h(n-M)} \tag{7.78}$$

of two different non-zero samples of the impulse response separated by M sampling periods. The factor g describes the decay constant per M samples. The decay constant per sampling period can

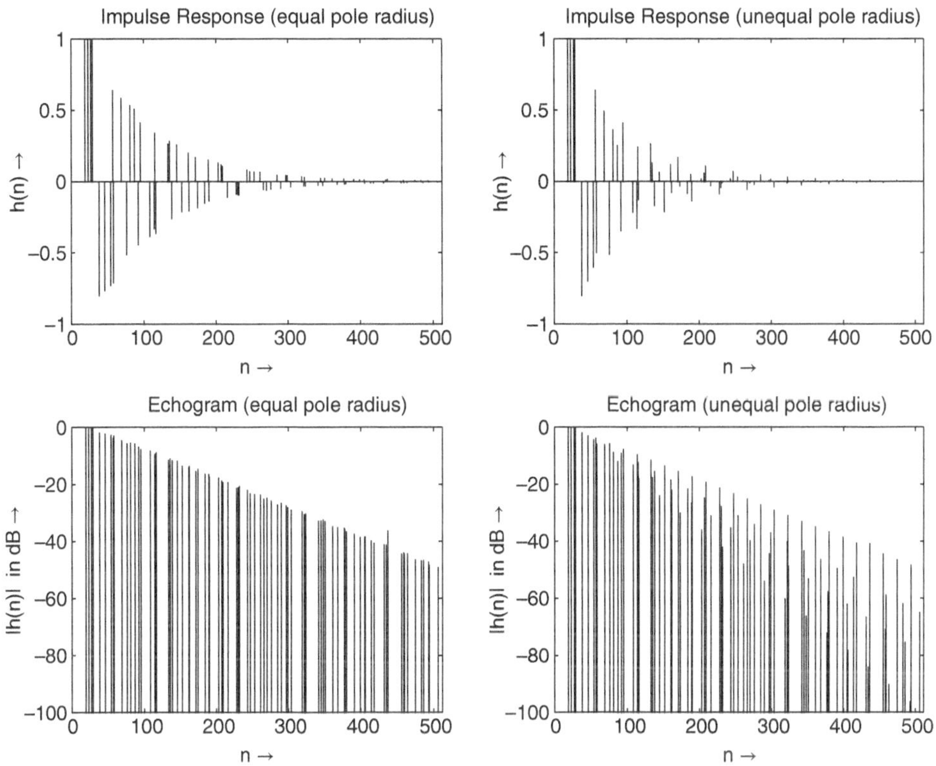

Figure 7.27 Impulse response and echogram.

be calculated from the pole radius $r = g^{1/M}$ and is defined as

$$r = \frac{h(n)}{h(n-1)}. \tag{7.79}$$

The relationship between feedback factor g and pole radius r can also be expressed using Eqs. (7.78) and (7.79) and is given by

$$g = \frac{h(n)}{h(n-M)} = \frac{h(n)}{h(n-1)} \cdot \frac{h(n-1)}{h(n-2)} \cdots \frac{h(n-(M-1))}{h(n-M)} = r \cdot r \cdot r \cdots r = r^M. \tag{7.80}$$

With the constant radius $r = g_p^{1/M_p}$ and the logarithmic parameters $R = 20\,\log_{10} r$ and $G_p = 20\,\log_{10} g_p$, the attenuation per sampling period is given by

$$R = \frac{G_p}{M_p}. \tag{7.81}$$

The reverberation time is defined as the decay time of the impulse response to $-60\,\text{dB}$. With $\frac{-60}{T_{60}} = \frac{R}{T_S}$, the reverberation time can be written as

$$T_{60} = -60\frac{T_S}{R} = -60\frac{T_S M_p}{G_p} = \frac{3}{\log_{10}|1/g_p|} M_p \cdot T_S. \tag{7.82}$$

The control of reverberation time can either be carried out with the feedback factor g or the delay parameter M. The increase of the reverberation time with factor g is responsible for a pole radius close to the unit circle and, hence, leads to an amplification of maxima of the frequency response (see Eq. (7.64)). This leads to a *coloring* of the sound impression. The increase of the delay parameter M, however, leads to an impulse response whose non-zero samples are far apart from each other so that individual echoes can be heard. The discrepancy between echo density and frequency density for a given reverberation time can be solved by a sufficient number of parallel comb filters.

Frequency-dependent Reverberation Time

The eigenfrequencies of rooms have a rapid decay for high frequencies. A frequency-dependent reverberation time can be implemented with a lowpass filter,

$$H_1(z) = \frac{1}{1 - az^{-1}}, \tag{7.83}$$

in the feedback loop of a comb filter. The modified comb filter in Fig. 7.28 has transfer function

$$H(z) = \frac{z^{-M}}{1 - gH_1(z)z^{-M}} \tag{7.84}$$

with the stability criterion

$$\frac{g}{1 - a} < 1. \tag{7.85}$$

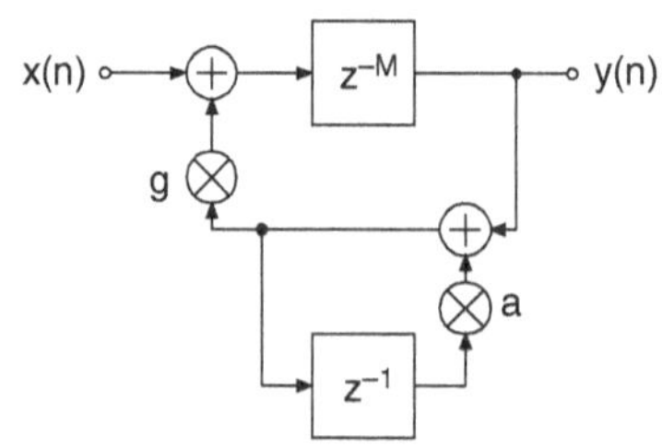

Figure 7.28 Modified lowpass comb filter.

The short-time spectra and the pole distribution of a parallel circuit with lowpass comb filters are presented in Fig. 7.29. Low eigenfrequencies decay slower than higher ones. The circular pole distribution becomes an elliptical distribution where the low-frequency poles are moved toward the unit circle.

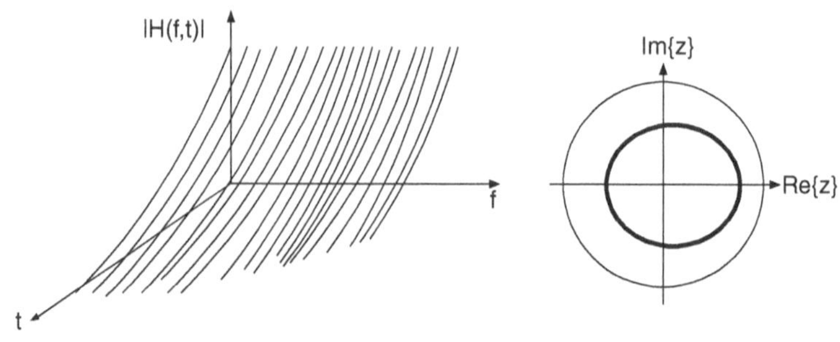

Figure 7.29 Short-time spectra of a parallel circuit of lowpass comb filters.

Stereo Room Simulation

An extension of the Schroeder algorithm was suggested by Moorer [Moo78]. In addition to a parallel circuit of comb filters in series with a cascade of allpass filters, a pattern of early reflections is generated. Figure 7.30 shows a room simulation system for a stereo signal. The generated room signals $e_L(n)$ and $e_R(n)$ are added to the direct signals $x_L(n)$ and $x_R(n)$. The input of the room simulation is the mono signal $x_M(n) = x_L(n) + x_R(n)$ (sum signal). This mono signal is added to the left and right room signals after going through a delay line DEL1. The total sum of all reflections is fed via another delay line DEL2 to a parallel circuit of comb filters which implements subsequent reverberation. To get a high-quality spatial impression, it is necessary to decorrelate the room signals $e_L(n)$ and $e_R(n)$ [Bla74, Bla85]. This can be achieved by taking left and right room signals at different points out of the parallel circuit of comb filters. These room signals are then fed to an allpass section to increase the echo density.

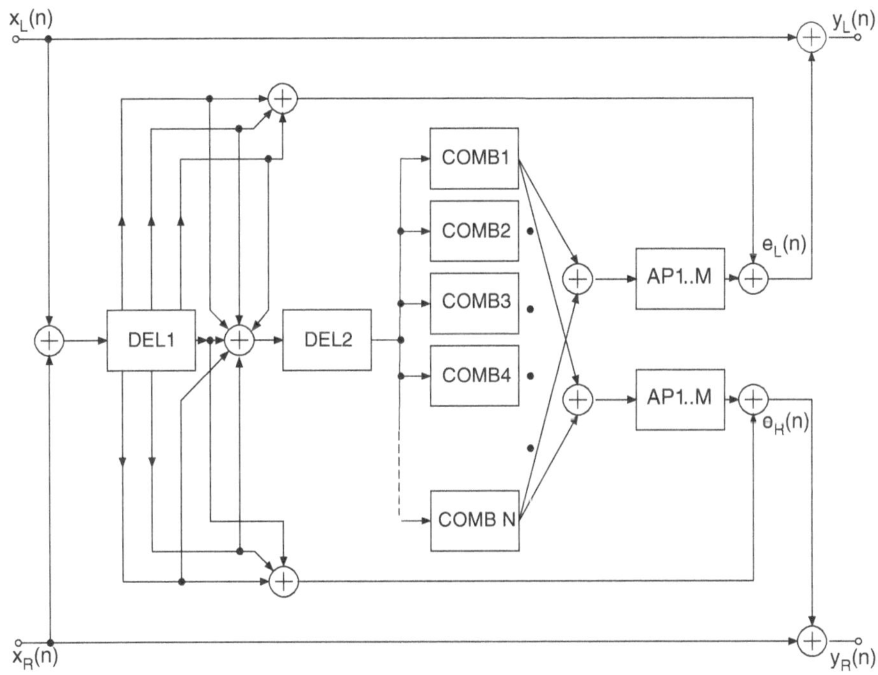

Figure 7.30 Stereo room simulation.

In addition to the described system for stereo room simulation in which the mono signal is processed with a room algorithm, it is also possible to perform complete stereo processing of $x_L(n)$ and $x_R(n)$, or to process a mono signal $x_M(n) = x_L(n) + x_R(n)$ and a side (difference) signal $x_S(n) = x_L(n) - x_R(n)$ individually.

7.3.2 General Feedback Systems

Further developments of the comb filter method by Schroeder tried to improve the acoustic quality of reverberation and especially the increase of echo density [Ger71, Ger76, Sta82, Jot91, Jot92, Roc95, Roc97a, RS97b]. With respect to [Jot91], the general feedback system in Fig. 7.31 is considered. For simplification, only three delay systems are shown. The feedback of output signals is carried out with the help of a matrix **A** which feeds back each of the three outputs to the three inputs.

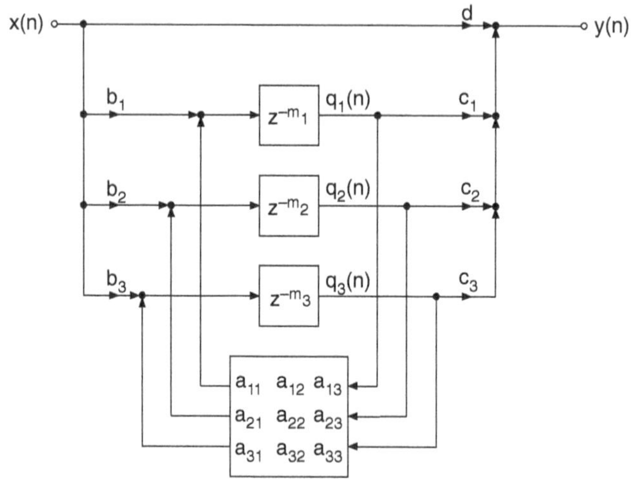

Figure 7.31 General feedback system.

In general, for N delay systems, we can write

$$y(n) = \sum_{i=1}^{N} c_i q_i(n) + dx(n), \tag{7.86}$$

$$q_j(n + m_j) = \sum_{i=1}^{N} a_{ij} q_i(n) + b_j x(n) \qquad 1 \leq j \leq N. \tag{7.87}$$

The Z-transform leads to

$$Y(z) = \mathbf{c}^T \mathbf{Q}(z) + d \cdot X(z), \tag{7.88}$$

$$\begin{aligned} &\mathbf{D}(z) \cdot \mathbf{Q}(z) = \mathbf{A} \cdot \mathbf{Q}(z) + \mathbf{b} \cdot X(z), \\ &\to \mathbf{Q}(z) = [\mathbf{D}(z) - \mathbf{A}]^{-1} \mathbf{b} \cdot X(z), \end{aligned} \tag{7.89}$$

with

$$\mathbf{Q}(z) = \begin{bmatrix} Q_1(z) \\ \vdots \\ Q_N(z) \end{bmatrix}, \quad \mathbf{b} = \begin{bmatrix} b_1 \\ \vdots \\ b_N \end{bmatrix}, \quad \mathbf{c} = \begin{bmatrix} c_1 \\ \vdots \\ c_N \end{bmatrix}, \tag{7.90}$$

and the diagonal delay matrix

$$\mathbf{D}(z) = \mathrm{diag}[z^{-m_1} \; \cdots \; z^{-m_N}]. \tag{7.91}$$

With Eq. (7.89), the Z-transform of the output is given by

$$Y(z) = \mathbf{c}^T [\mathbf{D}(z) - \mathbf{A}]^{-1} \mathbf{b} \cdot X(z) + d \cdot X(z) \tag{7.92}$$

and the transfer function by

$$H(z) = \mathbf{c}^T [\mathbf{D}(z) - \mathbf{A}]^{-1} \mathbf{b} + d. \tag{7.93}$$

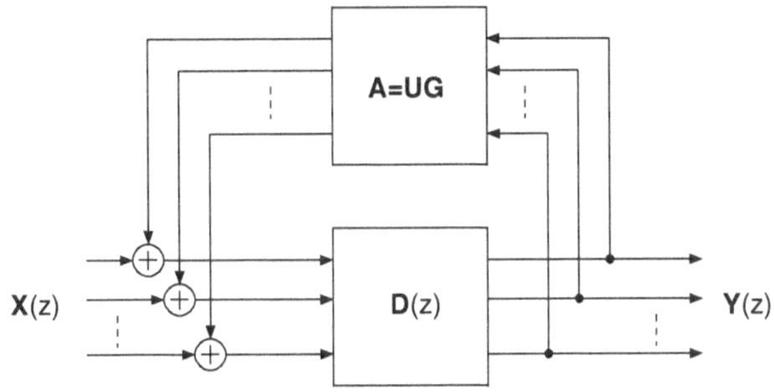

Figure 7.32 Feedback system.

The system is stable if the feedback matrix **A** can be expressed as a product of unitary matrix **U** ($\mathbf{U}^{-1} = \overline{\mathbf{U}}^T$) and a diagonal matrix with $g_{ii} < 1$ (derivation in [Sta82]). Figure 7.32 shows a general feedback system with input vector $\mathbf{X}(z)$, the output vector $\mathbf{Y}(z)$, a diagonal matrix $\mathbf{D}(z)$ consisting of purely delay systems z^{-m_i}, and a feedback matrix **A**. This feedback matrix consists of an orthogonal matrix **U** multiplied by the matrix **G**, which results in a weighting of the feedback matrix **A**.

If an orthogonal matrix **U** is chosen and the weighting matrix is equal to the unit matrix $\mathbf{G} = \mathbf{I}$, the system in Fig. 7.32 implements a white-noise random signal with Gaussian distribution when a pulse excitation is applied to the input. The time density of this signal slowly increases with time. If the diagonal elements of the weighting matrix **G** are less than one, a random signal with exponential amplitude decay results. With the help of the weighting matrix **G**, the reverberation time can be adjusted. Such a feedback system performs the convolution of an audio input signal with an impulse response of exponential decay.

The effect of the orthogonal matrix **U** on the subjective sound perception of subsequent reverberation is of particular interest. A relationship between the distribution of the eigenvalues of the matrix **U** on the unit circle and the poles of the system transfer function cannot be described analytically, owing to the high order of the feedback system. In [Her94], it is shown experimentally that the distribution of eigenvalues within the right-hand or left-hand complex plane produces a uniform distribution of poles of the system transfer function. Such a feedback matrix leads to an acoustically improved reverberation. The echo density rapidly increases to the maximum value of one sample per sampling period for a uniform distribution of eigenvalues. In addition to the feedback matrix, additional digital filtering is necessary to spectrally shape the subsequent reverberation and to implement frequency-dependent decay times (see [Jot91]). The following example illustrates the increase of the echo density.

Example: First, a system with only one feedback path per comb filter is considered. The feedback matrix is then given by

$$\mathbf{A} = \frac{g}{\sqrt{2}}\,\mathbf{I}. \tag{7.94}$$

Figure 7.33 shows the impulse response and the amplitude frequency response.

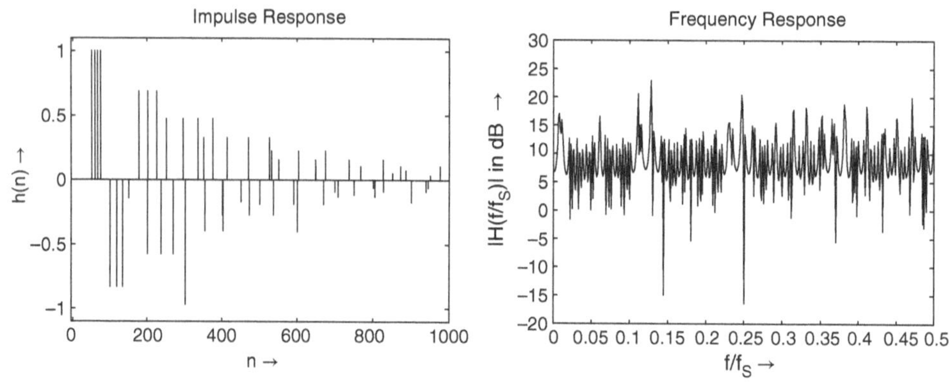

Figure 7.33 Impulse response and frequency response of 4-delay system with a unit matrix as unitary feedback matrix ($g = 0.83$).

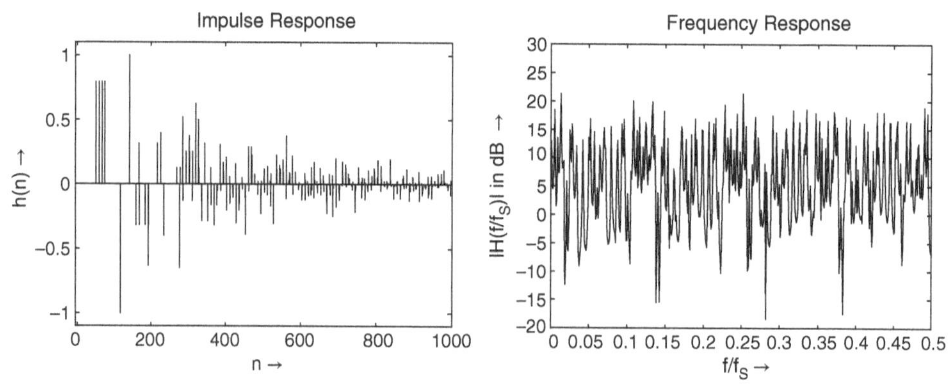

Figure 7.34 Impulse response and frequency response of a 4-delay system with unitary feedback matrix ($g = 0.63$).

With the feedback matrix

$$\mathbf{A} = \frac{g}{\sqrt{2}} \begin{bmatrix} 0 & 1 & 1 & 0 \\ -1 & 0 & 0 & -1 \\ 1 & 0 & 0 & -1 \\ 0 & 1 & -1 & 0 \end{bmatrix} \tag{7.95}$$

from [Sta82], the impulse response and the corresponding frequency response shown in Fig. 7.34 are obtained. In contrast to Fig. 7.33, an increase of the echo density of the impulse response is noticed.

7.3.3 Feedback Allpass Systems

In addition to the general feedback systems, simple delay systems with feedback have been used for room simulators (see Fig. 7.35). These simulators are based on a delay line, where single delays are fed back with L feedback coefficients to the input. The sum of input signal

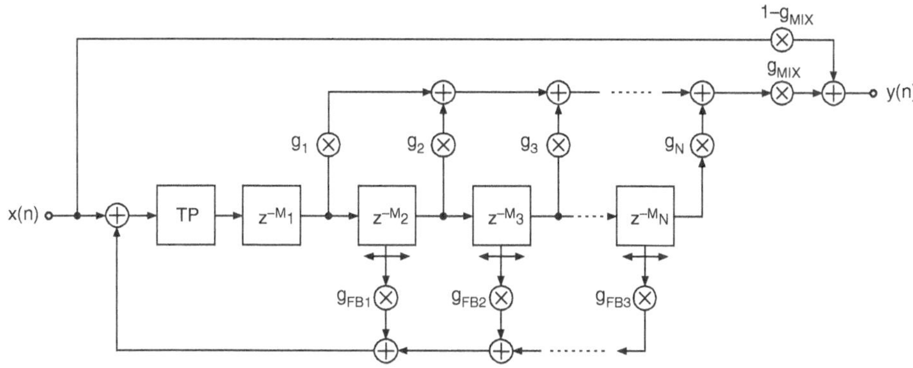

Figure 7.35 Room simulation with delay line and forward and backward coefficients.

and feedback signal is lowpass filtered or spectrally weighted by a low-frequency shelving filter and is then put to the delay line again. The first N reflections are extracted out of the delay line according to the reflection pattern of the simulated room. They are weighted and added to the output signal. The mixing between the direct signal and the room signal is adjusted by the factor g_{MIX}. The inner system can be described by a rational transfer function $H(z) = Y(z)/X(z)$. To avoid a low frequency density, the feedback delay lengths can be made time variant [Gri89, Gri91].

Increasing the echo density can be achieved by replacing the delays z^{-M_i} by frequency-dependent allpass systems $A(z^{-M_i})$. This extension was first proposed by Gardner in [Gar92a, Gar92b, Gar92c, Gar98]. In addition to the replacement of $z^{-M_i} \rightarrow A(z^{-M_i})$, the allpass systems can be extended by *embedded allpass systems* [Gar92a, Gar92b, Gar92c]. Figure 7.36 shows an allpass system (Fig. 7.36a), where the delay z^{-M} is replaced by a further allpass and a unit delay z^{-1} (Fig. 7.36b). The integration of a unit delay avoids delay free loops. In Fig. 7.36c, the inner allpass is replaced by a cascade of two allpass systems and a further delay z^{-M_3}. The resulting system is again an allpass system [Gar92a, Gar92b, Gar92c, Gar98]. A further modification of the general allpass system is shown in Fig. 7.36d [Dat97, Vää97, Dah00]. Here, a delay z^{-M} followed by a lowpass and a weighting coefficient is used. The resulting system is called an *absorbent allpass system*. With these embedded allpass systems, the room simulator shown in Fig. 7.35 is extended to a feedback allpass system, which is shown in Fig. 7.37 [Gar92a, Gar92b, Gar92c, Gar98]. The feedback is performed by a lowpass filter and a feedback coefficient g, which adjusts the decay behavior. The extension to a stereo room simulator is described in [Dat97, Dah00] and is depicted in Fig. 7.38 [Dah00]. The cascaded allpass systems $A_i(z)$ in the left and right channel can be a combination of embedded and absorbent allpass systems. Both output signals of the allpass chains are fed back to the input and added. In front of both allpass chains, a coupling of both channels with a weighted sum and difference is performed. The setup and parameters of such a system are discussed in [Dah00]. A precise adjustment of reverberation time and control of echo density can be achieved by the feedback coefficients of the allpasses. The frequency density is controlled by the scaling of the delay lengths of the inner allpass systems.

The original Schroeder comb and allpass reverberator was further improved by moving the allpass filters into the parallel arrangement of K allpass $A_k(z)$ and delay z^{-D_k} sections [Spin]. The principle structure is shown in Fig. 7.39.

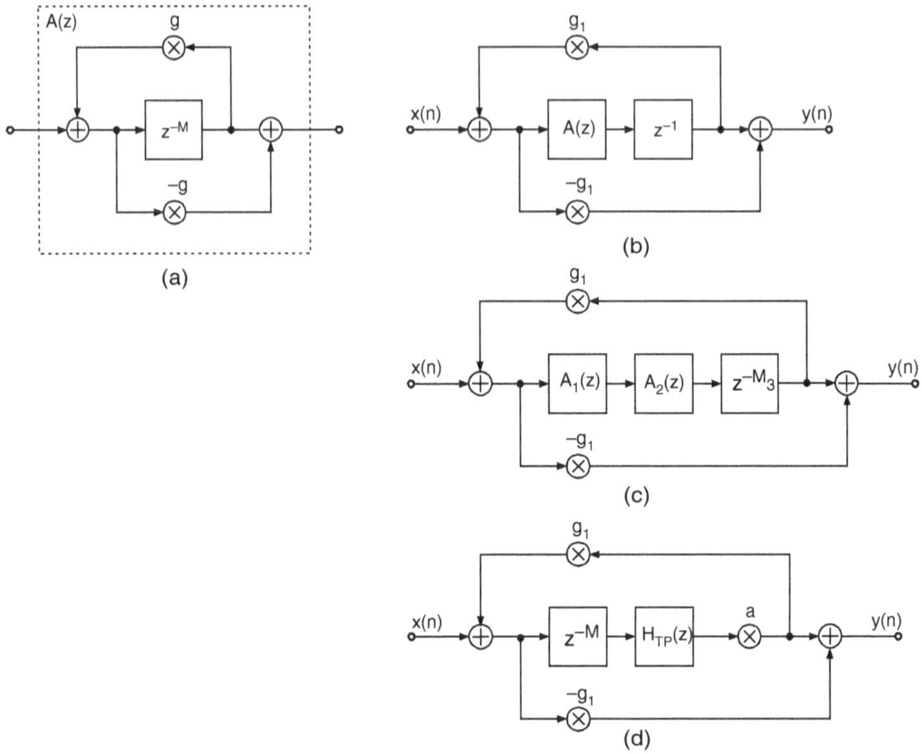

Figure 7.36 Embedded and absorbing allpass system [Gar92a, Gar92b, Gar92c, Gar98, Dat97, Vää97, Dah00].

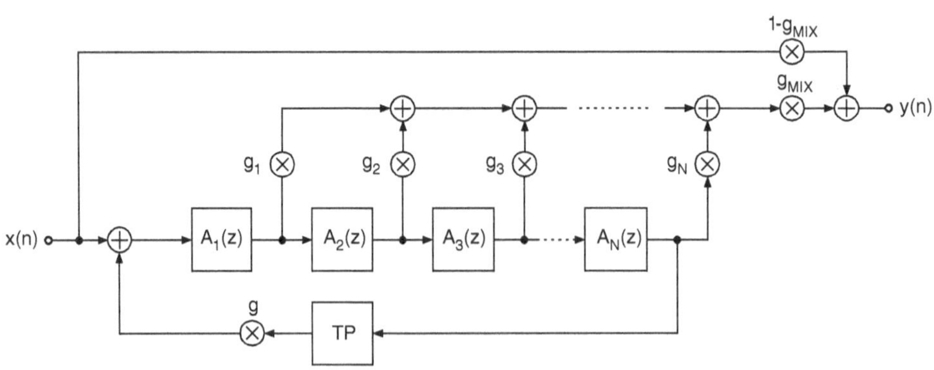

Figure 7.37 Room simulator with embedded allpass systems [Gar92a, Gar92b, Gar92c, Gar98].

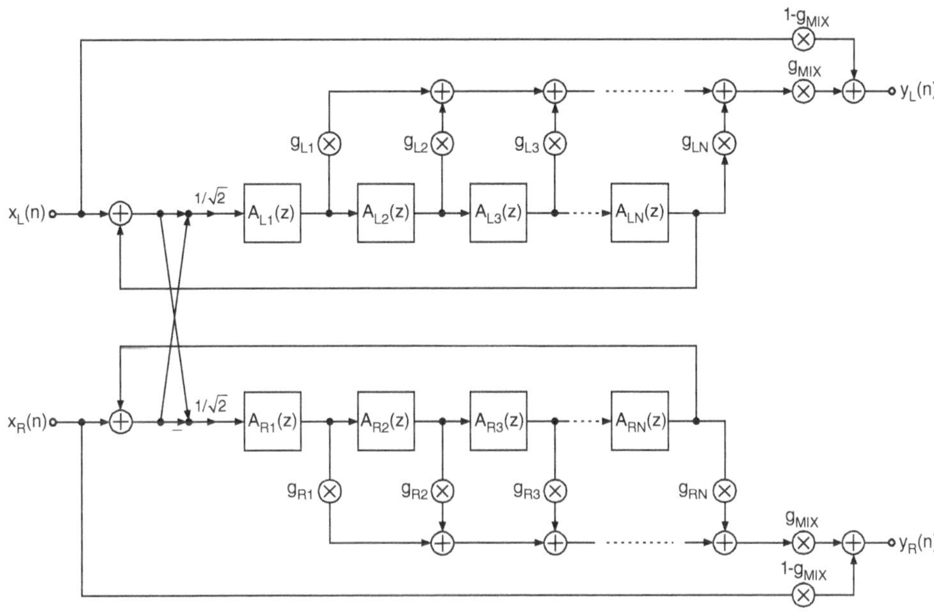

Figure 7.38 Stereo room simulator with absorbent allpass systems. Modified from [Dah00].

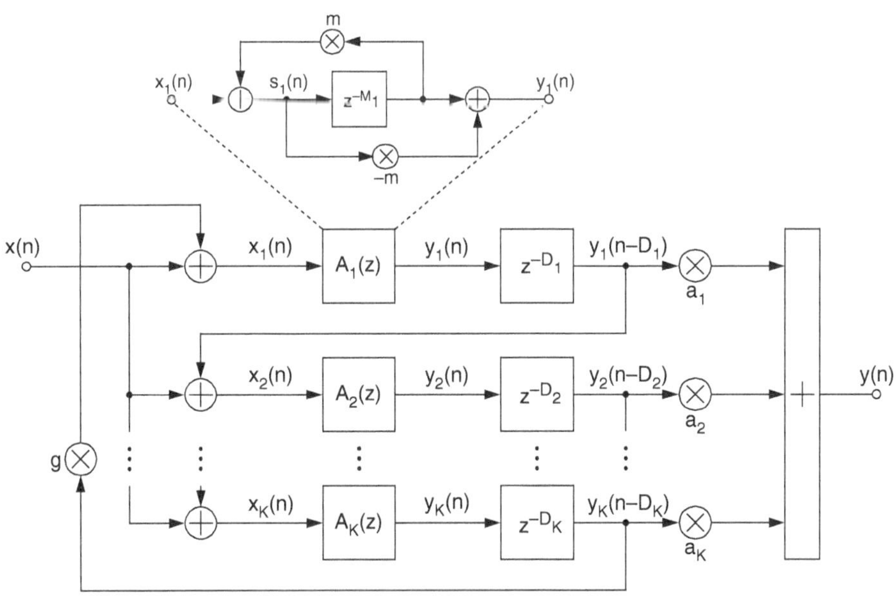

Figure 7.39 Simplified feedback delay network containing parallel allpass/delay branches.

The feedback of each allpass/delay branch into the next parallel allpass/delay and feeding the last allpass/delay back to the first via a feedback coefficient g leads to a simplified feedback delay network with a sparse feedback matrix. The input state variables to the allpasses are given by

$$x_k(n) = \begin{cases} x(n) + g \cdot y_K(n - D_k) & \text{for } k = 1, \\ x(n) + y_{k-1}(n - D_{k-1}) & \text{for } 2 \leq k \leq K, \end{cases} \tag{7.96}$$

and for the output of the allpasses according to

$$y_k(n) = -m_k \cdot x_k(n) + x_k(n - M_k) + m_k \cdot y_k(n - M_k), \tag{7.97}$$

where M_k and m_k define the delay and the coefficient inside the kth allpass, respectively. It is effectively a loop of several allpass/delays with a decay coefficient and a possible additional HF shelving filter for simulating air damping. The input is fed into each parallel section and the output is a weighted sum of all parallel sections given by

$$y(n) = \sum_{k=1}^{K} a_k \cdot y_k(n - D_k). \tag{7.98}$$

For a stereo output, a second weighted sum with orthogonal coefficients can be applied. The feedback coefficient and the HF shelving filter adjust the reverberation time. The room size can be adjusted by scaling the delays inside the allpasses and delays accordingly.

7.4 Approximation of Room Impulse Responses

In contrast to the systems for simulation of room impulse responses discussed up to this point, a method is now presented that measures and approximates the room impulse response in one step [Zöl90, Sch92, Sch93] (see Fig. 7.40). Moreover, it leads to a parametric representation of the

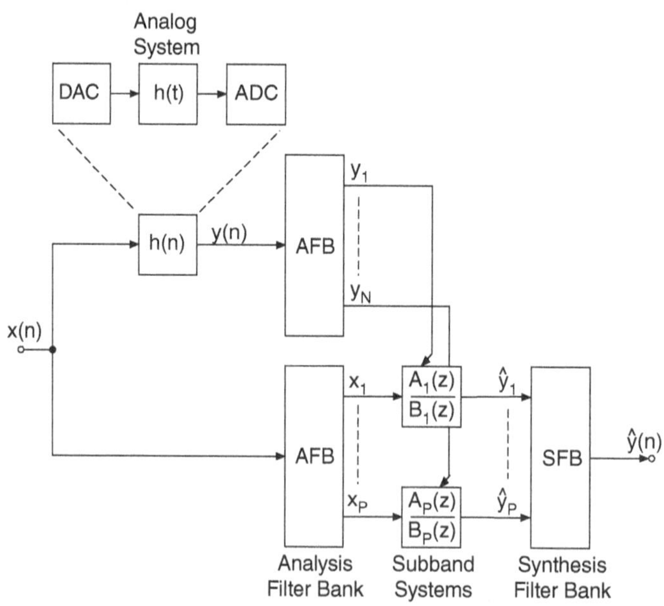

Figure 7.40 System measuring and approximating room impulse responses.

room impulse response. Because the decay times of room impulse responses decrease for high frequencies, use is made of multirate signal processing.

The analog system that is to be measured and approximated is excited with a binary pseudo-random sequence $x(n)$ via a DA converter. The resulting room signal gives a digital sequence $y(n)$ after AD conversion. The discrete-time sequence $y(n)$ and the pseudo-random sequence $x(n)$ are each decomposed by an analysis filter bank into sub-band signals $y_1, ..., y_P$ and $x_1, ..., x_P$, respectively. The sampling rate is reduced in accordance with the bandwidth of the signals. The sub-band signals $y_1, ..., y_P$ are approximated by adjusting the sub-band systems $H_1(z) = A_1(z)/B_1(z), ..., H_P(z) = A_P(z)/B_P(z)$. The outputs $\hat{y}_1, ..., \hat{y}_P$ of these sub-band systems give an approximation of the measured sub-band signals. With this procedure, the impulse response is given in parametric form (sub-band parameters) and can be directly simulated in the digital domain.

By suitably adjusting the analysis filter bank [Sch94], the sub-band impulse responses are obtained directly from the cross-correlation function

$$h_i \approx r_{x_i y_i}. \tag{7.99}$$

The sub-band impulse responses are approximated by a non-recursive filter and a recursive comb filter. The cascade of both filters leads to the transfer function

$$H_i(z) = \frac{b_0 + \dots + b_{M_i} z^{-M_i}}{1 - g_i z^{-N_i}} = \sum_{n_i=0}^{\infty} h_i(n_i) z^{-n_i}, \tag{7.100}$$

which is set equal to the impulse response in sub-band i. Multiplying both sides of Eq. (7.100) by the denominator $1 - g_i z^{-N_i}$ gives

$$(b_0 + \dots + b_{M_i} z^{-M_i}) = \left(\sum_{n_i=0}^{\infty} h_i(n_i) z^{-n_i} \right) (1 - g_i z^{-N_i}). \tag{7.101}$$

Truncating the impulse response of each sub-band to K samples and comparing the coefficients of powers of z on both sides of the equation, the following set of equations is obtained:

$$\begin{bmatrix} b_0 \\ b_1 \\ \vdots \\ b_M \\ 0 \\ \vdots \\ 0 \end{bmatrix} = \begin{bmatrix} h_0 & 0 & 0 & \cdots & 0 \\ h_1 & h_0 & 0 & \cdots & 0 \\ \vdots & \vdots & \vdots & & \vdots \\ h_M & h_{M-1} & h_{M-2} & \cdots & h_{M-N} \\ h_{M+1} & h_M & h_{M-1} & \cdots & h_{M-N+1} \\ \vdots & \vdots & \vdots & & \vdots \\ h_K & h_{K-1} & h_{K-2} & \cdots & h_{K-N} \end{bmatrix} \begin{bmatrix} 1 \\ 0 \\ \vdots \\ -g \end{bmatrix}. \tag{7.102}$$

The coefficients $b_0, \dots, b_M$ and g in the above equation are determined in two steps. First, the coefficient g of the comb filter is calculated from the exponentially decaying envelope of the measured sub-band impulse response. The vector $[1, 0, \dots, g]^T$ is then used to determine the coefficients $[b_0, b_1, \dots, b_M]^T$.

For the calculation of the coefficient g, we start with the impulse response of the comb filter $H(z) = 1/(1 - gz^{-N})$ given by

$$h(l = Nn) = g^l. \tag{7.103}$$

We further make use of the *integrated impulse response*

$$h_e(k) = \sum_{n=k}^{\infty} h(n)^2 \tag{7.104}$$

defined in [Schr65]. It describes the rest energy of the impulse response at time k. By taking the logarithm of $h_e(k)$, a straight line over time index k is obtained. From the slope of the straight line, we use

$$\ln g = N \cdot \frac{\ln h_e(n_1) - \ln h_e(n_2)}{n_1 - n_2} \qquad \text{with} \quad n_1 < n_2 \tag{7.105}$$

to determine the coefficient g [Sch94]. For $M = N$, the coefficients in Eq. (7.102) of the numerator polynomial are obtained directly from the impulse response

$$\begin{aligned} b_n &= h_n \qquad \text{for} \quad n = 0,1, \ldots, M-1, \\ b_M &= h_M - g h_0. \end{aligned} \tag{7.106}$$

Hence, the numerator polynomial of Eq. (7.100) is a direct reproduction of the first M samples of the impulse response (see Fig. 7.41). The denominator polynomial approximates the further exponentially decaying impulse response. This method is applied to each sub-band. The implementation complexity can be reduced by a factor of 10 compared with the direct implementation of the broadband impulse response [Sch94]. However, owing to the group delay caused by the filter bank, this method is not suitable for real-time applications.

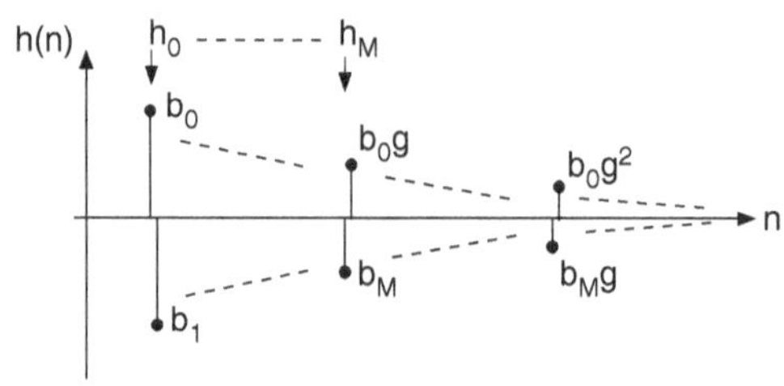

Figure 7.41 Determining model parameters from the measured impulse response.

7.5 JS Applet – Fast Convolution

The applet shown in Fig. 7.42 demonstrates audio effects resulting from a fast convolution algorithm. It is designed for a first insight into the perceptual effects of convolving an impulse response with an audio signal.

The applet generates an impulse response by modulating the amplitude of a random signal. The graphical interface presents the curve of the amplitude modulation, which can be manipulated with three control points. Two control points are used for the initial behavior of the amplitude modulation. The third control point is used for the exponential decay of the impulse response. You can choose between two predefined audio files from our web server (*audio1.wav* or *audio2.wav*) or your own local WAV file to be processed [Gui05].

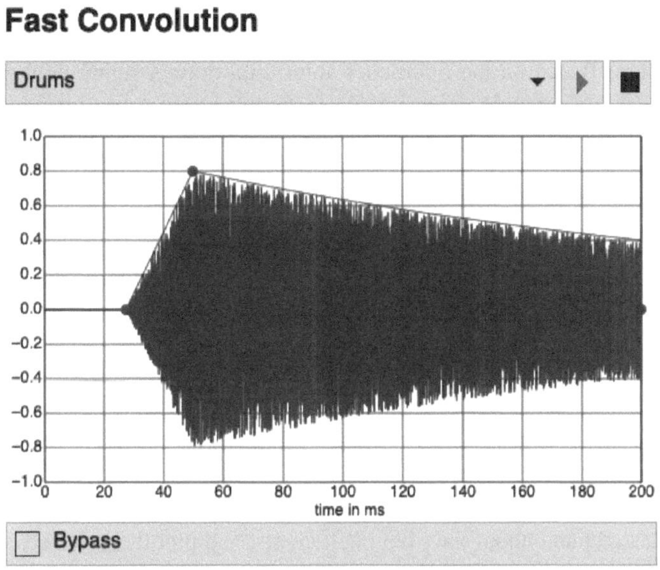

Figure 7.42 JS applet – fast convolution.

7.6 Exercises

1. Room Impulse Responses

1. How can we measure a room impulse response?
2. What kind of test signal is necessary?
3. How does the length of the impulse response affect the length of the test signal?

2. First Reflections

For a given sound (voice sound), calculate the delay time of a single first reflection. Write a Matlab program for the following computations.

1. Why do we have to choose this delay time? What coefficient should be used for this delay time?
2. Write an algorithm which performs the convolution of the input mono signal with two impulse responses which simulate a reflection to the left output $y_L(n)$ and a second reflection to the right output $y_R(n)$. Check the results by listening to the output sound.
3. Improve your algorithm to simulate two reflections which can be positioned to any angle inside the stereo mix.

3. Comb and Allpass Filters

1. **Comb Filters:** Based on the Schroeder algorithm, draw a signal flow graph for a comb filter consisting of a single delay line of M samples with a feedback loop containing an attenuation factor g.

   (a) Derive the transfer function of the comb filter.

   (b) Now, the attenuation factor g is in the feed-forward path and in the feedback loop no attenuation is applied. Why can we consider the impulse response of this model to be similar to the previous one?

   (c) In both cases, how should we choose the gain factor? What will happen if we do not respect that?

   (d) Calculate the reverberation time of the comb filter for $f_S = 44.1$ kHz, M=8, and g specified previously.

   (e) Make a statement about the filter coefficients, and plot the pole/zero locations and the frequency response of the filter.

2. **Allpass Filters:** Realize an allpass structure as suggested by Schroeder.

   (a) Why can we expect a better result with an allpass filter than with a comb filter? Write a Matlab function for a comb and allpass filter with $M = 8{,}16$.

   (b) Derive the transfer function and show the pole/zero locations, and the impulse, the magnitude, and phase responses.

   (c) Perform the filtering of an audio signal with the two filters and estimate the delay length M, which leads to a perception of a room impression.

4. Feedback Delay Networks

Write a Matlab program which realizes an FDN system.

1. What is the reason for a unitary feedback matrix?

2. What is the advantage of using a unitary circulant feedback matrix?

3. How do you control the reverberation time?

References

[Ant02] J. Antonio, L. Godinho, and A. Tadeu:*Reverberation times obtained using a numerical model versus those given by simplified formulas and measurements*. Acta Acustica, 88:25–261, 2002.

[All79] J.B. Allen, D.A. Berkeley: *Image Method for Efficient Simulating Small Room Acoustics*, J. Acoust. Soc. Am., Vol. 65, No. 4, pp. 943–950, 1979.

[And90] Y. Ando: *Concert Hall Acoustics*, Springer-Verlag, 1990.

[Bro01] S. Browne: *Hybrid Reverberation Algorithm Using Truncated Impulse Response Convolution and Recursive Filtering*, MSc Thesis, University of Miami, Coral Gables, Florida, June, 2001.

[Bar71] M. Barron: *The Subjective Effects of First Reflections in Concert Halls - The Need for Lateral Reflections*, J. Sound and Vibration 15, pp. 475–494, 1971.

[Bar81] M. Barron, A.H. Marschall: *Spatial Impression Due to Early Lateral Reflections in Concert Halls: The Derivation of a Physical Measure*, J. Sound and Vibration 77, pp. 211–232, 1981.

[Bel99] F.A. Beltran, J.R. Beltran, N. Holzem, and A. Gogu: *Matlab Implementation of Reverberation Algorithms*, Proc. Workshop on Digital Audio Effects (DAFx-99), pp. 91–96, Trondheim, Norway, 1999.

[Bla74] J. Blauert: *Räumliches Hören*, S. Hirzel Verlag, Stuttgart, 1974.

[Bla85] J. Blauert: *Räumliches Hören, Nachschrift-Neue Ergebnisse und Trends seit* 1972, S. Hirzel Verlag, Stuttgart, 1985.

[Ble01] B. Blesser: *An Interdisciplinary Synthesis of Reverberation Viewpoints*, J. Audio Eng. Soc., 49(10):867–903, October 2001.

[Cre78] L. Cremer, H.A. Müller: *Die wissenschaftlichen Grundlagen der Raumakustik - Bd. 1 u. 2*, S. Hirzel Verlag, Stuttgart, 1978/76.

[Cre03] L. Cremer, M. Möser: *Technische Akustik*, Springer-Verlag, Berlin, 2003.

[Dah00] L. Dahl, J.-M. Jot: *A Reverberator based on Absorbent All-pass Filters*, in Proceedings of the COST G-6 Conference on Digital Audio Effects (DAFX-00), Verona, Italy, December 7-9, 2000.

[Dat97] J. Dattorro: *Effect design - Part 1: Reverberator and other Filters*, J. Audio Eng. Soc., 45(19):660–684, September 1997.

[Ege96] G. P. M. Egelmeers and P. C. W. Sommen: *A new method for efficient convolution in frequency domain by nonuniform partitioning for adaptive filtering*, IEEE Trans. Signal Processing, Vol. 44, pp. 3123–3192, Dec. 1996.

[Far00] A. Farina: *Simultaneous Measurement of Impulse Response and Distortion with a Swept-sine Technique*, Proc. 108th AES Convention, Paris 18-22, February 2000.

[Far07] A. Farina: *Advancements in impulse response measurements by sine sweeps*. In Audio Engineering Society Convention 122, May 2007.

[Fre00] J. Frenette: *Reducing Artificial Reverberation Algorithm Requirements Using Time-varying Feddback Delay Networks*, MSc Thesis, University of Miami, Coral Gables, Florida, December 2000.

[Gar92a] W.G. Gardner: *A Realtime Multichannel Room Simulator*, J. Acoust. Soc. Am., Vol. 92 (4(A)):2395, 1992.

[Gar92b] W.G. Gardner: *Reverb - A Reverberator Design Tool for Audiomedia*, Proc. Int. Comp. Music Conf. (San Jose, CA), 1992.

[Gar92c] W.G. Gardner: *The Virtual Acoustic Room*, Master's Thesis, MIT Media Lab, 1992.

[Gar95] W. G. Gardner: *Efficient Convolution Without Input-output Delay*, J. Audio Eng. Soc., 43(3), pp. 127–136,1995.

[Gar98] W.G. Gardner: *Reverberation Algorithms*, In M. Kahrs and K. Brandenburg (eds), *Applications of Digital Signal Processing to Audio and Acoustics*, Kluwer Academic Publishers, p. 85–131, 1998.

[Ger71] M.A. Gerzon: Synthetic Stereo Reverberation, Studio Sound, no. 13, pp. 632–635, 1971 und no. 14, pp. 24–28, 1972.

[Ger76] M.A. Gerzon: *Unitary (Energy-Preserving) Multichannel Networks with Feedback*, Electronics Letters, Vol. 12, No. 11, pp. 278–279, 1976.

[Ger92] M.A. Gerzon: *The Design of Distance Panpots*, Proc. 92nd AES Convention, Preprint No. 3308, Vienna, 1992.

[Gri89] D. Griesinger: *Practical Processors and Programs for Digital Reverberation*, Proc. AES 7th Int. Conf., pp. 187–195, Toronto, 1989.

[Gri91] D. Griesinger: *Improving Room Acoustics Through Time-Variant Synthetic Reverberation*, in Proc. 90th Conv. Audio Eng. Soc., Preprint 3014, Feb. 1991.

[Gui05] M. Guillemard, C. Ruwwe, U. Zölzer: *J-DAFx - Digital Audio Effects in Java*, Proc. 8th Int. Conference on Digital Audio Effects (DAFx-05), Madrid, Spain, pp.161–166, 2005.

[Her94] T. Hertz: *Implementierung und Untersuchung von Rückkopplungssystemen zur digitalen Raumsimulation*, Diplomarbeit, TU Hamburg-Harburg, 1994.

[Hol09] M. Holters, T. Corbach, and U. Zölzer: *Impulse response measurement techniques and their applicability in the real world.* In Proc. of the 12th Int. Conference on Digital Audio Effects (DAFx-09), Como, Italy, September 1st-4th 2009.

[Iid95] K. Iida, K. Mizushima, Y. Takagi, and T. Suguta: *A New Method of Generating Artificial Reverberant Sound*, Proc. 99th AES Convention 1995, Preprint No. 4109, October 6-9, 1995.

[Joh00] M. Joho, G.S. Moschytz: *Connecting Partitioned Frequency-Domain Filters in Parallel or in Cascade*, IEEE Trans. CAS-II: Analog and Digital Signal Processing, Vol. 47, No. 8, pp. 685–698, Aug. 2000.

[Jot91] J.M. Jot, A. Chaigne: *Digital Delay Networks for Designing Artificial Reverberators*, Proc. 94th AES Convention, Preprint No. 3030, 1991.

[Jot92] J.M. Jot: *An Analysis/Synthesis Approach to Real-Time Artificial Reverberation*, Proc. ICASSP-92, pp. 221–224, San Francisco, 1992.

[Ken95a] G.S. Kendall: *A 3-D sound primer: directional hearing and stereo reproduction*, Computer Music J., 19(4):23–46, Winter 1995.

[Ken95b] G.S. Kendall. *The decorrelation of audio signals and its impact on spatial imagery*, Computer Music J., 19(4):71–87, Winter 1995.

[Kut91] H. Kuttruff: *Room Acoustics*, 3rd Edition, Elsevier Applied Sciences, London, 1991.

[Lee03a] W.-C. Lee, C.-M. Liu, C.-H. Yang, and J.-I. Guo, *Perceptual Convolution for Reverberation*, Proc. 115th Convention 2003, Preprint No. 5992, October 10-13, 2003

[Lee03b] W.-C. Lee, C.-M. Liu, C.-H. Yang, and J.-I. Guo, *Fast Perceptual Convolution for Reverberation*, In Proc. of the 6th Int. Conference on Digital Audio Effects (DAFX-03), London, UK, September 8-11, 2003.

[Leh08] E.A. Lehmann and A.M. Johansson: *Prediction of energy decay in room impulse responses simulated with an image-source model.* The Journal of the Acoustical Society of America, 124(1):269–277, June 2008.

[Lok01] T. Lokki, J. Hiipakka: *A Time-variant Reverberation Algorithm for Reverberation Enhancement Systems* In Proc. of the COST G-6 Conference on Digital Audio Effects (DAFX-01), Limerick, Ireland, December 6-8, 2001.

[Moo78] J.A. Moorer: *About this Reverberation Business*, Computer Music Journal, Vol. 3, No. 2, pp. 13–28, 1978.

[Mac76] F.J. MacWilliams, N.J.A. Sloane: *Pseudo-Random Sequences and Arrays*, IEEE Proceedings, Vol. 64, pp. 1715–1729, 1976.

[Mül01] S. Müller, P. Massarani: *Transfer-Function Measurement with Sweeps*, J. Audio Eng. Soc., Vol. 49, pp. 443–471, 2001.

[Pii98] E. Piirilä, T. Lokki, V. Välimäki, *Digital Signal Processing Techniques for Non-exponentially Decaying Reverberation*, in Proc. 1st COST-G6 Workshop on Digital Audio Effects (DAFX98) (Barcelona, Spain), 1998.

[Soo90] J.S. Soo, K.K. Pang: *Multidelay Block Frequency Domain Adaptive Filter*, IEEE Trans. ASSP, Vol. 38, No. 2, pp. 373–376, Feb. 1990.

[Rei95] A.J. Reijen, J.J. Sonke, and D. de Vries: *New Developments in Electro-Acoustic Reverberation Technology*, Proc. 98th AES Convention 1995, Preprint No. 3978, February 25-28, 1995.

[Roc95] D. Rocchesso: *The Ball within the Box: a sound-processing metaphor*, Computer Music J., 19(4):47–57, Winter 1995.

[Roc97a] D. Rocchesso: *Maximally-diffusive yet efficient feedback delay networks for artifcial reverberation*, IEEE Signal Processing Letters, 4(9):252–255, September 1997.

[RS97b] D. Rocchesso and J.O. Smith: *Circulant and elliptic feedback delay net- works for artificial reverberation*, IEEE Transactions on Speech and Audio Processing, 5(1):51–63, January 1997.

[Roc02] D. Rocchesso: *Spatial Effects*, In U. Zölzer (eds), DAFX -Digital Audio Effects, J. Wiley & Sons, p. 137–200, 2002.

[Sch92] M. Schönle, U. Zölzer, N. Fliege: *Modeling of Room Impulse Responses by Multirate Systems*, Proc. 93rd AES Convention, Preprint No. 3447, San Francisco 1992.

[Sch93] M. Schönle, N.J. Fliege, U. Zölzer: *Parametric Approximation of Room Impulse Responses by Multirate Systems*, Proc. ICASSP-93, Vol. 1, pp. 153–156, 1993.

[Sch94] M. Schönle: *Wavelet-Analyse und parametrische Approximation von Raumimpulsantworten*, Dissertation, TU Hamburg-Harburg, 1994.

[Schr61] M.R. Schroeder, B.F. Logan: *Colorless Artificial Reverberation*, J. Audio Eng. Soc., Vol. 9(3), pp. 192–197, 1961.

[Schr62] M.R. Schroeder: *Natural Sounding Artificial Reverberation*, J. Audio Eng. Soc., Vol. 10(3), pp. 219–223, 1962.

[Schr65] M.R. Schroeder: *New Method of Measuring Reverberation Time*, J. Acoust. Soc. Am., pp. 409–412, 1965.

[Schr70] M.R. Schroeder: *Digital Simulation of Sound Transmission in Reverberant Spaces*, J. Acoust. Soc. Am., Vol. 47, No. 2, pp. 424–431, 1970.

[Schr87] M.R. Schroeder: *Statistical Parameters of the Frequency Response Curves of Large Rooms*, J. Audio Eng. Soc., Vol. 35, No. 5, pp. 299–305, 1987.

[Smi85] J.O. Smith: *A new approach to digital reverberation using closed waveguide networks*, In Proc. International Computer Music Conference, pp. 47–53, Vancouver, Canada, 1985.

[Spin] Spin Semiconductor: *Audio effects*. http://www.spinsemi.com, accessed: 2021-05-20.

[Sta02] G.B. Stan, J.J. Embrechts, D. Archambeau: *Comparison of Different Impulse Response Measurement Techniques*, J. Audio Eng. Soc., Vol. 50, pp. 249–262, 2002.

[Sta82] J. Stautner, M. Puckette: *Designing Multi-Channel Reverberators*, Computer Music Journal, Vol. 6, No. 1, pp. 56–65, 1982.

[Vää97] R. Väänänen, V. Välimäki, and J. Huopaniemi: *Efficient and Parametric Reverberator for Room Acoustics Modeling*, in Proc. Int. Computer Music Conf. (ICMC'97), pp. 200–203, Thessaloniki, Greece, Sept. 1997.

[Väl12] V. Välimäki, J. D. Parker, L. Savioja, J. O. Smith, and J. S. Abel: *Fifty years of artificial reverberation*. IEEE Transactions on Audio, Speech, and Language Processing, 20(5):1421–1448, 2012.

[Zöl90] U. Zölzer, N.J. Fliege, M. Schönle, M. Schusdziarra: *Multirate Digital Reverberation System*, Proc. 89th AES Convention, Preprint No. 2968, Los Angeles 1990.

Chapter 8

Dynamic Range Control

U. Zölzer and E. Gerat

The dynamic range of a signal is defined as the logarithmic ratio of maximum to minimum signal amplitude and is given in decibels. The dynamic range of an audio signal lies between 40 and 120 dB. Dynamic range control of audio signals is used in many applications to match the dynamic behavior of the audio signal to different requirements. While recording, dynamic range control protects the AD converter from overload or it is used in the signal path to optimally use the full amplitude range of a recording system. For suppressing low-level noise, so-called noise gates are used so that the audio signal is passed through only from a certain level onward. While reproducing music and speech in a car, a shopping center, restaurant, or inside a disco, the dynamics have to match the special noise characteristic of the environment. Therefore the signal level is measured from the audio signal and a control signal is derived which then changes the signal level to control the loudness of the audio signal. This loudness control is adaptive to the input level. The combination of level measurement and adaptive signal level adjustment is called *dynamic range control*.

8.1 Basics

Figure 8.1 shows a block diagram of a system for dynamic range control. After measuring the input level $X_{\mathrm{dB}}(n)$, the output level $Y_{\mathrm{dB}}(n)$ is affected by multiplying the delayed input signal $x(n)$ by a factor $g(n)$ according to

$$y(n) = g(n) \cdot x(n - D). \tag{8.1}$$

The delay of the signal $x(n)$ compared with the control signal $g(n)$ allows predictive control of the output signal level. This multiplicative weighting is carried out with corresponding attack and release time. Multiplication leads, in terms of a logarithmic level representation of the

Digital Audio Signal Processing, Third Edition. Udo Zölzer.
© 2022 John Wiley & Sons Ltd. Published 2022 by John Wiley & Sons Ltd.

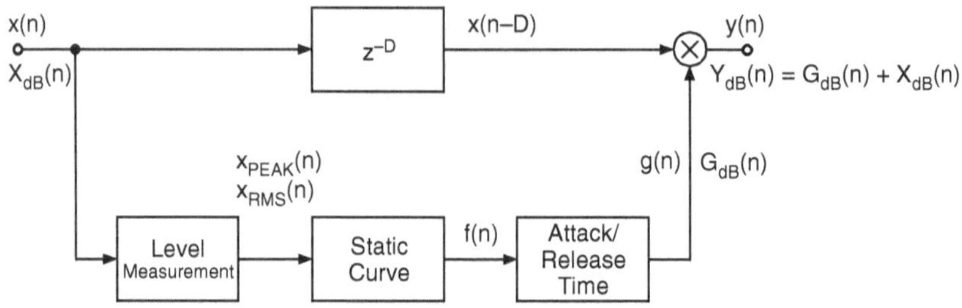

Figure 8.1 System for dynamic range control.

corresponding signals, to the addition of the weighting level $G_{\mathrm{dB}}(n)$ to the input level $X_{\mathrm{dB}}(n)$ giving the output level

$$Y_{\mathrm{dB}}(n) = X_{\mathrm{dB}}(n) + G_{\mathrm{dB}}(n). \tag{8.2}$$

8.2 Static Curve

The relationship between input level and weighting level is defined by a static level curve $G_{\mathrm{dB}}(n) = f(X_{\mathrm{dB}}(n))$. An example of such a static curve is given in Fig. 8.2. Here, the output level and the weighting level are given as functions of the input level.

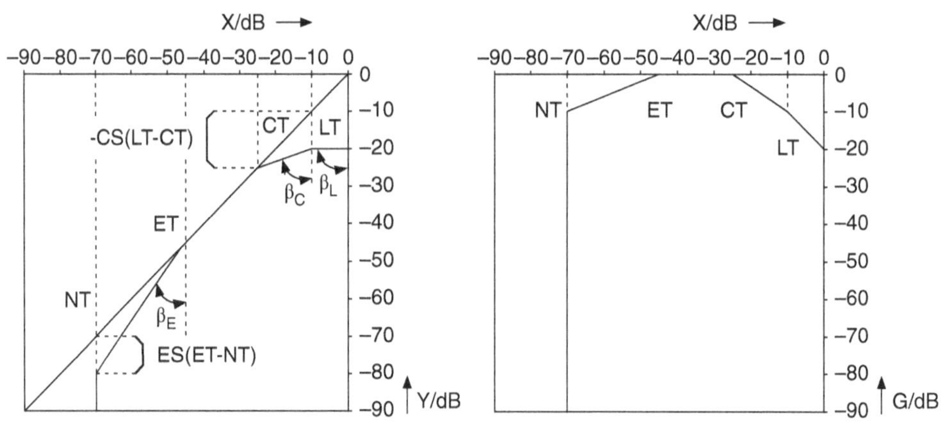

Figure 8.2 Static curve with the parameters: LT = *limiter threshold*; CT = *compressor threshold*; ET = *expander threshold*; and NT = *noise gate threshold*.

With the help of a limiter, the output level is limited when the input level exceeds the limiter threshold (LT). All input levels above this threshold lead to a constant output level. The compressor maps a change of input level on a certain smaller change of output level. In contrast to a limiter, the compressor increases the loudness of the audio signal. The expander increases changes in the input level to larger changes in the output level. With this, an increase of the dynamics for low levels is achieved. The noise gate is used to suppress low-level signals, for noise reduction, and is also used for sound effects like truncating the decay of room reverberation. Every threshold used in particular parts of the static curve is defined as the lower limit for the limiter and compressor and upper limit for the expander and noise gate.

In the logarithmic representation of the static curve, the compression factor R (*ratio*) is defined by the ratio of the input level change ΔP_I to the output level change ΔP_O, as given by

$$\mathrm{R} = \frac{\Delta P_I}{\Delta P_O}. \tag{8.3}$$

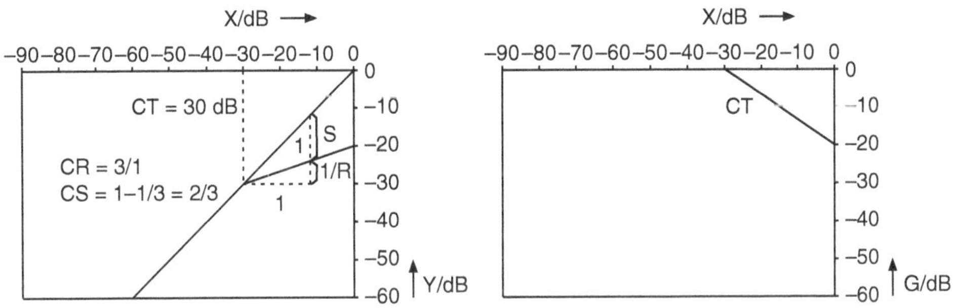

Figure 8.3 Compressor curve (compressor ratio CR/compressor slope CS).

With the help of Fig. 8.3, the straight line equation $Y_{\mathrm{dB}}(n) = \mathrm{CT} + \frac{1}{\mathrm{R}}(X_{\mathrm{dB}}(n) - \mathrm{CT})$ and the compression factor

$$\mathrm{R} = \frac{X_{\mathrm{dB}}(n) - \mathrm{CT}}{Y_{\mathrm{dB}}(n) - \mathrm{CT}} = \tan \beta_C \tag{8.4}$$

are obtained, where the angle β is defined as shown in Fig. 8.2. The relationship between the *ratio* R and the *slope* S can also be derived from Fig. 8.3 and is expressed as

$$\mathrm{S} = 1 - \frac{1}{\mathrm{R}} \tag{8.5}$$

$$\mathrm{R} = \frac{1}{1 - \mathrm{S}}. \tag{8.6}$$

Typical compression factors are

$$\begin{array}{rcll} \mathrm{R} & = & \infty & \text{limiter;} \\ \mathrm{R} & > & 1 & \text{compressor (CR: compressor ratio);} \\ 0 < \mathrm{R} & < & 1 & \text{expander (ER: expander ratio);} \\ \mathrm{R} & = & 0 & \text{noise gate.} \end{array} \tag{8.7}$$

The transition from logarithmic to linear representation leads, from Eq. (8.4), to

$$\mathrm{R} = \frac{\log_{10} \frac{\hat{x}(n)}{c_T}}{\log_{10} \frac{\hat{y}(n)}{c_T}}, \tag{8.8}$$

where $\hat{x}(n)$ and $\hat{y}(n)$ are the linear levels and c_T denotes the linear compressor threshold. Rewriting Eq. (8.8) gives the linear output level

$$\frac{\hat{y}(n)}{c_T} = 10^{\frac{1}{R}\log_{10}\left(\frac{\hat{x}(n)}{c_T}\right)} = \left(\frac{\hat{x}(n)}{c_T}\right)^{\frac{1}{R}},$$

$$\hat{y}(n) = c_T^{1-\frac{1}{R}} \cdot \hat{x}^{\frac{1}{R}}(n), \tag{8.9}$$

as a function of input level. The control factor $g(n)$ can be calculated by the quotient

$$g(n) = \frac{\hat{y}(n)}{\hat{x}(n)} = \left(\frac{\hat{x}(n)}{c_T}\right)^{\frac{1}{R}-1}. \tag{8.10}$$

With the help of tables and interpolation methods, it is possible to determine the control factor without taking logarithms and antilogarithms. The implementation described as follows, however, makes use of the logarithm of the input level and calculates the control level $G_{\mathrm{dB}}(n)$ with the help of the straight line equation. The antilogarithm leads to the value $f(n)$ which gives the control factor $g(n)$ with corresponding attack and release times (see Fig. 8.1).

To further smoothen the transition between the compressed (or expanded) and uncompressed portions of the static curve at the threshold level, the width W of the knee (angle at the threshold) can be adjusted. A width set to zero will give a *hard knee* while a larger width will set what is called a *soft knee*. It consists of a range centered on the threshold level with a variable width, as shown in Fig. 8.4. This allows a more transparent dynamic range control operation. The knee portion of the static curve is replaced by a piecewise continuous function

$$\mathrm{Y_{dB}} = \begin{cases} \mathrm{X_{dB}} & 2(\mathrm{X_{dB}} - \mathrm{CT}) < -\mathrm{W}, \\ \mathrm{X_{dB}} + \frac{1-\mathrm{R}}{2WR}\left(\mathrm{X_{dB}} - \mathrm{CT} + \frac{\mathrm{W}}{2}\right)^2 & 2|(\mathrm{X_{dB}} - \mathrm{CT})| \le \mathrm{W}, \\ \mathrm{CT} + \mathrm{R}^{-1}(\mathrm{X_{dB}} - \mathrm{CT}) & 2(\mathrm{X_{dB}} - \mathrm{CT}) > \mathrm{W}. \end{cases} \tag{8.11}$$

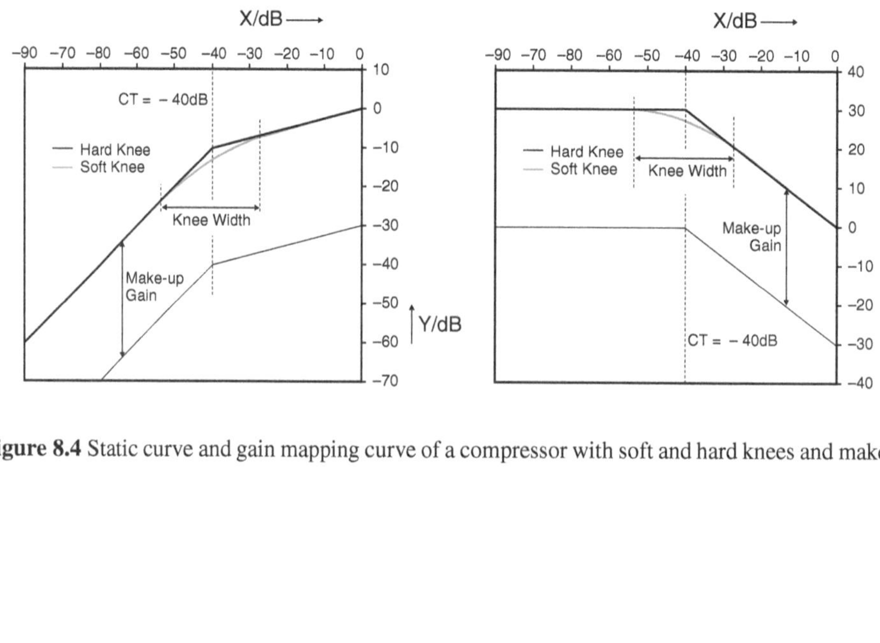

Figure 8.4 Static curve and gain mapping curve of a compressor with soft and hard knees and make-up gain.

The *make-up gain* is an additional gain used to compensate the level loss produced by the compression operation. It allows the generation of a sound perceived to be louder without the distortions induced by the clipping of the audio signal.

8.3 Dynamic Behavior

In addition to the static curve of dynamic range control, the dynamic behavior in terms of attack and release times plays a significant role in sound quality. The rapidity of dynamic range control depends also on the measurement of PEAK and RMS values [McN84, Sti86].

8.3.1 Level Measurement

Level measurements [McN84] can be performed with the systems shown in Figs. 8.5 and 8.9. For PEAK measurement, the absolute value of the input is compared with the peak value $x_{\mathrm{PEAK}}(n)$. If the absolute value is greater than the peak value, the difference is weighted with the coefficient AT (attack time) and added to $(1-\mathrm{AT})\cdot x_{\mathrm{PEAK}}(n-1)$. For this attack case $|x(n)| > x_{\mathrm{PEAK}}(n-1)$, we get the difference equation (see Fig. 8.5)

$$x_{\mathrm{PEAK}}(n) = (1-\mathrm{AT})\cdot x_{\mathrm{PEAK}}(n-1) + \mathrm{AT}\cdot |x(n)| \tag{8.12}$$

with the transfer function

$$H(z) = \frac{\mathrm{AT}}{1-(1-\mathrm{AT})z^{-1}}. \tag{8.13}$$

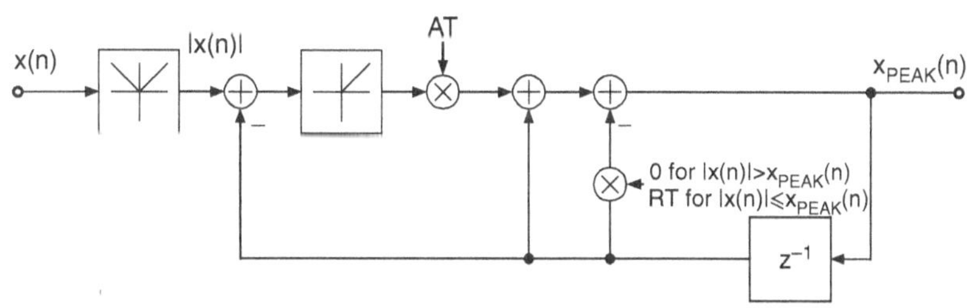

Figure 8.5 PEAK measurement.

If the absolute value of the input is smaller than the peak value $|x(n)| \leq x_{\mathrm{PEAK}}(n-1)$ (the release case), the new peak value is given by

$$x_{\mathrm{PEAK}}(n) = (1-\mathrm{RT})\cdot x_{\mathrm{PEAK}}(n-1) \tag{8.14}$$

with the release time coefficient RT. The difference signal of the input will be muted by the non-linearity such that the difference equation for the peak value is given according to Eq. (8.14). For the release case, the transfer function

$$H(z) = \frac{1}{1-(1-\mathrm{RT})z^{-1}} \tag{8.15}$$

is valid. For the attack case, the transfer function (8.13) with coefficient AT is used, and for the release case, the transfer function (8.15) with the coefficient RT is used. The coefficients (see Section 8.3.3) are given by

$$\mathrm{AT} = 1 - \exp\left(\frac{-2.2T_S}{t_a/1000}\right), \tag{8.16}$$

$$\mathrm{RT} = 1 - \exp\left(\frac{-2.2T_S}{t_r/1000}\right), \tag{8.17}$$

where the attack time t_a and the release time t_r are given in msec (T_S sampling interval). With this switching between filter structures, one achieves fast attack responses for increasing input signal amplitudes and slow decay responses for decreasing input signal amplitudes.

A first variation for peak detection is the *branched* peak detector that has two different branches of operation, as shown in Fig. 8.6. For the attack case AC when $|x(n)| > x_{\mathrm{PEAK}}(n-1)$ and the release case RC when $|x(n)| \leq x_{\mathrm{PEAK}}(n-1)$, the operations are given by

$$x_{\mathrm{PEAK}}(n) = \begin{cases} (1-\mathrm{AT}) \cdot x_{\mathrm{PEAK}}(n-1) + \mathrm{AT} \cdot |x(n)|, & \text{for AC,} \\ (1-\mathrm{RT}) \cdot x_{\mathrm{PEAK}}(n-1), & \text{for RC.} \end{cases} \tag{8.18}$$

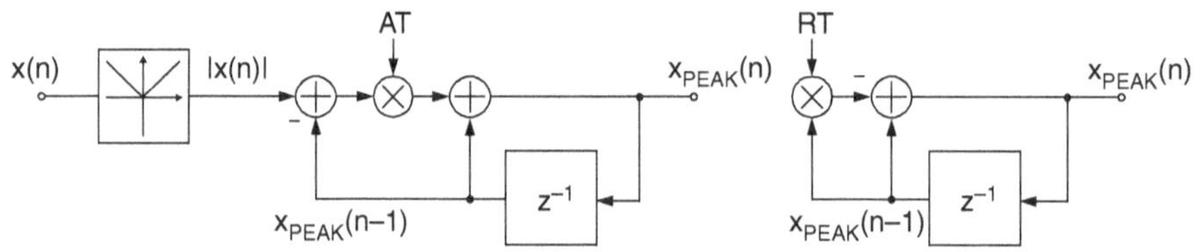

Figure 8.6 Block diagrams of the attack and release parts of the branched peak level detector.

One further variation of the peak detector is the so-called *decoupled* peak level detector, which is presented in Fig. 8.7. It computes an auxiliary signal $x_1(n)$ based on the release time and then feeds this signal to an attack filter, as given by

$$x_1(n) = \max\left(|x(n)|, (1-\mathrm{RT}) \cdot x_1(n-1) + \mathrm{RT} \cdot |x(n)|\right), \tag{8.19}$$

$$x_{\mathrm{PEAK}}(n) = (1-\mathrm{AT}) \cdot x_{\mathrm{PEAK}}(n-1) + \mathrm{AT} \cdot x_1(n). \tag{8.20}$$

This detector adds a small onset to the release time, as shown in Fig. 8.8. This leads to the release time approximately equal to $t_{\mathrm{AT}} + t_{\mathrm{RT}}$ and is therefore always bigger than the attack time. This can be useful to avoid artifacts arising from release times that are too short [Gia12].

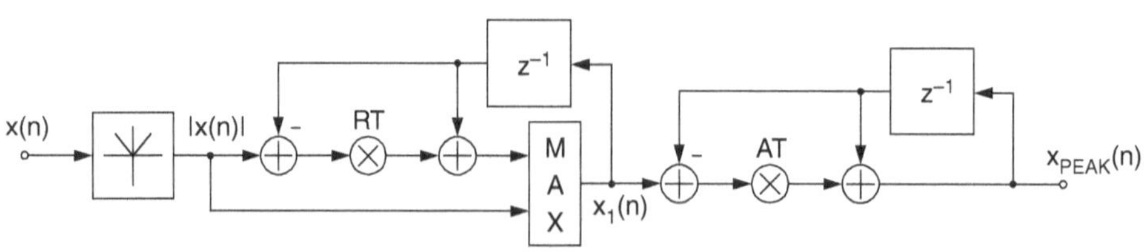

Figure 8.7 Block diagram of the decoupled peak level detector.

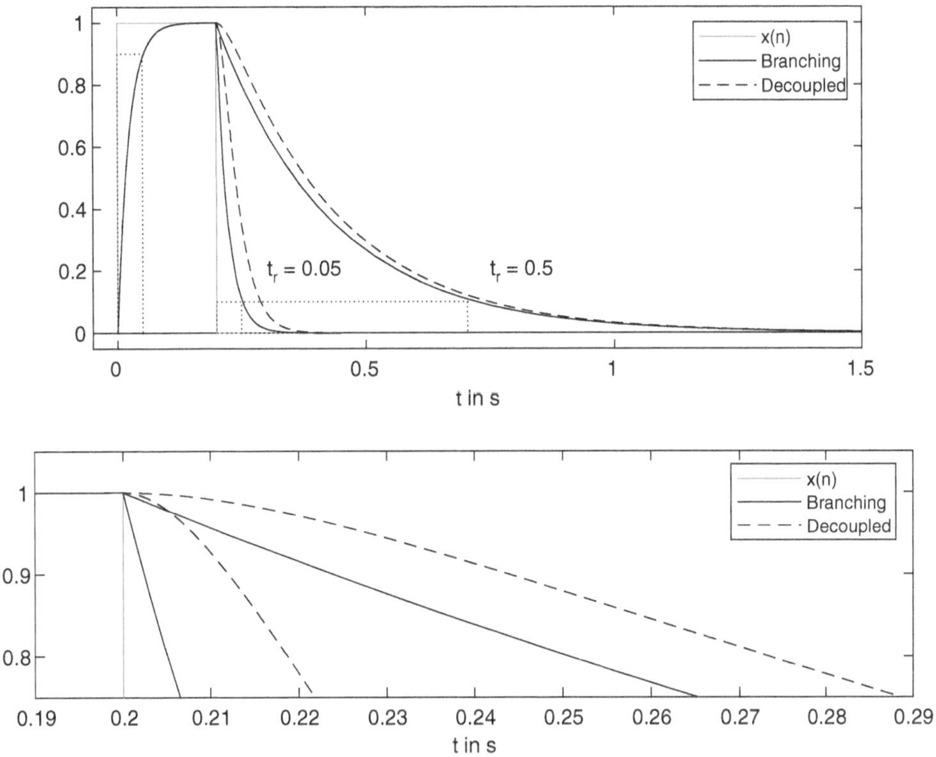

Figure 8.8 Attack and release behavior of branching and decoupled peak level detectors.

The computation of the RMS value

$$x_{\mathrm{RMS}}(n) = \sqrt{\frac{1}{N}\sum_{i=0}^{N-1} x^2(n-i)} \tag{8.21}$$

over N input samples can be achieved by a recursive formulation. The RMS measurement shown in Fig. 8.9 uses the square of the input and performs averaging with a first-order lowpass filter. The averaging coefficient

$$\mathrm{TAV} = 1 - \exp\left(\frac{-2.2T_A}{t_M/1000}\right) \tag{8.22}$$

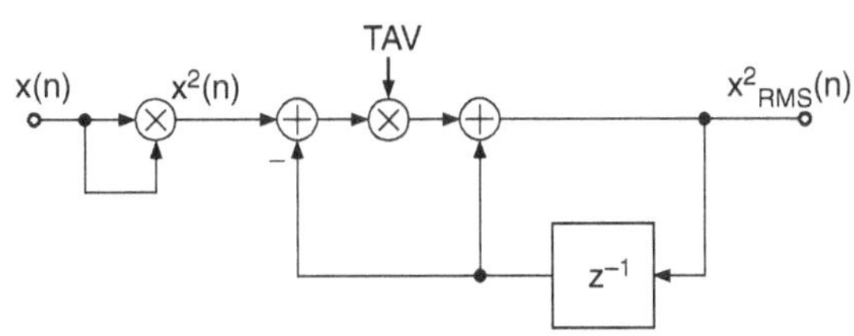

Figure 8.9 RMS measurement (TAV = averaging coefficient).

is determined according to the time constant calculation discussed in Section 8.3.3, where t_M is the averaging time in msec. The difference equation is given by

$$x^2_{\mathrm{RMS}}(n) = (1 - \mathrm{TAV}) \cdot x^2_{\mathrm{RMS}}(n-1) + \mathrm{TAV} \cdot x^2(n) \tag{8.23}$$

with the transfer function

$$H(z) = \frac{\mathrm{TAV}}{1 - (1 - \mathrm{TAV})z^{-1}}. \tag{8.24}$$

8.3.2 Gain Factor Smoothing

Attack and release times can be implemented by the system shown in Fig. 8.10 [McN84]. The attack coefficient AT or release coefficient RT is obtained by comparing the input control factor with the previous one. A small hysteresis curve determines whether the control factor is in the attack or release status and hence gives the coefficient AT or RT. The system also serves to smooth the control signal. The difference equation is given by

$$g(n) = (1 - k) \cdot g(n-1) + k \cdot f(n), \tag{8.25}$$

with $k = \mathrm{AT}$ or $k = \mathrm{RT}$ and the corresponding transfer function leads to

$$H(z) = \frac{k}{1 - (1 - k)z^{-1}}. \tag{8.26}$$

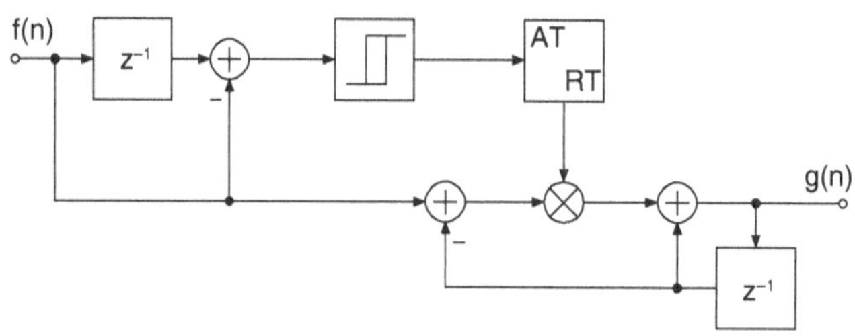

Figure 8.10 Implementing attack and release times or gain factor smoothing.

8.3.3 Time Constants

If the step response of a continuous-time system is

$$g(t) = 1 - e^{-t/\tau} \qquad \tau = \text{time constant}, \tag{8.27}$$

then sampling (step-invariant transform) the step response gives the discrete-time step response

$$g(nT_S) = \varepsilon(nT_S) - e^{-nT_S/\tau} = 1 - z_\infty^n \qquad \text{with} \quad z_\infty = e^{-T_S/\tau}. \tag{8.28}$$

The Z-transform leads to

$$\begin{aligned} G(z) &= \frac{z}{z-1} - \frac{1}{1 - z_\infty z^{-1}} \\ &= \frac{1 - z_\infty}{(z-1)(1 - z_\infty z^{-1})}. \end{aligned} \tag{8.29}$$

With the definition of attack time $t_a = t_{90} - t_{10}$, we derive

$$0.1 = 1 - e^{-t_{10}/\tau} \quad \leftarrow t_{10} = 0.1\tau, \tag{8.30}$$

$$0.9 = 1 - e^{-t_{90}/\tau} \quad \leftarrow t_{90} = 0.9\tau. \tag{8.31}$$

The relationship between attack time t_a and the time constant τ of the step response is obtained as follows:

$$0.9/0.1 = e^{(t_{90}-t_{10})/\tau}$$

$$\ln(0.9/0.1) = (t_{90} - t_{10})/\tau$$

$$t_a = t_{90} - t_{10} = 2.2\tau. \tag{8.32}$$

Hence, the pole is calculated as

$$\boxed{z_\infty = e^{-2.2T_S/t_a}}. \tag{8.33}$$

A system for implementing the given step response is obtained by the relationship between the Z-transform of the impulse response and the Z-transform of the step response:

$$H(z) = \frac{z-1}{z} G(z). \tag{8.34}$$

The transfer function can now be written as

$$H(z) = \frac{(1 - z_\infty)z^{-1}}{(1 - z_\infty z^{-1})} \tag{8.35}$$

with the pole $z_\infty = e^{-2.2T_S/t_a}$ adjusting the attack, release, or averaging time. For the coefficients of the corresponding time constant filters, the attack case is given by Eq. (8.16), the release case by Eq. (8.17), and the averaging case by Eq. (8.22). Figure 8.11 shows an example where the dotted lines mark the t_{10} and t_{90} times.

8.4 Implementation

The programming of a system for dynamic range control is described in the following sections.

8.4.1 Limiter

The block diagram of a limiter is presented in Fig. 8.12. The signal $x_{\text{PEAK}}(n)$ is determined from the input with variable attack and release times. The logarithm to the base 2 of this peak signal is taken and compared with the limiter threshold. If the signal is above the threshold, the difference is multiplied by the negative slope of the limiter LS. After this, the antilogarithm of the result is taken. The obtained control factor $f(n)$ is then smoothed with a first-order lowpass filter (SMOOTH). If the signal $x_{\text{PEAK}}(n)$ lies below the limiter threshold, the signal $f(n)$ is set to $f(n) = 1$. The delayed input $x(n - D_1)$ is multiplied by the smoothed control factor $g(n)$ to give the output $y(n)$.

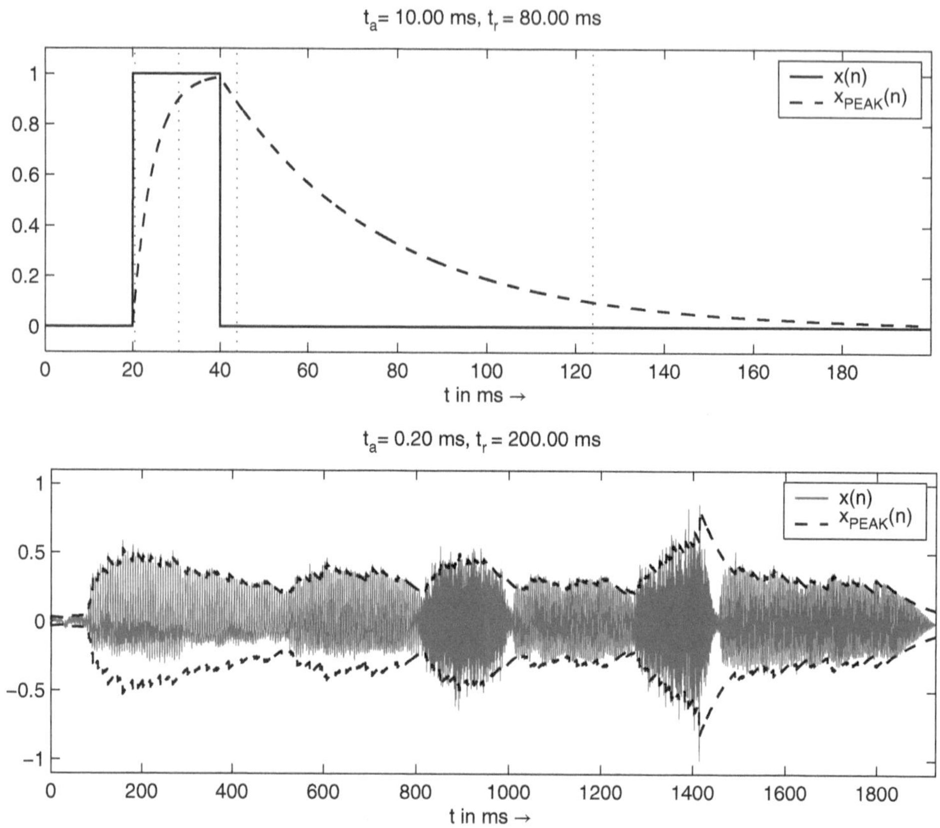

Figure 8.11 Attack and release behavior for time constant filters.

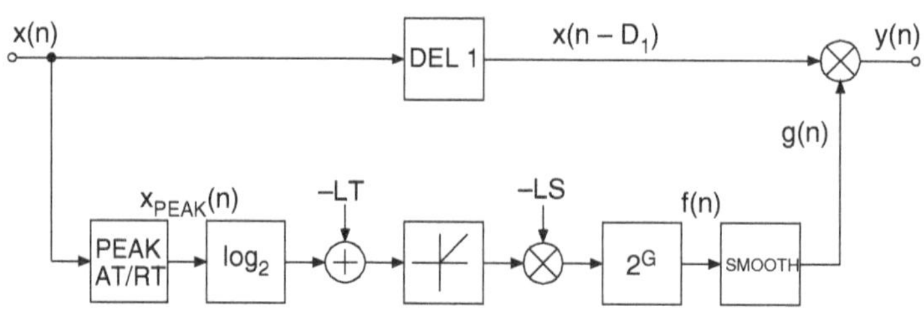

Figure 8.12 Limiter.

8.4.2 Compressor

Feedback implementation

A DRC system can be implemented in a feedback form, as shown in Fig. 8.13, where level sensing is performed on the output signal $y(n)$. A delay has to be introduced in the feedback side-chain path. The gain is then applied to the input signal $x(n)$.

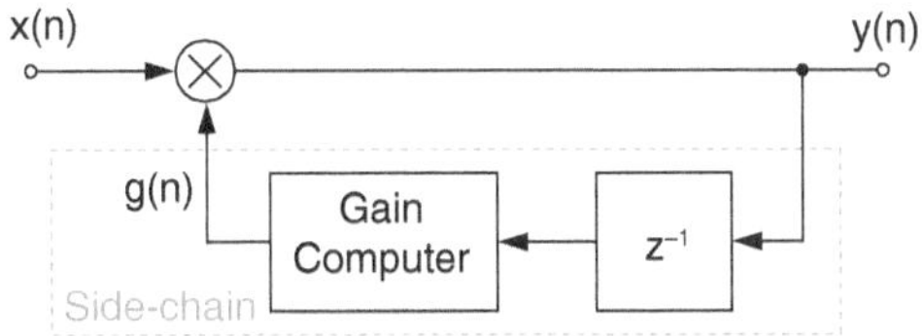

Figure 8.13 Feedback DRC System.

Ducking

Ducking consists of using a second signal $x_D(n)$ as input to the side-chain of the DRC system, as shown in Fig. 8.14. This has various applications, for example, in the case of background music playing and an announcement has to be made. The level of the music can be automatically reduced by sensing the level of the speaker and thus applying a negative gain to the music (compression). This effect is also widely used in modern music production to give a stronger feeling of energy to the kick drum. In this case, the kick drum is used as a side-chain signal to periodically attenuate the rest of the instruments.

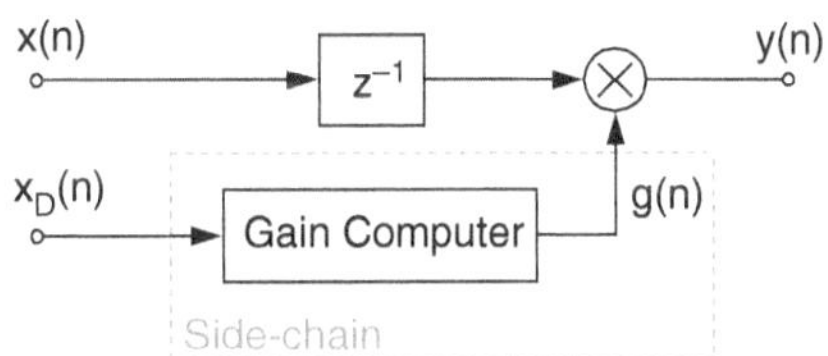

Figure 8.14 Ducking DRC system.

Lookahead

The lookahead consists of introducing a delay in the direct path to sense the level prior to applying it, as shown in Fig. 8.15. It is useful when applied on transients or fast changing signals, because it predicts the variations and reduces (or avoids) the time needed to react to changes in the signal dynamic. When realized offline, the delay introduced can be compensated, but in real-time, the delay is equal to the lookahead time. Given a lookahead time t_{LH} in ms, the number of samples is given by $N = \frac{t_{LH} \cdot f_S}{1000}$.

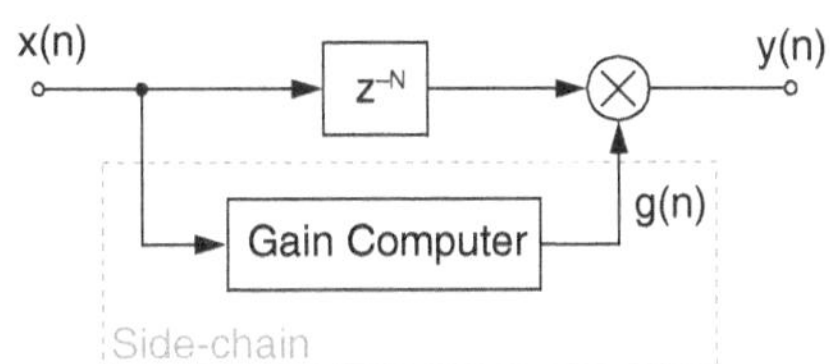

Figure 8.15 DRC system with a lookahead of N samples.

8.4.3 Compressor, Expander, Noise Gate

The block diagram of a compressor/expander/noise gate is shown in Fig. 8.16. The basic structure is similar to the limiter. In contrast to the limiter, the logarithm of the signal $x_{RMS}(n)$ is taken and multiplied by 0.5. The obtained value is compared with three thresholds to determine the operating range of the static curve. If one of the three thresholds is crossed, the resulting difference is multiplied by the corresponding slope (CS, ES, NS) and the antilogarithm of the result is taken. A following first-order lowpass filter provides the attack and release times.

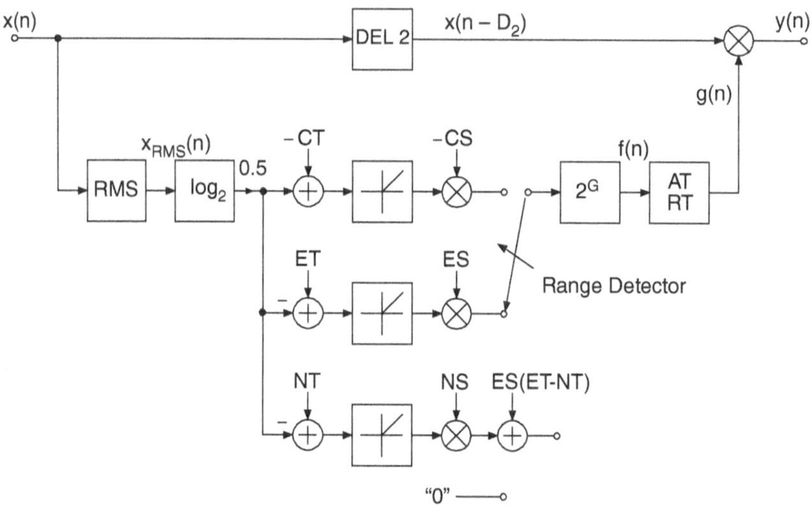

Figure 8.16 Compressor/expander/noise gate.

8.4.4 Combination System

A combination of a limiter that uses PEAK measurement and a compressor/expander/noise gate that is based on RMS measurement, is presented in Fig. 8.17. The PEAK and RMS values are measured simultaneously. If the linear threshold of the limiter is crossed, the logarithm of the peak signal $x_{PEAK}(n)$ is taken and the upper path of the limiter is used to calculate the characteristic curve. If the limiter threshold is not crossed, the logarithm of the RMS value is taken and one of the three lower paths is used. The additive terms in the limiter and noise gate paths result from the static curve. After going through the range detector, the antilogarithm is taken. The sequence $f(n)$ is smoothed with a SMOOTH filter in the limiter case, or weighted with corresponding attack and release times of the relevant operating range (compressor, expander, or noise gate). By limiting the maximum level, the dynamic range is reduced. As a consequence, the overall static curve can be shifted up by a gain factor. Figure 8.18 demonstrates this with a gain factor equal to 10 dB. This static parameter value is directly included in the control factor $g(n)$.

As an example, Fig. 8.19 illustrates the input $x(n)$, the output $y(n)$, and the control factor $g(n)$ of a compressor/expander system. It is observed that signals with high amplitude are compressed and those with low amplitude are expanded. An additional gain of 12 dB shows the maximum

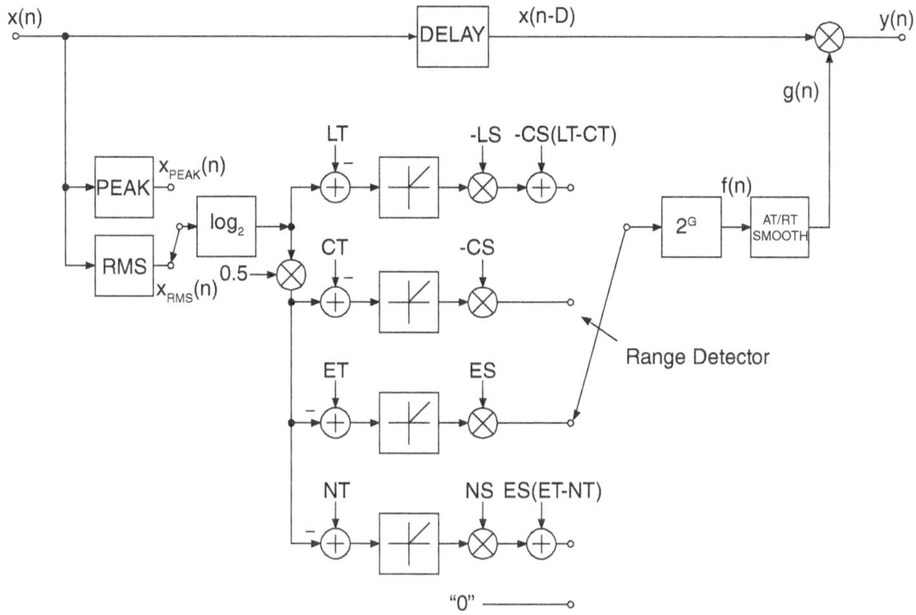

Figure 8.17 Limiter/compressor/expander/noise gate.

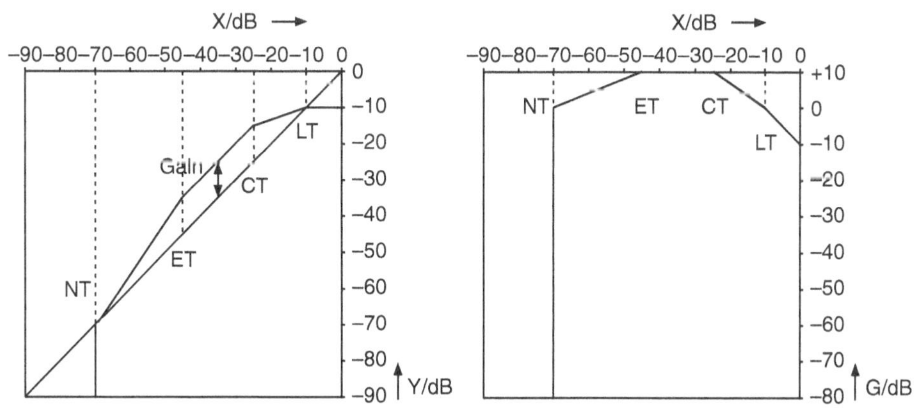

Figure 8.18 Shifting the static curve by a gain factor.

value of 4 for the control factor $g(n)$. The compressor/expander system operates in the linear region of the static curve if the control factor is equal to 4. If the control factor is between 1 and 4, the system operates as a compressor. For control factors lower than 1, the system works as an expander ($3500 < n < 4500$ and $6800 < n < 7900$). The compressor is responsible for increasing the loudness of the signal whereas the expander increases the dynamic range for signals of small amplitude.

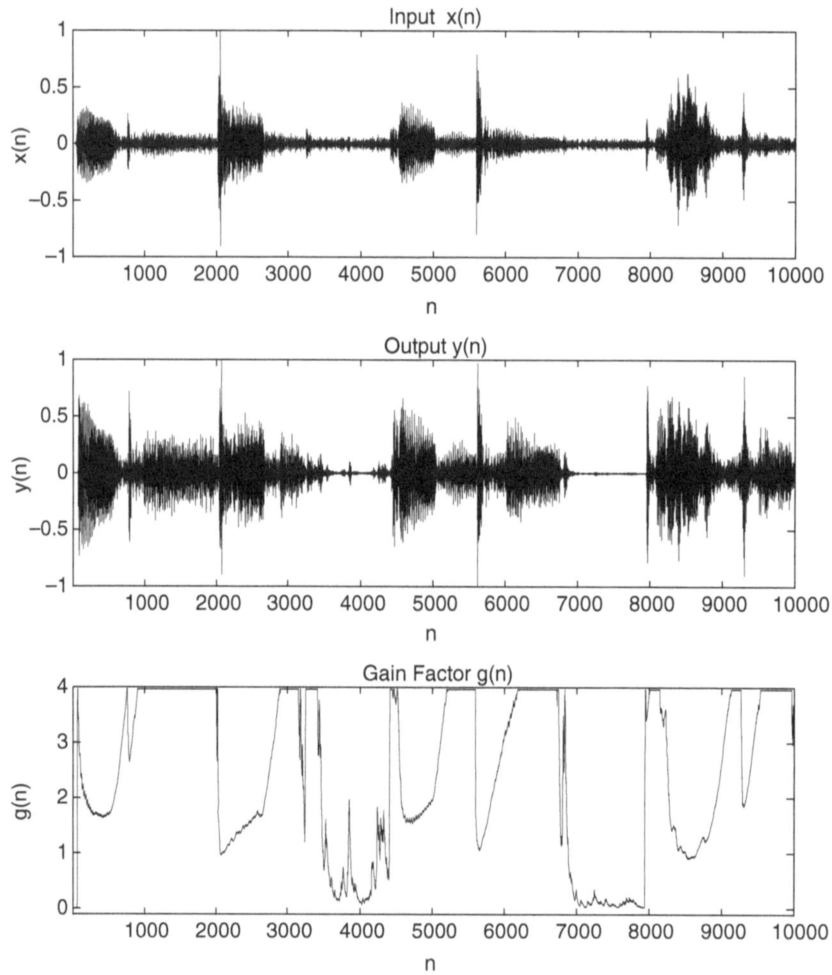

Figure 8.19 Signals $x(n)$, $y(n)$, and $g(n)$ for dynamic range control.

8.5 Realization Aspects

8.5.1 Sampling Rate Reduction

To reduce the computational complexity, downsampling can be carried out after calculating the PEAK/RMS value (see Fig. 8.20). As the signals $x_{\mathrm{PEAK}}(n)$ and $x_{\mathrm{RMS}}(n)$ are already band limited, they can be directly downsampled by taking every second or fourth value of the sequence. This downsampled signal is then processed by taking its logarithm, calculating the static curve, taking the antilogarithm, and filtering with corresponding attack and release time with reduced sampling rate. A following upsampling by a factor of 4 is achieved by repeating the output value four times. This procedure is equivalent to upsampling by a factor 4 followed by a sample-and-hold transfer function.

The nesting and spreading of partial program modules over four sampling periods is shown in Fig. 8.21. The modules PEAK/RMS (i.e. PEAK/RMS calculation) and MULT (delay of input and

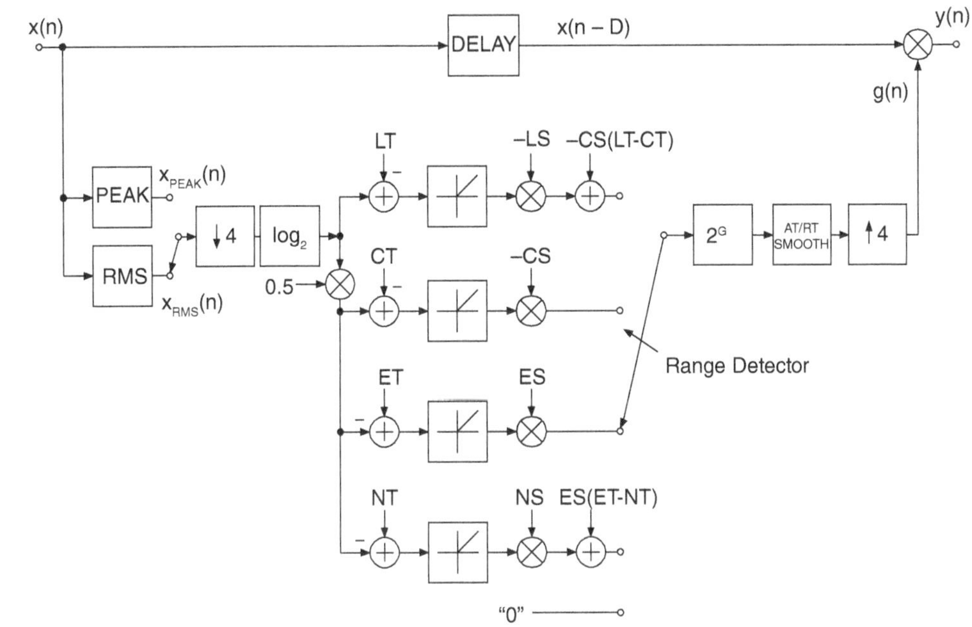

Figure 8.20 Dynamic system with sampling rate reduction.

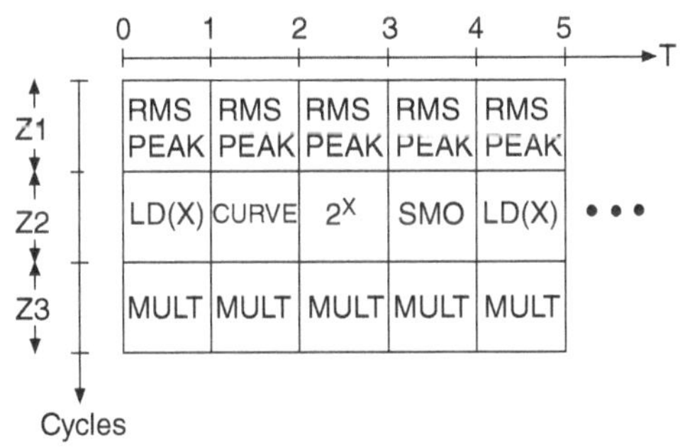

Figure 8.21 Nesting technique.

multiplication with $g(n)$) are performed every input sampling period. The number of processor cycles for PEAK/RMS and MULT are denoted by Z1 and Z3, respectively. The modules LD(x), CURVE, 2^x, and SMO have a maximum number of processor cycles of Z2 and are processed consecutively in the given order. This procedure is repeated every four sampling periods. The total number of processor cycles per sampling period for the complete dynamics algorithm results from the sum of all three modules.

8.5.2 Curve Approximation

In addition to taking the logarithm and antilogarithm, other simple operations like comparisons and addition/multiplication occur in calculating the static curve. The logarithm of the PEAK/RMS value is taken as follows:

$$x = M \cdot 2^E, \tag{8.36}$$

$$\mathrm{ld}(x) = \mathrm{ld}(M) + E. \tag{8.37}$$

First, the mantissa is normalized and the exponent is determined. The function ld(M) is then calculated by a series expansion. The exponent is simply added to the result.

The logarithmic weighting factor G and the antilogarithm 2^G are given by

$$G = -E - M, \tag{8.38}$$

$$2^G = 2^{-E} \cdot 2^{-M}. \tag{8.39}$$

Here, E is a natural number and M is a fractional number. The antilogarithm 2^G is calculated by expanding the function 2^{-M} in a series and multiplication by 2^{-E}. A reduction of computational complexity can be achieved by directly using tables for log and antilog.

8.5.3 Stereo Processing

For stereo processing, a common control factor $g(n)$ is needed. If different control factors are used for both channels, limiting or compressing one of the two stereo signals causes a displacement of the stereo balance. Figure 8.22 shows a stereo dynamic system in which the sum of the two signals is used to calculate a common control factor $g(n)$. The following processing steps of measuring the PEAK/RMS value, downsampling, taking logarithm, calculating static curve, taking antilogarithm attack and release time, and upsampling with a sample-and-hold function remain the same. The delay (DEL) in the direct path must be the same for both channels.

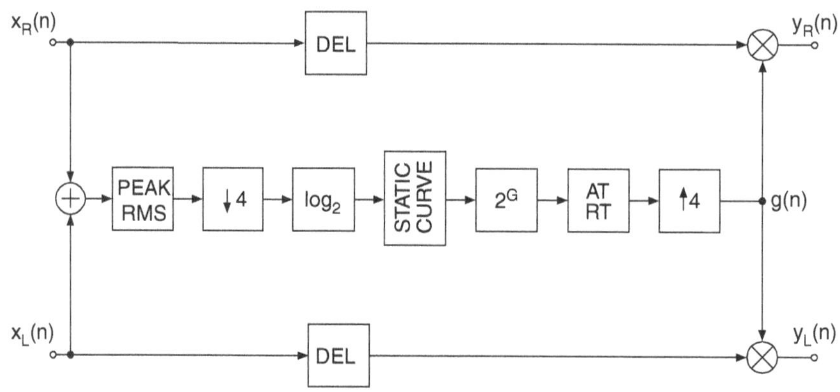

Figure 8.22 Stereo dynamic system.

8.6 Multiband DRC

A multiband DRC system consists of several DRC devices applied on different portions of the frequency range. The signal is split into bands (usually 3 to 5) with a complementary filter bank and each band is treated separately with it own DRC device (usually a compressor), as shown in

Fig. 8.23. Multiband compressors are useful when only a specific frequency region of the signal has to be treated. Especially at the mastering stage where the mix is already done, each instrument can not be treated on its own anymore. It also prevents common DRC artifacts that occur over a wider frequency range. The level is measured on each band and is used for the gain computation of the corresponding DRC device [Dut11]. A problem of multiband compressors is that the signal is split into bands even when the device is not active. This can cause filtering problems, such as phase cancellation or shifting, making the effect not fully transparent even when deactivated. Also important to notice is that such systems are relatively expensive to compute and can lead to CPU overload when heavily used on systems that are too weak.

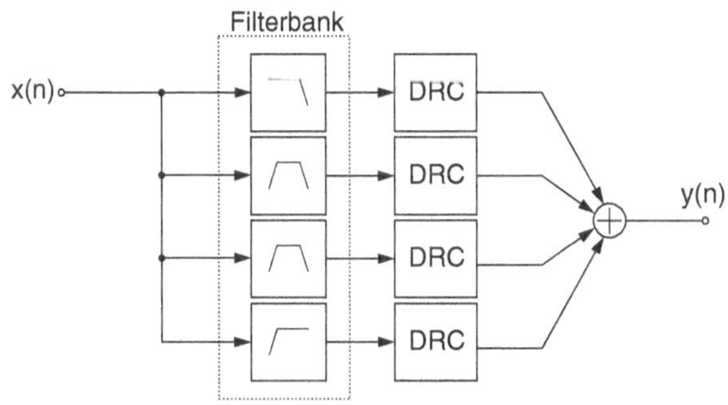

Figure 8.23 Multiband DRC system.

8.7 Dynamic Equalizers

Dynamic equalizers are very similar to multiband DRC systems, but are in their own way more transparent. The filters for a dynamic equalizer are placed in series like a regular equalizer, but the sensing can be performed in parallel, as shown in Fig. 8.24. The levels of several band-limited signals are sensed and are used to control the gain of a peak (or shelving) filter. The control can be positive and work as an expander (the gain of the filter is proportional to the gain which exceeds the threshold), or can also be negative and work as a compressor (the filter gain is inverse

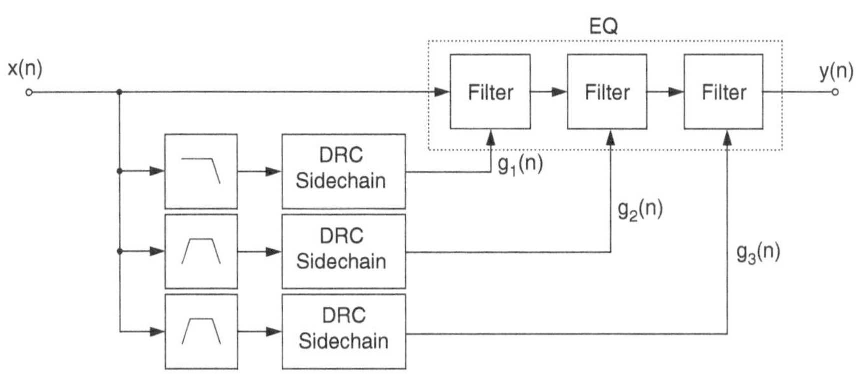

Figure 8.24 Dynamic EQ system.

proportional to the gain exceeding the threshold) [Wis09, Väl16]. A corresponding frequency response at a specific time instant n is shown in Fig. 8.25.

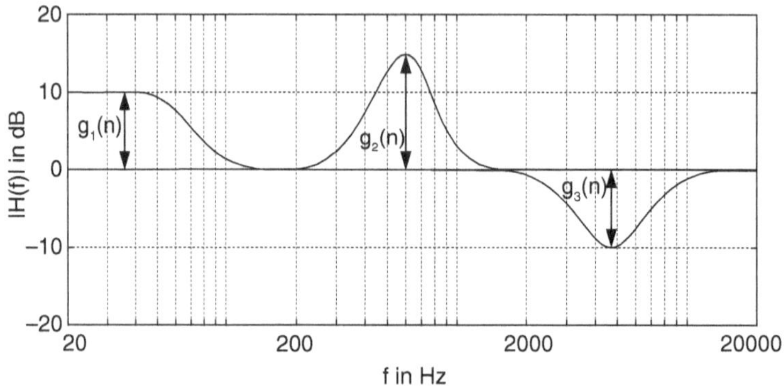

Figure 8.25 Frequency response of a dynamic equalizer at sample n.

A further approach, shown in Figs. 8.26 and 8.27, implements a dynamic shelving and peak filter based on an allpass filter decomposition for parametric equalization [Zöl08]. This

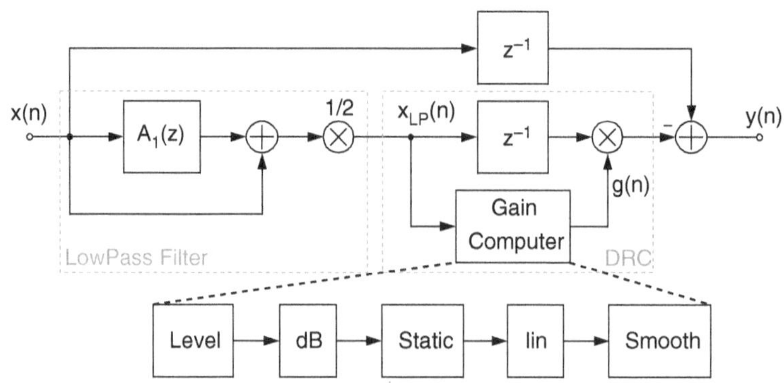

Figure 8.26 Block diagram of a low-shelving dynamic filter.

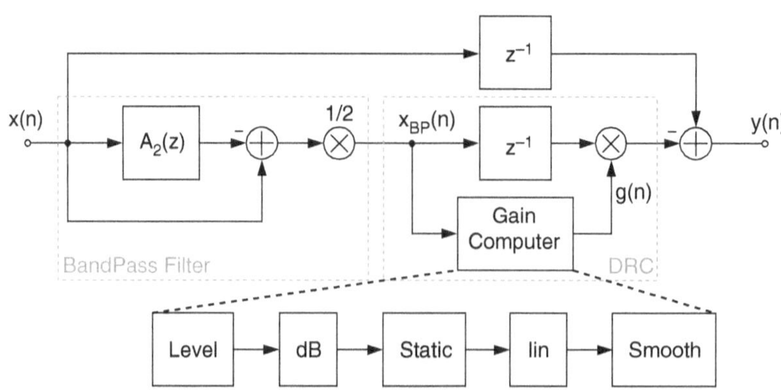

Figure 8.27 Block diagram of a peak dynamic filter.

implementation takes advantage of the lowpass and bandpass signal generated at the first stage to compute the level used later at the DRC stage.

8.8 Source-filter DRC

8.8.1 Introduction

Source-filter separation and processing [Arf11] has been extensively used to extract the so-called spectral envelope (filter coefficients) and the source signal from an audio signal. When applying the spectral envelope to the source signal again, the filtering operation will perfectly reconstruct the original. In the case of a speech signal, the spectral envelope has the advantage to provide a good representation of its formants. The idea of using the extracted formants to re-synthesize voice has been expressed for a long time in [Sla62] and is nowadays used in applications such as Vocoders [Nag09] or artificial speech models [Kad07].

The system proposed here will apply some dynamic processing on the error signal (source signal) to affect the re-synthesized result. However, too drastic modifications of the source signal can generate artifacts at the re-synthesis stage. One of the applications presented here aims to reduce background noise and is especially effective when applied to speech recordings.

Source-filter separation [Arf11], as shown in Fig. 8.28, uses linear predictive coding and decoding (LPC) by generating a prediction $\hat{x}(n)$ of the input signal $x(n)$ with an adaptive FIR filter $P(z)$ of order p. This filter estimates a spectral envelope of the input $x(n)$. The difference between the predicted signal and the input is called the prediction error $e(n) = x(n) - \hat{x}(n)$ and represents the source signal. The prediction error $e(n)$ can be seen as a whitened version of $x(n)$. For speech signals, the prediction error (source signal) represents an excitation signal similar to the sound emitted by the vocal cords, which will be filtered by the vocal tract to give the spectral shape of the speech signal. Further dynamic range processing of the prediction error will lead to a processed source/error $\tilde{e}(n)$, which is then fed to the LPC decoding as shown in Fig. 8.28 to reconstruct the processed output $y(n)$.

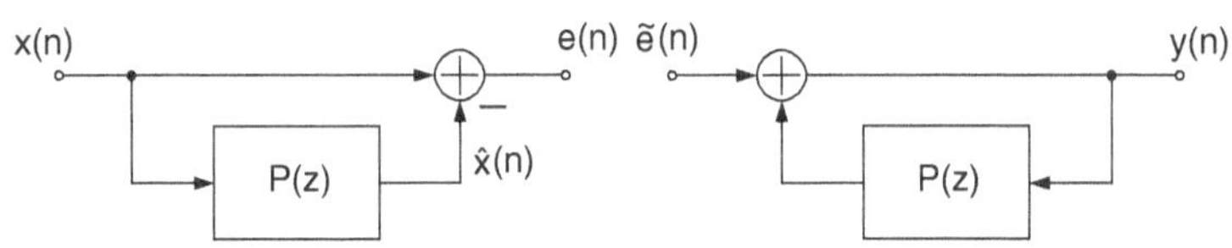

Figure 8.28 Source-filter separation and processing using linear predictive coding (analysis) and decoding (synthesis).

The prediction filter $P(z)$ of order p used in the analysis and synthesis is updated at every new incoming sample using a buffer of the p past samples from $x(n)$. The filter coefficients a_k are updated according to the least-mean-square (LMS) method given by

$$a_k(n+1) = a_k(n) + \mu e(n)x(n-k); \tag{8.40}$$

for $k = 1, \dots, p$, where k is the filter coefficient index and μ corresponds to the step size of the adaption. The predicted signal $\hat{x}(n)$ is calculated using

$$\hat{x}(n) = \sum_{k=1}^{p} a_k x(n-k), \tag{8.41}$$

which represents FIR filtering of the input $x(n)$ with the previously estimated filter coefficients. The transfer function for coding (analysis) is given by

$$H_C(z) = \frac{E(z)}{X(z)} = 1 - P(z). \tag{8.42}$$

The decoder for the re-synthesis uses the inverse transfer function of the coder given by

$$H_D(z) = \frac{Y(z)}{\tilde{E}(z)} = \frac{1}{1 - P(z)}. \tag{8.43}$$

This filter uses the coefficients calculated by the prediction filter $P(z)$ and performs as an all-pole filter. From an implementation point of view, this is simply using the filter in a feedback loop, as depicted in Fig. 8.28.

8.8.2 Combination with DRC

In this particular system, the error signal will be processed by a DRC system to influence the re-synthesized signal $y(n)$, as depicted in Fig. 8.29. DRC produces generally a quite transparent effect. Therefore, the processed error signal will not be affected too drastically, so the reconstructed signal will be less prompt to artifacts arising from imperfect reconstruction.

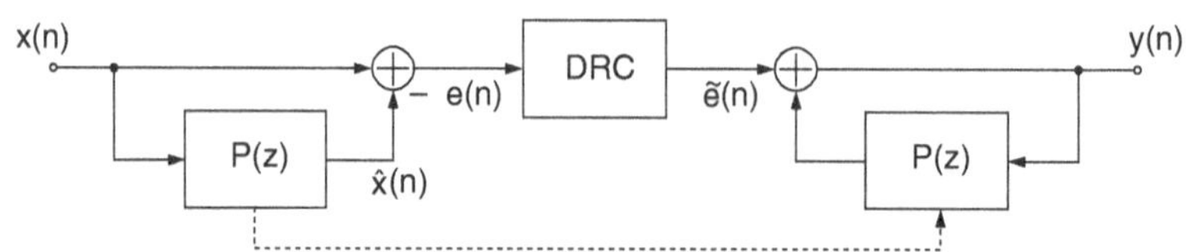

Figure 8.29 Block diagram of the combined systems LPC and DRC.

8.8.3 Applications

Denoising

Dynamic range control has been one of the first tools used to denoise audio signals. As an example, a noise gate or expander attenuates the input signal when its level goes below a defined threshold. By setting this threshold right above the background noise level, the parts where no signal of interest is present will see their level reduced, thus denoised. This denoising method is, by definition, effective only when the energy of the input signal is low in the case of a speech signal when the speaker is not speaking or pausing between words.

In this denoising application, an expander is used as a DRC block, because it tends to be less prompt to artifacts than a noise gate. The expander chosen in this application is composed of a peak level detector [Gia12]. Because the error signal carries out most of the background noise, the expander threshold parameter can be set to a few dBs above the noise level (see Fig. 8.30). The pulse train remains, whereas the noise in between pulses is slightly reduced. Moreover, the silent part sees its background noise reduced as well, because the level is located below the expander threshold. For this application, the time constants of the expander have been set to a very fast

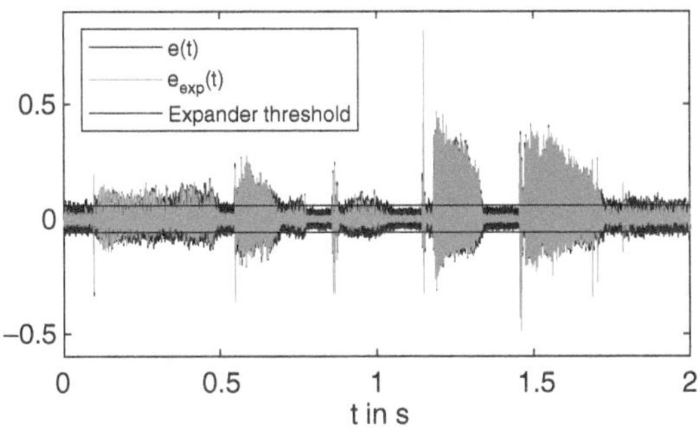

Figure 8.30 Error signals $e(t)$ and $e_{exp}(t)$ before and after dynamic range expansion, respectively.

attack AT = 0.1 ms to leave the transients and pulses untouched and a slower release RT = 1 ms to avoid artifacts. A ratio ER = 40 has been chosen to drastically reduce the quieter sounds, thus the background noise. In this particular case, the threshold has been set up manually using prior knowledge of the noise level. However, an adaptive threshold may be applied for non-stationary noise situations. It is worth mentioning that for high noise levels (that cover the speech), the presented algorithm will see its performances reduced drastically because its fundamental principle is to increase the level differences between parts of the signal. As mentioned in [Arf11], LPC can be parameterized so that the error signal resembles a whispering effect. This means that the resonant part of the voice is removed resulting in a combination of plosives, noise and pulse sequences. Removing the background noise from such a signal results in denoising the re-synthesized signal. To obtain an excitation signal with such properties, filter orders in the range $p = [80, 150]$ are most suitable (with $f_S = 44.1$ kHz).

The top left plot in Fig. 8.31 shows the spectrogram of a female singing voice with white noise (30 dB SNR). The top right plot spectrogram in Fig. 8.31 shows the effect of an expander applied as denoiser. In the case of the presence of stationary low-colored background noise in a signal, it will be contained in the error signal $e(n)$ as well, as the separation by the LPC is whitening the signal. By denoising the error signal with an expander, the re-synthesized signal $y(n)$ will also be denoised in parts where voicing is happening. This effect is visible on the spectrogram shown in the lower plot of Fig. 8.31, where an attenuation of the background noise is occurring without altering the original signal.

Transient Control

The error signal obtained by the source–filter separation contains the transients of the original sound. Because transients are sudden burst of energy, the prediction filter is not fast enough to predict them. This arises from to the weighted update of the filter coefficients (μ). Moreover, transients are generally broadband. This character makes them prompt not to be modeled by the filter, and therefore are present in the error signal.

Once transients are extracted, it is possible to choose the amount to re-inject into the error signal before the re-synthesis. For this matter, a gain G_{tr} controlling the amount of transient to add or cancel (by phase inversion) is placed, as shown in Fig. 8.32. One of the problems occurring frequently when recording speech is the presence of too strong transients or clipping.

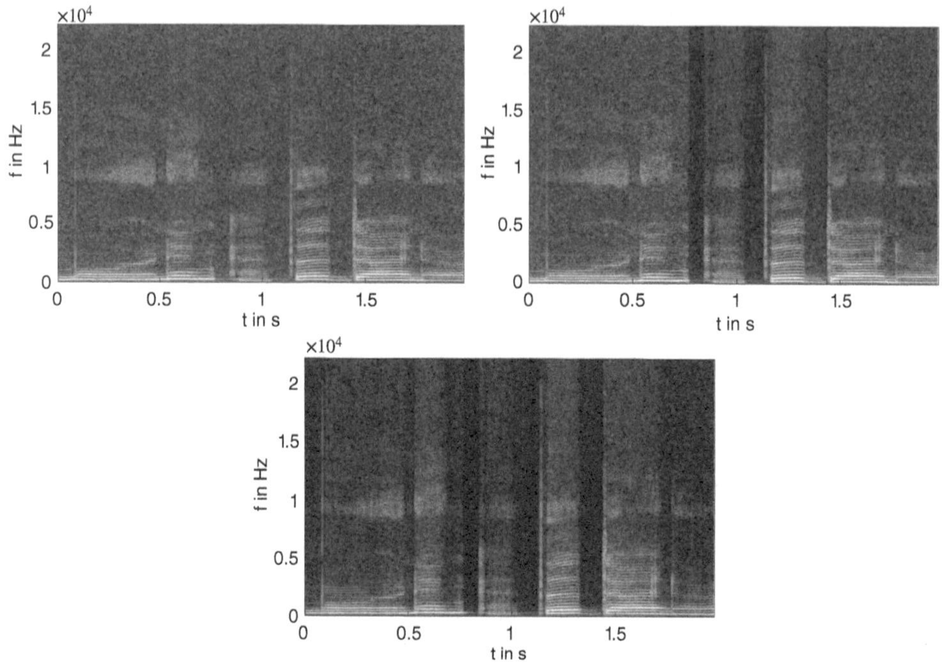

Figure 8.31 Spectrograms of a female singing voice with white noise (30 dB SNR; top left, input; top right, expanded input; bottom, denoised output).

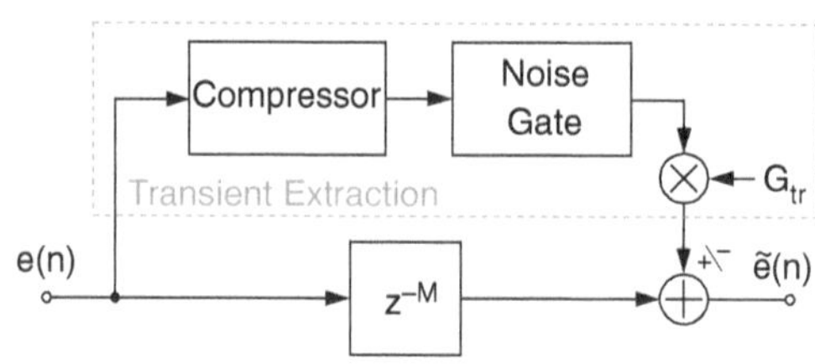

Figure 8.32 Block diagram of the DRC block for transient control.

This can arise from the air flow hitting the microphone membrane when pronouncing a plosive. Reducing transients may be used to reduce this effect (pop filtering). Traditionally, compressors are used to overcome such problems. Nevertheless, it is also considered that the attack portion of sounds conveys brightness character, so increasing the transients may also give a brighter effect to the re-synthesized signal. Depending on the desired effect, being able to control the amount of transients in a sound is useful. This property makes the error a good pre-processing step to extract transients. Processing it with a compressor can achieve a complete transient extraction.

An example of extracted transients is shown in Fig. 8.33 and the corresponding spectrograms are visible in Fig. 8.34. Selecting a slower attack time will allow for the transient portion of the sound to pass through before the compressor starts clamping. However, if the attack time is too slow, then too big a portion of the transient may pass through. A noise gate is cascaded after the expander to fully remove the non-transients parts (see Fig. 8.32).

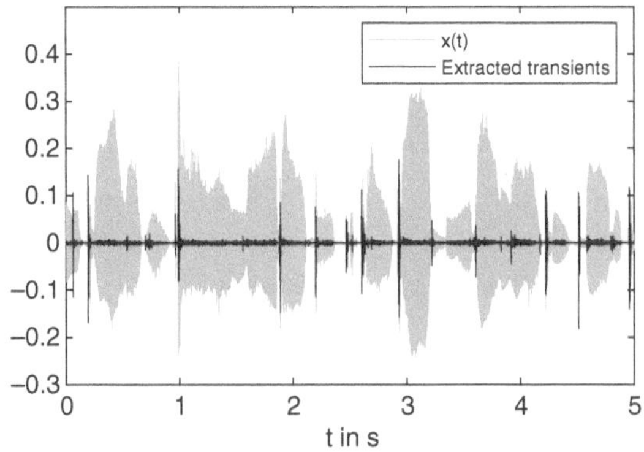

Figure 8.33 Input signal $x(t)$ and the corresponding extracted transients.

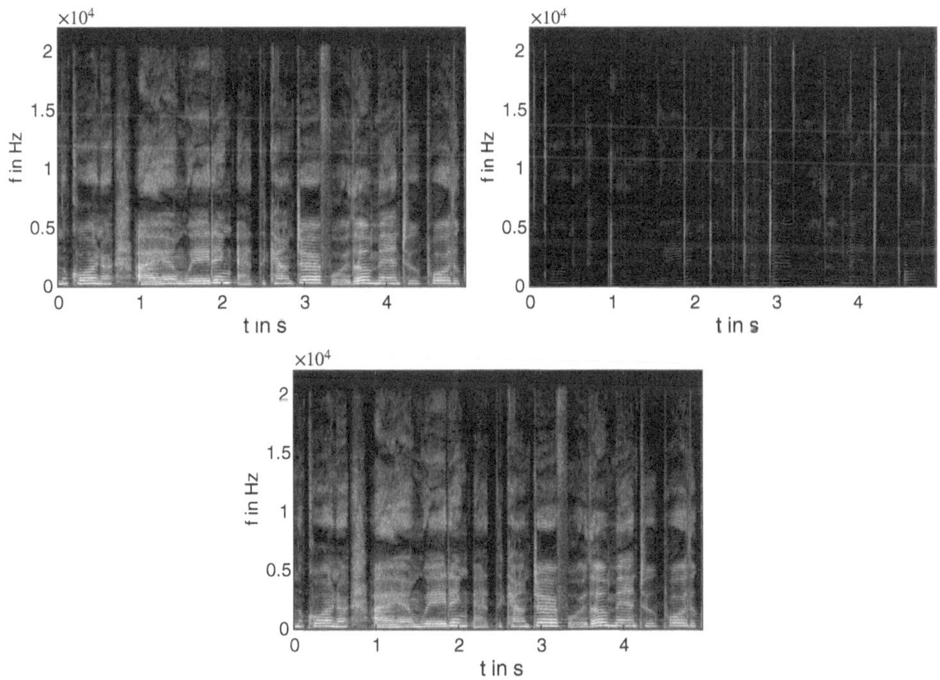

Figure 8.34 Spectrogram of a female singing voice (top left, input; top right, extracted transients; bottom, output with attenuated transients).

8.9 JS Applet – Dynamic Range Control

The applet shown in Fig. 8.35 demonstrates dynamic range control. It is designed for a first insight into the perceptual effects of dynamic range control of an audio signal. You can adjust the characteristic curve with two control points. You can choose between two predefined audio files from our web server (*audio1.wav* or *audio2.wav*) or your own local WAV file to be processed [Gui05].

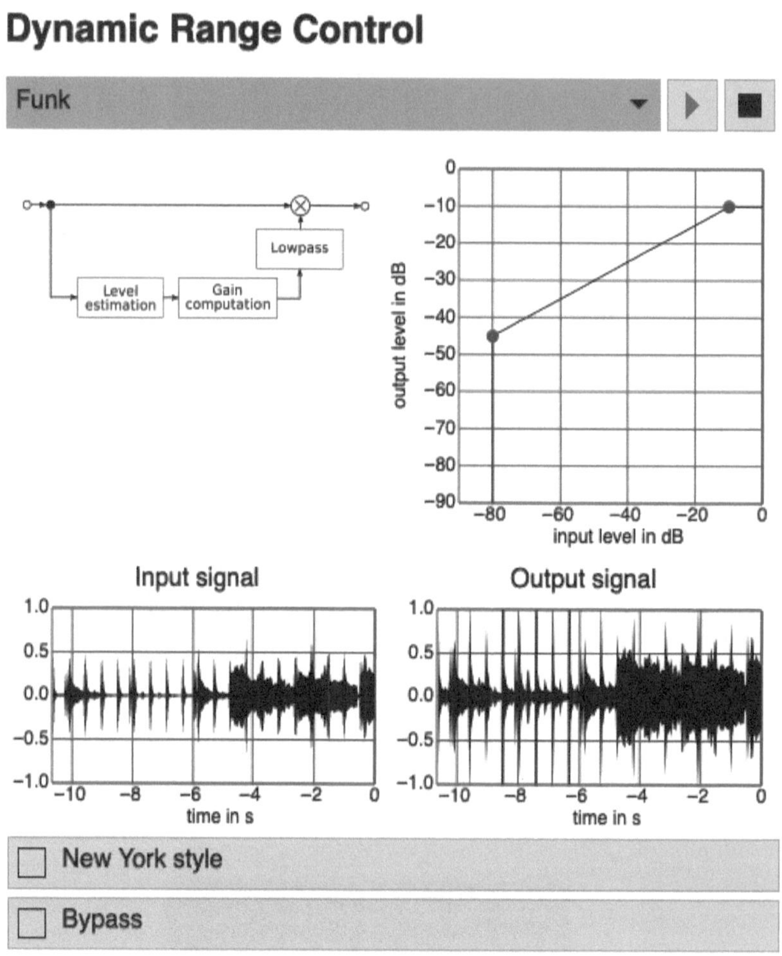

Figure 8.35 JS applet – dynamic range control.

8.10 Exercises

1. Lowpass Filtering for Envelope Detection

Generally, envelope computation is performed by lowpass filtering the input signal's absolute value or its square.

1. Sketch the block diagram of a recursive first-order lowpass $H(z) = \frac{\lambda}{1-(1-\lambda)z^{-1}}$.
2. Sketch its step response. What characteristic measure of the envelope detector can be derived from the step response and how?
3. Typically, the lowpass filter is modified to use a non-constant filter coefficient λ. How does λ depend on the signal? Sketch the response to a rect-signal of the lowpass filter thus modified.

2. Discrete-time Specialties of Envelope Detection

Taking absolute value or squaring are nonlinear operations. Hence care must be taken when using them in discrete-time systems as they introduce harmonics the frequency of which may violate the Nyquist bound. This can lead to unexpected results, as a simple example shall illustrate. Consider the input signal $x(n) = \sin\left(\frac{\pi}{2}n + \varphi\right), \varphi \in [0,2\pi]$.

1. Sketch $x(n)$, $|x(n)|$, and $x^2(n)$ for different values of φ.
2. Determine the value of the envelope after perfect lowpass filtering, i.e. averaging, $|x(n)|$. Note: As the input signal is periodical, it is sufficient to consider one period, e.g.

$$\overline{x} = \frac{1}{4}\sum_{n=0}^{3}|x(n)|.$$

3. Similarly, determine the value of the envelope after averaging $x^2(n)$.

3. Dynamic Range Processors

Sketch the characteristic curves mapping input level to output level and input level to gain for, and describe briefly the application of:

1. limiter;
2. compressor;
3. expander; and
4. noise gate.

References

[Arf11] D. Arfib, F. Keiler, U. Zölzer, and V. Verfaille: *Source-Filter Processing*, chapter 8, pages 279–320. John Wiley and Sons, Ltd, 2011.

[Dut11] P. Dutilleux, K. Dempwolf, M. Holters, and U. Zölzer: *Nonlinear Processing*, chapter 4, pages 101–138. John Wiley and Sons, Ltd, 2011.

[Gia12] D. Giannoulis, M. Massberg, and J.D. Reiss: *Digital dynamic range compressor design - a tutorial and analysis*. Journal of the Audio Engineering Society, 60(6):399–408, 2012.

[Gui05] M. Guillemard, C. Ruwwe, U. Zölzer: *J-DAFx - Digital Audio Effects in Java*, Proc. 8th Int. Conference on Digital Audio Effects (DAFx-05), Madrid, Spain, pp.161–166, 2005.

[Kad07] Manjunath D. Kadaba: *Artificial speech synthesis using LPC* In Audio Engineering Society Convention 122, May 2007.

[McN84] G.W. McNally: *Dynamic Range Control of Digital Audio Signals*, J. Audio Eng. Soc., Vol. 32, No. 5, pp. 316–327, 1984.

[Nag09] F. Nagel, S. Disch, and N. Rettelbach: *A phase vocoder driven bandwidth extension method with novel transient handling for audio codecs*. In Audio Engineering Society Convention 126, May 2009.

[Sla62] F.H. Slaymaker and R.A. Houde. *Speech compression by analysis-synthesis*. J. Audio Eng. Soc, 10(2):144–148, 1962.

[Sti86] E. Stikvoort: *Digital Dynamic Range Compressor for Audio*, J. Audio Eng. Soc., Vol. 34, No. 1/2, pp. 3–9, 1986.

[Väl16] V. Välimäki and J.D. Reiss: *All about audio equalization: Solutions and frontiers*. Applied Sciences, 6(5), 2016.

[Wis09] D.K. Wise: *Concept, design, and implementation of a general dynamic parametric equalizer*. Journal of the Audio Engineering Society, 57(1/2):1–28, January 2009.

[Zöl08] U. Zölzer: *Digital Audio Signal Processing*. John Willey and Sons, 2008.

Chapter 9

Audio Coding

U. Zölzer and P. Bhattacharya

For transmission and storage of audio signals, different methods for compressing data have been investigated in addition to the pulse-code modulation (PCM) representation. The requirements of different applications have resulted in a variety of audio coding methods which have become international standards. In this chapter, the basic principles of audio coding will be introduced and then the most important audio coding standards will be discussed. Audio coding can be divided into two types: *lossless* and *lossy* audio coding. Lossless audio coding is based on a statistical model of the signal amplitudes and coding of the audio signal (audio coder). The reconstruction of the audio signal at the receiver allows a lossless resynthesis of the signal amplitudes of the original audio signal (audio decoder). Alternatively, lossy audio coding makes use of a psychoacoustic model of the human acoustic perception for quantizing and coding the audio signal. In this case, only the acoustic relevant parts of the signal are coded and reconstructed at the receiver. The samples of the original audio signal are not exactly reconstructed. The objective of both audio coding methods is a data rate reduction or data compression for transmission or storage compared with the original PCM signal.

9.1 Lossless Audio Coding

Lossless audio coding is based on linear prediction followed by entropy coding [Jay84], as shown in Fig. 9.1.

- Linear Prediction. A quantized set of coefficients P for a block of M samples is determined, which leads to an estimate $\hat{x}(n)$ of the input sequence $x(n)$. The aim is to minimize the power of the difference signal $d(n)$ without any additional quantization errors, i.e. the word length of the signal $\hat{x}(n)$ must be equal to the word length of the input. An alternative approach [Han98, Han01] quantizes the prediction signal $\hat{x}(n)$ such that the word length of the difference signals $d(n)$ remains the same as the input signal word length. Figure 9.2

Digital Audio Signal Processing, Third Edition. Udo Zölzer.
© 2022 John Wiley & Sons Ltd. Published 2022 by John Wiley & Sons Ltd.

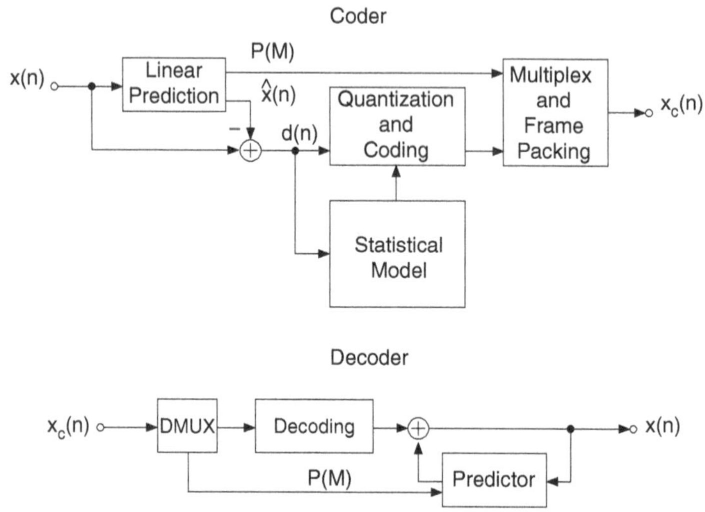

Figure 9.1 Lossless audio coding based on linear prediction and entropy coding.

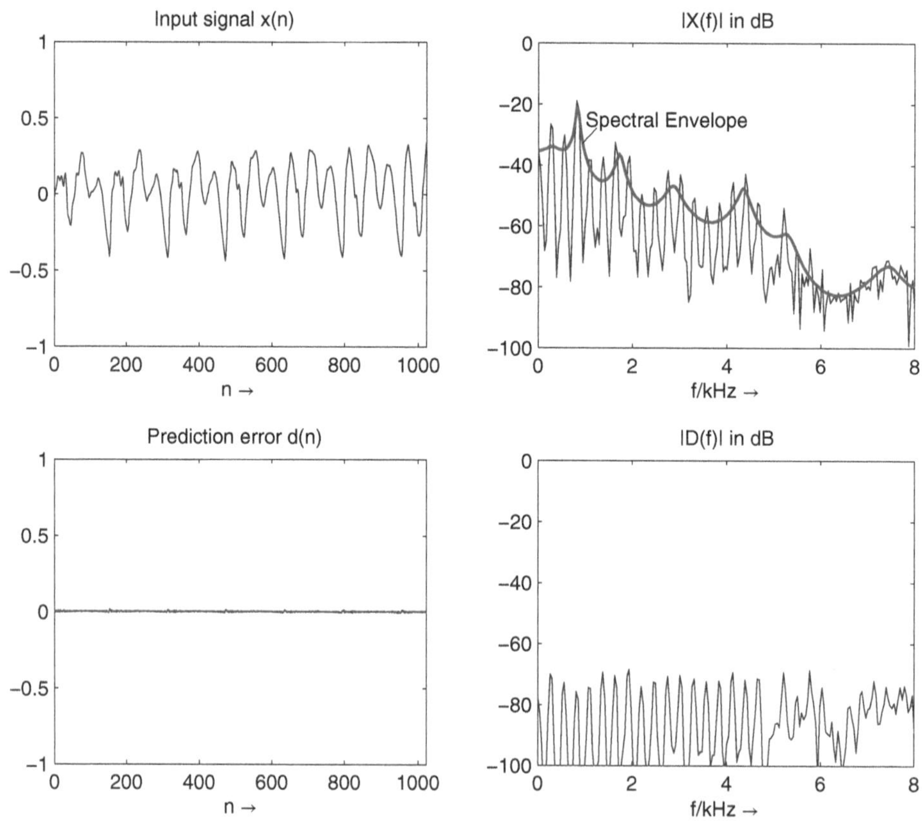

Figure 9.2 Signals and spectra for linear prediction.

shows a signal block $x(n)$ and the corresponding spectrum $|X(f)|$. Filtering the input signal with the predictor filter transfer function $P(z)$ delivers the estimate $\hat{x}(n)$. Subtracting input and prediction signals yields the prediction error $d(n)$, which is also shown in Fig. 9.2 and which has a considerably lower power compared with the input power. The spectrum of this prediction error is nearly white (see Fig. 9.2 bottom right figure). The prediction can be represented as a filter operation with an analysis transfer function $H_A(z) = 1 - P(z)$ on the coder side.

- Entropy Coding. Quantization of signal $d(n)$ arising from the probability density function of the block. Samples $d(n)$ of higher probability are coded with shorter data words whereas samples $d(n)$ of lower probability are coded with longer data words [Huf52].

- Frame Packing. The frame packing uses the quantized and coded difference signal and the coding of the M coefficients of the predictor filter $P(z)$ of order M.

- Decoder. On the decoder side, the inverse synthesis transfer function $H_S(z) = H_A^{-1}(z) = [1 - P(z)]^{-1}$ reconstructs the input signal with the coded difference samples and the M filter coefficients. The frequency response of this synthesis filter represents the spectral envelope shown in the top right part of Fig. 9.2. The synthesis filter shapes the white spectrum of the difference (prediction error) signal with the spectral envelope of the input spectrum.

The attainable compression rates depend on the statistics of the audio signal and allow a compression rate of up to two [Bra92, Cel93, Rob94, Cra96, Cra97, Pur97, Han98, Han01, Lie02, Raa02, Sch02]. Figure 9.3 illustrates examples of the necessary word length for lossless audio coding [Blo95, Sqa88]. In addition to the local entropy of the signal (entropy computed over a block length of 256), results for linear prediction followed by Huffman coding [Huf52] are presented. Huffman coding is carried out with a fixed code table [Pen93] and a power-controlled choice of adapted code tables. It is observed from Fig. 9.3 that for high signal powers, a reduction of word length is possible if the choice is made from several adapted code tables. Lossless compression methods are used for storage media with limited word length (16 bit), which are used for recording audio signals of higher word lengths (>16 bit). Further applications are in the transmission and archiving of audio signals.

9.2 Lossy Audio Coding

Significantly higher compression rates (of factor four to eight) can be obtained with *lossy* coding methods. Psychoacoustic phenomena of human hearing are used for signal compression. The fields of application have a wide range, from professional audio like source coding for audio transmission to home entertainment applications.

An outline of the coding methods [Bra94] is standardized in an international specification ISO/IEC 11172-3 [ISO92], which is based on the following processing (see Fig. 9.4) steps.

- Sub-band decomposition with filter banks of short latency time.

- Calculation of psychoacoustic model parameters based on short-time fast Fourier transform (FFT).

- Dynamic bit allocation owing to psychoacoustic model parameters (signal-to-mask ratio SMR).

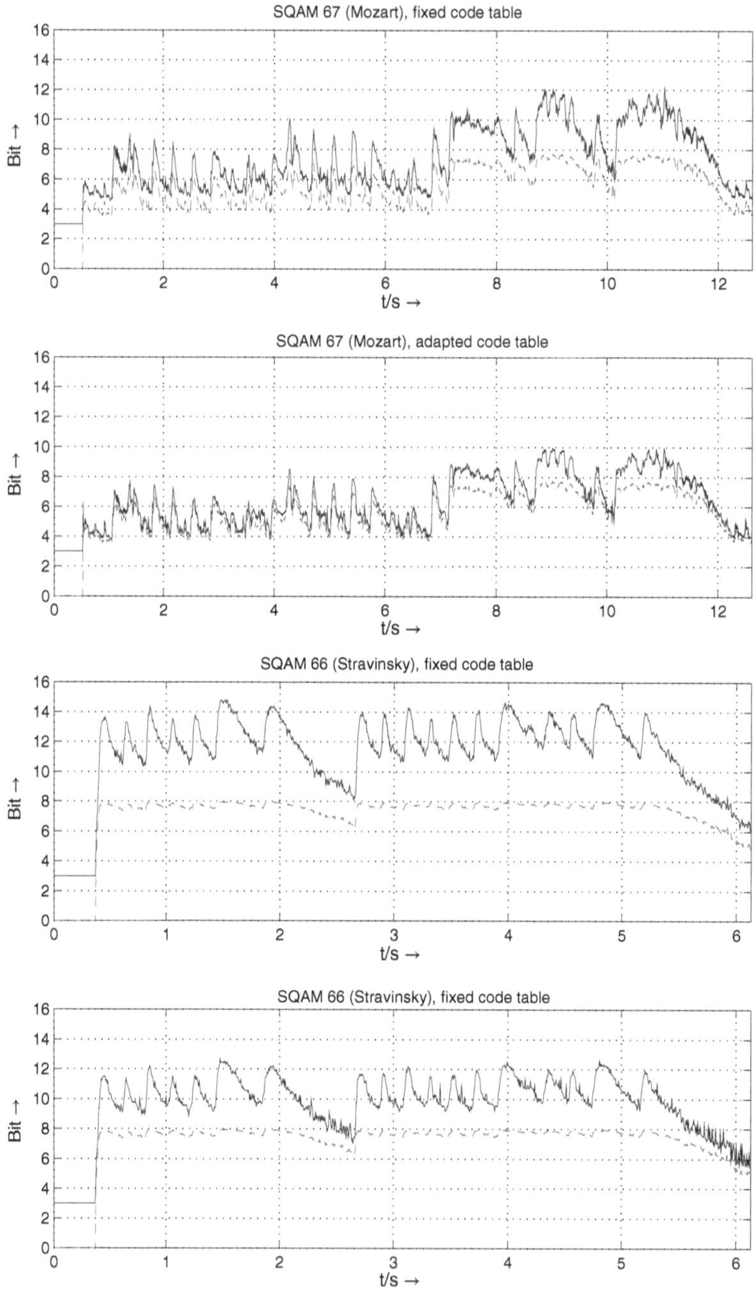

Figure 9.3 Lossless audio coding (Mozart, Stravinsky): word length in bits versus time (entropy - -, linear prediction with Huffman coding –).

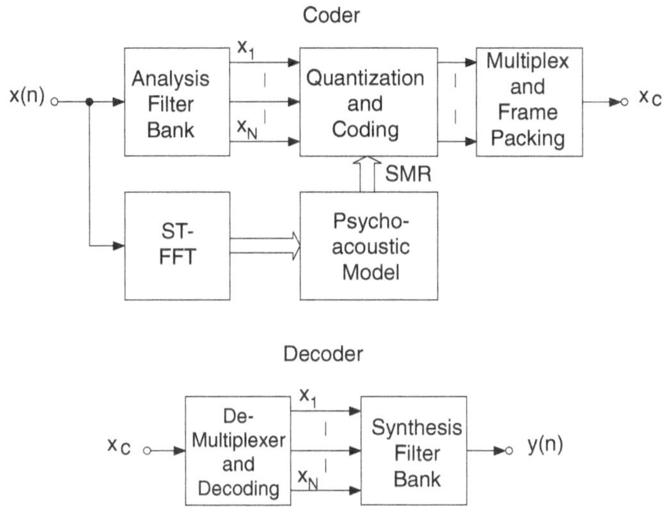

Figure 9.4 Lossy audio coding based on sub-band coding and psychoacoustic models.

- Quantization and coding of sub-band signals.
- Multiplex and frame packing.

Owing to lossy audio coding, post-processing of such signals or several coding and decoding steps is associated with some additional problems. The high compression rates justify the use of lossy audio coding techniques in applications like transmission.

9.3 Psychoacoustics

In this section, the basic principles of psychoacoustics are presented. The results of psychoacoustic investigations by Zwicker [Zwi82, Zwi90] form the basis for audio coding based on models of human perception. These coded audio signals have a significantly reduced data rate compared with the linearly quantized PCM representation. The human auditory system analyzes broadband signals in so-called critical bands. The aim of psychoacoustic coding of audio signals is to decompose the broadband audio signal into sub-bands, which are matched to the critical bands, and then perform quantization and coding of these sub-band signals [Joh88a, Joh88b, Thei88]. Because the perception of sound below the absolute threshold of hearing is not possible, sub-band signals below this threshold need neither to be coded nor transmitted. In addition to the perception in critical bands and the absolute threshold, the effects of signal masking in human perception play an important role in signal coding. These will be explained in the following and their application to psychoacoustic coding will be discussed.

9.3.1 Critical Bands and Absolute Threshold

Critical Bands. Critical bands, as investigated by Zwicker [Zwi82], are listed in Table 9.1.

Table 9.1 Critical bands as given by Zwicker. Modified from [Zwi82].

z/Bark	f_l/Hz	f_u/Hz	f_B/Hz	f_c/Hz
0	0	100	100	50
1	100	200	100	150
2	200	300	100	250
3	300	400	100	350
4	400	510	110	450
5	510	630	120	570
6	630	770	140	700
7	770	920	150	840
8	920	1080	160	1000
9	1080	1270	190	1170
10	1270	1480	210	1370
11	1480	1720	240	1600
12	1720	2000	280	1850
13	2000	2320	320	2150
14	2320	2700	380	2500
15	2700	3150	450	2900
16	3150	3700	550	3400
17	3700	4400	700	4000
18	4400	5300	900	4800
19	5300	6400	1100	5800
20	6400	7700	1300	7000
21	7700	9500	1800	8500
22	9500	1200	2500	10500
23	12000	15500	3500	13500
24	15500			

A transformation of the linear frequency scale into a hearing adapted scale is given by Zwicker [Zwi90] (units of z in Bark)

$$\frac{z}{\text{Bark}} = 13\arctan\left(0.76\frac{f}{\text{kHz}}\right) + 3.5\arctan\left(\frac{f}{7.5\text{ kHz}}\right)^2. \tag{9.1}$$

The individual critical bands have the following bandwidths:

$$\Delta f_B = 25 + 75\left(1 + 1.4\left(\frac{f}{\text{kHz}}\right)^2\right)^{0.69}. \tag{9.2}$$

Absolute Threshold. The absolute threshold L_{T_q} (threshold in quiet) denotes the curve of sound pressure level L [Zwi82] versus frequency, which leads to the perception of a sinusoidal tone. The absolute threshold is given by [Ter79]

$$\frac{L_{T_q}}{\text{dB}} = 3.64\left(\frac{f}{\text{kHz}}\right)^{-0.8} - 6.5\exp\left(-0.6\left(\frac{f}{\text{kHz}} - 3.3\right)^2\right) + 10^{-3}\left(\frac{f}{\text{kHz}}\right)^4. \tag{9.3}$$

Below the absolute threshold, no perception of signals is possible. Figure 9.5 shows the absolute threshold versus frequency. Band splitting in critical bands and the absolute threshold allow for

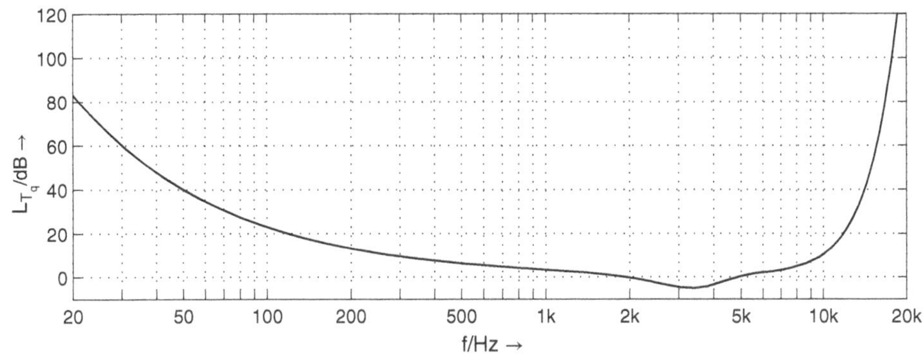

Figure 9.5 Absolute threshold (threshold in quiet).

the calculation of an offset between the signal level and the absolute threshold for every critical band. This offset is responsible for choosing appropriate quantization steps per critical band.

9.3.2 Masking

For audio coding, the use of sound perception in critical bands and absolute threshold only is not sufficient for high compression rates. The bases for further data reduction are the masking effects investigated by Zwicker [Zwi82, Zwi90]. For band-limited noise or a sinusoidal signal, frequency-dependent masking thresholds can be given. These thresholds perform masking of frequency components if these components are below a masking threshold (see Fig. 9.6). The application of masking for perceptual coding is described in the following.

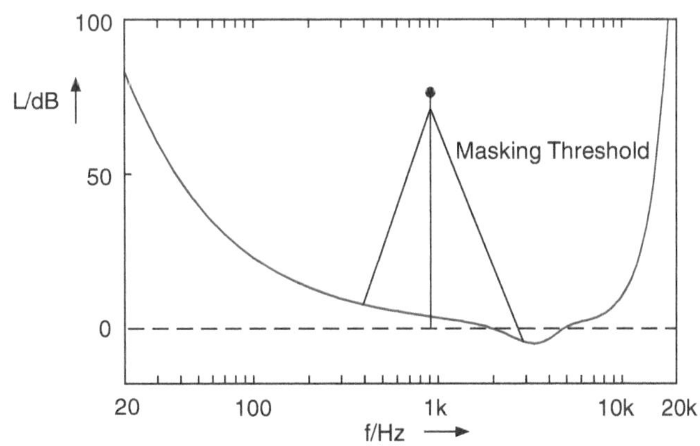

Figure 9.6 Masking threshold of band-limited noise.

Calculation of Signal Power in Band *i*. First, the sound pressure level within a critical band is calculated. The short-time spectrum $X(k) = \text{DFT}[x(n)]$ is used to calculate the power density spectrum

$$S_p\left(e^{j\Omega}\right) = S_p\left(e^{j\frac{2\pi k}{N}}\right) = X_R^2\left(e^{j\frac{2\pi k}{N}}\right) + X_I^2\left(e^{j\frac{2\pi k}{N}}\right) \tag{9.4}$$

$$S_p(k) = X_R^2(k) + X_I^2(k) \quad 0 \le k \le N-1, \tag{9.5}$$

with the help of an N-point FFT. The signal power in band i is calculated by the sum

$$S_p(i) = \sum_{\Omega=\Omega_{li}}^{\Omega_{ui}} S_p(k) \tag{9.6}$$

from the lower frequency up to the upper frequency of critical band i. The sound pressure level in band i is given by $L_S(i) = 10\log_{10} S_p(i)$.

Absolute Threshold. The absolute threshold is set such that a 4-kHz signal with peak amplitude ±1 LSB for a 16-bit representation lies at the lower limit of the absolute threshold curve. Every masking threshold calculated in individual critical bands, which lies below the absolute threshold, is set to a value equal to the absolute threshold in the corresponding band. Because the absolute threshold within a critical band varies for low and high frequencies, it is necessary to make use of the mean absolute threshold within a band.

Masking Threshold. The offset between signal level and the masking threshold in critical band i (see Fig. 9.7) is given by [Hel72]

$$\frac{O(i)}{\text{dB}} = \alpha(14.5 + i) + (1 - \alpha)a_v, \tag{9.7}$$

where α denotes the tonality index and a_v is the masking index. The masking index [Kap92] is given by

$$a_v = -2 - 2.05\ \arctan\left(\frac{f}{4\ \text{kHz}}\right) - 0.75\ \arctan\left(\frac{f^2}{2.56\ \text{kHz}^2}\right). \tag{9.8}$$

As an approximation,

$$\frac{O(i)}{\text{dB}} = \alpha(14.5 + i) + (1 - \alpha)5.5 \tag{9.9}$$

can be used [Joh88a, Joh88b]. If a tone is masking a noise-like signal ($\alpha = 1$), the threshold is set $14.5 + i$ dB below the value of $L_S(i)$. If a noise-like signal is masking a tone ($\alpha = 0$), the threshold is set $5.5 + i$ dB below $L_S(i)$. To recognize a tonal or noise-like signal within a certain number of samples, the *spectral flatness measure* (SFM) is estimated. The SFM is defined by the ratio of the

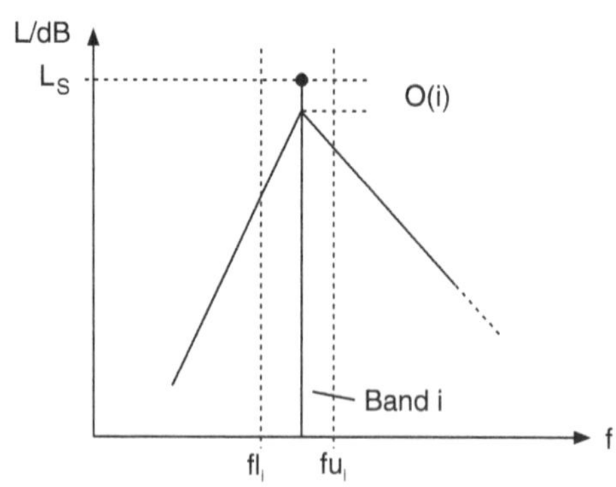

Figure 9.7 Offset between signal level and masking threshold.

geometric to arithmetic mean value of $S_p(i)$ according to

$$\mathrm{SFM} = 10\log_{10}\left(\frac{\left[\prod_{k=1}^{\frac{N}{2}} S_p\left(e^{j\frac{2\pi k}{N}}\right)\right]^{1/\frac{N}{2}}}{\frac{1}{N/2}\sum_{k=1}^{\frac{N}{2}} S_p\left(e^{j\frac{2\pi k}{N}}\right)} i\right). \tag{9.10}$$

The SFM is compared with the SFM of a sinusoidal signal (definition $\mathrm{SFM}_{\max} = -60\,\mathrm{dB}$) and the tonality index is calculated [Joh88a, Joh88b] by

$$\alpha = MIN\left(\frac{\mathrm{SFM}}{\mathrm{SFM}_{\max}}, 1\right). \tag{9.11}$$

An SFM = 0 dB corresponds to a noise-like signal and leads to $\alpha = 0$, whereas an SFM = 75 dB gives a tone-like signal ($\alpha = 1$). With the sound pressure level $L_S(i)$ and the offset $O(i)$, the masking threshold is given by

$$T(i) = 10^{[L_S(i)-O(i)]/10}. \tag{9.12}$$

Masking Across Critical Bands. Masking across critical bands can be carried out with the help of the Bark scale. The masking threshold is of a triangular form which decreases with S_1 dB per Bark for the lower slope and with S_2 dB per Bark for the upper slope, depending on the sound pressure level L_i and the center frequency f_{c_i} in band i (see [Ter79]), according to

$$S_1 = 27 \qquad \mathrm{dB/Bark} \tag{9.13}$$

$$S_2 = 24 + 0.23\left(\frac{f_{c_i}}{\mathrm{kHz}}\right)^{-1} - 0.2\frac{L_S(i)}{\mathrm{dB}} \qquad \mathrm{dB/Bark}. \tag{9.14}$$

An approximation of the minimum masking within a critical band can be made using Fig. 9.8 [Thei88, Sauv90]. Masking at the upper frequency f_{u_i} in the critical band i is responsible for masking the quantization noise with approximately 32 dB using the lower masking threshold that

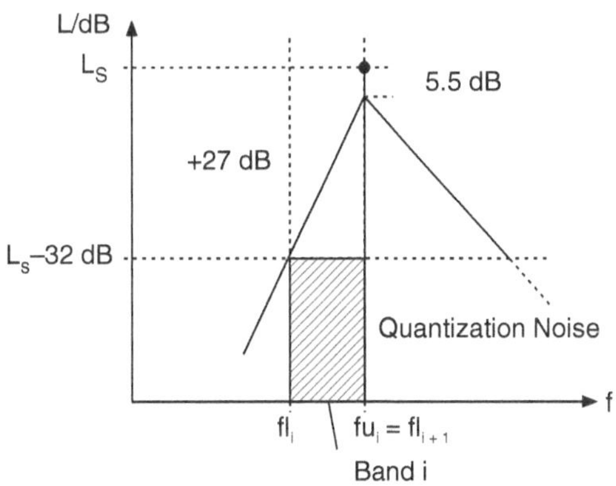

Figure 9.8 Masking within a critical band. Based on [Thei88] and [Sauv90].

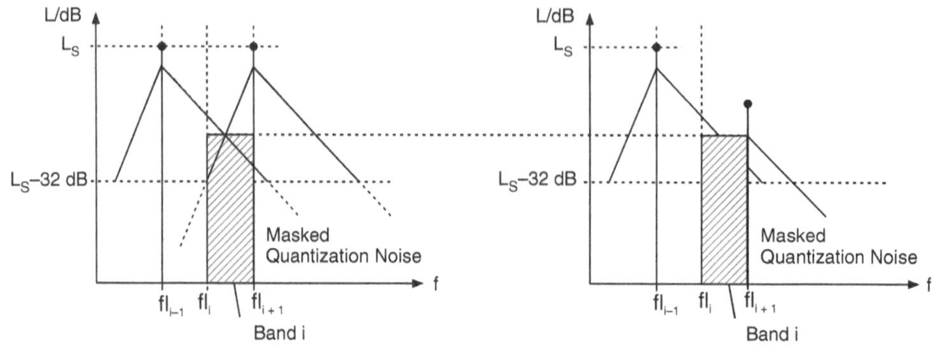

Figure 9.9 Masking across critical bands.

decreases by 27 dB/Bark. The upper slope has a steepness which depends on the sound pressure level. This steepness is lower than the steepness of the lower slope. Masking across critical bands is presented in Fig. 9.9. The masking signal in critical band $i - 1$ is responsible for masking the quantization noise in critical band i as well as the masking signal in critical band i. This kind of masking across critical bands further reduces the number of quantization steps within critical bands.

An analytical expression for masking across critical bands [Schr79] is given by

$$10\log_{10}[B(\Delta i)] = 15.81 + 7.5(\Delta i + 0.474) - 17.5[1 + (\Delta i + 0.474)^2]^{\frac{1}{2}}. \tag{9.15}$$

Here, Δi denotes the distance between two critical bands in Bark. Expression (9.15) is called the *spreading* function. With the help of this *spreading* function, masking of critical band i by critical band j can be calculated [Joh88a, Joh88b] with $\mathrm{abs}(i - j) \leq 25$ such that

$$S_m(i) = \sum_{j=0}^{24} B(i-j) \cdot S_p(i). \tag{9.16}$$

The masking across critical bands can therefore be expressed as a matrix operation given by

$$\begin{bmatrix} S_m(0) \\ S_m(1) \\ \vdots \\ S_m(24) \end{bmatrix} = \begin{bmatrix} B(0) & B(-1) & B(-2) & \cdots & B(-24) \\ B(1) & B(0) & B(-1) & \cdots & B(-23) \\ \vdots & \vdots & \vdots & & \vdots \\ B(24) & B(23) & B(22) & \cdots & B(0) \end{bmatrix} \begin{bmatrix} S_p(0) \\ S_p(1) \\ \vdots \\ S_p(24) \end{bmatrix}. \tag{9.17}$$

A renewed calculation of the masking threshold with Eq. (9.16) leads to the global masking threshold

$$T_m(i) = 10^{\log_{10} S_m(i) - O(i)/10}. \tag{9.18}$$

For a clarification of the single steps for a psychoacoustic-based audio coding, we will summarize the operations with exemplified analysis results.

- Calculation of the signal power $S_p(i)$ in critical bands
  $\rightarrow L_S(i)$ in dB (Fig. 9.10a).
- Calculation of masking across critical bands $T_m(i)$
  $\rightarrow L_{T_m}(i)$ in dB (Fig. 9.10b).

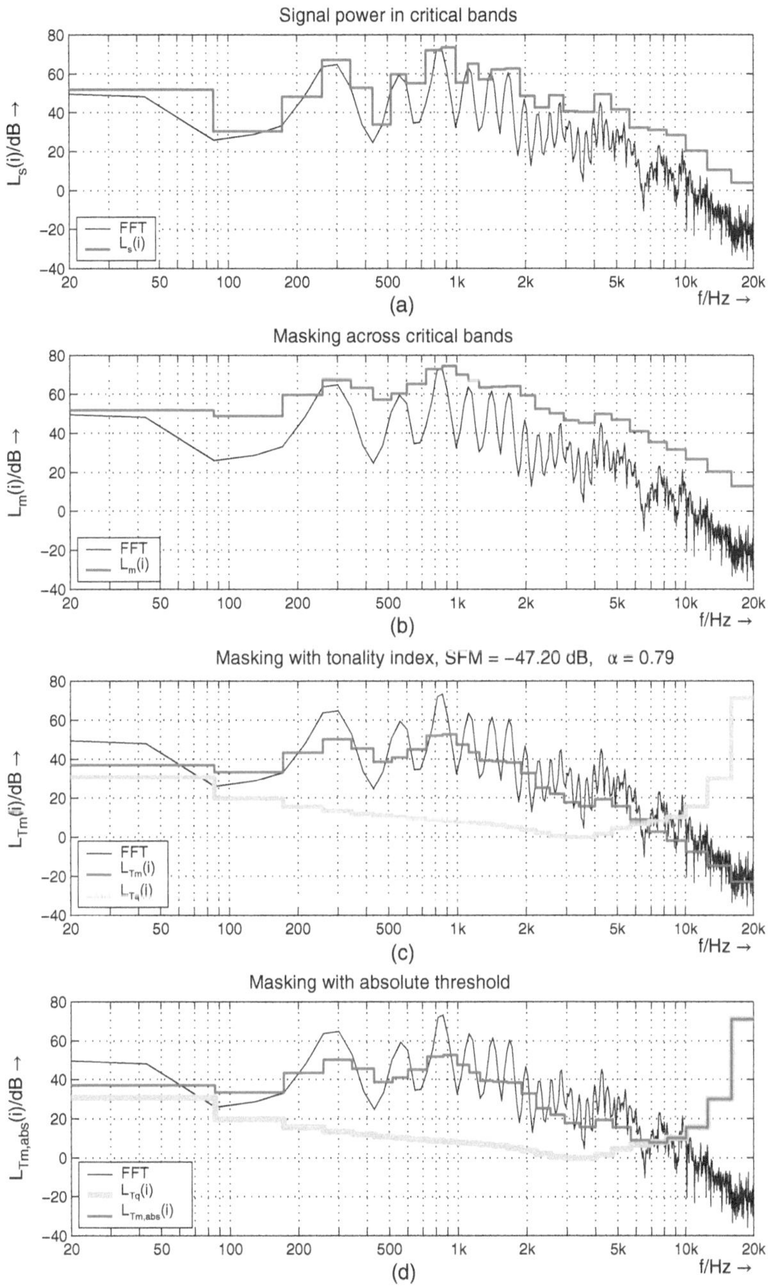

Figure 9.10 Stepwise calculation of psychoacoustic model.

- Masking with tonality index
  $\rightarrow L_{T_m}(i)$ in dB (Fig. 9.10c).
- Calculation of global masking threshold with respect to threshold in quiet L_{T_q}
  $\rightarrow L_{T_{m,abs}}(i)$ in dB (Fig. 9.10d).

With the help of the global masking threshold $L_{T_{m,abs}}(i)$, we calculate the signal-to-mask ratio (SMR)

$$\mathrm{SMR}(i) = L_S(i) - L_{T_{m,abs}}(i) \quad \text{in dB} \tag{9.19}$$

per Bark band. This SMR defines the necessary number of bits per critical band, such that masking of quantization noise is achieved. For the given example, the signal power and the global masking threshold are shown in Fig. 9.11a. The resulting signal-to-mask ratio SMR(i) is shown in Fig. 9.11b. As soon as SMR(i) > 0, one has to allocate bits to the critical band i. For SMR(i) < 0, the corresponding critical band will not be transmitted. Figure 9.12 shows the masking thresholds in critical bands for a sinusoid of 440 Hz. Compared to the first example, the influence of masking thresholds across critical bands is easier to note and to interpret.

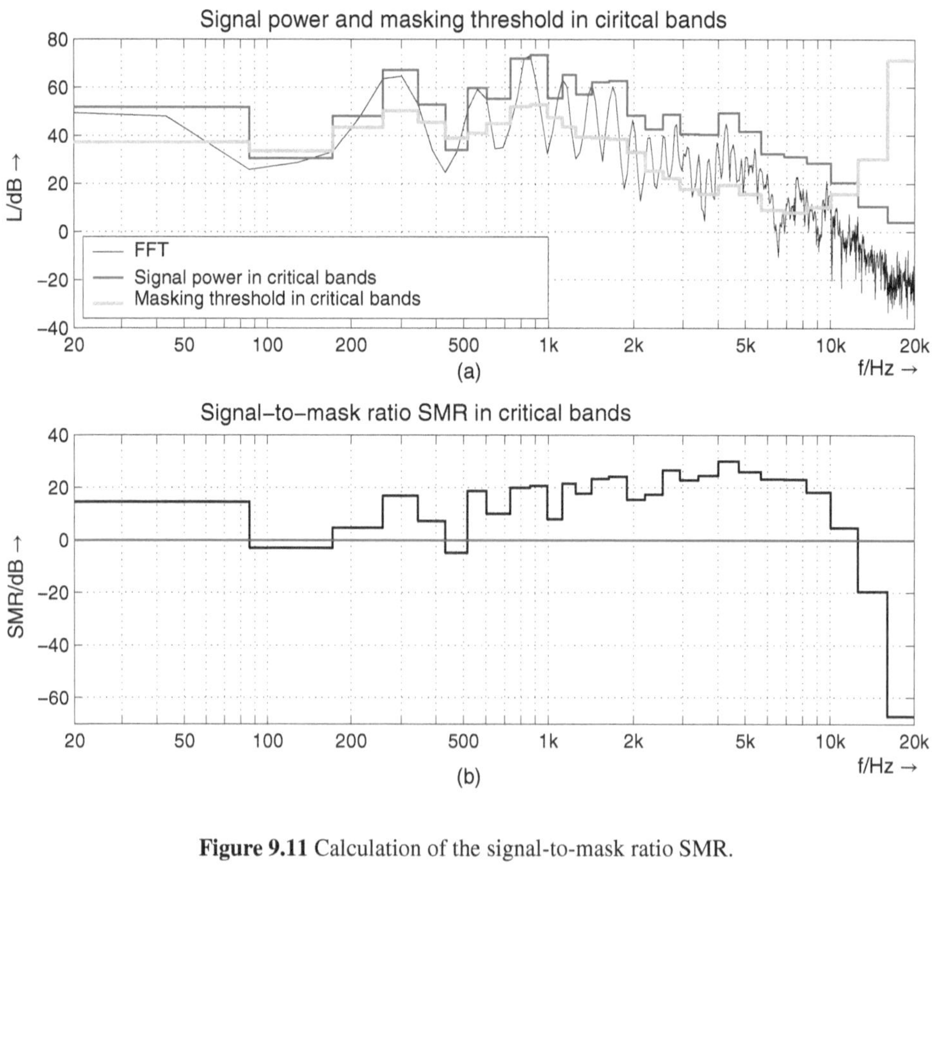

Figure 9.11 Calculation of the signal-to-mask ratio SMR.

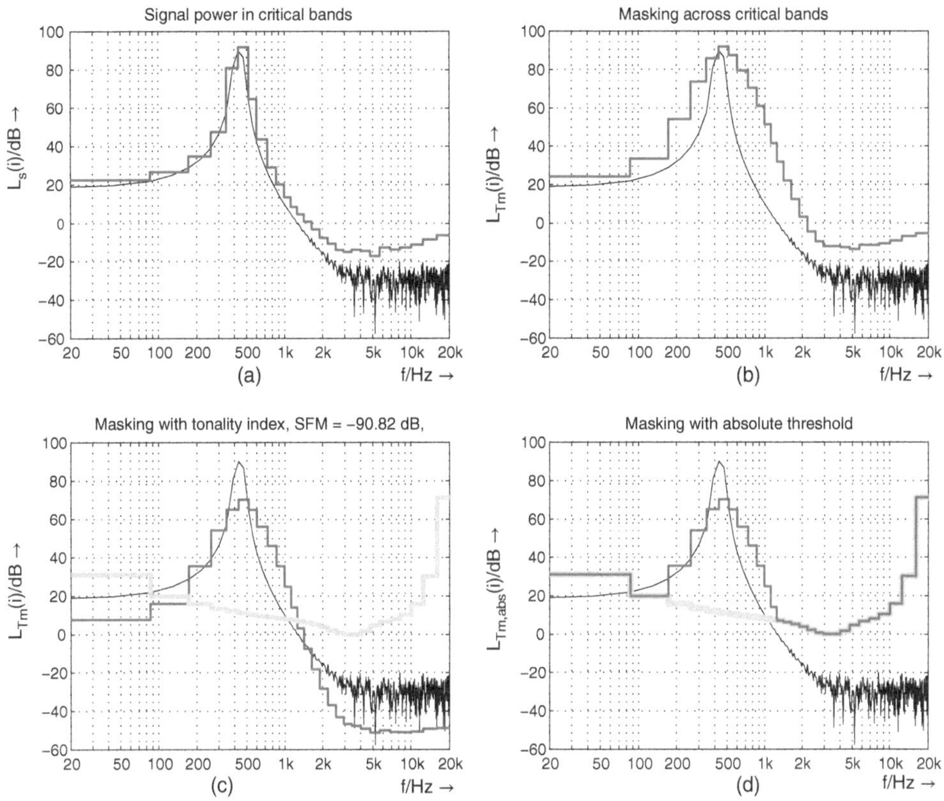

Figure 9.12 Calculation of psychoacoustic model for a pure sinusoid with 440 Hz.

9.4 ISO-MPEG1 Audio Coding

In this section, the coding method for digital audio signals is described, which is specified in the standard ISO/IEC 11172-3 [ISO92]. The filter banks used for sub-band decomposition, the psychoacoustic models, dynamic bit allocation, and coding are discussed. A simplified block diagram of the coder for implementing layers I and II of the standard is shown in Fig. 9.13. The corresponding decoder is shown in Fig. 9.14. It uses the information from the ISO-MPEG1 frame and feeds the decoded sub-band signals to a synthesis filter bank for reconstructing the broadband PCM signal. The complexity of the decoder is significantly lower compared with that of the coder. Prospective improvements of the coding method are being made entirely for the coder.

9.4.1 Filter Banks

The sub-band decomposition is done with a pseudo-quadrature mirror filter (QMF) bank (see Fig. 9.15). The theoretical background is found in the related literature [Rot83, Mas85, Vai93]. The decomposition of the broadband signal is made into M uniformly spaced sub-bands. The sub-bands are processed further after a sampling rate reduction by a factor of M. The implementation of an ISO-MPEG1 coder is based on $M = 32$ frequency bands. The individual bandpass

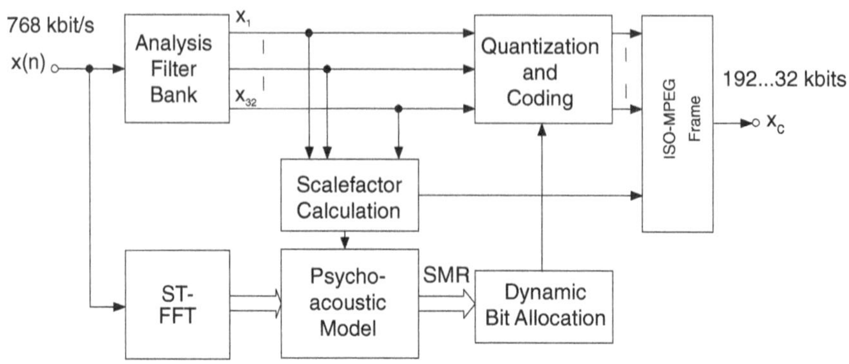

Figure 9.13 Simplified block diagram of an ISO-MPEG1 coder.

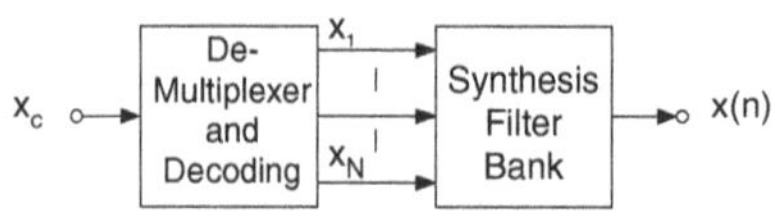

Figure 9.14 Simplified block diagram of an ISO-MPEG1 decoder.

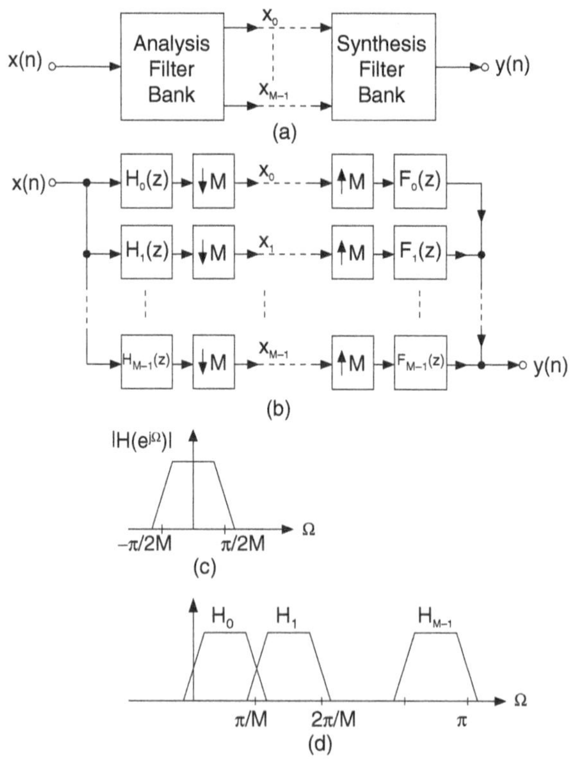

Figure 9.15 Pseudo-QMF bank.

filters $H_0(z) \dots H_{M-1}(z)$ are designed using a prototype lowpass filter $H(z)$ and frequency-shifted versions. The frequency shifting of the prototype with cutoff frequency $\pi/2M$ is done by modulating the impulse response $h(n)$ with a cosine term [Bos02] according to

$$h_k(n) = h(n) \cdot \cos\left(\frac{\pi}{32}(k + 0{,}5)(n - 16)\right), \tag{9.20}$$

$$f_k(n) = 32 \cdot h(n) \cdot \cos\left(\frac{\pi}{32}(k + 0{,}5)(n + 16)\right), \tag{9.21}$$

with $k = 0, \dots, 31$ and $n = 0, \dots, 511$. The bandpass filters have bandwidth π/M. For the synthesis filter bank, corresponding filters $F_0(z) \dots F_{M-1}(z)$ give outputs which are added together resulting in a broadband PCM signal. The prototype impulse response with 512 taps, the modulated bandpass impulse responses, and the corresponding magnitude responses are shown in Fig. 9.16. The magnitude responses of all 32 bandpass filters are also shown. The overlap of neighboring bandpass filters is limited to the lower and upper filter bands. This overlap reaches up to the center frequency of the neighboring bands. The resulting aliasing after downsampling in each sub-band will be canceled in the synthesis filter bank. The pseudo-QMF bank can be implemented by the combination of a polyphase filter structure followed by a discrete cosine transform [Rot83, Vai93, Kon94].

For increasing the frequency resolution, layer III of the standard decomposes each of the 32 sub-bands further into a maximum of 18 uniformly spaced sub-bands (see Fig. 9.17). The decomposition is carried out with the help of an overlapped transform of windowed sub-band samples. The method is based on a modified discrete cosine transform (MDCT), also known as the TDAC filter bank (time domain aliasing cancellation) and MLT (modulated lapped transform). An exact description is found in [Pri87, Mal92]. This extended filter bank is denoted as the polyphase/MDCT hybrid filter bank [Bra94]. The higher frequency resolution enables a higher coding gain but has the disadvantage of having a worse time resolution. This is observed for impulse-like signals. To minimize these artifacts, the number of sub-bands per sub-band can be altered from 18 down to 6. Sub-band decompositions that are matched to the signal can be obtained by specially designed window functions with overlapping transforms [Edl89, Edl95]. The equivalence of overlapped transforms and filter banks is found in [Mal92, Glu93, Vai93, Edl95, Vet95].

9.4.2 Psychoacoustic Models

Two psychoacoustic models have been developed for layers I to III of the ISO-MPEG1 standard. Both models can be used independently of each other for all three layers. Psychoacoustic model 1 is used for layers I and II whereas model 2 is used for layer III. Owing to the numerous applications of layers I and II, we will discuss psychoacoustic model 1 in the following.

Psychoacoustic Model 1. Bit allocation in each of the 32 sub-bands is carried out using the signal-to-mask ratio SMR(i). This is based on the minimum masking threshold and the maximum signal level within a sub-band. To calculate this ratio, the power density spectrum is estimated with the help of a short-time FFT in parallel with the analysis filter bank. As a consequence, a higher frequency resolution is obtained to estimate the power density spectrum in contrast to the frequency resolution of the 32-band analysis filter bank. The signal-to-mask ratio for every sub-band is determined as follows.

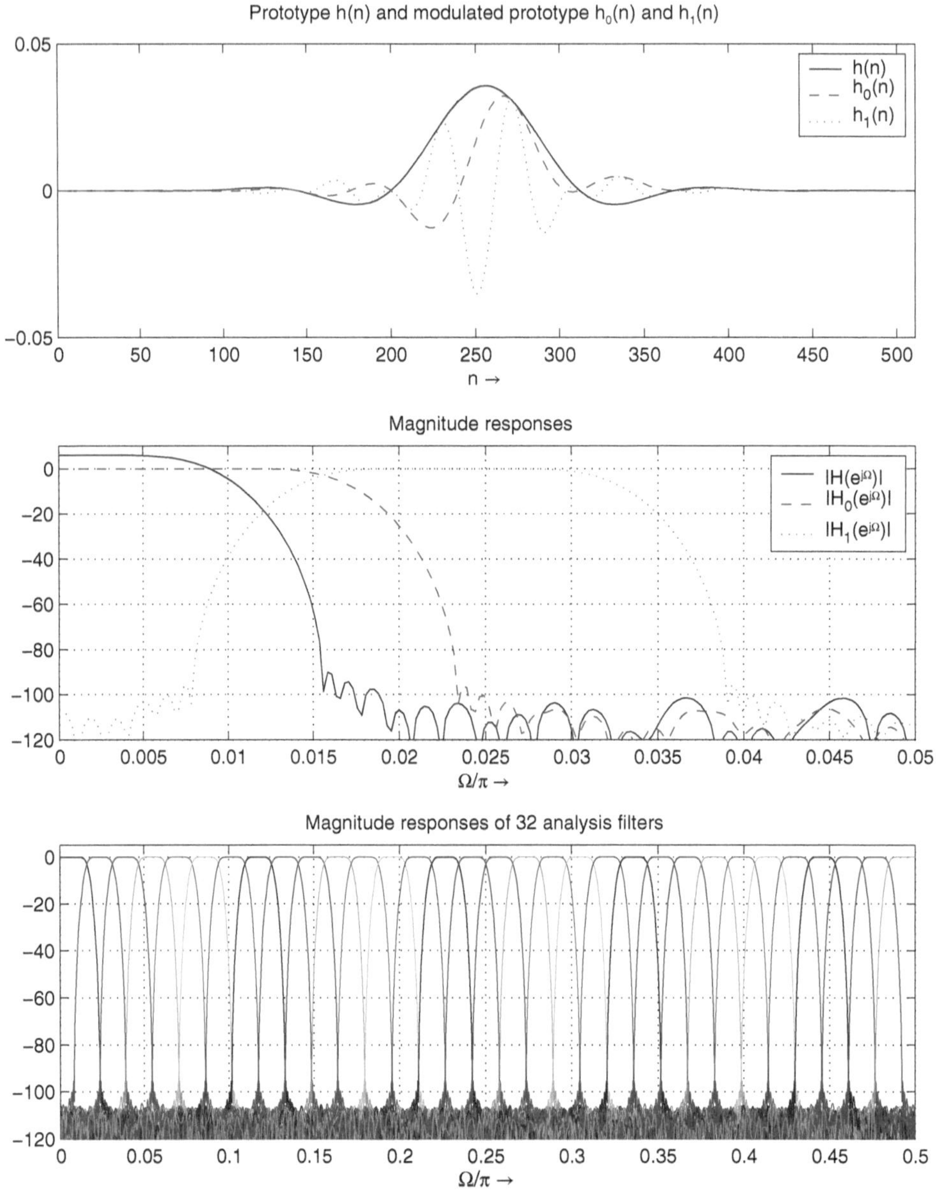

Figure 9.16 Impulse responses and magnitude responses of pseudo-QMF bank.

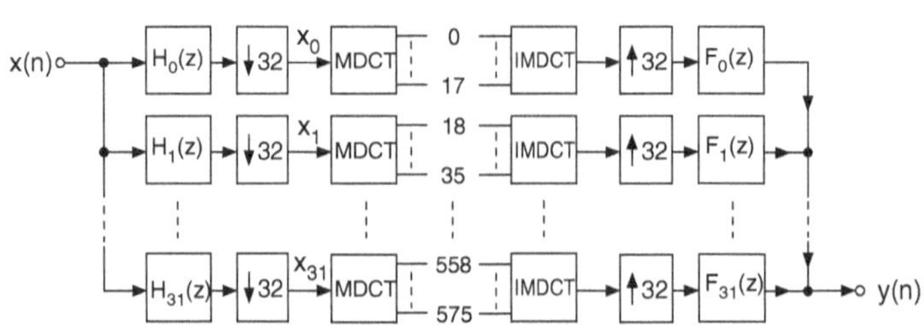

Figure 9.17 Polyphase/MDCT hybrid filter bank.

1. Calculating the power density spectrum of a block of N samples using FFT. After windowing a block of $N = 512$ ($N = 1024$ for layer II) input samples, the power density spectrum

$$X(k) = 10 \log_{10} \left| \frac{1}{N} \sum_{n=0}^{N-1} h(n)x(n)e^{-jnk2\pi/N} \right|^2 \quad \text{in dB} \tag{9.22}$$

is calculated. After this, the window $h(n)$ is displaced by 384 ($12 \cdot 32$) samples and the next block is processed.

2. Determination of sound pressure level in every sub-band. The sound pressure level is derived from the calculated power density spectrum and by calculating a scaling factor in the corresponding sub-band, as given by

$$L_S(i) = MAX[X(k), 20 \log_{10}[\text{SCF}_{max}(i) \cdot 32768] - 10] \qquad \text{in dB.} \tag{9.23}$$

For $X(k)$, the maximum of the spectral lines in a sub-band is used. The scaling factor SCF(i) for sub-band i is calculated from the absolute value of the maximum of 12 consecutive sub-band samples. A nonlinear quantization to 64 levels is carried out (layer I). For layer II, the sound pressure level is determined by choosing the largest of the three scaling factors from $3 \cdot 12$ sub-band samples.

3. Considering the absolute threshold. The absolute threshold $LT_q(m)$ is specified for different sampling rates in [ISO92]. The frequency index m is based on a reduction of $N/2$ relevant frequencies with the FFT of index k (see Fig. 9.18). The sub-band index is still i.

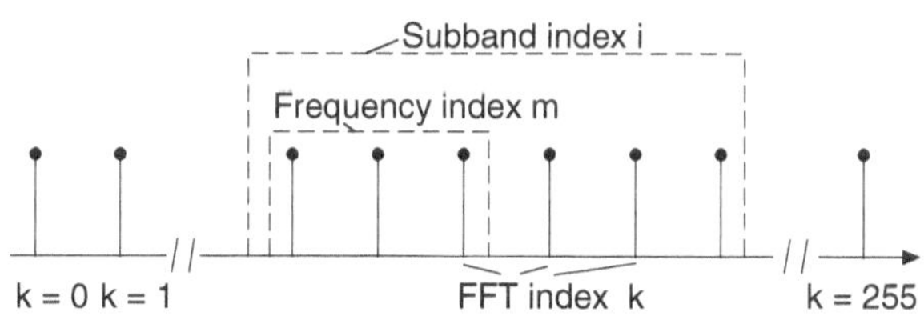

Figure 9.18 Nomenclature of frequency indices.

4. Calculating tonal $X_{tm}(k)$ or non-tonal $X_{nm}(k)$ masking components and determining relevant masking components (for details, see [ISO92]). These masking components are denoted as $X_{tm}[z(j)]$ and $X_{nm}[z(j)]$. With the index j, tonal and non-tonal masking components are labeled. The variable $z(m)$ is listed for reduced frequency indices m in [ISO92]. It allows a finer resolution of the 24 critical bands with the frequency group index z.

5. Calculating the individual masking thresholds. For masking thresholds of tonal and non-tonal masking components $X_{tm}[z(j)]$ and $X_{nm}[z(j)]$, the following calculation is performed:

$$LT_{tm}[z(j), z(m)] = X_{tm}[z(j)] + a_{v_{tm}}[z(j)] + v_f[z(j), z(m)] \quad \text{in dB,} \tag{9.24}$$

$$LT_{nm}[z(j), z(m)] = X_{nm}[z(j)] + a_{v_{nm}}[z(j)] + v_f[z(j), z(m)] \quad \text{in dB.} \tag{9.25}$$

The masking index for tonal masking components is given by

$$a_{v_{tm}} = -1.525 - 0.275 \cdot z(j) - 4.5 \quad \text{in dB} \tag{9.26}$$

and the masking index for non-tonal masking components is

$$a_{v_{nm}} = -1.525 - 0.175 \cdot z(j) - 0.5 \quad \text{in dB.} \tag{9.27}$$

The masking function $v_f[z(j), z(m)]$ with distance $\Delta z = z(m) - z(j)$ is given by

$$v_f = \begin{cases} 17 \cdot (\Delta z + 1) - (0.4 \cdot X[z(j)] + 6) & -3 \leq \Delta z < -1, \\ (0.4 \cdot X[z(j)] + 6) \cdot \Delta z & -1 \leq \Delta z < 0, \\ -17 \cdot \Delta z & 0 \leq \Delta z < 1, \\ -(\Delta z - 1) \cdot (17 - 0.15 \cdot X[z(j)]) - 17 & 1 \leq \Delta z < 8, \\ \quad \text{in dB} & \quad \text{in Bark.} \end{cases}$$

This masking function $v_f[z(j), z(m)]$ describes the masking of the frequency index $z(m)$ by the masking component $z(j)$.

6. Calculating the global masking threshold. For frequency index m, the global masking threshold is calculated as the sum of all contributing masking components according to

$$LT_g(m) = 10\log_{10}\left[10^{LT_q(m)/10} + \sum_{j=1}^{T_m} 10^{LT_{tm}[z(j),z(m)]/10} + \sum_{j=1}^{R_m} 10^{LT_{nm}[z(j),z(m)]/10}\right] \quad \text{in dB.} \tag{9.28}$$

The total number of tonal and non-tonal masking components are denoted as T_m and R_m, respectively. For a given sub-band i, only masking components that lie in the range -8 to $+3$ Bark will be considered. Masking components outside this range are neglected.

7. Determination of the minimum masking threshold in every sub-band:

$$LT_{min}(i) = MIN[LT_g(m)] \quad \text{in dB.} \tag{9.29}$$

Several masking thresholds $LT_g(m)$ can occur in a sub-band as long as m lies within the sub-band i.

8. Calculation of the signal-to-mask ratio SMR(i) in every sub-band:

$$\text{SMR}(i) = L_S(i) - LT_{min}(i) \quad \text{in dB.} \tag{9.30}$$

The signal-to-mask ratio determines the dynamic range that has to be quantized in the particular sub-band so that the level of quantization noise lies below the masking threshold. The signal-to-mask ratio is the basis for the bit allocation procedure for quantizing the sub-band signals.

9.4.3 Dynamic Bit Allocation and Coding

Dynamic Bit Allocation. Dynamic bit allocation is used to determine the number of bits that are necessary for the individual sub-bands so that a transparent perception is possible. The minimum number of bits in sub-band i can be determined from the difference between scaling factor SCF(i) and the absolute threshold $LT_q(i)$ as $b(i) = \mathrm{SCF}(i) - LT_q(i)$. With this, quantization noise remains under the masking threshold. Masking across critical bands is used for the implementation of the ISO-MPEG1 coding method.

For a given transmission rate, the maximum possible number of bits B_m for coding sub-band signals and scaling factors is calculated as

$$B_m = \sum_{i=1}^{32} b(i) + \mathrm{SCF}(i) + \text{additional information.} \tag{9.31}$$

The bit allocation is performed within an allocation frame consisting of 12 sub-band samples ($384 = 12 \cdot 32$ PCM samples) for layer I and 36 sub-band samples ($1152 = 36 \cdot 32$ PCM samples) for layer II.

The dynamic bit allocation for the sub-band signals is carried out as an iterative procedure. At the beginning, the number of bits per sub-band is set to zero. First, the mask-to-noise ratio,

$$\mathrm{MNR}(i) = \mathrm{SNR}(i) - \mathrm{SMR}(i), \tag{9.32}$$

is determined for every sub-band. The signal-to-mask ratio SMR(i) is the result of the psychoacoustic model. The signal-to-noise ratio SNR(i) is defined by a table in [ISO92], in which for every number of bits, a corresponding signal-to-noise ratio is specified. The number of bits must be increased as long as the mask-to-noise ratio MNR is less than zero.

The iterative bit allocation is performed by the following steps.

1. Determination of the minimum MNR(i) of all sub-bands.

2. Increasing the number of bits of these sub-bands on to the next stage of the MPEG1 standard. Allocation of 6 bits for the scaling factor of the MPEG1 standard when the number of bits is increased for the first time.

3. New calculation of MNR(i) in this sub-band.

4. Calculation of the number of bits for all sub-bands and scaling factors and comparison with the maximum number B_m. If the number of bits is smaller than the maximum number, the iteration starts again with step 1.

Quantization and Coding of Sub-band Signals. The quantization of the sub-band signals is done with the allocated bits for the corresponding sub-band. The 12 (36) sub-band samples are divided by the corresponding scaling factor and then linearly quantized and coded (for details see [ISO92]). This is followed by a frame packing. In the decoder, the procedure is reversed. The decoded sub-band signals with different word lengths are reconstructed to a broadband PCM signal with a synthesis filter bank (see Fig. 9.14). MPEG-1 audio coding has a one or a two channel stereo mode with sampling frequencies of 32, 44.1, and 48 kHz and a bit rate of 128 kbit/s per channel.

9.5 MPEG-2 Audio Coding

The aim of the introduction of MPEG-2 audio coding was the extension of MPEG-1 to lower sampling frequencies and multichannel coding [Bos97]. Backward compatibility to existing MPEG-1 systems is achieved through the version MPEG-2 BC (Backward Compatible) and the introduction toward lower sampling frequencies of 32, 22.05, 24 kHz with version MPEG-2 LSF (Lower Sampling Frequencies). The bit rate for a five-channel MPEG-2 BC coding with full bandwidth of all channels is 640–896 kBit/s.

9.6 MPEG-2 Advanced Audio Coding

To improve the coding of mono, stereo, and multichannel audio signals, the MPEG-2 AAC (Advanced Audio Coding) standard is specified. This coding standard is not backward compatible with the MPEG-1 standard and forms the kernel for new extended coding standards such as MPEG-4. The achievable bit rate for a five-channel coding is 320 kBit/s. In the following, the main signal processing steps for MPEG-2 AAC will be introduced and the principle functionalities will be explained. An extensive explanation can be found in [Bos97, Bra98, Bos02]. An MPEG-2 AAC coder is shown in Fig. 9.19. The corresponding decoder performs the functional units in reverse order with corresponding decoder functionalities.

Pre-processing: the input signal will be band limited according to the sampling frequency. This step is used only in the *scalable sampling rate profile* [Bos97, Bra98, Bos02].

Filter bank: the time-frequency decomposition into $M = 1024$ sub-bands with an overlapped MDCT [Pri86, Pri87] is based on blocks of $N = 2048$ input samples. A step-wise explanation of the implementation will be introduced. A graphical representation of the single steps is depicted in Fig. 9.20. The single steps are as follows.

1. Partitioning of the input signal $x(n)$ with time index n into overlapped blocks

$$x_m(r) = x(mM + r) \quad r = 0, \ldots, N-1; -\infty \leq m \leq \infty \tag{9.33}$$

   of length N with an overlap (hop size) of $M = \frac{N}{2}$. The time index inside a block is denoted by r. The variable m denotes the block index.

2. Windowing of blocks with window function $w(r) \to x_m(r) \cdot w(r)$.

3. MDCT

$$X(m,k) = \sqrt{\frac{2}{M}} \sum_{r=0}^{N-1} x_m(r) w(r) \cos\left(\frac{\pi}{M}\left(k + \frac{1}{2}\right)\left(r + \frac{M+1}{2}\right)\right), \quad k = 0, \ldots, M-1 \tag{9.34}$$

   yields every M input samples $M = N/2$ spectral coefficients from N windowed input samples.

4. Quantization of spectral coefficients $X(m,k)$ leads to quantized spectral coefficients $X_Q(m,k)$ based on a psychoacoustic model.

5. IMDCT

$$\hat{x}_m(r) = \sqrt{\frac{2}{M}} \sum_{k=0}^{M-1} X_Q(m,k) \cos\left(\frac{\pi}{M}\left(k + \frac{1}{2}\right)\left(r + \frac{M+1}{2}\right)\right), \quad r = 0, \ldots, N-1 \tag{9.35}$$

   yields every M input samples N output samples in block $\hat{x}_m(r)$.

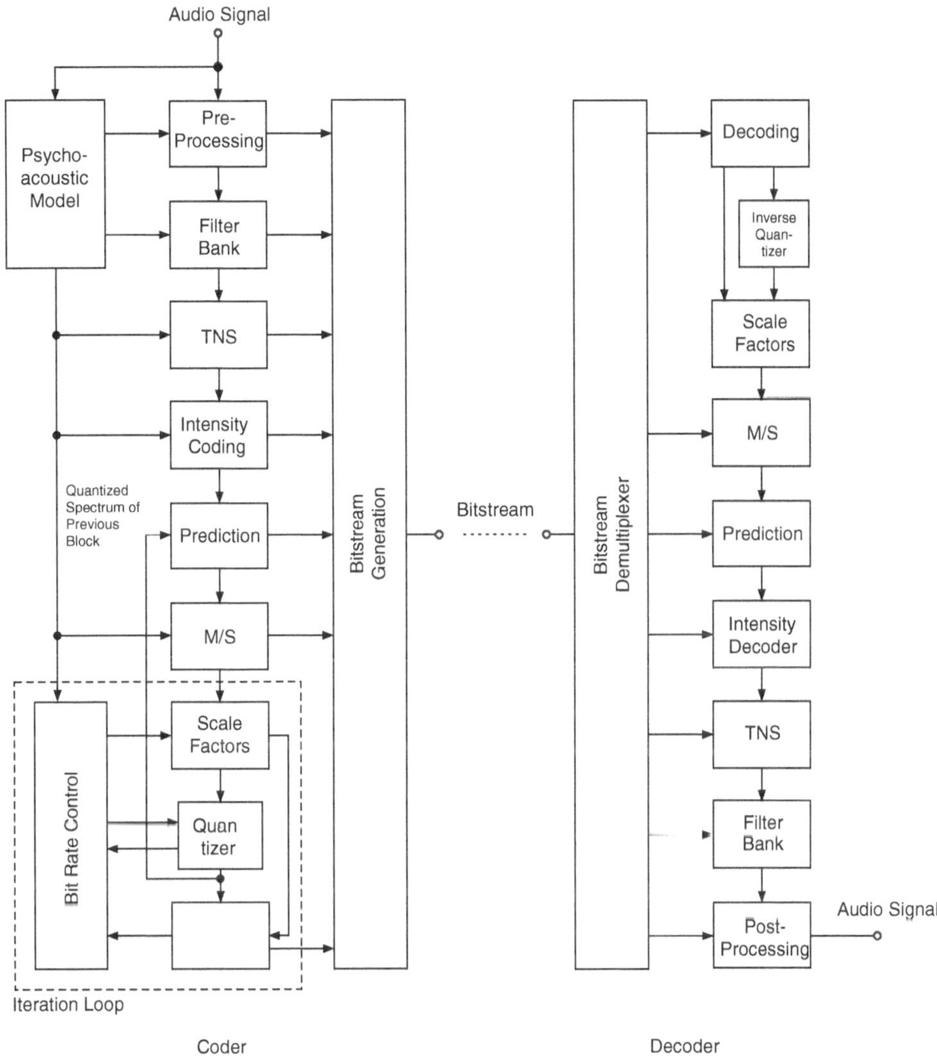

Figure 9.19 MPEG-2 AAC coder and decoder.

6. Windowing of inverse transformed block $\hat{x}_m(r)$ with window function $w(r)$.
7. Reconstruction of output signal $y(n)$ by overlap and add operation according to

$$y(n) = \sum_{m=-\infty}^{\infty} \hat{x}_m(r)w(r), \quad r = 0, \ldots, N-1 \tag{9.36}$$

with overlap M.

To explain the procedural steps, we consider the MDCT/IMDCT of a sine pulse shown in Fig. 9.21. The left column shows top down the input signal and partitions of the input signal of block length $N = 256$. The window function is a sine window. The corresponding MDCT coefficients of length $M = 128$ are shown in the middle column. The IMDCT delivers the signals in the right column. One can notice that the inverse transforms with the IMDCT do

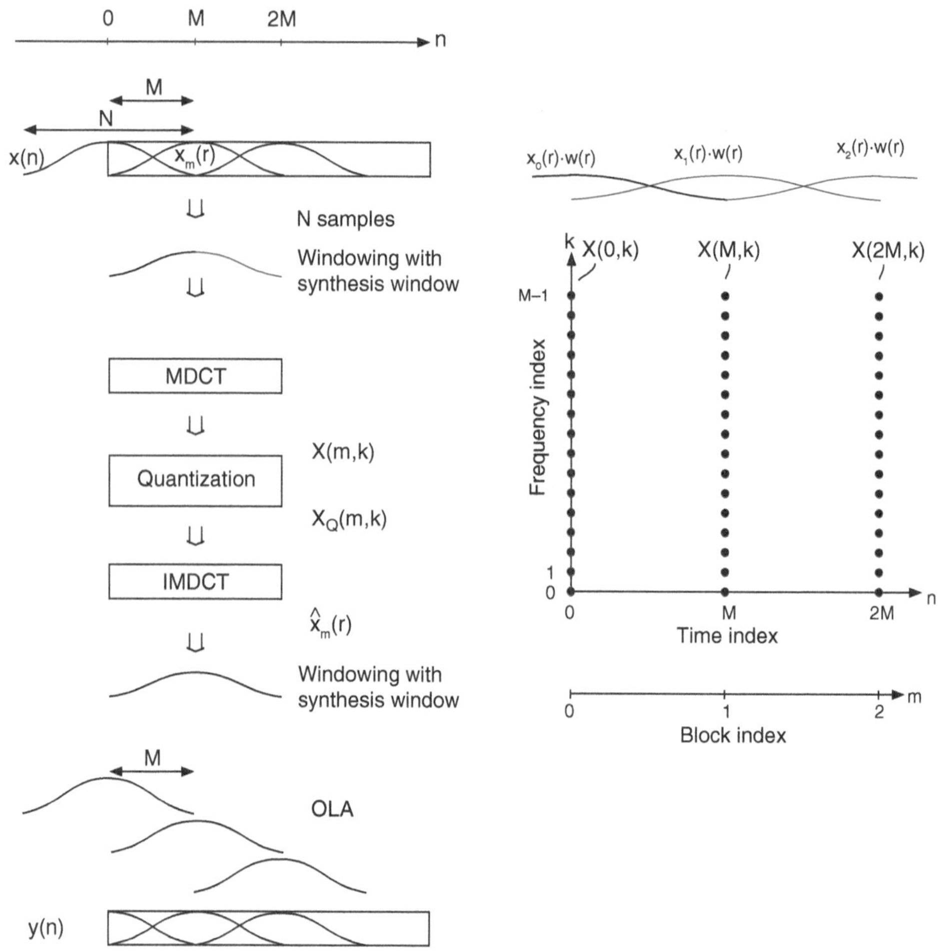

Figure 9.20 Time-frequency decompostion with MDCT/inverse modified discrete cosine transform (IMDCT).

not exactly reconstruct the single input blocks. Moreover, each output block consists of an input block and a special superposition of a time-reversed and circular shifted by $M = N/2$ input block, which is denoted by time domain aliasing [Pri86, Pri87, Edl89]. The overlap and add operation of the single output blocks perfectly recovers the input signal, which is shown in the top signal of the right column (Fig. 9.21). For a perfect reconstruction of the output signal, the window function of the analysis and synthesis step has to fulfill the condition $w^2(r) + w^2(r+M) = 1, \quad r = 0, \ldots, M-1$. The Kaiser–Bessel-derived window [Bos02] and a sine window $h(n) = \sin\left((n+\frac{1}{2})\frac{\pi}{N}\right)$ with $n = 0, \ldots, N-1$ [Mal92] are applied. Figure 9.22 shows both window functions with $N = 2048$ and the corresponding magnitude responses for a sampling frequency of $f_S = 44100$ Hz. The sine window has a smaller passband width but slower falling side lobes. In contrast, the Kaiser–Bessel-derived window shows a wider passband and a faster decay of the side lobes. To demonstrate the filter bank properties and, in particular, the

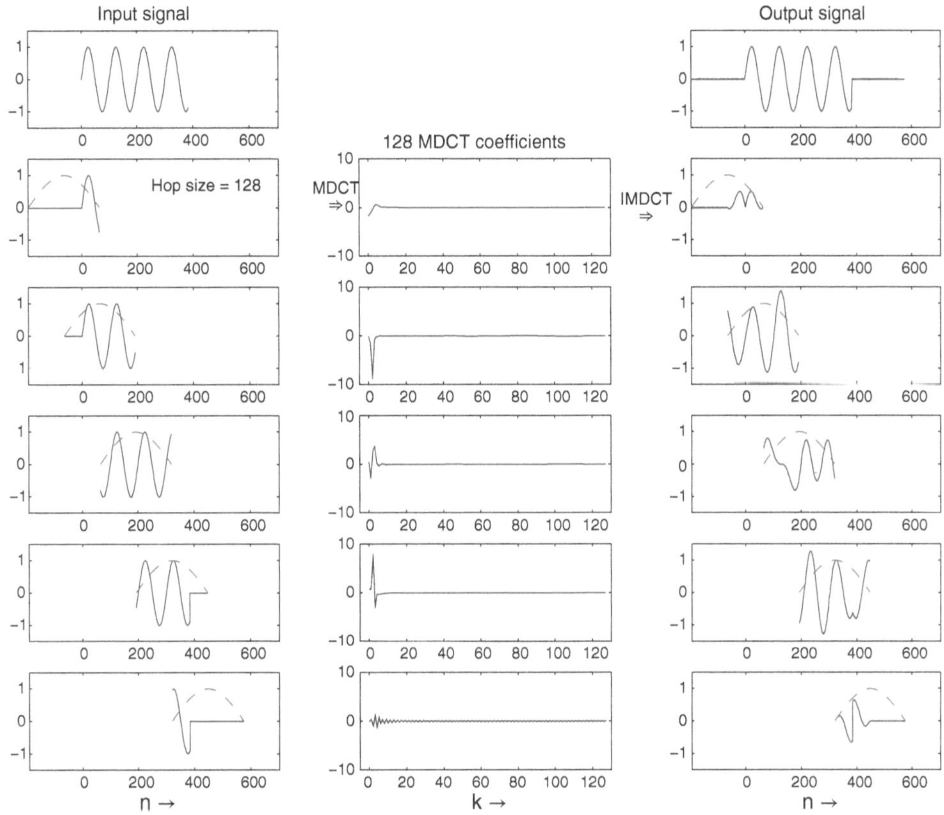

Figure 9.21 Signals of MDCT/IMDCT.

frequency decomposition of MDCT, we derive the modulated bandpass impulse responses of the window functions (prototype impulse response $w(n) = h(n)$) according to

$$h_k(n) = 2 \cdot h(n) \cdot \cos\left(\frac{\pi}{M}\left(k+\frac{1}{2}\right)\left(n+\frac{M+1}{2}\right)\right) \tag{9.37}$$

$$k = 0, \ldots, M-1; n = 0, \ldots, N-1.$$

Figure 9.23 shows the normalized prototype impulse response of the sine window and the first two modulated bandpass impulse responses $h_0(n)$ and $h_1(n)$, and accordingly, the corresponding magnitude responses are depicted. In addition to the increased frequency resolution with $M = 1024$ bandpass filters, a reduced stopband attenuation can be observed. A comparison of this magnitude responses of the MDCT with the frequency resolution of the pseudo-QMF bank with $M = 32$ in Fig. 9.16 points out the different properties of both sub-band decompositions.

To adjust the time and frequency resolution to the properties of an audio signal, several methods have been investigated. Signal-adaptive audio coding based on the wavelet transform can be found in [Sin93, Ern00]. Window switching can be applied to achieve a time-variant time-frequency resolution for MDCT and IMDCT applications. For stationary signals, a high frequency resolution and a low time resolution are necessary. This leads to long windows with $N = 2048$. Coding of attacks of instruments needs a high time resolution (reduction of window

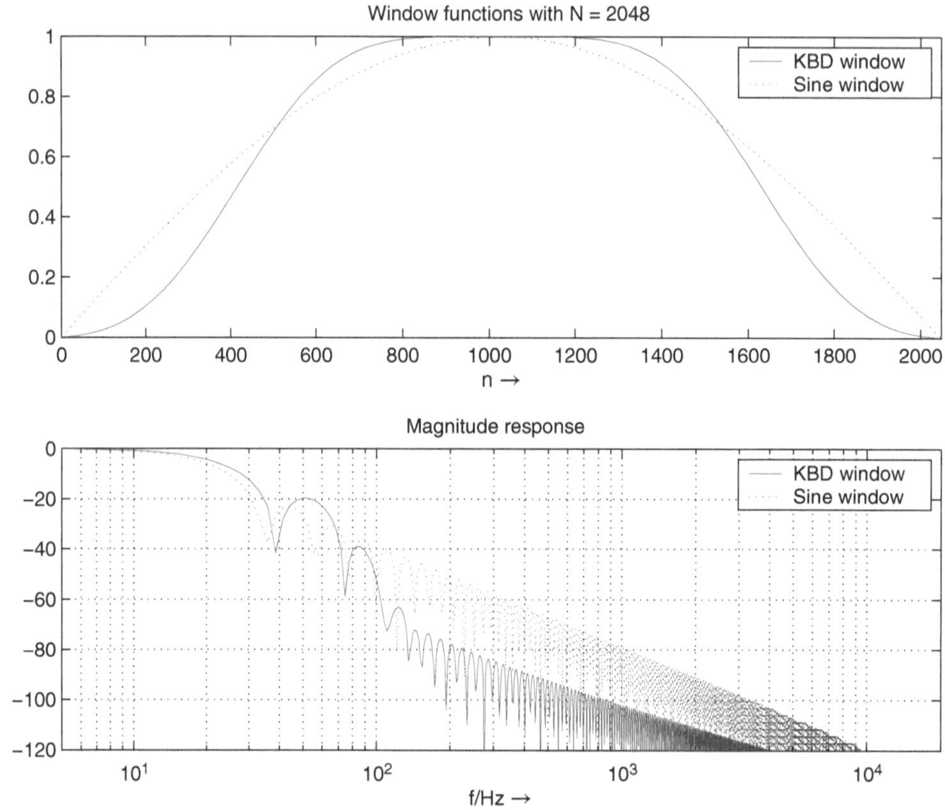

Figure 9.22 Kaiser–Bessel-derived window and sine window for $N = 2048$ and magnitude responses of the normalized window functions.

length to $N = 256$) and thus reduces the frequency resolution (reduction of number of spectral coefficients). A detailed description of switching between time–frequency resolution with the MDCT/IMDCT can be found in [Edl89, Bos97, Bos02]. Examples of switching between different window functions and windows of different length are shown in Fig. 9.24.

Temporal Noise Shaping. A further method for adapting the time–frequency resolution of a filter bank and here a MDCT/IMDCT to the signal characteristic is based on linear prediction along the spectral coefficients in the frequency domain [Her96, Her99]]. This method is called *temporal noise shaping* (TNS) and is a weighting of the temporal envelope of the time domain signal. Such weighting of the temporal envelope is demonstrated in Fig. 9.25.

Figure 9.25a shows a signal from a castanet attack. Making use of the discrete cosine transform (DCT, [Rao90]),

$$X^{C(2)}(k) = \sqrt{\frac{2}{N}} c_k \sum_{n=0}^{N-1} x(n) \cos\left(\frac{(2n+1)k\pi}{2N}\right), \quad k = 0, \dots, N-1, \tag{9.38}$$

and the IDCT,

$$x(n) = \sqrt{\frac{2}{N}} \sum_{k=0}^{N-1} c_k X^{C(2)}(k) \cos\left(\frac{(2n+1)k\pi}{2N}\right), \quad n = 0, \dots, N-1 \tag{9.39}$$

$$\text{with } c_k = \begin{cases} 1/\sqrt{(2)} & k = 0, \\ 1 & \text{otherwise,} \end{cases}$$

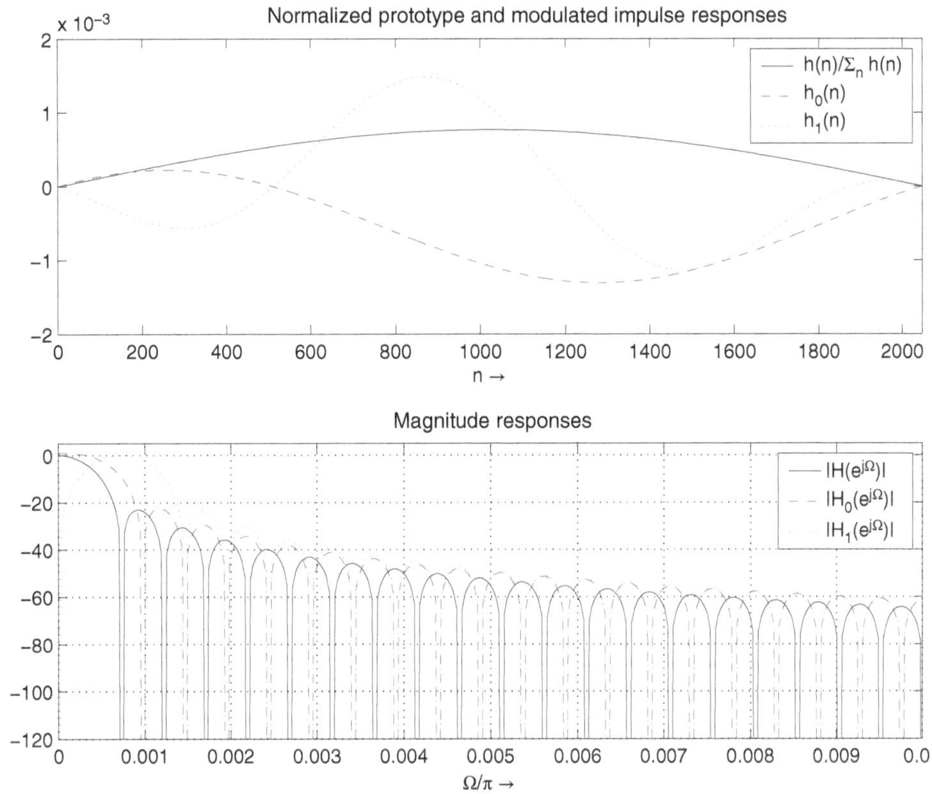

Figure 9.23 Normalized impulse responses of sine window for $N = 2048$, modulated bandpass impulse responses, and magnitude responses.

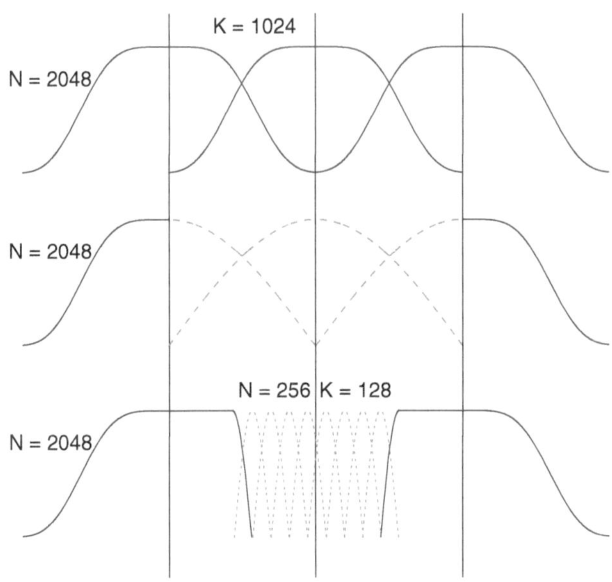

Figure 9.24 Switching of window functions.

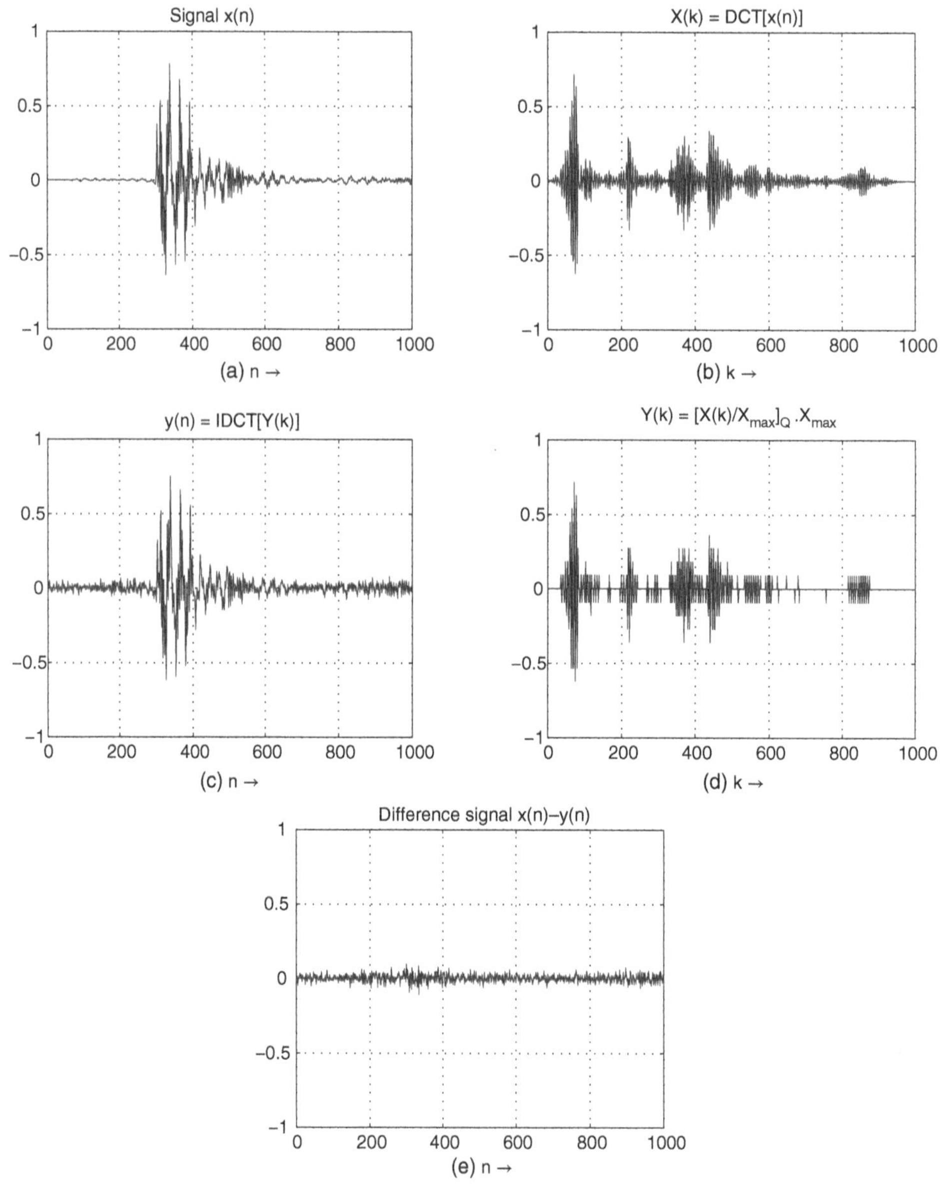

Figure 9.25 Attack of castanet and spectrum.

the spectral coefficients of the DCT of this castanet attack are represented in Fig. 9.25b. After quantization of these spectral coefficients $X(k)$ to 4 bit (Fig. 9.25d) and IDCT of the quantized spectral coefficients, the time-domain signal in Fig. 9.25c and the difference signal in Fig. 9.25e between input and output results. One can notice in the output and difference signal that the error is spread along the entire block length. This means that before the attack of the castanet happens, the error signal of the block is perceptible. The time-domain masking, the so-called pre-masking [Zwi90], is not sufficient. Ideally, the spreading of the error signal should follow the time-domain envelope of the signal itself. From the forward linear prediction in the time domain, it is known that the power spectral density of the error signal after coding and decoding is weighted by the envelope of the power spectral density of the input signal [Var06]. Performing a forward linear prediction along the frequency axis in the frequency domain and quantization and coding leads to an error signal in the time domain where the temporal envelope of the error signal follows the time-domain envelope of the input signal [Her96]. To point out the temporal weighting of the error signal, we consider the forward prediction in the time domain using Fig. 9.26a. For coding the input signal, $x(n)$ is predicted by an impulse response $p(n)$. The output of the predictor is subtracted from the input signal $x(n)$ and delivers the signal $d(n)$, which is then quantized to a reduced word length. The quantized signal $d_Q(n) = x(n) * a(n) + e(n)$ is the sum of the convolution of $x(n)$ with the impulse response $a(n)$ and the additive quantization error $e(n)$. The power spectral density of the coder output is $S_{D_Q D_Q}(e^{j\Omega}) = S_{XX}(e^{j\Omega}) \cdot |A(e^{j\Omega})|^2 + S_{EE}(e^{j\Omega})$. The decoding operation performs the convolution of $d_Q(n)$ with the impulse response $h(n)$ of the inverse system to the coder. Therefore, $a(n) * h(n) = \delta(n)$ must hold and thus $H(e^{j\Omega}) = 1/A(e^{j\Omega})$. Hereby, the output signal $y(n) = x(n) + e(n) * h(n)$ is derived with the corresponding discrete Fourier transform $Y(k) = X(k) + E(k) \cdot H(k)$. The power spectral density of the decoder out signal is given by $S_{YY}(e^{j\Omega}) = S_{XX}(e^{j\Omega}) + S_{EE}(e^{j\Omega}) \cdot |H(e^{j\Omega})|^2$. Herein, one can notice the spectral weighting of the quantization error with the spectral envelope of the input signal which is represented by $|H(e^{j\Omega})|$. The same kind of forward prediction will now be applied in the frequency domain to the spectral coefficients $X(k) = \mathrm{DCT}[x(n)]$ for a block of input samples $x(n)$ shown in Fig. 9.26b. The output of the decoder is then given by $Y(k) = X(k) + E(k) * H(k)$ with

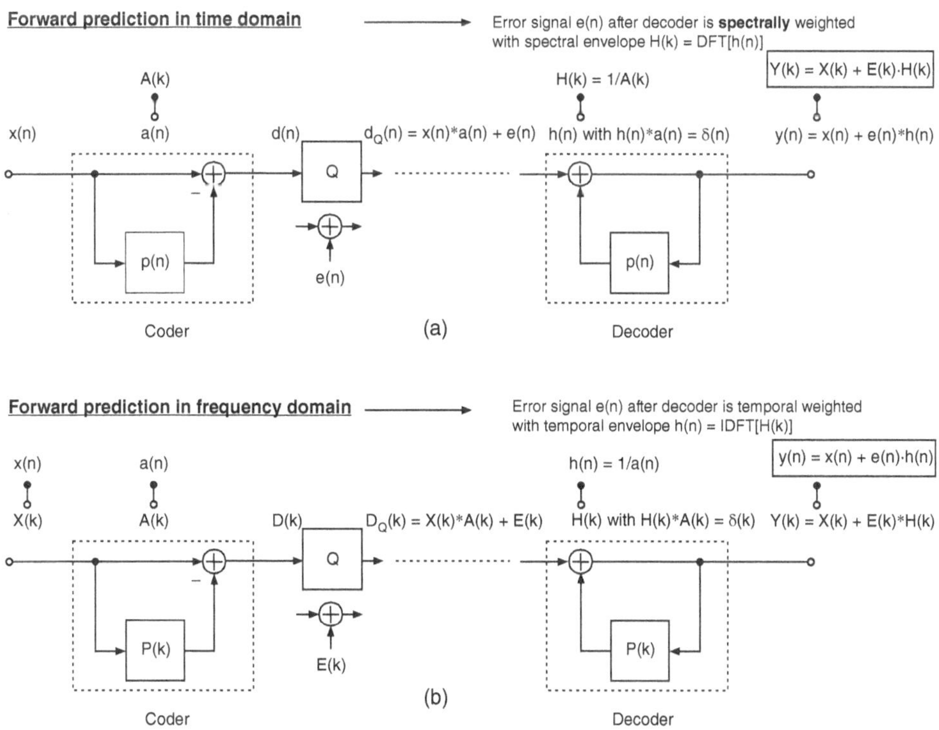

Figure 9.26 Forward prediction in time and frequency domains.

$A(k) * H(k) = \delta(k)$. Thus, the corresponding time-domain signal is $y(n) = x(n) + e(n) \cdot h(n)$, where the temporal weighting of the quantization error with the temporal envelope of the input signal is clearly evident. The temporal envelope is represented by the absolute value $|h(n)|$ of the impulse response $h(n)$. The relation between the temporal signal envelope (absolute value of the analytical signal) and the auto-correlation function of the analytical spectrum is discussed in [Her96]. The dualities between forward linear predictions in the time and frequency domains are summarized in Table 9.2. Figure 9.27 demonstrates the operations for temporal noise shaping in the coder, where the prediction is performed along the spectral coefficients. The coefficients of the forward predictor have to be transmitted to the decoder, where inverse filtering is performed along the spectral coefficients.

Table 9.2 Forward prediction in the time and frequency domains.

Prediction in time domain	Prediction in frequency domain
$y(n) = x(n) + e(n) * h(n)$	$y(n) = x(n) + e(n) \cdot h(n)$
$Y(k) = X(k) + E(k) \cdot H(k)$	$Y(k) = X(k) + E(k) * H(k)$

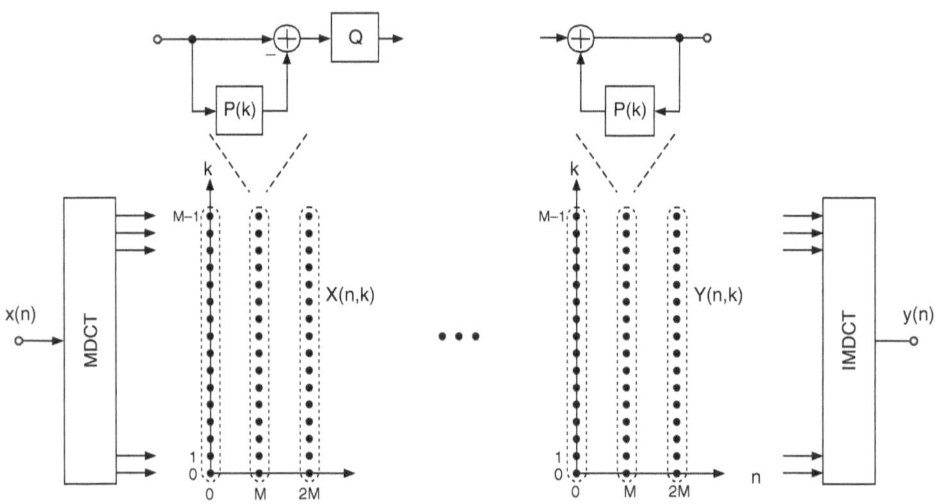

Figure 9.27 Temporal noise shaping with forward prediction in the frequency domain.

The temporal weighting is finally demonstrated in Fig. 9.28, where the corresponding signals with the forward prediction in the frequency domain are shown. Figure 9.28a/b shows the castanet signal $x(n)$ and its corresponding spectral coefficients $X(k)$ of the applied DCT. The forward prediction delivers $D(k)$ in Fig. 9.28d and the quantized signal $D_Q(k)$ in Fig. 9.28f. After the decoder, the signal $Y(k)$ in Fig. 9.28h is reconstructed by the inverse transfer function. The IDCT of $Y(k)$ finally results in the output signal $y(n)$ in Fig. 9.28e. The difference signal $x(n) - y(n)$ in Fig. 9.28g demonstrates the temporal weighting of the error signal with the temporal envelope from Fig. 9.28c. For this example, the order of the predictor is chosen to be 20 [Bos97] and the prediction along the spectral coefficients $X(k)$ is performed by the Burg method. The prediction gain for this signal in the frequency domain is $G_p = 16\,\text{dB}$ (see Fig. 9.28d).

Frequency Domain Prediction. A further compression of the bandpass signals is possible by using linear prediction. A backward prediction [Var06] of the bandpass signals is applied on the

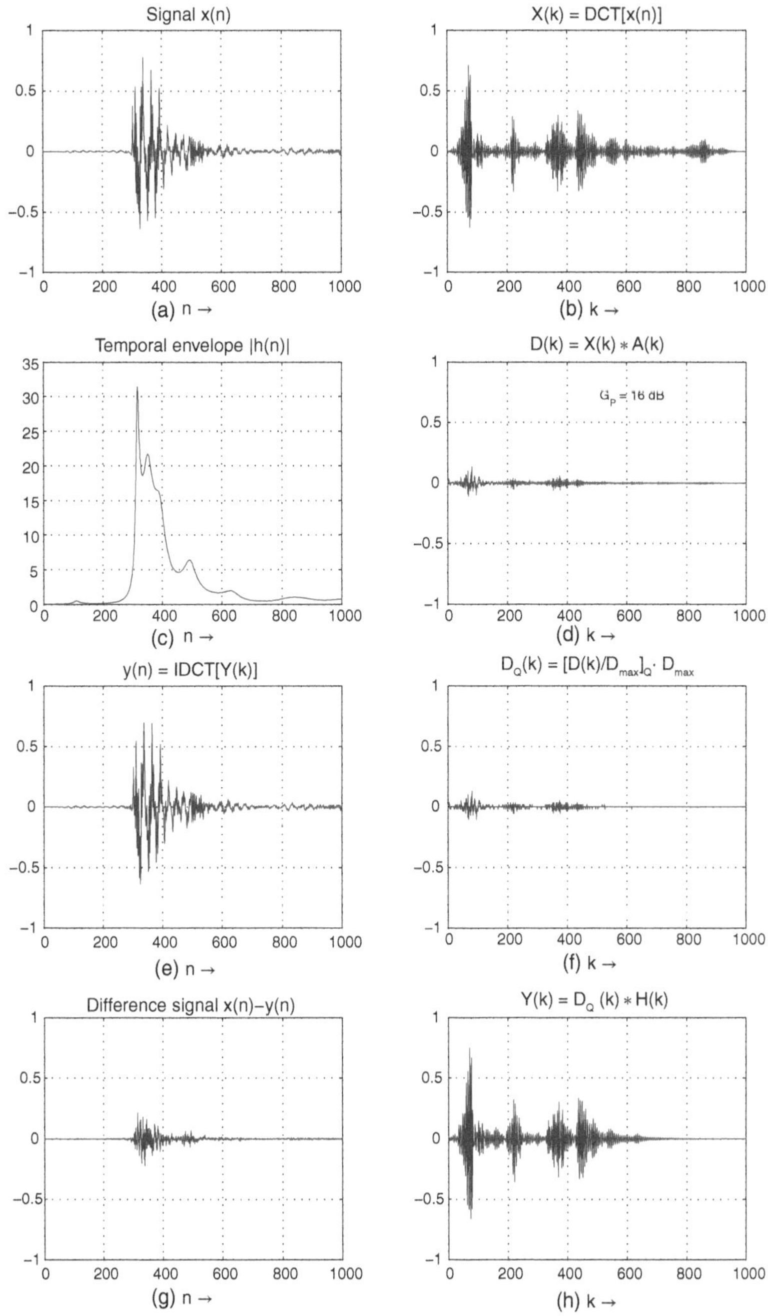

Figure 9.28 Temporal noise shaping: attack of castanet and spectrum.

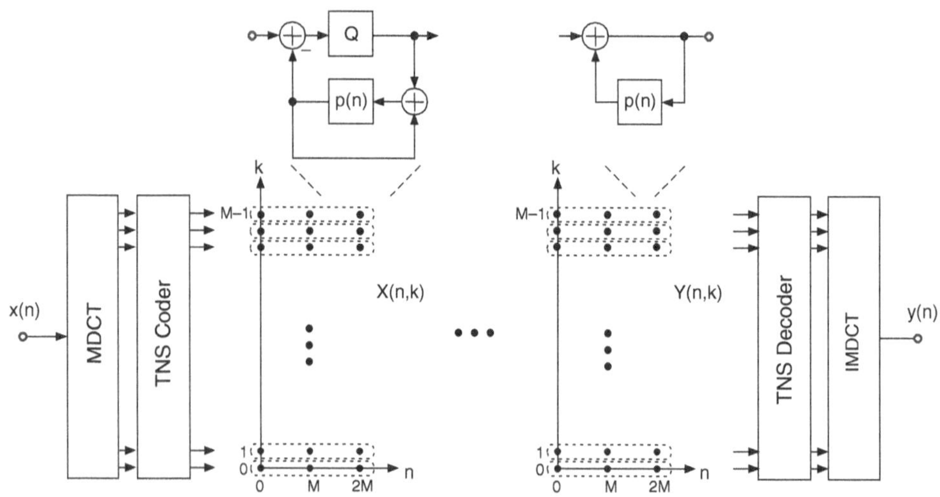

Figure 9.29 Backward prediction of bandpass signals.

coder side (see Fig. 9.29). In using a backward prediction, the predictor coefficients need not be coded and transmitted to the decoder, because the estimate of the input sample is based on the quantized signal. The decoder derives the predictor coefficients $p(n)$ in the same way from the quantized input. A second-order predictor is sufficient, because the bandwidth of the bandpass signals is very low [Bos97].

Mono/Side Coding. Coding of stereo signals with left and right signals $x_L(n)$ and $x_R(n)$ can be achieved by coding a mono signal (M) $x_M(n) = (x_L(n) + x_R(n))/2$ and a side (S, difference) signal $x_S(n) = (x_L(n) - x_R(n))/2$ (M/S coding). Because the power of the side signal is reduced for highly correlated left and right signals, a reduction of the bit rate for this signal can be achieved. The decoder can reconstruct the left signal $x_L(n) = x_M(n) + x_S(n)$ and the right signal $x_R(n) = x_M(n) - x_S(n)$, if no quantization and coding is applied to the mono and side signals. This M/S coding is carried out for MPEG-2 AAC [Bra98, Bos02] with the spectral coefficients of a stereo signal (see Fig. 9.30).

Intensity Stereo Coding. For intensity stereo (IS) coding, a mono signal $x_M(n) = x_L(n) + x_R(n)$ and two temporal envelopes $e_L(n)$ and $e_R(n)$ of the left and right signals are coded and transmitted.

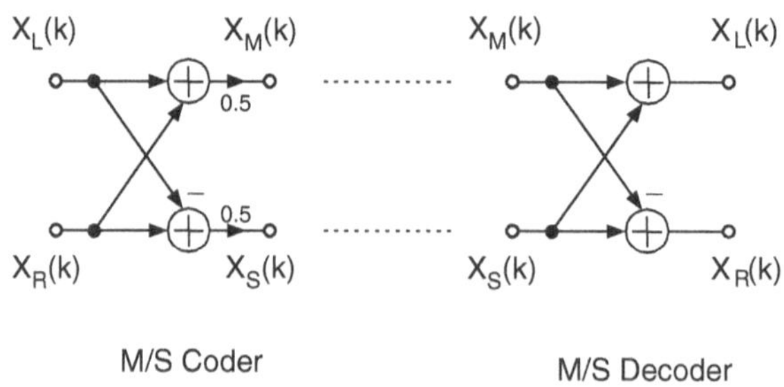

Figure 9.30 M/S coding in the frequency domain.

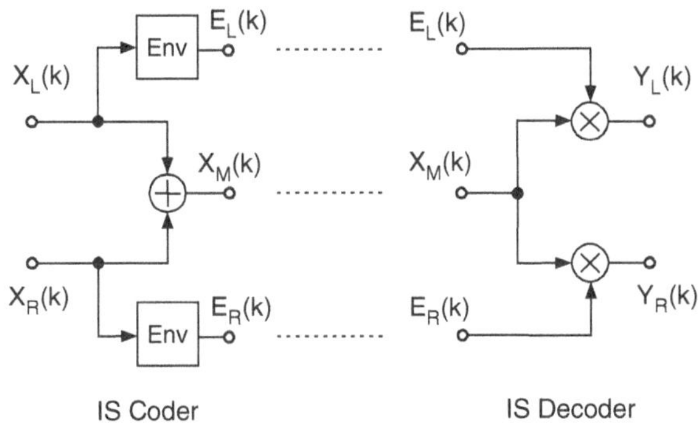

Figure 9.31 Intensity stereo coding in the frequency domain.

On the decoding side, the left signal is reconstructed by $y_L(n) = x_M(n) \cdot e_L(n)$ and the right signal by $y_R(n) = x_M(n) \cdot e_R(n)$. This reconstruction is lossy. The intensity stereo coding of MPEG-2 AAC [Bra98] is performed by the summation of spectral coefficients of both signals and by coding of scale factors which represent the temporal envelope of both signals (see Fig. 9.31). This type of stereo coding is only useful for higher frequency bands, because the human perception for phase shifts is non-sensitive for frequencies above 2 kHz.

Quantization and Coding. During the last coding step, the quantization and coding of the spectral coefficients take place. The quantizers, which are used in the figures for prediction along spectral coefficients in the frequency direction (Fig. 9.27) and prediction in the frequency domain along bandpass signals (Fig. 9.29), are now combined to a single quantizer per spectral coefficient. This quantizer performs nonlinear quantization similar to a floating-point quantizer presented in Chapter 2, such that a nearly constant SNR over a wide amplitude range is achieved. This floating-point quantization with a so-called scale factor is applied to several frequency bands, in which several spectral coefficients use a common scale factor derived from an iteration loop (see Fig. 9.19). Finally, a Huffman coding of the quantized spectral coefficients is performed. An extensive presentation can be found in [Bos97, Bra98, Bos02].

9.7 MPEG-4 Audio Coding

The MPEG-4 audio coding standard consists of a family of audio and speech coding methods for different bit rates and a variety of multimedia applications [Bos02, Her02]. In addition to a higher coding efficiency, new functionalities such as scalability, object-oriented representation of signals, and interactive synthesis of signals at the decoder are integrated. The MPEG-4 coding standard is based on the following speech and audio coders.

- Speech coders
  - CELP: Code-excited linear prediction (bit rate 4–24 kBit/s).
  - HVXC: Harmonic vector excitation coding (bit rate 1.4–4 kBit/s).
- Audio coders
  - Parametric audio: representation of a signal as a sum of sinusoids, harmonic components, and residual components (bit rate 4–16 kBit/s).

- Structured audio: synthetic signal generation at decoder (extension of the MIDI standard[1] (0.2–4 kBit/s).
- Generalized audio: extension of MPEG-2 AAC with additional methods in the time-frequency domain. The basic structure is depicted in Fig. 9.19 (bit rate 6–64 kBit/s).

The basics of speech coders can found in [Var06]. The specified audio coders allow coding with lower bit rates (parametric audio and structured audio) and coding with higher quality at lower bit rates compared with MPEG-2 AAC.

Compared with the so far introduced coding methods, such as MPEG-1 and MPEG-2, parametric audio coding is of special interest as an extension to the filter bank methods [Pur99, Edl00]. A parametric audio coder is shown in Fig. 9.32. The analysis of the audio signal leads to a decomposition into sinusoidal, harmonic, and noise-like signal components and the quantization and coding of these signal components is based on psychoacoustics [Pur02a]. According to an analysis/synthesis approach [McA86, Ser89, Smi90, Geo92, Geo97, Rod97, Mar00a] shown in Fig. 9.33, the audio signal is represented in a parametric form given by

$$x(n) = \sum_{i=1}^{M} A_i(n) \cos\left(2\pi \frac{f_i(n)}{f_A} n + \varphi_i(n)\right) + x_n(n). \tag{9.40}$$

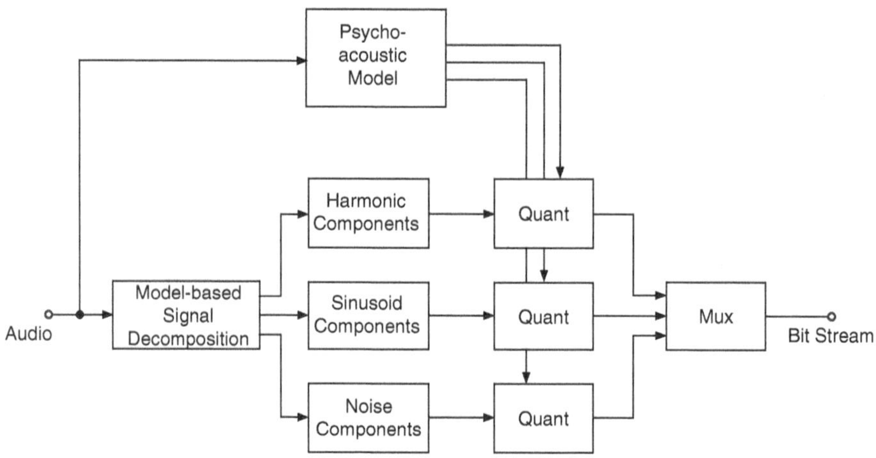

Figure 9.32 MPEG-4 parametric coder.

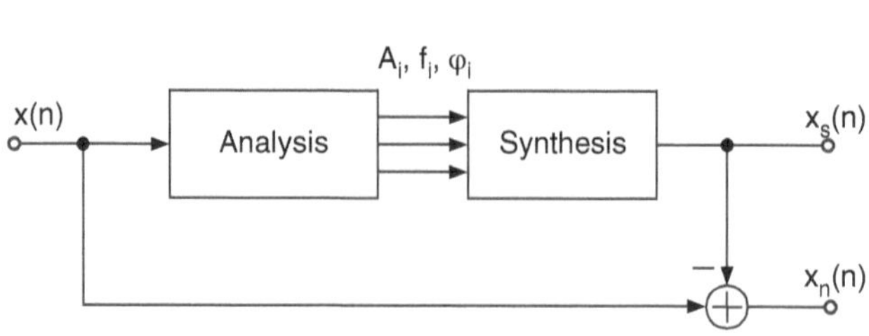

Figure 9.33 Parameter extraction with analysis/synthesis.

[1] `http://www.midi.org/`

The first term describes a sum of sinusoids with time-varying amplitudes $A_i(n)$, frequencies $f_i(n)$, and phases $\varphi_i(n)$. The second term consists of a noise-like component $x_n(n)$ with a time-varying temporal envelope. This noise-like component $x_n(n)$ is derived by subtracting the synthesized sinusoidal components from the input signal. With the help of a further analysis step, harmonic components with a fundamental frequency and multiples of this fundamental frequency are identified and grouped to harmonic components. The extraction of deterministic and stochastic components from an audio signal can be found in [Alt99, Hai03, Kei01, Kei02, Mar00a, Mar00b, Lag02, Lev98, Lev99, Pur02b]. In addition to the extraction of sinusoidal components, the modeling of noise-like components and transient components is of specific importance [Lev98, Lev99]. Figure 9.34 exemplifies the decomposition of an audio signal into a sum of sinusoids $x_s(n)$ and a noise-like signal $x_n(n)$. The spectrogram shown in Fig. 9.35 represents the short-time spectra of the sinusoidal components. The extraction of the sinusoids has been achieved by a modified FFT method [Mar00a] with an FFT length of $N = 2048$ and an analysis hop size of $R_A = 512$.

The corresponding parametric MPEG-4 decoder is shown in Fig. 9.36 [Edl00, Mei02]. The synthesis of the three signal components can be achieved by inverse FFT and overlap and add methods or can be directly performed by time-domain methods [Rod97, Mei02]. A significant advantage of parametric audio coding is the direct access at the decoder to the three main signal components which allows effective post-processing for the generation of a variety of audio

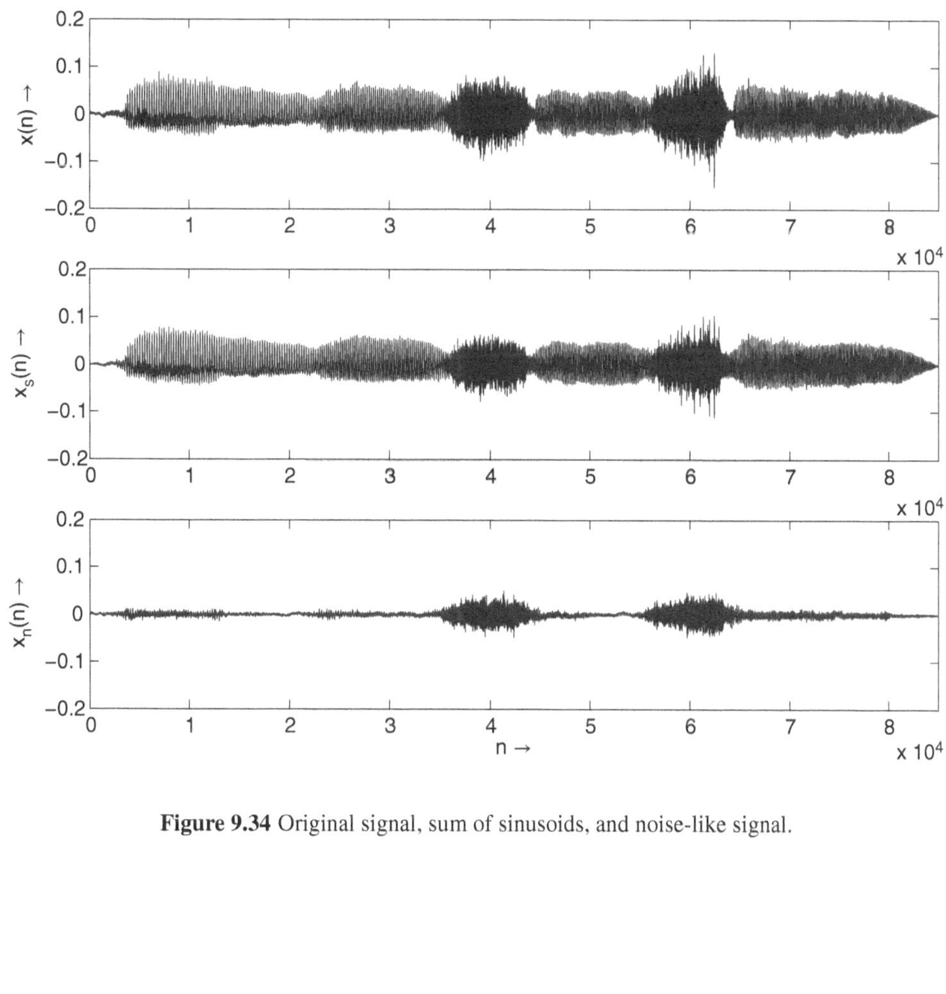

Figure 9.34 Original signal, sum of sinusoids, and noise-like signal.

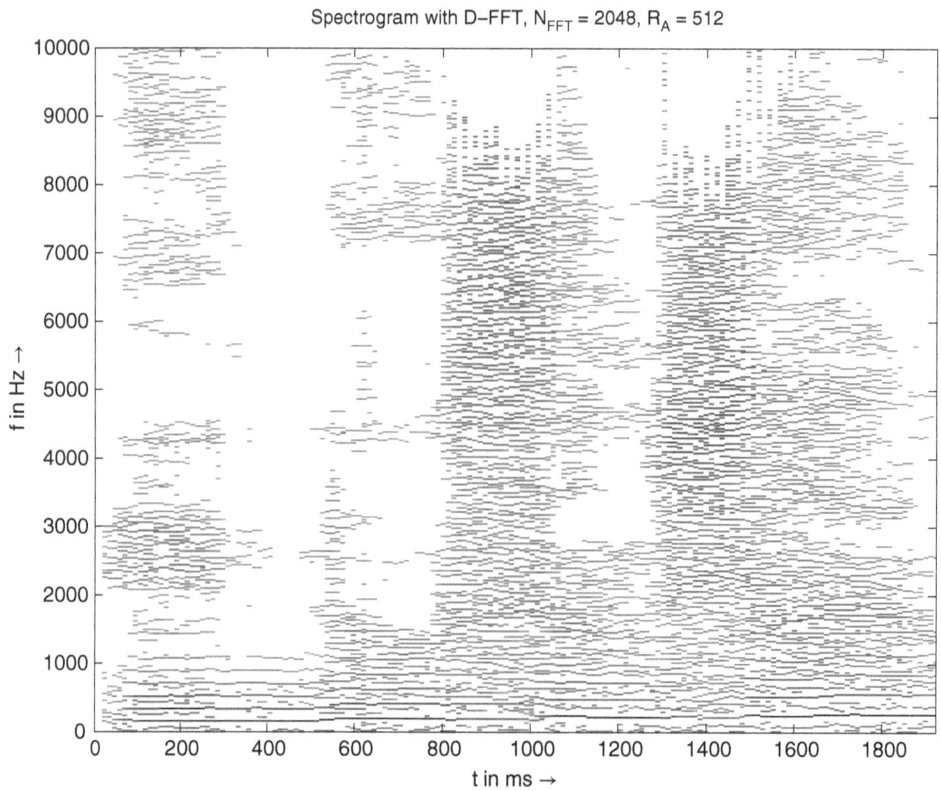

Figure 9.35 Spectrogram of sinusoidal components.

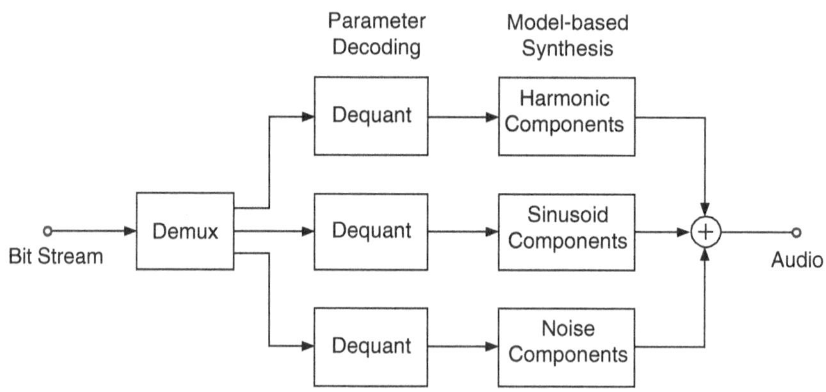

Figure 9.36 MPEG-4 parametric decoder.

effects [Zöl11]. Effects such as time and pitch scaling, virtual sources in three-dimensional spaces, and cross-synthesis of signals (karaoke) are just a few examples for interactive sound design on the decoding side.

9.8 Spectral Band Replication

To further reduce the bit rate, an extension of MPEG-1 layer 3 with the name MP3pro was introduced [Die02, Zie02]. The underlying method, called *spectral band replication* (SBR) performs a lowpass and highpass decomposition of the audio signal, where the lowpass filtered part is coded by a standard coding method (e.g. MPEG-1 Layer 3) and the highpass part is represented by a spectral envelope and a difference signal [Eks02, Zie03]. Figure 9.37 shows the functional units of an SBR coder. For the analysis of the difference signal, the highpass part (HP generator) is reconstructed from the lowpass part and compared with the actual highpass part. The difference is coded and transmitted. For decoding (see Fig. 9.38), the decoded lowpass part of a standard decoder is used by the HP generator to reconstruct the highpass part. An additional coded difference signal is added at the decoder. An equalizer provides the spectral envelope shaping for the highpass part. The spectral envelope of the highpass signal can be achieved by a filter bank and computing the root-mean-square values of each bandpass signal [Eks02, Zie03]. The reconstruction of the highpass part (HP generator) can also be achieved by a filter bank and substituting the bandpass signals by using the lowpass parts [Schu96, Her98]. To code the difference signal of the highpass part, additive sinusoidal models can be applied such as the parametric methods of the MPEG-4 coding approach.

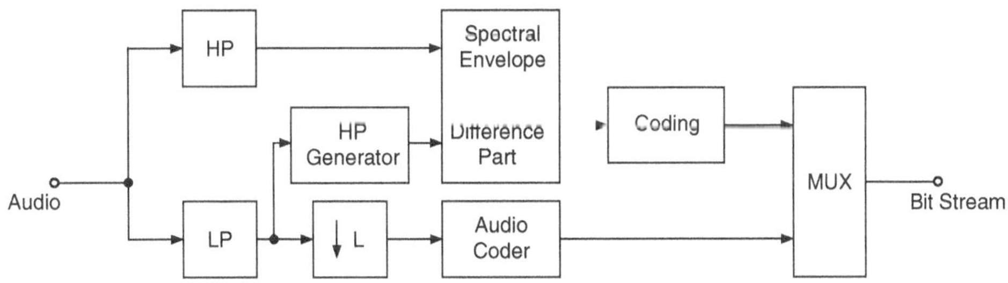

Figure 9.37 SBR coder.

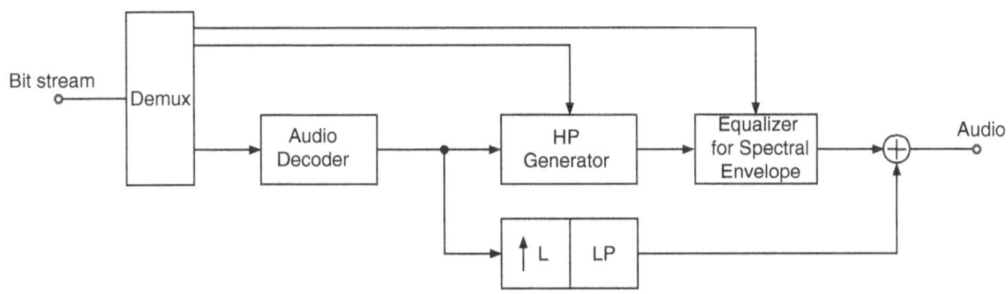

Figure 9.38 SBR decoder. Based on [Edl00] and [Mei02].

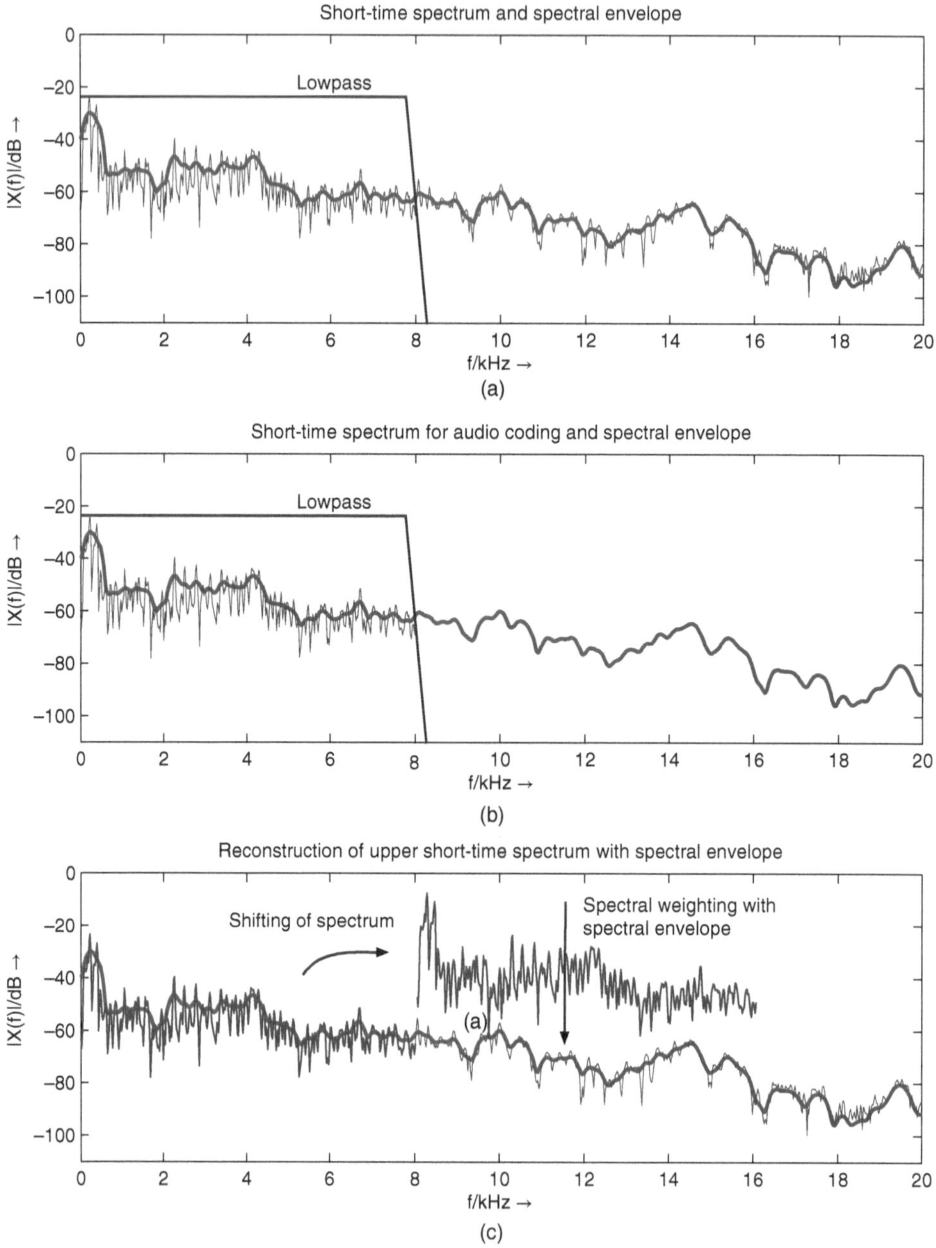

Figure 9.39 Functional units of the SBR method.

Figure 9.39 shows the functional units of the SBR method in the frequency domain. First, the short-time spectrum is used to calculate the spectral envelope (Fig. 9.39a). The spectral envelope can be derived from an FFT, a filter bank, the cepstrum, or by linear prediction [Zöl11]. The band-limited lowpass signal can be downsampled and coded by a standard coder which operates at a reduced sampling rate. In addition, the spectral envelope has to be coded

(Fig. 9.39b). On the decoding side, the reconstruction of the upper spectrum is achieved by frequency shifting of the lowpass part, or even specific lowpass parts, and applying the spectral envelope onto this artificial highpass spectrum (Fig. 9.39c). An efficient implementation of a time-varying spectral envelope computation (at the coder side) and spectral weighting of the highpass signal (at the decoder side) with a complex-valued QMF bank is described in [Eks02].

9.9 Constrained Energy Lapped Transform – Gain and Shape Coding

Another audio coding system using overlapped MDCT and a gain-shape quantization scheme in the frequency domain was introduced in [Val09, Val10], which is lossy in nature. The block diagram in Fig. 9.40 shows a simple version of the so-called constrained energy lapped transform (CELT) encoder and decoder. In this section, a simplified version of the CELT codec is described that excludes some refinements such as pre- and post-filtering, bit allocation for pitch period filter, and filter gain, which are described in [Val13].

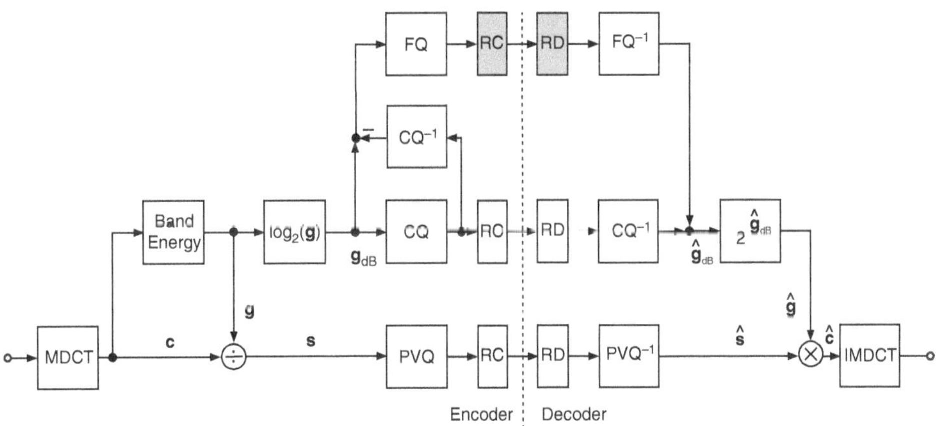

Figure 9.40 A simplified illustration of the CELT codec.

In the CELT codec, the MDCT with a given window length N is performed on a pre-processed audio input to generate the coefficients $\mathbf{c}$. From the coefficients, the energies in 20 Bark bands are computed as

$$E_i = \sqrt{\mathbf{c}_i^T \cdot \mathbf{c}_i}, \tag{9.41}$$

where $i = 0,1, \ldots , 19$ denote the band indices. These band energies are called gain coefficients and they are given by $g_i = E_i$. Division of the Bark band coefficient vectors $\mathbf{c}_i$ by these gain coefficients g_i leads to the shape coefficient vectors given by

$$\mathbf{s}_i = \frac{\mathbf{c}_i}{g_i}. \tag{9.42}$$

The mapping from DCT coefficients $\mathbf{c}$ to Bark bands $\mathbf{c}_i$ is described in Table IV of [Val10]. The corresponding frequency for each DCT coefficient c_k is given by $f_k = \frac{k}{2N} \cdot f_S$ in Hz with $k = 0,1 \ldots, 255$. The sequential operations for computing gain and shape are given by

$$\mathbf{c} = \begin{bmatrix} \mathbf{c}_0 & = \begin{bmatrix} c_0 \\ c_1 \\ c_2 \end{bmatrix} \\ \mathbf{c}_1 & = \begin{bmatrix} c_3 \\ c_4 \\ c_5 \end{bmatrix} \\ \vdots & \\ \mathbf{c}_{19} & = \begin{bmatrix} c_{180} \\ \vdots \\ c_{232} \end{bmatrix} \end{bmatrix} \rightarrow \mathbf{g} = \begin{bmatrix} g_0 = \sqrt{\mathbf{c}_0^T \cdot \mathbf{c}_0} \\ g_1 = \sqrt{\mathbf{c}_1^T \cdot \mathbf{c}_1} \\ \vdots \\ g_{19} = \sqrt{\mathbf{c}_{19}^T \cdot \mathbf{c}_{19}} \end{bmatrix} \rightarrow \mathbf{s} = \begin{bmatrix} \mathbf{s}_0 = \mathbf{c}_0/g_0 \\ \mathbf{s}_1 = \mathbf{c}_1/g_1 \\ \vdots \\ \mathbf{s}_{19} = \mathbf{c}_{19}/g_{19} \end{bmatrix}. \tag{9.43}$$

Figure 9.41 shows the individual spectrograms representing the energies per Bark band in dB, the corresponding gain coefficients in dB, and the shape coefficient vectors. The values of elements in each shape vector lie between −1 and 1. In the next step, quantization of the gain and shape

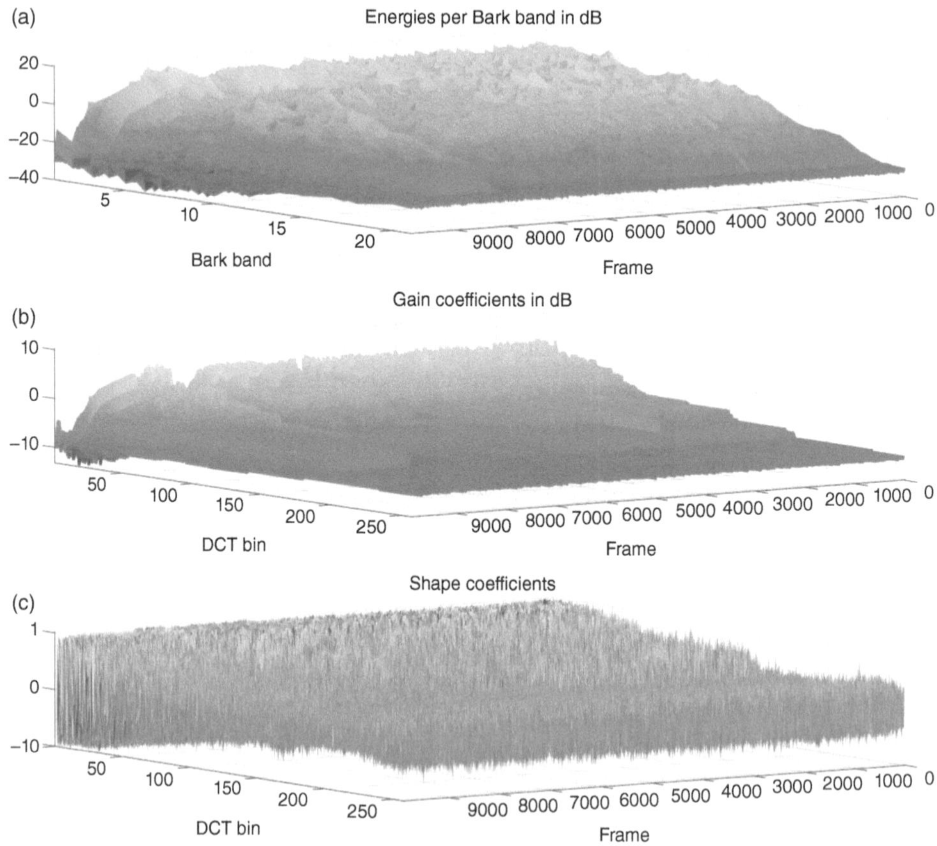

Figure 9.41 CELT spectrograms for a) energies per Bark band, b) gain coefficients in dB, and c) shape coefficients.

coefficients takes place. The gain coefficients use scalar quantization and the shape coefficients use a form of vector quantization, which usually performs a dimension reduction and leads to compression.

9.9.1 Gain Quantization

CELT uses a strategy to quantize the gain in the log2 domain that includes two steps – coarse and fine quantization. For coarse quantization, gain coefficients from each band can be directly quantized with a pre-defined resolution. However, direct quantization of the coefficients will result in the usage of a high number of bits during encoding if the variance among the coefficients or the entropy of their distribution is high. When CELT is used for real-time audio coding over a network as part of the OPUS codec [Val12], the available bandwidth governs the bit allocation. Hence, the CELT encoder eliminates the redundancy between the gain coefficients across time and frequency by performing a prediction. The corresponding details of the predictive method are provided in [Val10]. The method leads to a set of predictive gain coefficients, denoted by $\mathbf{g}^{\mathrm{p}}$, with lower variance or entropy than the original gain coefficients. Figure 9.42 shows the mean gain vector compared with the mean predictive gain vector, averaged across all bands for 100 overlapping frames of an audio signal. The variance of the mean gain vector is 0.088 while the variance of the average predictive gain vector is 0.017. The coarse quantization (CQ) is then performed as

$$\mathbf{g}^{\mathrm{p}}_{\mathrm{cq}} = \mathrm{CQ}(\mathbf{g}^{\mathrm{p}}_{\mathrm{db}}) = \mathrm{CQ}\left(\log_2(\mathbf{g}^{\mathrm{p}})\right) = \begin{bmatrix} \mathrm{CQ}\left(\log_2\left(g^{\mathrm{p}}_0\right)\right) \\ \mathrm{CQ}\left(\log_2\left(g^{\mathrm{p}}_1\right)\right) \\ \vdots \\ \mathrm{CQ}\left(\log_2\left(g^{\mathrm{p}}_{19}\right)\right) \end{bmatrix} \tag{9.44}$$

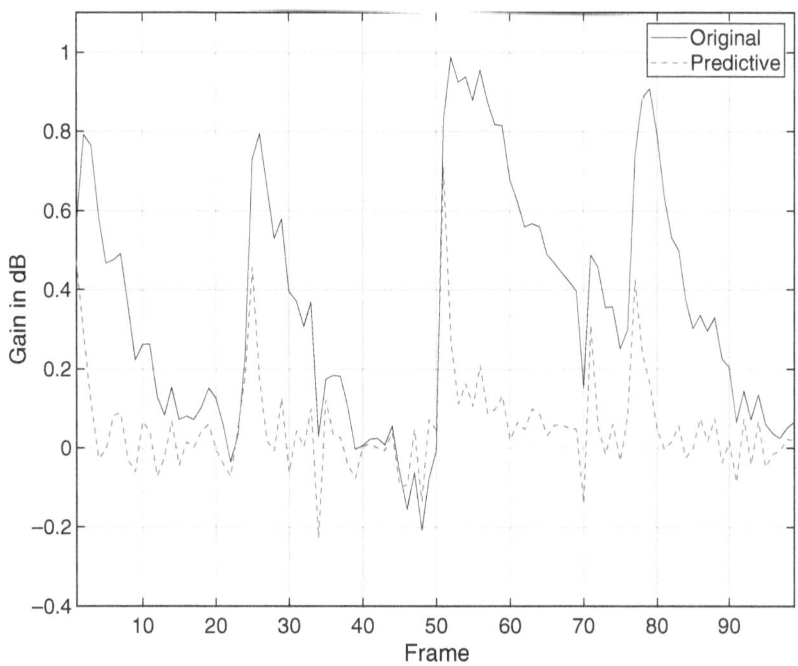

Figure 9.42 Mean gain and mean predictive gain coefficients, averaged over all bands.

and the fine quantization (FQ) is performed as

$$\mathbf{g}^{\mathrm{p}}_{\mathrm{fq}} = \mathrm{FQ}\left(\mathbf{g}^{\mathrm{p}}_{\mathrm{db}} - \mathbf{g}^{\mathrm{p}}_{\mathrm{cq}}\right) = \mathrm{FQ}\left(\mathbf{e}_{\mathrm{cq}}\right), \tag{9.45}$$

where $\mathbf{g}^{\mathrm{p}}_{\mathrm{db}}$ denote the predictive gain coefficients expressed in decibels and $\mathbf{e}_{\mathrm{cq}}$ denotes the coarse quantization error. The range of the rounded values after coarse quantization varies between frames and depends on the quantization resolution q_r. The quantized values or indices are transformed into positive integers for encoding. The coarse quantization symbols are usually distributed in such a way that it can be modeled by a generalized Gaussian or Laplace distribution. An example of the real symbol distribution over 100 overlapping frames is provided in Fig. 9.43 along with the ideal Laplace distribution that can be used during entropy coding. The coarse quantization error usually lies between $-0.5 \times q_r$ and $0.5 \times q_r$. Fine quantization maps the error value to an integral symbol which represents a sub-interval of the error range. The resolution for fine quantization depends on its bit allocation. The fine quantization symbols per frame are usually not distributed well enough to be modeled by any standard distribution. However, if many frames are taken into account, its distribution approaches a uniform distribution, as shown in Fig. 9.44.

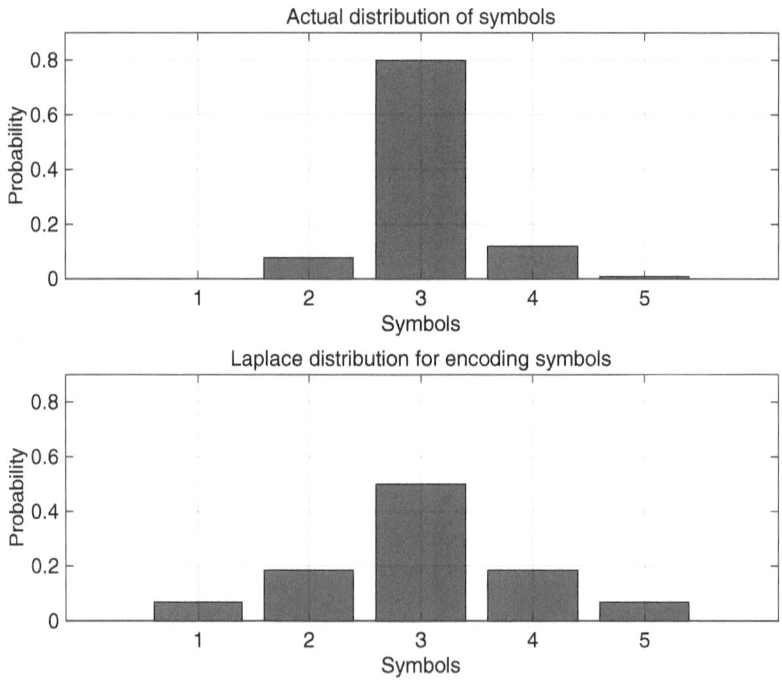

Figure 9.43 Actual and ideal distribution of the coarse quantization symbols for 100 frames.

9.9.2 Shape Quantization

Quantization of the shape vectors is performed by pyramid vector quantization (PVQ) [Fis86, Say12, Val13]. In PVQ, a codebook containing reference vectors of length N is created, where N denotes the length of the shape vector to be quantized. The codebook is made of integers,

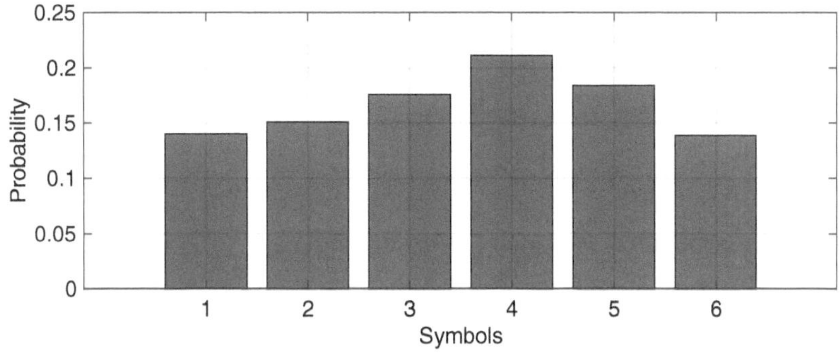

Figure 9.44 Distribution of the fine quantization symbols for 100 frames.

whose absolute values add up to a predefined constant K referred to as the number of pulses. The codebook vectors are normalized to produce unit vectors which are ordered and assigned with numeric indices. The shape vector to be encoded receives the index that corresponds to its nearest unit vector in the codebook. The L_1 distance is used to find the appropriate unit vector in a codebook search. Each band is uniquely encoded based on its bit allocation and the low- and mid-frequency bands usually have a better bit resolution. A shape vector $\mathbf{s}_i$ of the ith band is mapped to a codebook index $s_{pvq,i}$ according to

$$\mathbf{s}_{\mathrm{pvq}} = \mathrm{PVQ}(\mathbf{s}) = \begin{bmatrix} s_{pvq,0} = \mathrm{PVQ}\left(\mathbf{s}_0\right) \\ s_{pvq,1} = \mathrm{PVQ}\left(\mathbf{s}_1\right) \\ \vdots \\ s_{pvq,19} = \mathrm{PVQ}\left(\mathbf{s}_{19}\right) \end{bmatrix}. \tag{9.46}$$

The distribution of the shape quantization symbols per frame is usually sparse and similar to the fine quantization symbols, it cannot be modeled by any standard distribution. When many frames are considered, it approaches a uniform distribution for each band. The inverse operation maps the codebook index $s_{pvq,i}$ to a codebook vector $\hat{\mathbf{s}}_i$ as given by

$$\hat{\mathbf{s}} = \mathrm{PVQ}^{-1}(\mathbf{s}_{\mathrm{pvq}}) = \begin{bmatrix} \hat{\mathbf{s}}_0 = \mathrm{PVQ}^{-1}\left(s_{pvq,0}\right) \\ \hat{\mathbf{s}}_1 = \mathrm{PVQ}^{-1}\left(s_{pvq,1}\right) \\ \vdots \\ \hat{\mathbf{s}}_{19} = \mathrm{PVQ}^{-1}\left(s_{pvq,19}\right) \end{bmatrix}. \tag{9.47}$$

9.9.3 Range Coding

Range coding [Nig79, Say12] is an entropy coding method that performs a progressive encoding of symbols by narrowing down an initial range, based on the probability distribution of the symbols, until the final range and codeword are determined. The final codeword is decoded with the help of the probability distribution of the symbols as well. It is very similar to arithmetic encoding, except that the encoding is done with digits in any base instead of with bits. However, CELT uses range coding as a bit packer. It performs an iterative compression of a sequence of 20 symbols corresponding to 20 bands with a pre-defined probability density function and outputs a single codeword. From this codeword, the iterative decompression delivers the 20 input symbols. This range coding and decoding is usually performed for the course gain symbols only while the shape and fine quantization symbols are encoded as raw and decoded directly. However, if the standalone CELT codec is used for non-real-time compression of audio after a look-ahead of multiple frames, entropy coding is able to compress the shape and fine quantization symbols as well with the use of a uniform distribution function.

9.9.4 CELT Decoding

Figure 9.40 shows a simplified operation of the CELT decoder. Once the bit streams are available to the decoder, range decoding is applied to retrieve the coarse quantization symbols while the raw bits are decoded to retrieve the fine and shape quantization symbols. The inverse quantization of the symbols takes place in the next step as denoted by FQ^{-1}, GQ^{-1}, and PVQ^{-1} to get the respective coefficients. The reconstructed gain error is added to the reconstructed predictive gain coefficients and subsequently the reconstructed gain coefficients $\hat{\mathbf{g}}_{db}$ are calculated based on the inverse prediction method. The gain coefficients are transformed to the linear domain as $\hat{\mathbf{g}} = 2^{\hat{\mathbf{g}}_{db}}$ and the frequency domain coefficients are reconstructed as

$$\hat{\mathbf{c}}_i = \hat{g}_i \cdot \hat{\mathbf{s}}_i. \tag{9.48}$$

The coefficients are transformed using IMDCT to get the reconstructed time-domain signal. Figure 9.45 shows an original and reconstructed signal along with the reconstruction error for 100 frames of a test audio file using a quantization resolution of 6 dB. For the above example, CELT encoding results in an average rate of 348 bits per frame while the average rate for coarse quantization after entropy coding is nearly 26 bits per frame. The overall compression achieved for the corresponding audio snippet is nearly 25 % of the original.

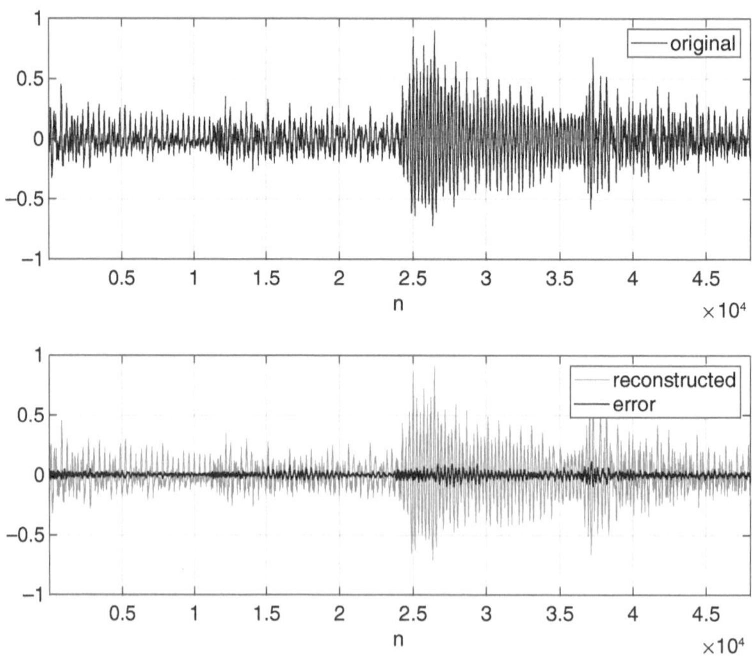

Figure 9.45 Performance of the CELT codec on an audio snippet.

9.10 JS Applet – Psychoacoustics

The applet shown in Fig. 9.46 demonstrates psychoacoustic audio masking effects [Gui05]. It is designed for a first insight into the perceptual experience of masking a sinusoidal signal with band-limited noise.

You can choose between two predefined audio files from our web server ("Audio1.wav" or "Audio2.wav"). These are band-limited noise signals with different frequency ranges. A sinusoidal signal is generated by the applet, and two sliders can be used to control its frequency and magnitude values.

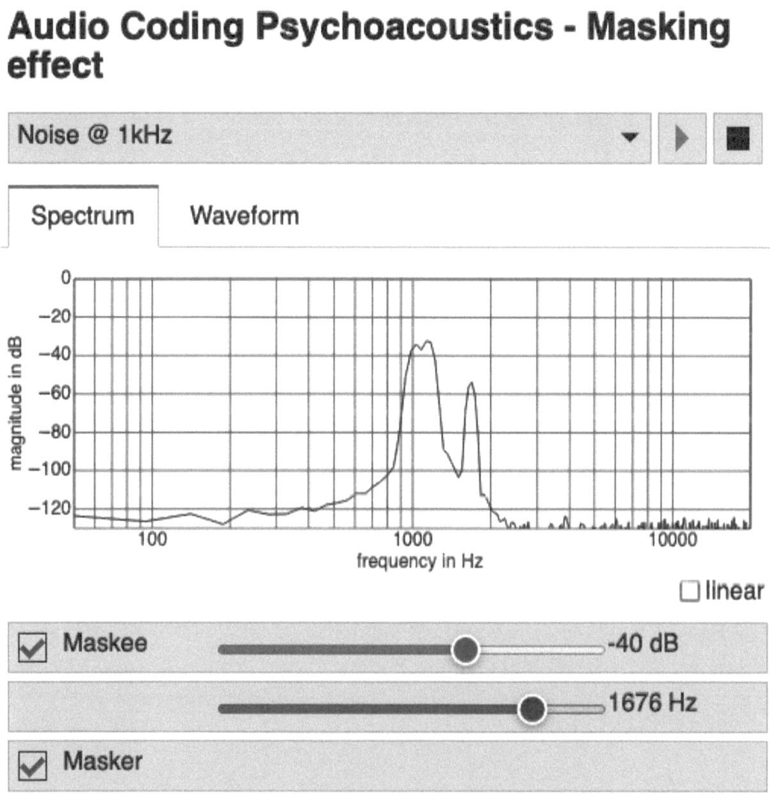

Figure 9.46 JS applet – psychoacoustics.

9.11 Exercises

1. Psychoacoustics

1. Human hearing

   (a) In which frequency range is a human able to perceive sounds?

   (b) What is the frequency range of speech?

(c) In the above specified range, where is human hearing most sensitive?

(d) How is the absolute threshold of hearing determined?

2. Masking

(a) What is frequency-domain masking?

(b) What is a critical band and why is it needed for frequency masking phenomena?

(c) Consider a_i and f_i to be respectively the amplitude and the frequency of a partial at index i and $V(a_i)$ to be the corresponding volume in dB. The difference between the level of the masker and the masking threshold is -10 dB. The masking curves toward lower and higher frequencies are described respectively by a left slope (27 dB/Bark) and a right slope (15 dB/Bark). Explain the main steps of frequency masking in this case and show with plots how this masking phenomena is achieved.

(d) What are the psychoacoustic parameters used for lossy audio coding?

(e) How can we explain temporal masking and what is its duration after stopping the active masker?

2. Audio coding

1. How do a lossless coder and decoder work.
2. What is the achievable compression factor for lossless coding?
3. How do an MPEG-1 Layer 3 coder and decoder work.
4. How do an MPEG-2 AAC coder and decoder work.
5. What is temporal noise shaping?
6. How do an MPEG-4 coder and decoder work.
7. What is the benefit of SBR?

References

[Alt99] R. Althoff, F. Keiler, U. Zölzer: *Extracting Sinusoids from Harmonic Signals*, Proc. DAFX-99 Workshop on Digital Audio Effects, pp. 97–100, Trondheim, Norway, 1999.

[Blo95] T. Block: *Untersuchung von Verfahren zur verlustlosen Datenkompression von digitalen Audiosignalen*, Studienarbeit, TU Hamburg-Harburg, 1995.

[Bos97] M. Bosi, K. Brandenburg, S. Quackenbush, L. Fielder, K. Akagiri, H. Fuchs, M. Dietz, J. Herre, G. Davidson, Y. Oikawa: *ISO/IEC MPEG-2 Advanced Audio Coding*, J. Aud. Eng. Soc., Vol. 45, No. 10, pp. 789–814, October 1997.

[Bos02] M. Bosi, R.E. Goldberg: *Introduction to Digital Audio Coding and Standards*, Kluwer Academic Publishers, 2002.

[Bra92] K. Brandenburg, J. Herre: *Digital Audio Compression for Professional Applications*, Proc. 92nd AES Convention, Preprint No. 3330, Vienna 1992.

[Bra94] K. Brandenburg, G. Stoll: *The ISO-MPEG-1 Audio: A Generic Standard for Coding of High Quality Digital Audio*, J. Aud. Eng. Soc., Vol. 42, No. 10, pp. 780–792, October 1994.

[Bra98] K. Brandenburg: *Perceptual Coding of High Quality Digital Audio*, in M. Kahrs, K. Brandenburg: *Applications of Digital Signal Processing to Audio and Acoustics*, Kluwer Academic Publishers, 1998.

[Cel93] C. Cellier, P. Chenes, M. Rossi: *Lossless Audio Data Compression for Real-Time Applications*, Proc. 95th AES Convention, Preprint No. 3780, New York 1993.

[Cra96] P. Craven, M. Gerzon: *Lossless coding for audio discs*, J. Audio Eng. Soc., Vol. 44, No. 9, pp.706–720, 1996.

[Cra97] P. Craven, M. Law, and J. Stuart: *Lossless compression using IIR prediction*, Proc. 102nd AES Convention, Preprint No. 4415, Munich, Germany, 1997.

[Die02] M. Dietz, L. Liljeryd, K. Kjörling, O. Kunz: *Spectral Band Replication: A Novel Approach in Audio Coding* Proc. 112th AES Convention, Preprint No. 5553, Munich 2002.

[Edl89] B. Edler: *Codierung von Audiosignalen mit überlappender Transformation und adaptiven Fensterfunktionen*, Frequenz, Vol. 43, pp. 252–256, 1989.

[Edl00] B. Edler, H. Purnhagen: *Parametric Audio Coding*, 5th International Conference on Signal Processing (ICSP 2000), Beijing, August 2000.

[Edl95] B. Edler: *Äquivalenz von Transformation und Teilbandzerlegung in der Quellencodierung*, Dissertation, Universität Hannover, 1995.

[Eks02] P. Ekstrand: *Bandwidth Extension of Audio Signals by Spectral Band Replication*, Proc. 1st IEEE Benelux Workshop on Model-based Processing and Coding of Audio (MPCA-2002), Leuven, Belgium, 2002.

[Ern00] M. Erne: *Signal Adpative Audio Coding Using Wavelets and Rate Optimization*, Dissertation, ETH Zrich, 2000.

[Fis86] T.H. Fischer: *A pyramid vector quantizer*. IEEE Trans. on Information Theory, 32:56–583, 1986.

[Geo92] E.B. George, M.J.T. Smith: *Analysis-by-Synthesis/Overlap-Add Sinusoidal Modeling applied to the Analysis and Synthesis of Musical Tones*, Journal of the Audio Engineering Society, Vol. 40, No. 6, pp. 497–516, June 1992.

[Geo97] E.B. George, M.J.T. Smith: *Speech Analysis/Synthesis and Modification using an Analysis-by-Synthesis/Overlap-Add Sinusoidal Model*, IEEE Transactions on Speech and Audio Processing, Vol. 5, No. 5, pp.389–406, September 1997.

[Glu93] R. Gluth: *Beiträge zur Beschreibung und Realisierung digitaler, nichtrekursiver Filterbänke auf der Grundlage linearer diskreter Transformationen*, Dissertation, Ruhr-Universität Bochum, 1993.

[Gui05] M. Guillemard, C. Ruwwe, U. Zölzer: *J-DAFx - Digital Audio Effects in Java*, Proc. 8th Int. Conference on Digital Audio Effects (DAFx-05), Madrid, Spain, pp.161–166, 2005.

[Hai03] S. Hainsworth, M. Macleod: *On Sinusoidal Parameter Estimation*, Proc. DAFX-03 Conference on Digital Audio Effects, London, UK, September 2003.

[Han98] M. Hans: *Optimization of Digital Audio for Internet Transmission*, PhD Thesis, Georgia Inst. Technol., Atlanta, 1998.

[Han01] M. Hans, R.W. Schafer: *Lossless Compression of Digital Audio*, IEEE Signal Processing Magazine, Vol. 18, No. 4, pp. 21–32, July 2001.

[Hel72] R.P. Hellman: *Asymmetry in Masking Between Noise and Tone*, Perception and Psychophys., Vol. 11, pp. 241–246, 1972.

[Her96] J. Herre, J.D. Johnston: *Enhancing the Performance of Perceptual Audio Coders by Using Temporal Noise Shaping (TNS)*, Proc. 101st AES Convention, Preprint No. 4384, Los Angeles, 1996.

[Her98] J. Herre, D. Schultz: *Extending the MPEG-4 AAC Codec by Perceptual Noise Substitution*, Proc. 104th AES Convention, Preprint No. 4720, Amsterdam, 1998.

[Her99] J. Herre: *Temporal Noise Shaping, Quantization and Coding Methods In Perceptual Audio Coding: A Tutorial Introduction*, Proc. AES 17th International Conference on High Quality Audio Coding, Florence, September 1999.

[Her02] J. Herre, B. Grill: *Overview of MPEG-4 Audio and its Applications in Mobile Communications*, Proc. 112th AES Convention, Preprint No. 5553, Munich 2002.

[Huf52] D.A. Huffman: *A Method for the Construction of Minimum-"-Redundancy Codes*, Proc. of the IRE, pp. 1098–1101, 1952.

[ISO92] ISO/IEC 11172-3: *Coding of Moving Pictures and Associated Audio for Digital Storage Media at up to 1.5 Mbits/s - Audio Part*, International Standard, 1992.

[Jay84] N.S. Jayant, P. Noll: *Digital Coding of Waveforms*, Prentice-Hall, New Jersey, 1984.

[Joh88a] J.D. Johnston: *Transform Coding af Audio Signals Using Perceptual Noise Criteria*, IEEE Journal on Selected Areas in Communications, Vol. 6, No. 2, pp. 314–323, February 1988.

[Joh88b] J.D. Johnston: *Estimation of Perceptual Entropy Using Noise Masking Criteria*, Proc. ICASSP-88, pp. 2524–2527, 1988.

[Kap92] R. Kapust: *A Human Ear Related Objective Measurement Technique Yields Audible Error and Error Margin*, Proc. 11th Int. AES Conference - Test&Measurement, Portland, pp. 191–202, 1992.

[Kei01] F. Keiler, U. Zölzer: *Extracting Sinusoids from Harmonic Signals*, Journal of New Music Research, Special Issue: "Musical Applications of Digital Signal Processing", Guest Editor: Mark Sandler, Vol. 30, No. 3, pp. 243–258, September 2001.

[Kei02] F. Keiler, S. Marchand: *Survey on Extraction of Sinusoids in Stationary Sounds*, Proc. DAFX-02 Conference on Digital Audio Effects, pp. 51–58, Hamburg, 2002.

[Kon94] K. Konstantinides: *Fast Subband Filtering in MPEG Audio Coding*, IEEE Signal Processing Letters, Vol. 1, No. 2, pp. 26–28, February 1994.

[Lag02] M. Lagrange, S. Marchand, J.-B. Rault: *Sinusoidal Parameter Extraction and Component Selection in a Non Stationary Model*, Proc. DAFX-02 Conference on Digital Audio Effects, pp. 59–64, Hamburg, 2002.

[Lev98] S. Levine: *Audio Representations for Data Compression and Compressed Domain Processing*, PhD Thesis, Stanford University, 1998.

[Lev99] S. Levine, J.O. Smith: *Improvements to the Switched Parametric & Transform Audio Coder*, Proc. 1999 IEEE Workshop on Applicatrions of Signal Proceassing to Audio and Acoustics, New Paltz, October 1999.

[Lie02] T. Liebchen: *Lossless Audio Coding Using Adaptive Multichannel Prediction*, Proc. 113th AES Convention, Preprint No. 5680, Los Angeles, 2002.

[Mar00a] S. Marchand: *Sound Models for Computer Music*, PhD Thesis, University of Bordeaux, October 2000.

[Mar00a] S. Marchand: *Compression of Sinusoidal Modeling Parameters*, Proc. DAFX-00 Conference on Digital Audio Effects, pp. 273–276, Verona, Italy, December 2000.

[Mal92] H.S. Malvar: *Signal Processing with Lapped Transforms*, Artech House, Norwood, 1992.

[McA86] R. McAulay, T. Quatieri: *Speech Transformations Based in a Sinusoidal Representation*, IEEE Trans. Acoustics, Speech, Signal Processing, Vol. 34, No. 4, pp. 744–754, 1989.

[Mas85] J. Masson, Z. Picel: *Flexible Design of Computationally Efficient Nearly Perfect QMF Filter Banks*, Proc. ICASSP-85, pp. 541–544, 1985.

[Mei02] N. Meine, H. Purnhagen: *Fast Sinusoid Synthesis For MPEG-4 HILN Parametric Audio Decoding*, Proc. DAFX-02 Conference on Digital Audio Effects, pp. 239–244, Hamburg, September 2002.

[Nig79] G. Nigel and N. Martin: *Range encoding: An algorithm for removing redundancy from a digitized message*. In Proc. Video and Data Recording Conference, Oct 1979.

[Pen93] W.B. Pennebaker, J.L. Mitchell: *JPEG Still Image Data Compression Standard*, Van Nostrand Reinhold, New York, 1993.

[Pri86] J.P. Princen, A.B. Bradley: *Analysis/Synthesis Filter Bank Design Based on Time Domain Aliasing Cancellation*, IEEE Trans. on Acoustics, Speech, and Signal Processing, Vol. 34, No. 5, pp. 1153–1161, October 1986.

[Pri87] J.P. Princen, A.W. Johnston, A.B. Bradley: *Subband/Transform Coding Using Filter Bank Designs Based on Time Domain Aliasing Cancellation*, Proc. ICASSP-87, pp. 2161–2164, 1987.

[Pur97] M. Purat, T. Liebchen, and P. Noll: *Lossless Transform Coding of Audio Signals*, Proc. 102nd AES Convention, Preprint No. 4414, Munich, Germany, 1997.

[Pur99] H. Purnhagen: *Advances in Parametric Audio Coding*, Proc. 1999 IEEE Workshop on Applicatrions of Signal Proceassing to Audio and Acoustics, New Paltz, October 1999.

[Pur02a] H. Purnhagen, N. Meine, B. Edler: *Sinusoidal Coding Using Loudness-Based Component Selection*, Proc. ICASSP-2002, May 13-17, Orlando, USA, 2002.

[Pur02b] H. Purnhagen: *Parameter Estimation and Tracking For Time-Varying Sinusoids*, Proc. 1st IEEE Benelux Workshop on Model Based Processing and Coding of Audio, Leuven, Belgium, November 2002.

[Raa02] M. Raad, A. Mertins: *From Lossy to Lossless Audio Coding Using SPIHT*, Proc. DAFX-02 Conference on Digital Audio Effects, pp. 245–250, Hamburg, 2002.

[Rao90] K.R. Rao, P. Yip: *Discrete Cosine Transform – Algorithms, Advantages, Applications*, Academic Press, Inc., San Diego, 1990.

[Rob94] T. Robinson: *SHORTEN: Simple lossless and near-lossless waveform compression*, Technical Report CUED/F-INFENG/TR.156, Cambridge University Engineering Department, Cambridge, UK, December 1994.

[Rod97] X. Rodet: *Musical Sound Signals Analysis/Synthesis: Sinusoidal+Residual and Elementary Waveform Models*, Proceedings of the IEEE Time-Frequency and Time-Scale Workshop (TFTS-97), University of Warwick, Coventry, UK, August 1997.

[Rot83] J.H. Rothweiler: *Polyphase Quadrature Filters - A New Subband Coding Technique*, Proc. ICASSP-87, pp. 1280–1283, 1983.

[Sauv90] U. Sauvagerd: *Bitratenreduktion hochwertiger Musiksignale unter Verwendung von Wellendigitalfiltern*, VDI-Verlag, Düsseldorf 1990.

[Say12] K. Sayood: *Introduction to Data Compression*. Morgan Kaufmann, 2012.

[Schr79] M.R. Schroeder, B.S. Atal, J.L. Hall: *Optimizing Digital Speech Coders by Exploiting Masking Properties of the Human Ear*, J. Acoust. Soc. Am., Vol. 66, No. 6, pp. 1647–1652, December 1979.

[Sch02] G.D.T. Schuller, Bin Yu, Dawei Huang, B. Edler: *Perceptual Audio Coding Using Adaptive Pre- and Post-Filters and Lossless Compression*, IEEE Transactions on Speech and Audio Processing, Vol. 10, No. 6, pp. 379–390, Sept. 2002.

[Schu96] D. Schulz: *Improving Audio Codecs by Noise Substitution*, J. Aud. Eng. Soc., Vol. 44, No. 7/8, pp. 593–598, July/August 1996.

[Ser89] X. Serra: *A System for Sound Analysis/Transformation/Synthesis based on a Deterministic plus Stochastic Decomposition*, PhD Thesis, Stanford University, 1989.

[Sin93] D. Sinha, A.H. Tewfik: *Low bit rate Transparent Audio Compression Using Adapted Wavelets*, IEEE Transactions on Signal Processing, Vol. 41, pp.3463–3479, 1993.

[Smi90] J.O. Smith, X. Serra: *Spectral Modeling Synthesis: A Sound Analysis/Synthesis System based on a Deterministic plus Stochastic Decomposition*, Computer Music Journal, Vol. 14, No. 4, pp. 12–24, 1990.

[Sqa88] EBU-SQAM: *Sound Quality Assessment Material*, Recordings for Subjective Tests, CompactDisc, 1988.

[Ter79] E. Terhardt: *Calculating Virtual Pitch*, Hearing Res., Vol. 1, pp. 155–182, 1979.

[Thei88] G. Theile, G. Stoll, M. Link: *Low Bit-Rate Coding of High-quality Audio Signals*, EBU Review, No. 230, pp. 158–181, August 1988.

[Vai93] P.P. Vaidyanathan: *Multirate Systems and Filter Banks*, Prentice-Hall, Englewood Cliffs, 1993.

[Val09] J.-M. Valin, T.B. Terriberry, and G. Maxwell: *A full-bandwidth audio codec with low complexity and very low delay*. In Proc. EUSIPCO 2009, pages 125–1258, 2009.

[Val10] J.-M. Valin, T. B. Terriberry, C. Montgomery, and G. Maxwell: *A high-quality speech and audio codec with less than 10 ms delay*. IEEE Trans. Audio, Speech and Language Processing, 18(1):5–67, 2010.

[Val12] J.-M. Valin, K. Vos, and T. Terriberry: *Definition of the Opus Audio Codec*. RFC 6716, Sep 2012.

[Val13] J.-M. Valin, T. B. Terriberry, C. Montgomery, and K. Vos: *Highquality, low-delay coding in the opus codec*. In Audio Engineering Society Convention 135, Oct 2013.

[Var06] P. Vary, R. Martin: *Digital Speech Transmission. Enhancement, Coding and Error Concealment*, J. Wiley & Sons, Chichester, 2006.

[Vet95] M. Vetterli, J. Kovacevic: *Wavelets and Subband Coding*, Prentice-Hall, Englewood Cliffs, 1995.

[Zie02] T. Ziegler, A. Ehret, P. Ekstrand, M. Lutzky: *Enhancing mp3 with SBR: Features and Capabilities of the new mp3PRO Algorithm*, Proc. 112th AES Convention, Preprint No. 5560, Munich 2002.

[Zie03] T. Ziegler, M. Dietz, K. Kjörling, A. Ehret: *aacPlus-Full Bandwidth Audio Coding for Broadcast and Mobile Applications*, International Signal Processing Conference, Dallas, 2003.

[Zöl11] U. Zölzer (Ed.): *DAFX - Digital Audio Effects*, 2nd edition, J. Wiley & Sons, Chichester, 2011.

[Zwi82] E. Zwicker: *Psychoakustik*, Springer-Verlag, Berlin, 1982.

[Zwi90] E. Zwicker, H. Fastl: *Psychoacoustics*, Springer-Verlag, Berlin, 1990.

Chapter 10

Nonlinear Processing

M. Holters and L. Köper

10.1 Fundamentals

Linear systems – filters – play an important role in audio signal processing. Linearity here means that the superposition property holds: denote the output of a system excited with input signals $x_1(n)$ and $x_2(n)$ with $y_1(n)$ and $y_2(n)$, respectively. If the system maps the superposition of the input signals $x(n) = a_1x_1(n) + a_2x_2(n)$ to the corresponding superposition $y(n) = a_1y_1(n) + a_2y_2(n)$ at the output (for any x_1, x_2, a_1, a_2), it is considered linear. Systems that do not fulfill this property are accordingly called nonlinear. One important consequence of linearity, when combined with time invariance, is that systems can be fully described in the frequency domain by their transfer function. Therefore, frequency components present in the input signal can be amplified or attenuated and phase-shifted, but no new components can appear in the output. This is not the case for nonlinear systems, where in particular, components at multiples of the input frequency (harmonics) can appear. Figure 10.1 visualizes this important difference.

We have already seen nonlinear systems, namely the dynamic range controllers of Chapter 8. However, when only considering short time spans (in relation to the attack and release times), they can be approximated by simply scaling the signal, a linear operation. This chapter, however, will focus on nonlinear systems where such an approximation is not fruitful.

An important distinction between nonlinear systems is whether they only act on the current input signal value (static, memoryless, or stateless system) or also contain an internal state (dynamic or stateful system). The latter can exhibit arbitrary behavior, while the former are more restricted and allow some general observations. We therefore first focus on static nonlinear systems.

A static nonlinear system can be described by a single function f mapping input to output by

$$y(n) = f\,(x(n))\,. \tag{10.1}$$

Digital Audio Signal Processing, Third Edition. Udo Zölzer.
© 2022 John Wiley & Sons Ltd. Published 2022 by John Wiley & Sons Ltd.

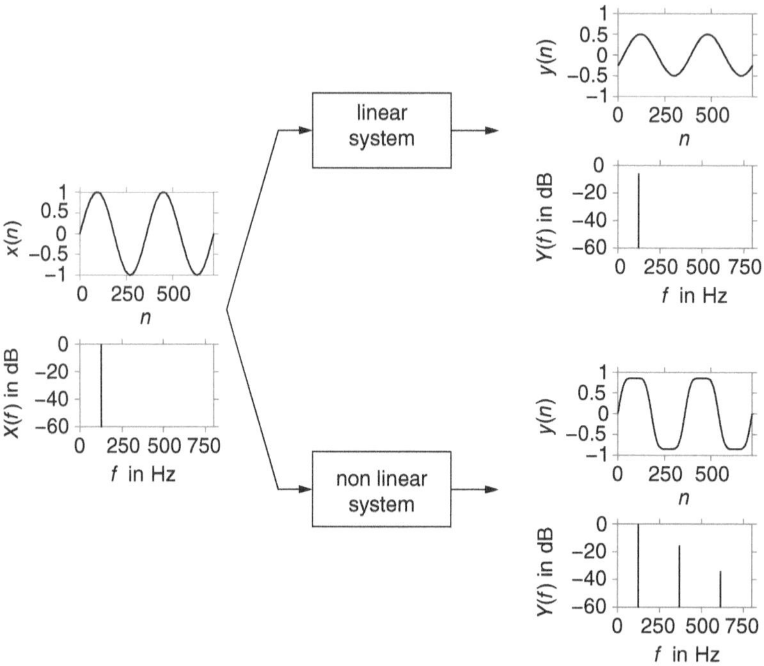

Figure 10.1 Comparison of a linear system and a nonlinear system when excited with a single sinusoid.

Now obviously, if $x(n)$ is periodic with period N_0, then so is $y(n)$. In particular, for a sinusoidal input $x(n) = \sin(\Omega_0 n)$ of frequency $\Omega_0 = 2\pi/N_0$, because the output $y(n)$ has period N_0, from Fourier theory we obtain that it can be decomposed as

$$y(n) = A_0 + A_1 \sin(\Omega_0 n - \varphi_1) + A_2 \sin(2\Omega_0 n - \varphi_2) + A_3 \sin(3\Omega_0 n - \varphi_3) + \cdots , \quad (10.2)$$

i.e. it consists of sinusoids at multiples of the input frequency Ω_0. While the DC component A_0 is usually ignored, the component at Ω_0 is referred to as the fundamental and the remaining components are called its harmonics.

In contexts where a linear system is desired, the effect of a nonlinearity is therefore also called harmonic distortion. It is commonly measured by the total harmonic distortion (THD)

$$\text{THD} = \sqrt{\frac{A_2^2 + A_3^2 + \cdots}{A_1^2 + A_2^2 + A_3^2 + \cdots}}, \quad (10.3)$$

the ratio of the combined power of all harmonics to the total signal power (without the DC component). Usually, the THD is given in dB. While it quantifies the deviation from linearity for systems that should ideally be linear, it is of little use to characterize systems which are intentionally nonlinear, as the relative levels of the different harmonics have a strong influence on the resulting timbre.

Typical audio signals of interest comprise more than a single sinusoid. Single note tones of musical instruments consist of a fundamental and its harmonics themselves. Any new components

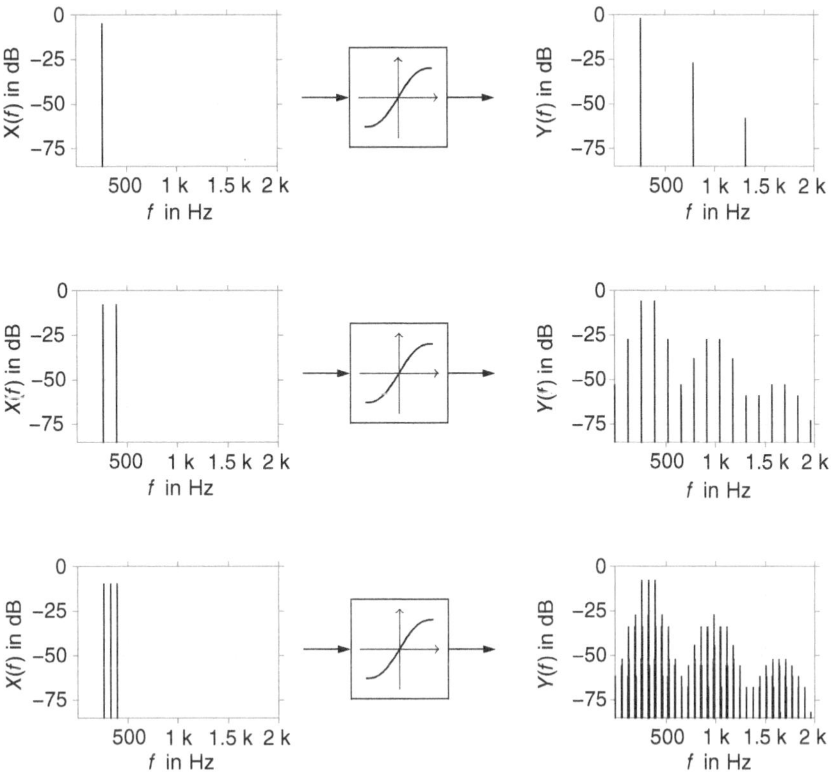

Figure 10.2 Effect of a static nonlinearity on a single sinusoid, two sinusoids, and three sinusoids.

introduced by a static nonlinear system will still be harmonics, so the tonal quality remains, the appearance of new harmonics or amplification of existing ones will only make the sound brighter or even harsher. For input signals containing more than a single sinusoid and its harmonics, however, additional components appear spaced by the differences of the input signal frequencies. An example is given in Fig. 10.2, where the same static nonlinear system is excited with a single sinusoid at 261.63 Hz (C4, top), two sinusoids at 261.63 Hz and 392 Hz (C4 and G4, middle), and 261.63 Hz, 329.63 Hz and 392 Hz (C4, E4, and G4, i.e. Cmajor, bottom). As is apparent, additional components in the input signal quickly yield a very dense output spectrum. For that reason, intentional nonlinear processing is usually limited to single notes or chords with very few notes (e.g. power chords on an electric guitar).

10.2 Overdrive, Distortion, Clipping

Overdrive, distortion, and clipping effects are heavily used in many kinds of musical devices. The terms overdrive and distortion are by no means exactly defined. However, most musicians will agree by defining overdrive as a soft saturating amplification of low-level signals, which results in an operating point in the linear as well as nonlinear regions of the characteristic curve of the overdrive. A distortion effect will mainly be used in its nonlinear region. This leads to

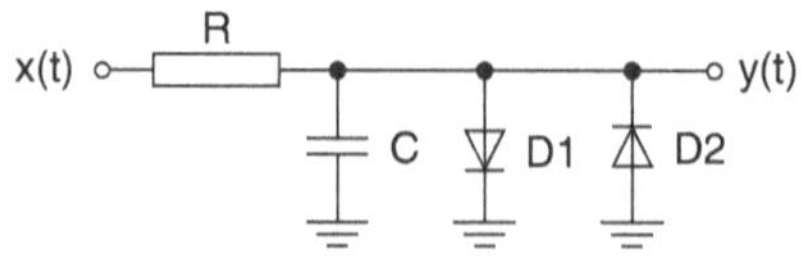

Figure 10.3 First-order diode clipper.

a harsher sound with a lot of harmonic frequency content. To design and analyze such effects, an understanding of the underlying nonlinear processing is crucial. A very basic clipping circuit can be obtained by connecting antiparallel diodes to the signal path. Figure 10.3 illustrates a soft-clipping lowpass filter using only two additional diodes.

By disregarding the diodes, the transfer function of the linear lowpass filter can simply be found to be

$$H(s) = \frac{1}{1 + RCs} \tag{10.4}$$

with the corresponding differential equation

$$\dot{y}(t) = \frac{1}{RC}\left(x(t) - y(t)\right). \tag{10.5}$$

However, including the nonlinearity, which is introduced by the diodes, to the transfer function is not straightforward. The nonlinear relation of the voltage over and the current through the diode can be obtained using Shockley's law:

$$I_d = I_s\left(e^{\frac{V_d}{\eta v_t}} - 1\right), \tag{10.6}$$

with I_d as the current through, V_d the voltage over the diode, I_s as the reverse saturation current, v_t as thermal voltage, and η as quality factor. Applying Kirchoff's voltage and current laws to the circuit from Fig. 10.3 and assuming identical diodes, a first-order nonlinear differential equation,

$$\dot{y}(t) = \frac{1}{RC}\left[x(t) - y(t) - 2RI_s \sinh\left(\frac{y(t)}{v_t}\right)\right], \tag{10.7}$$

can be derived.

Although this equation describes the system in its entirety, implementing it into a digital processing unit cannot be done easily, because the nonlinear differential equation needs to be solved numerically for each and every time step. This computational effort even increases for more complicated systems. Hence, the question arises whether the system can be somehow separated into a linear and a nonlinear part. Looking at the system from Fig. 10.3, the most intuitive idea is to apply first a linear lowpass filter and afterwards a nonlinear mapping function. Figure 10.4 shows such a system. The discrete input signal $x(n)$ is filtered by a linear lowpass filter and afterwards

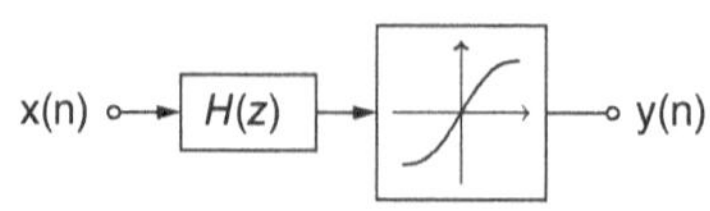

Figure 10.4 Combination of linear filtering and nonlinear static mapping.

fed into the nonlinearity. Note that this separation into a linear stateful filter and a static nonlinear mapping function does not perfectly recreate the dynamic behavior of nonlinear stateful filters such as in Eq. (10.7). However, many nonlinear systems can be sufficiently approximated by approaches based on Fig. 10.4.

As shown for the simple example in Fig. 10.3, nonlinear processing of such dynamic systems is very tedious. Hence, we will focus on the use of static nonlinear mapping functions to create overdrive and distortion effects. A simple soft clipping characteristic curve [Sche80] is given by the equation

$$f(x) = \begin{cases} 2x & \text{for } 0 \leq x \leq 1/3, \\ \frac{3-(2-3x)^2}{3} & \text{for } 1/3 \leq x \leq 2/3, \\ 1 & \text{for } 2/3 \leq x \leq 1. \end{cases} \tag{10.8}$$

A corresponding hard clipping characteristic curve

$$f(x) = \begin{cases} 2x & \text{for } 0 \leq x \leq 1/3, \\ 1 & \text{for } 2/3 \leq x \leq 1, \end{cases} \tag{10.9}$$

can be obtained by removing the quadratic term from the soft-clipping equation. Figure 10.5 shows both soft- and hard clipping mapping functions corresponding to Eq. (10.8) and Eq. (10.9) in symmetrical application $f(x) = -f(-x)$ for $x < 0$. The output signal for such a system given a sinusoidal input of 1 kHz and an amplitude of 0.8 V is depicted in Fig. 10.6. Applying such symmetrical nonlinear functions to a sinusoidal input signal resolves into an output signal which contains only odd harmonics of the original signal. Furthermore, hard clipping nonlinearities will produce a higher harmonic frequency content compared with the smoother soft-clipping nonlinearity. However, if the characteristic curve is asymmetric, e.g. Eq. (10.8) with $f(x) = x$ for $x < 0$, the output signal will contain both even and odd harmonics. Stateful nonlinear filters might also add non-harmonic frequency content to the signal. This is, for example, the case if looking at self-oscillating systems.

Another nonlinear processing technique similar to clipping is called wavefolding. It also uses a nonlinear static mapping function, but instead of saturating the input signal, some parts of the input signal are folded back onto itself. The name wavefolding has its origin in the early days of electronic sound synthesis. A wavefolder takes a sine or cosine wave as an input and produces an output wave with high harmonic frequency content [Roa79]. The wavefolding process is depicted

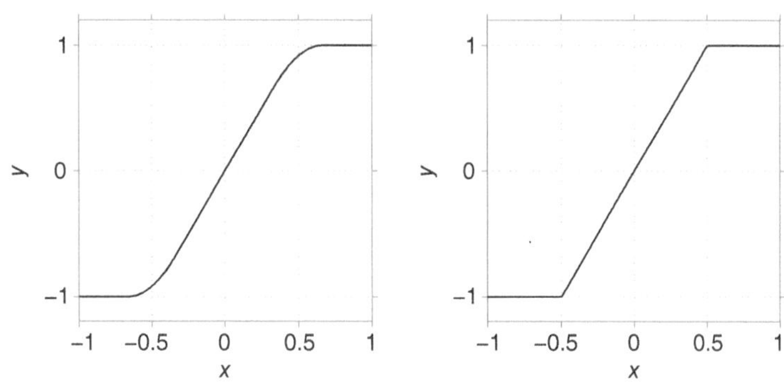

Figure 10.5 Static characteristic curve of a symmetrical soft-clipping (left) and hard clipping (right) nonlinearity.

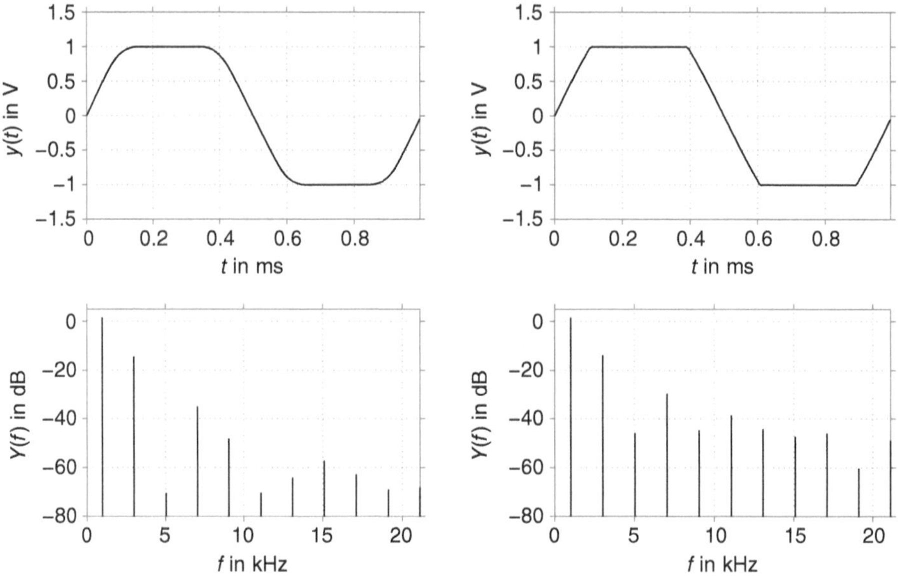

Figure 10.6 Output signal for sinusoidal input with soft-clipping (left) and hard clipping (right) nonlinearity.

in Fig. 10.7. Using the upper linear mapping function, the signal can be inverted. Applying this inversion only to certain parts of the input signal results in an output signal similar to that shown at the bottom of Fig. 10.7. The resulting output signal has a high amount of harmonic frequency content. Because the input signal to such wavefolders is typically a sinusoid with no harmonic frequency content, this technique is often used for additive synthesis.

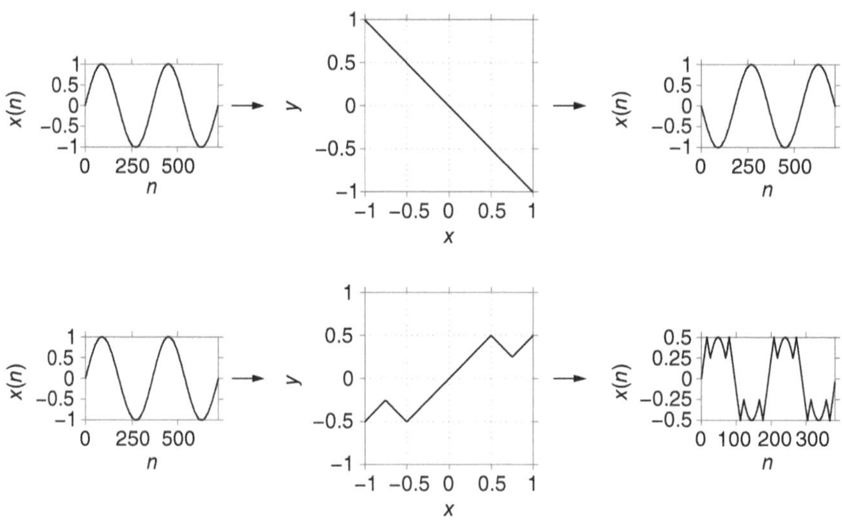

Figure 10.7 Wavefolding using nonlinear static mapping functions.

10.3 Nonlinear Filters

Describing nonlinear systems as a combination of linear filters and static nonlinear mapping functions is sufficient for many applications. However, sometimes it is necessary to implement the nonlinearities directly into the filter, which results in a stateful nonlinear filter. For example, by adding saturating nonlinearities, a digital filter can be constructed that has a more ‘analog’-like sound [Ros92]. The stability of these nonlinear systems cannot be as easily determined as for linear systems. We have to take into account the fact that the pole locations are now dependent on the state of the filter. Therefore, stability analysis will make use of the instantaneous pole locations of the filter. We will investigate a stateful nonlinear filter by looking at a second-order filter [Cho20], as depicted in Fig. 10.8. The difference equation for the output can be directly constructed from the block diagram with

$$y(n) = b_0u(n) + b_1u(n-1) + b_2u(n-2) - a_1y(n-1) - a_2y(n-2). \tag{10.10}$$

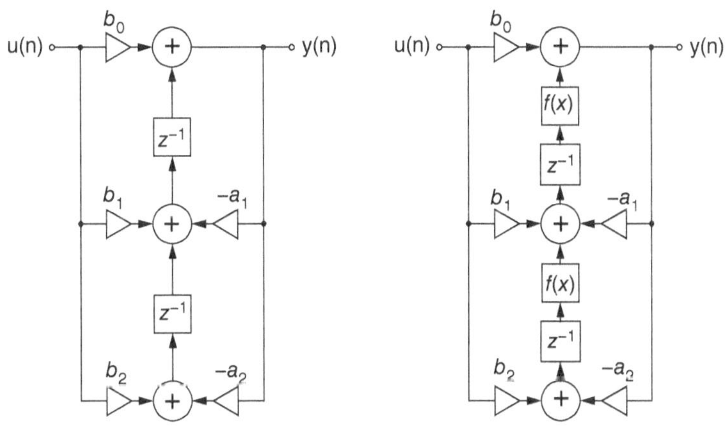

Figure 10.8 Linear (left) and nonlinear (right) second-order filter.

Transforming the difference equation into the frequency domain, the transfer function

$$H(z) = \frac{b_0z^2 + b_1z + b_2}{z^2 + a_1z + a_2} \tag{10.11}$$

is obtained. This leads to the pole locations

$$p_{1,2} = -\frac{a_1}{2} \pm \sqrt{\frac{a_1^2}{4} - a2}. \tag{10.12}$$

For further analysis, it is helpful to bring the system into a state-space representation. Therefore we define the states to be the outputs of the delay blocks. This leads to the discrete-time state-space system:

$$\begin{bmatrix} x_1(n+1) \\ x_2(n+1) \end{bmatrix} = \begin{bmatrix} -a_1 & 1 \\ -a_2 & 0 \end{bmatrix} \begin{bmatrix} x_1(n) \\ x_2(n) \end{bmatrix} + \begin{bmatrix} b_1 - a_1b_0 \\ b_2 - a_2b_0 \end{bmatrix} u(n) \tag{10.13}$$

$$y(n) = \begin{bmatrix} 1 & 0 \end{bmatrix} \begin{bmatrix} x_1(n) \\ x_2(n) \end{bmatrix} + b_0u(n).$$

Note that the stability of this linear system requires all eigenvalues of the system matrix to have a magnitude smaller than one. A nonlinear version of this filter can be achieved by inserting nonlinear mapping functions, as mentioned in Section 10.2, into the filter. The nonlinear blocks are put directly behind the delay blocks. The output equation changes to

$$y(n) = b_0u(n) + f\left(b_1u(n-1) - a_1y(n-1) + f\left(b_2u(n-2) - a_2y(n-2)\right)\right). \tag{10.14}$$

The nonlinear state-space model results in

$$\begin{bmatrix} x_1(n+1) \\ x_2(n+1) \end{bmatrix} = h\left(\begin{bmatrix} x_1(n) \\ x_2(n) \end{bmatrix}\right) + \begin{bmatrix} b_1 - a_1b_0 \\ b_2 - a_2b_0 \end{bmatrix} u(n)$$
$$y(n) = f(x_1(n)) + b_0u(n) \tag{10.15}$$

with

$$h_1\left(x_1(n), x_2(n)\right) = -a_1f\left(x_1(n)\right) + f\left(x_2(n)\right), \tag{10.16}$$
$$h_2\left(x_1(n), x_2(n)\right) = -a_2f\left(x_1(n)\right). \tag{10.17}$$

Because the poles of this nonlinear system are dependent on the state of the system, stability can be analyzed by looking at the instantaneous pole locations of the filter. Stability can be assured if all possible instantaneous pole locations have a magnitude strictly smaller than one. Therefore, the Lyapunov stability [Che04] of the system can be analyzed. Note that the Lyapunov stability is more restrictive than the BIBO stability, because a Lyapunov stable system will only have poles inside the unit circle for any given point in time. A system is considered Lyapunov stable if the eigenvalues for the Jacobian of the discrete-time system matrix have magnitudes strictly smaller than one. The Jacobian of the nonlinear state-space system yields

$$J = \begin{bmatrix} -a_1f'(x_1(n)) & f'(x_2(n)) \\ -a_2f'(x_1(n)) & 0 \end{bmatrix}. \tag{10.18}$$

Consequently, there are two restrictions for the system to be stable. The backward coefficients have the same restriction as for the linear system with $a_1 + a_2 > -1 \cap a_1 - a_2 < 1$ for real poles and $a_1 < \sqrt{2(a_2 + 1)}$ for conjugate complex poles. Furthermore, the first derivative of the nonlinear mapping function must not exceed a value of one at all operating points. Many nonlinear saturating functions fulfill this requirement. For example, a hyperbolic tangent, as in Fig. 10.9, can be used.

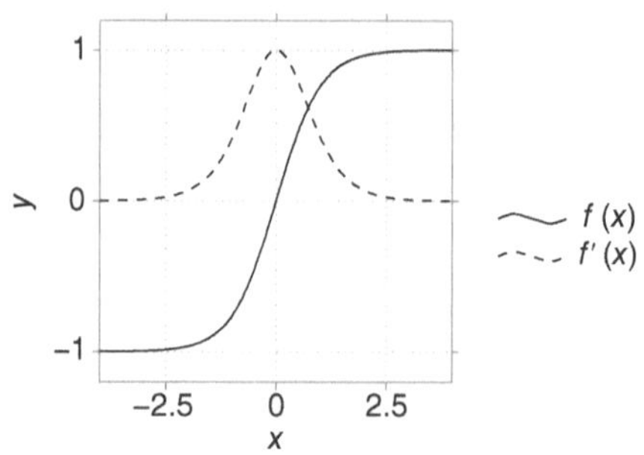

Figure 10.9 Hyperbolic tangent mapping function.

To illustrate the characteristics of such nonlinear filters, we take the second-order filter from Fig. 10.8 as an example. The filter coefficients are set to $a_1 = -1.89$, $a_2 = 0.9$, $b_0 = b_2 = 0.02$, and $b_1 = 0.04$, which results in a lowpass configuration of the filter. The frequency response of the linear filter can be seen in Fig. 10.10. In this example, we will use a linear sinesweep with amplitude 1 and a frequency range from 50 Hz to 4000 Hz as an input signal to the filters. The output signal over time and frequency can be seen in the waterfall representations shown in Fig. 10.11 and Fig. 10.12. The lowpass behavior of the filters can be observed in both figures because the higher frequencies of the sinesweep receive a higher attenuation. The nonlinear filter additionally

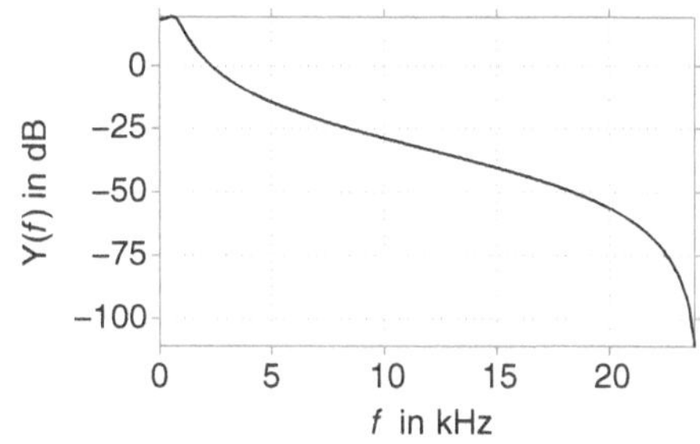

Figure 10.10 Frequency response of linear second-order filter.

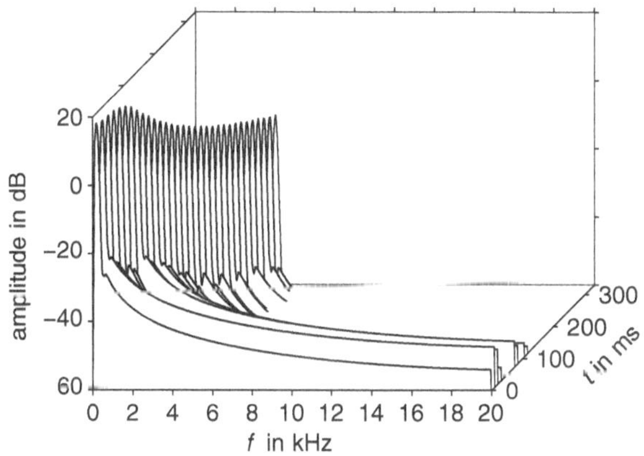

Figure 10.11 Waterfall presentation of a linear filtered sinesweep.

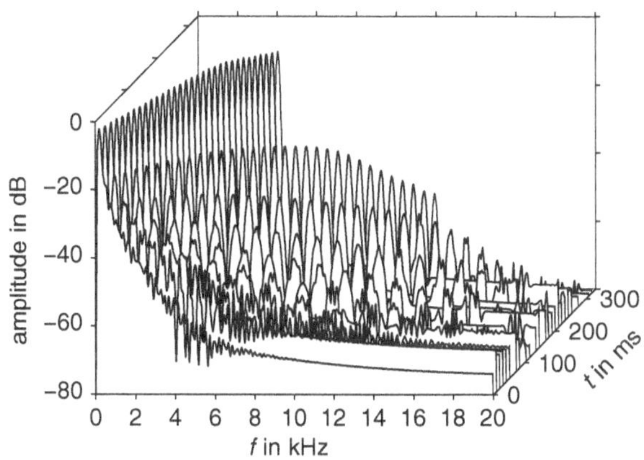

Figure 10.12 Waterfall presentation of a nonlinear filtered sinesweep.

produces filtered harmonics of the input signal. This feature can be used to create sounds with a more analog-like feel like saturated transistor or tube stages.

10.4 Aliasing and its Mitigation

So far, we have ignored one important aspect of nonlinear processing: the newly introduced signal components, especially higher harmonics, may exceed the Nyquist limit at half the sampling frequency. This will result in aliasing distortion, which is typically undesired. We may first observe that when considering a continuous-time input signal $\overline{x}(t)$, the order of sampling and applying a static nonlinear mapping f may be reversed: First applying f gives $\overline{y}(t) = f\left(\overline{x}(t)\right)$, which, after sampling with sampling rate f_s, results in $y(n) = \overline{y}(n/f_\mathrm{s}) = f\left(\overline{x}(n/f_\mathrm{s})\right)$. However, by first sampling, we get $x(n) = \overline{x}(n/f_\mathrm{s})$, which is then mapped to the same $y(n) = f(x(n)) = f\left(\overline{x}(n/f_\mathrm{s})\right)$. This holds true even if $\overline{y}(t)$ is not band limited to $f_\mathrm{s}/2$.

As an extreme case, we consider the mapping function $f(x) = \mathrm{sgn}(x)$, which corresponds to applying an infinite gain and then hard clipping to the range $[-1,1]$. Applied to a sinusoidal input, the output becomes a square wave. Figure 10.13a depicts the corresponding spectrum for an input frequency of 1318.5 Hz (the note E6). As the harmonics only decay slowly with frequency, their level above the Nyquist limit of 22.05 kHz (marked by a vertical line) for the common sampling rate of $f_\mathrm{s} = 44.1$ kHz is still significant. Consequently, the sampled signal of Fig. 10.13b contains many and strong aliased components. These will be audible as both a noise floor and inharmonic tones.

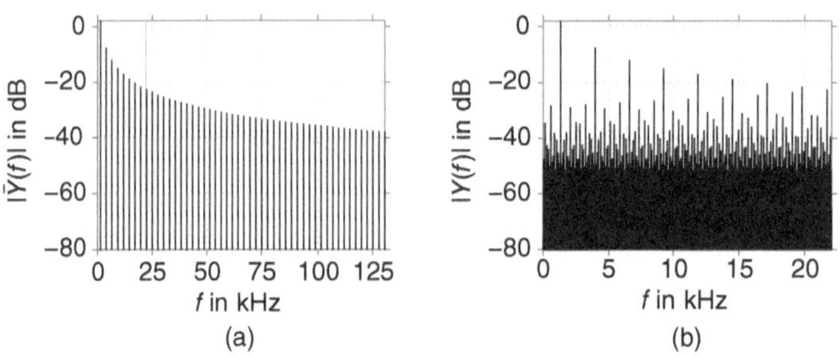

Figure 10.13 Spectra of (a) continuous-time signal $\overline{y}(t) = \mathrm{sgn}(\sin(2\pi \cdot 1318.5\ \mathrm{Hz} \cdot t))$ with a marker at the Nyquist frequency 22.05 kHz and (b) sampled signal $y(n) = \overline{y}(n/44.1\ \mathrm{kHz})$.

The conceptually simplest approach to reducing this aliasing distortion is to increase the sampling rate. Typically, the input signal will be upsampled before and downsampled to the original sampling rate after the nonlinearity, where the resampling includes appropriate interpolation and decimation filters. A corresponding system is shown in Fig. 10.14. The effectiveness of this method depends on how fast the harmonics decay with frequency and obviously the oversampling factor L. For the example above, the spectra obtained by oversampling with different factors L are shown in Fig. 10.15.

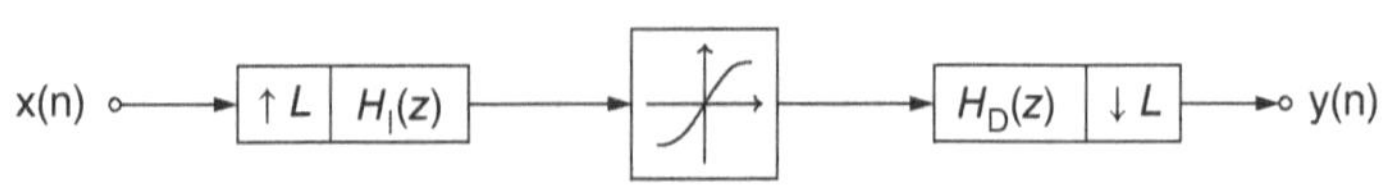

Figure 10.14 Operation of a nonlinear system at a sampling frequency increased by factor L, where $H_\mathrm{I}(z)$ and $H_\mathrm{D}(z)$ denote the interpolation and decimation filter, respectively.

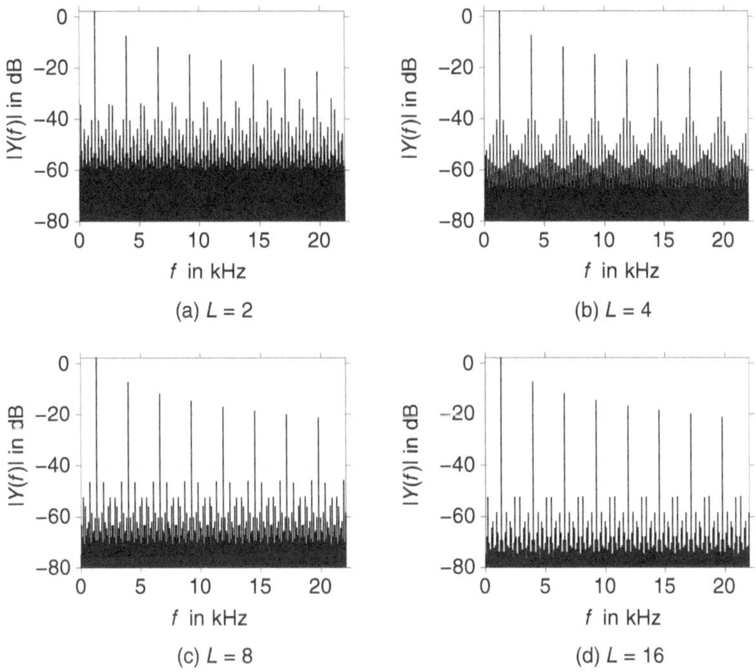

Figure 10.15 Spectra of $y(t) = \text{sgn}\,(\sin(2\pi \cdot 1318.5\text{ Hz} \cdot n/(L \cdot 44.1\text{ kHz})))$ for different values of L.

It is clearly visible that with increasing L, the aliasing distortion gets reduced. For less extreme nonlinear systems, the harmonics typically decay faster and the effect of oversampling will be even more pronounced.

However, owing to increasing computational costs, the oversampling factor usually cannot be made arbitrarily large. Therefore, various other strategies to reduce the aliasing distortion have been developed, e.g. [Esq15, Esq16a, Esq16b, Mul17]. In the case of a memoryless nonlinear system, an attractive approach is based on approximating a continuous-time system [Par16, Bil17a, Bil17b], which will be explained in the following. Perfect aliasing suppression can be obtained by carrying out the processing in the continuous-time domain, i.e. by converting the input signal to its continuous representation, applying the nonlinearity to it, and then sampling it after appropriate lowpass filtering. While theoretically perfect, this is clearly impractical. However, by crude approximation of this process, we can obtain a practical implementation that can considerably lower the aliasing distortion.

First, we replace the exact continuous-time representation of the input signal by a piecewise linear approximation:

$$\tilde{x}(t) = x(n) + (tf_s - n) \cdot (x(n+1) - x(n)) \quad \text{where } n = \lfloor tf_s \rfloor. \tag{10.19}$$

To this we apply the nonlinear mapping to obtain

$$\tilde{y}(t) = f(\tilde{x}(t)). \tag{10.20}$$

Before sampling this, we need to apply a lowpass to suppress frequency content above the Nyquist limit. In the simplest case, we average over one sampling interval (corresponding to convolution with a rect), i.e.

$$y(n) = f_s \int_{(n-1)/f_s}^{n/f_s} \tilde{y}(t)\mathrm{d}t = f_s \int_{(n-1)/f_s}^{n/f_s} f(\tilde{x}(t))\mathrm{d}t. \tag{10.21}$$

Now we apply integration by substitution and rewrite as

$$y(n) = f_s \int_{(n-1)/f_s}^{n/f_s} f(\tilde{x}(t))\mathrm{d}t = f_s \int_{x(n-1)}^{x(n)} f(\tilde{x})\frac{\mathrm{d}t}{\mathrm{d}\tilde{x}}\mathrm{d}\tilde{x} \tag{10.22}$$

and exploit the fact that, thanks to the piecewise linear approximation,

$$\frac{\mathrm{d}t}{\mathrm{d}\tilde{x}} = \left(\frac{\mathrm{d}\tilde{x}}{\mathrm{d}t}\right)^{-1} = \frac{1}{f_s \cdot (x(n) - x(n-1))} \tag{10.23}$$

is easy to compute. We thus obtain

$$y(n) = \frac{1}{x(n) - x(n-1)} \int_{x(n-1)}^{x(n)} f(\tilde{x})\mathrm{d}\tilde{x}. \tag{10.24}$$

Finally, by the fundamental theorem of calculus, we may rewrite to

$$y(n) = \frac{F(x(n)) - F(x(n-1))}{x(n) - x(n-1)}, \tag{10.25}$$

where F denotes the antiderivative of f. The method is therefore also referred to as antiderivative antialiasing. If the antiderivative cannot be derived in closed form, it can be precomputed numerically and tabulated.

Assuming evaluation of F to be approximately as expensive as evaluation of f and when memorizing $F(x(n))$ to be used as $F(x(n-1))$ in the next time step, the antialiased system only requires two additional subtractions and one additional division compared with the non-antialiased system. There is one minor complication though, namely, when $x(n) \approx x(n-1)$, the denominator of Eq. (10.25) becomes nearly (or even exactly) zero, which will result in numerical problems. In the limit $x(n) \to x(n-1)$, Eq. (10.25) reduces to $f(x(n))$ or equally $f(x(n-1))$. Thus, one could use either of those in case $x(n) \approx x(n-1)$, but as will be explained momentarily, their mean $\frac{1}{2}(f(x(n)) + f(x(n-1)))$ is the most consistent choice. So to summarize, antiderivative antialiasing is given by

$$y(n) = \begin{cases} \frac{1}{2}(f(x(n)) + f(x(n-1))) & \text{if } x(n) \approx x(n-1), \\ \frac{F(x(n))-F(x(n-1))}{x(n)-x(n-1)} & \text{otherwise.} \end{cases} \tag{10.26}$$

While the approach thus derived is attractive owing to its simplicity, the interpolation and decimation filters are far from the ideal brick-wall filters. This will lead to both an unwanted lowpass filtering of the desired signal components and an imperfect suppression of the image spectra. To first analyze the unwanted lowpass filtering, we consider the effect of applying the antiderivative antialiasing to a linear system, choosing in particular $f(x) = x$. With $F(x) = \frac{1}{2}x^2$, we find

$$\frac{F(x(n)) - F(x(n-1))}{x(n) - x(n-1)} = \frac{\frac{1}{2}(x(n))^2 - \frac{1}{2}(x(n-1))^2}{x(n) - x(n-1)}$$
$$= \frac{\frac{1}{2}(x(n) - x(n-1))(x(n) + x(n-1))}{x(n) - x(n-1)} = \frac{1}{2}(x(n) + x(n-1)), \quad (10.27)$$

which equals our choice for the $x(n) \approx x(n-1)$ case, thereby justifying it. We notice that the antialiasing introduces a half-sample delay and the expected lowpass filtering. Concerning the imperfect suppression of the image spectra, note that the frequency response of the rect filter has zeros at the multiples of the sampling rate f_s, i.e. those components that would be aliased to DC are suppressed perfectly. Components that are aliased to low frequencies still see very high suppression. However, components just above the Nyquist limit are only attenuated by a meager 3 dB, leaving strong aliased components at high frequencies.

These effects can be seen in Fig. 10.16a, where the same example of $f(x) = \text{sgn}(x)$ excited with a sine at 1318.5 Hz as before is now subject to antiderivative antialiasing. Compared with Fig. 10.13b, one clearly sees that the aliasing distortion at low frequencies is greatly reduced, while the strongest components at high frequencies see almost no reduction. At the same time, the desired signal components undergo a slight lowpass filtering, as can be seen from the comparison to their desired levels marked by crosses.

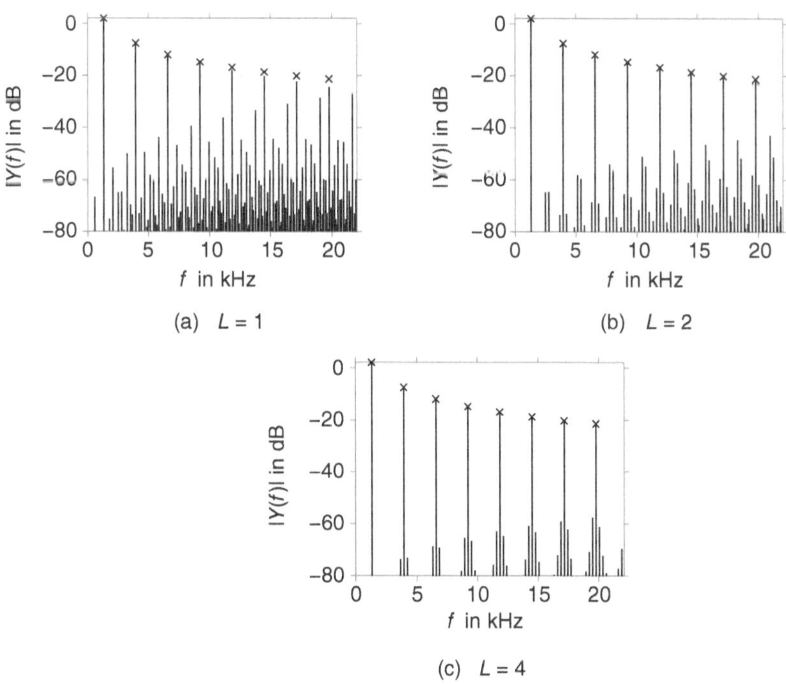

Figure 10.16 Spectra of output obtained with antiderivative antialiasing for different values of L, where crosses mark the desired harmonics.

As the antiderivative antialiasing works especially well at low frequencies, it is beneficial to combine it with oversampling to effectively broaden the frequency range of satisfactory aliasing suppression in addition to the aliasing reduction by the oversampling itself. The results obtained from two- and four-times oversampling are shown in Fig. 10.16b and Fig. 10.16c. Compared with the case of oversampling alone (Fig. 10.15a and Fig. 10.15b), the effectiveness of antiderivative antialiasing is obvious.

While the piecewise linear interpolation is required to make the integration by substitution feasible in Eq. (10.22), alternatives for the decimation filter are possible. In [Par16], a tri filter is explored, while [Bil17a, Bil17b] view Eq. (10.25) as a discrete-time approximation of differentiating the antiderivative $F(x)$ and explore using higher-order antiderivatives and different differentiation schemes. This allows to strike different balances between computational complexity, aliasing suppression, and alteration of the desired signal components.

Unfortunately, antiderivative antialiasing is only applicable to memoryless nonlinear systems. In [Hol20], an extension to a class of stateful systems is proposed. However, it is limited to systems which can be written in a particular way where all nonlinear functions only depend on a single scalar input. An alternative approach is discussed in [Mul17], where the system output is calculated not only based on the system state, but also its derviative, which enables an additional smoothing that can reduce aliasing. In particular, it can achieve high suppression of the harmonics around the sampling rate but has little effect on even higher harmonics. It is therefore mainly effective for nonlinear systems with relatively quickly decaying harmonics.

10.5 Virtual Analog Modeling

Constructing nonlinear systems directly in the digital domain is a sound and flexible method. However, oftentimes there is a need to create a digital model taken from existing analog circuits. The field of virtual analog modeling provides many different approaches and techniques for creating suitable digital models from analog reference circuits. There are a variety of different approaches, which all can be categorized into three different model types: blackbox models, graybox models, and whitebox models. A blackbox model is based solely on the input and output data of the analog circuit. In whitebox modeling approaches, all information regarding the analog circuit is known and can be used. This includes input and output data as well as voltage and current relations on each and every point in the circuit. A graybox model is something in between these two by having access to more information than just input and output data, but limited nonetheless. In this chapter, we will give a brief introduction of the most commonly used whitebox modeling approaches i.e. wave digital filters and state-space modeling. Although we will not give a detailed analysis of the graybox and blackbox modeling approaches, we will provide a brief reference to some of the most used techniques. A commonly applied graybox modeling approach is the use of Wiener–Hammerstein models. These models divide the overall model into several linear and nonlinear parts. This approach is similar to that shown in Fig. 10.4, where we separated a nonlinear filter into a linear filter with a subsequent static nonlinearity. This and similar graybox modeling approaches are applied successfully in [Kem08, Eic16, Fra13]. In the domain of blackbox modeling, only input and output data are available for the model construction. To model such kinds of time-series data, certain artificial neural network structures can be used. The most common one is the recurrent neural network, which is fruitfully used in [Wri19]. In [Par19], a neural network is combined with a whitebox state-space modeling approach.

10.5.1 Wave Digital Filters

Wave digital filters can be used to transform an analog circuit from the Kirchhoff domain to the so-called wave domain. This representation allows efficient and precise modeling of linear as well as nonlinear analog circuits. A circuit element is described in the wave domain by an incident and a reflected wave with a corresponding port resistance. A port is characterized by its port voltage v_0 and its port current i_0. The corresponding wave variables are constructed with a linear combination of port voltage and current with the port resistance as a parameter. The incident and reflected wave are defined as [Fet86]

$$\begin{aligned} a_0 &= v_0 + R_0 i_0, \\ b_0 &= v_0 - R_0 i_0. \end{aligned} \tag{10.28}$$

This can be written into matrix notation as

$$\begin{bmatrix} a_0 \\ b_0 \end{bmatrix} = \begin{bmatrix} 1 & R_0 \\ 1 & -R_0 \end{bmatrix} \begin{bmatrix} v_0 \\ i_0 \end{bmatrix}. \tag{10.29}$$

Solving now Eq. (10.29) for v_0 and i_0, the port voltages and currents can respectively be obtained from the wave variables with

$$v_0 = \frac{1}{2} a_0 + \frac{1}{2} b_0, \tag{10.30a}$$

$$i_0 = \frac{1}{2R_0} a_0 - \frac{1}{2R_0} b_0. \tag{10.30b}$$

With these definitions, circuit elements can now be transformed from Kirchhoff to wave domain. Rather than covering all possible circuit elements in the wave domain, we restrict ourselves to the most common ones to give a brief overview and a basic understanding. We start with the simple example of transforming a resistor into the wave domain. A resistor is described in Kirchhoff domain by Ohm's law.

$$v_0 = R \cdot i_0. \tag{10.31}$$

Combining this with Eq. (10.30a) and Eq. (10.30b) and solving for the reflected wave b_0 yields

$$b_0 = \frac{R - R_0}{R + R_0} a_0. \tag{10.32}$$

This is the so-called unadapted form of a resistor in the wave domain. Most elements can be adapted by parameterizing the port resistance R_0 to a suitable value. In this case, we can set $R_0 = R$ and we obtain $b_0 = 0$ as the adapted form of a resistor in the wave domain [Wer16].

Similar to this derivation, we can obtain the wave-domain representation of a capacitor. Starting with the differential equation

$$i_0 = C \frac{dv_0}{dt}, \tag{10.33}$$

we can use Eqs. (10.30a) and (10.30b) to obtain a differential equation depending on the wave variables with

$$\frac{d}{dt} \left\{ a_0(t) + b_0(t) \right\} = \frac{1}{R_0 C} \left(a_0(t) - b_0(t) \right). \tag{10.34}$$

Going further, the continuous time differential equation needs to be discretized. Many discretization schemes might by used here, all with certain advantages and drawbacks. Here we choose one of the most common ones with the trapezoidal rule:

$$y(n) = y(n-1) + \frac{T}{2}\left(f(t_n, y_n) + f(t_{n-1}, y_{n-1})\right), \tag{10.35}$$

where $y' = f(t, y)$ is the corresponding differential equation and T the sampling interval. After discretization with the trapezoidal rule, the difference equation for the reflected wave in the time domain yields [Wer16]

$$b_0(n) = -\frac{T - 2R_0C}{T + 2R_0C} b_0(n-1) + \frac{T - 2R_0C}{T + 2R_0C} a_0(n) + a_0(n-1). \tag{10.36}$$

This unadapted form can be adapted as well by setting the port resistance to $R_0 = \frac{T}{2C}$, which results in the adapted wave-domain representation

$$b_0(n) = a_0(n-1). \tag{10.37}$$

Note that instead of applying the trapezoidal rule in the time domain, we could also transform the differential equation to the frequency domain and use the bilinear transform for discretization, which will ultimately lead to the same result.

Other algebraic or reactive elements can be derived similarly. Table 10.1 comprises the wave-domain representation of the most common linear circuit elements.

Table 10.1 Wave domain representation of common linear circuit elements

Element	Port resistance	Wave equation
R 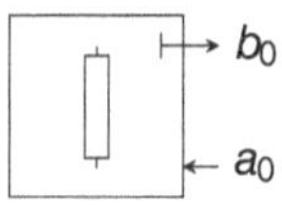	$R_0 = R$	$b_0 = 0$
C 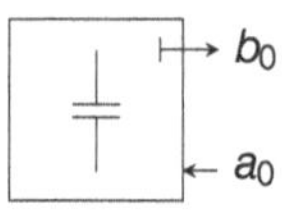	$R_0 = \frac{T}{2C}$	$b_0(n) = a_0(n-1)$
L 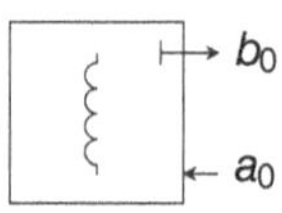	$R_0 = \frac{2L}{T}$	$b_0(n) = -a_0(n-1)$
e 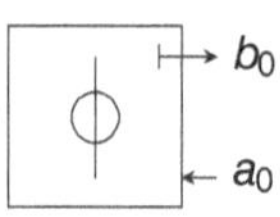	not adaptable	$b_0(n) = 2e(n) - a_0(n)$
j 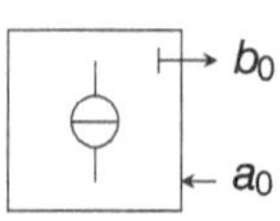	not adaptable	$b_0(n) = 2R_0 j(n) - a_0(n)$

As an example for a nonlinear circuit element, we will derive a wave-domain representation of a simple diode. An ideal diode can be described by Shockley's law given in Eq. (10.6). For the sake of simplicity, we will set the ideality factor to $\eta = 1$. Transforming Shockley's law into the wave domain, we obtain

$$\frac{1}{2R_0}a_0 - \frac{1}{2R_0}b_0 = I_s\left(e^{(a_0+b_0)/2v_t} - 1\right). \tag{10.38}$$

For use in a wave digital filter, the wave equation should be given in explicit form. However, owing to the exponential, Eq. (10.38) cannot be simply brought into an explicit formulation. Luckily we can bring Eq. (10.38) into a form which can be explicitly solved for the reflected wave b_0 if we make use of the Lambert w function $\omega(x)$. The resulting wave equation yields

$$b_0 = a_0 + 2R_0 I_s - 2v_t\omega\left\{\frac{R_0 I_s}{v_t} e^{\frac{a_0+R_0 I_s}{v_t}}\right\} \tag{10.39}$$

with ω as the Lambert w function[Wer16].

Another important element of wave digital filters is the adaptor. Adaptors are used to connect wave digital filter elements with each other. This can be done in a series or a parallel connection. In a two-port parallel connection, the voltage and current relations can be easily obtained with

$$v_0 = v_1, \tag{10.40a}$$

$$i_0 = -i_1. \tag{10.40b}$$

Inserting again the definition of the wave variables, we can construct an unadapted wave equation for a two-port parallel adaptor

$$\begin{bmatrix} b_0 \\ b_1 \end{bmatrix} = \begin{bmatrix} -\frac{R_0-R_1}{R_0+R_1} & \frac{2R_0}{R_0+R_1} \\ \frac{2R_1}{R_0+R_1} & \frac{R_0-R_1}{R_0+R_1} \end{bmatrix} \begin{bmatrix} a_0 \\ a_1 \end{bmatrix}. \tag{10.41}$$

Choosing $R_0 = R_1$ for both port resistances, an adapted form can be found with

$$\begin{bmatrix} b_0 \\ b_1 \end{bmatrix} = \begin{bmatrix} 0 & 1 \\ 1 & 0 \end{bmatrix} \begin{bmatrix} a_0 \\ a_1 \end{bmatrix}. \tag{10.42}$$

The wave equation for a two-port series adaptor can be obtained similarly by using $v_0 = -v_1$ and $i_0 = i_1$ as voltage and current relations.

For many applications, adaptors will be needed which have more than two ports. Most commonly used are three-port adaptors. Furthermore, three-port adaptors can also be used as building blocks for N-port adaptors [Wer16]. Consequently, we will only deal with the derivation of three-port series or parallel adaptors.

The voltage and current relations of a three-port parallel adaptor are given by

$$v_0 = v_1 = v_2, \tag{10.43a}$$

$$i_0 + i_1 + i_2 = 0. \tag{10.43b}$$

This can be brought similarly to the two-port adaptor in an adapted wave equation

$$\begin{bmatrix} b_0 \\ b_1 \\ b_2 \end{bmatrix} = \begin{pmatrix} 0 & \frac{R_2}{R1+R_2} & \frac{R_1}{R_1+R_2} \\ 1 & -\frac{R_1}{R1+R_2} & \frac{R_1}{R_1+R2} \\ 1 & \frac{R_2}{R_1+R_2} & -\frac{R_2}{R_1+R_2} \end{pmatrix} \begin{bmatrix} a_0 \\ a_1 \\ a_2 \end{bmatrix} \tag{10.44}$$

Table 10.2 Parallel and series adaptors

Element	Port resistance	Wave equation
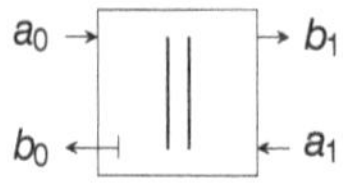	$R_0 = R_1$	$\begin{bmatrix} b_0 \\ b_1 \end{bmatrix} = \begin{bmatrix} 0 & 1 \\ 1 & 0 \end{bmatrix} \begin{bmatrix} a_0 \\ a_1 \end{bmatrix}$
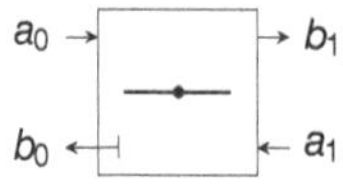	$R_0 = R_1$	$\begin{bmatrix} b_0 \\ b_1 \end{bmatrix} = \begin{bmatrix} 0 & -1 \\ -1 & 0 \end{bmatrix} \begin{bmatrix} a_0 \\ a_1 \end{bmatrix}$
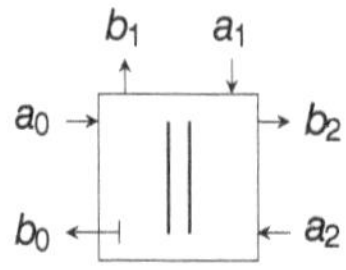	$R_0 = \frac{R_1 R_2}{R_1+R_2}$	$\begin{bmatrix} b_0 \\ b_1 \\ b_2 \end{bmatrix} = \begin{bmatrix} 0 & \frac{R_2}{R1+R_2} & \frac{R_1}{R_1+R_2} \\ 1 & -\frac{R_1}{R1+R_2} & \frac{R_1}{R_1+R2} \\ 1 & \frac{R_2}{R_1+R_2} & -\frac{R_2}{R_1+R_2} \end{bmatrix} \begin{bmatrix} a_0 \\ a_1 \\ a_2 \end{bmatrix}$
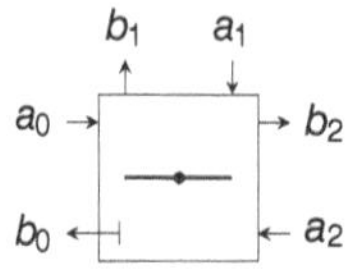	$R_0 = R_1 + R_2$	$\begin{bmatrix} b_0 \\ b_1 \\ b_2 \end{bmatrix} = \begin{bmatrix} 0 & -1 & -1 \\ -R_1 & \frac{R_2}{R_1+R_2} & \frac{R_1}{R_1+R2} \\ -R_2 & -\frac{R_2}{R_1+R_2} & \frac{R_1}{R_1+R_2} \end{bmatrix} \begin{bmatrix} a_0 \\ a_1 \\ a_2 \end{bmatrix}$

with the adapted port resistance $R_0 = \frac{R_1 R_2}{R_1+R_2}$. All adapted wave equations for two-port as well as for three-port series and parallel adaptors can be found in Table 10.2.

We will conclude this section with a simple example of how to construct a wave digital filter model from a given circuit schematic. An easy nonlinear circuit can be seen in Fig. 10.17. It comprises an asymmetrical second-order diode clipper. The wave digital filter structure of this circuit can be seen on the left-hand side of Fig. 10.18. Constructing such a wave digital filter structure underlies certain restrictions and building rules to make the filter realizable. These restrictions can be best understood if we introduce the concept of connection tree structures for wave digital filters. The structure of Fig. 10.18 can be transformed into a tree-based topology. This structure comprises several elements, namely the root, the leaves, and adaptors. The corresponding connection tree of the wave digital filter can be seen on the right-hand side of Fig. 10.18. The root of a wave digital filter has no upward facing connections to other elements, an adaptor has one upward facing connection and one or more downward faced ports, and a leaf is an element containing only one upwards facing connection. A non-adaptable element, such like a voltage source, should always be the root of the connection tree. It is not allowed to be placed as a leaf or adaptor. This can lead to complications if the analog circuit contains more than one non-adaptable element. For voltage and current sources, this can be solved by combining the source with its

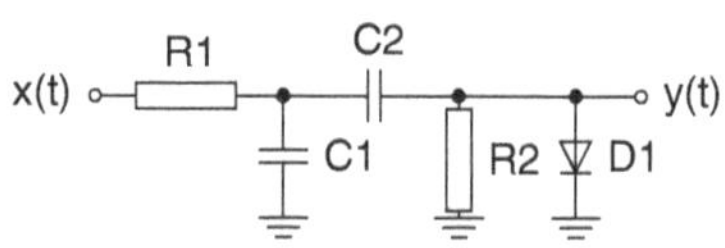

Figure 10.17 Second-order diode clipper.

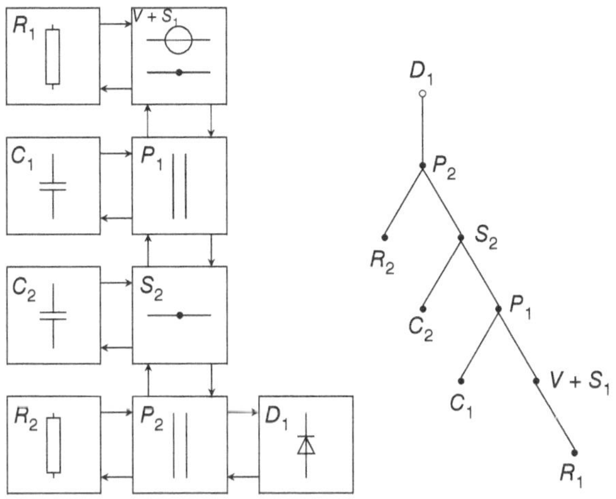

Figure 10.18 Wave digital filter structure of second-order diode clipper with corresponding connection tree.

adjacent adaptor. The voltage source can be absorbed into a series connection resulting into a two-port block with the wave equation

$$\begin{bmatrix} b_0 \\ b_1 \end{bmatrix} = \begin{bmatrix} -\frac{R_0-R_1}{R_0+R_1} & -\frac{2R_0}{R_0+R_1} \\ -\frac{2R_1}{R_0+R_1} & \frac{R_0-R_1}{R_0+R_1} \end{bmatrix} \begin{bmatrix} a_0 \\ b_0 \end{bmatrix} + \begin{bmatrix} -2\frac{R_0}{R_0+R_1} \\ -2\frac{R_1}{R_0+R_1} \end{bmatrix} e, \tag{10.45}$$

where e is the source voltage. This wave equation can be adapted by setting $R_0 = R_1$ yielding the adapted wave equation

$$\begin{bmatrix} b_0 \\ b_1 \end{bmatrix} = \begin{bmatrix} 0 & -1 \\ -1 & 0 \end{bmatrix} \begin{bmatrix} a_0 \\ a_1 \end{bmatrix} - \begin{bmatrix} 1 \\ 1 \end{bmatrix} e, \tag{10.46}$$

which can now be used as an adaptor in the connection tree. Going further, the nonlinear elements are also restricting the construction of the wave digital filter. Note that we restricted ourselves in this example to a circuit with only one nonlinear element. The reason for this is that a nonlinear element should always be placed at the root of the wave digital filter. Constructing a filter with more nonlinear elements runs into several problems, whose solutions are a major research area in wave digital filter design. One approach of dealing with multiple nonlinear or non-adaptable elements is the use of so-called R-type adaptors, which have successfully been implemented to model these kinds of circuits [Wer16]. For the sake of simplicity, we will stick with our simple example in Fig. 10.17. One major advantage of using binary connection trees, like that from Fig. 10.18, is that there is no possibility the graph can have a delay free loop. This property directly assures realizable wave digital filter structures. From the tree structure in Fig. 10.18, a realizable signal flow graph can be derived, which can finally be used to compute the output signal of the wave digital filter.

10.5.2 State-space Approaches

One of the most used approaches in the domain of virtual analog modeling is to derive a non-linear state-space model from any given circuit schematic. This can be done very systematically by using, for example, the nodal DK method proposed by David Yeh in [Yeh10]. The resulting nonlinear state-space model has the form

$$x(n) = Ax(n-1) + Bu(n) + Ci(n), \tag{10.47a}$$

$$y(n) = Dx(n-1) + Eu(n) + Fi(n), \tag{10.47b}$$

$$v(n) = Gx(n-1) + Hu(n) + Ki(n), \tag{10.47c}$$

$$i(n) = f(v(n)), \tag{10.47d}$$

with $u(n)$ as input vector, $y(n)$ as output vector, and $v(n)$ and $i(n)$ as the voltage over and current through the nonlinear circuit elements. The matrices A, B, C, D, E, F, G, H, K describe the dynamics of the system and f is a nonlinear function containing all voltage–current relations of the nonlinear elements. In comparison to the nodal-DK method, we will derive the state-space model with an approach introduced in [Hol15]. We will start with the description of individual circuit elements. Any circuit element will be described with the equation

$$M_{v,e}v_e + M_{i,e}i_e + M_{x,e}x_e + M_{\dot{x},e}\dot{x}_e + M_{q,e}q_e = u_e \tag{10.48a}$$

$$f_e(q_e) = 0, \tag{10.48b}$$

where v_e, i_e are the port voltages and currents, x_e, $\dot{x}_e$ are the element's state and state derivative vectors, respectively, q_e is the auxiliary vector, u_e the source vector, and f_e the element's nonlinear voltage–current relationship. The coefficient matrices of common circuit elements are given in Table 10.3.

A description of the whole circuit can be now achieved by combining all individual coefficient matrices, nonlinear functions as well as voltage, current, state, state derivative, and source vectors into one system. The vectors are simply stacked like $v = \begin{pmatrix} v_{e,1}^T & v_{e,1}^T & \dots & v_{e,N}^T \end{pmatrix}^T$. The coefficient matrices are combined into one block diagonal matrix with the form

$$M_v = \begin{pmatrix} M_{v,e,1} & 0 & \dots & 0 \\ 0 & M_{v,e,2} & \dots & 0 \\ \vdots & \vdots & \ddots & \vdots \\ 0 & 0 & \dots & M_{v,e,N} \end{pmatrix}. \tag{10.49}$$

Table 10.3 Coefficient matrices and nonlinear functions of common circuit elements

Element	$M_{v,e}$	$M_{i,e}$	$M_{x,e}$	$M_{\dot{x},e}$	$M_{q,e}$	u_e	$f_e(q_e)$
Voltage source v_s	(1)	(0)	$()$	$()$	$()$	(v_s)	
Current source i_s	(0)	(-1)	$()$	$()$	$()$	(i_s)	
Resistor R	(-1)	(R)	$()$	$()$	$()$	(0)	
Capacitor C	$\begin{pmatrix} C \\ 0 \end{pmatrix}$	$\begin{pmatrix} 0 \\ 1 \end{pmatrix}$	$\begin{pmatrix} -1 \\ 0 \end{pmatrix}$	$\begin{pmatrix} 0 \\ -1 \end{pmatrix}$	$()$	$\begin{pmatrix} 0 \\ 0 \end{pmatrix}$	
Inductor L	$\begin{pmatrix} 1 \\ 0 \end{pmatrix}$	$\begin{pmatrix} 0 \\ L \end{pmatrix}$	$\begin{pmatrix} 0 \\ -1 \end{pmatrix}$	$\begin{pmatrix} -1 \\ 0 \end{pmatrix}$	$()$	$\begin{pmatrix} 0 \\ 0 \end{pmatrix}$	
Diode D	$\begin{pmatrix} 1 \\ 0 \end{pmatrix}$	$\begin{pmatrix} 0 \\ 1 \end{pmatrix}$	$()$	$()$	$\begin{pmatrix} -1 & 0 \\ 0 & -1 \end{pmatrix}$	$\begin{pmatrix} 0 \\ 0 \end{pmatrix}$	$\left(I_s \cdot \left(e^{q_{e,1}/v_t} - 1\right) - q_{e,2}\right)$

The nonlinear functions are collected in the vector

$$f\begin{pmatrix} q_{e,1} \\ q_{e,2} \\ \vdots \\ q_{e,N} \end{pmatrix} = \begin{pmatrix} f_{e,1}(q_{e,1}) \\ f_{e,2}(q_{e,2}) \\ \vdots \\ f_{e,N}(q_{e,N}) \end{pmatrix}. \tag{10.50}$$

Going further, all constraints introduced by the circuit's elements can be described by

$$M_v v + M_i i + M_x x + M_{\dot{x}} \dot{x} + M_q q = u, \tag{10.51a}$$

$$f(q) = 0. \tag{10.51b}$$

Applying now Kirchhoff's voltage and current laws, the circuit topology can be incorporated with the additional equations $T_v v = 0$ and $T_i i = 0$, where the matrices T_v and T_i are derived using standard network analysis techniques. This leads to the nonlinear differential equation system:

$$\begin{pmatrix} M_v & M_i & M_x & M_{\dot{x}} & M_q \\ T_v & 0 & 0 & 0 & 0 \\ 0 & T_i & 0 & 0 & 0 \end{pmatrix} \begin{pmatrix} v \\ i \\ x \\ \dot{x} \\ q \end{pmatrix} = \begin{pmatrix} u \\ 0 \\ 0 \end{pmatrix} \tag{10.52a}$$

$$f(q) = 0. \tag{10.52b}$$

From this point, it is possible to derive a continuous-time state-space model, and with subsequent discretization, a discrete-time model. However, the equation system can also be directly discretized. For that, we decide to use the trapezoidal integration rule for time discretization:

$$\hat{x}(n) = \hat{x}(n-1) + \frac{T}{2}\left(\hat{\dot{x}}(n) + \hat{\dot{x}}(n-1)\right), \tag{10.53}$$

where T is the sampling interval, $\dot{x}$ the discrete approximation of the state at time nT, and $\hat{\dot{x}}$ the exact solution for $\dot{x}$. With the introduction of canonical states,

$$\overline{x}(n) = \hat{x}(n) + \frac{T}{2}\hat{\dot{x}}(n), \tag{10.54}$$

we can make use of the substitutions

$$\hat{\dot{x}}(n) = \frac{1}{T}\left(\overline{x}(n) - \overline{x}(n-1)\right), \tag{10.55a}$$

$$\hat{x}(n) = \frac{1}{2}\left(\overline{x}(n) + \overline{x}(n-1)\right), \tag{10.55b}$$

and the definitions $\overline{M}_{x'} = \frac{1}{T}M_{\dot{x}} + \frac{1}{2}M_x$ and $\overline{M}_x = \frac{1}{T}M_{\dot{x}} - \frac{1}{2}M_x$ to construct a discrete-time system of the form

$$\begin{pmatrix} M_v & M_i & \overline{M}_{x'} & M_q \\ T_v & 0 & 0 & 0 \\ 0 & T_i & 0 & 0 \end{pmatrix} \begin{pmatrix} \overline{v}(n) \\ \overline{i}(n) \\ \overline{x}(n) \\ \overline{q}(n) \end{pmatrix} = \begin{pmatrix} \overline{M}_x \overline{x}(n-1) \\ 0 \\ 0 \end{pmatrix} + \begin{pmatrix} \overline{u}(n) \\ 0 \\ 0 \end{pmatrix} \tag{10.56a}$$

$$f(\overline{q}(n)) = 0, \tag{10.56b}$$

where $\overline{v}(n), \overline{i}(n), \overline{q}(n)$, and $\bar{u}(n)$ are the discrete-time values at time nT. This discrete-time equation system cannot be solved uniquely; however, a general solution can be obtained with

$$\begin{pmatrix} \overline{v}(n) \\ \overline{i}(n) \\ \overline{x}(n) \\ \overline{q}(n) \end{pmatrix} = \begin{pmatrix} D_v \\ D_i \\ A \\ D_q \end{pmatrix} \overline{x}(n-1) + \begin{pmatrix} E_v \\ E_i \\ B \\ E_q \end{pmatrix} \bar{u}(n) + \begin{pmatrix} F_v \\ F_i \\ C \\ F_q \end{pmatrix} z(n). \tag{10.57}$$

Here, $z(n)$ is an arbitrary vector depending on the chosen solution with as many entries as $f(\overline{q}(n))$. Finally, we can extract from $\overline{v}$ and $\overline{i}$ only the quantities of interest resulting into a nonlinear state-space system like in Eqs. (10.47a) to (10.47d):

$$\overline{x}(n) = A\overline{x}(n-1) + B\bar{u}(n) + Cz(n), \tag{10.58a}$$

$$y(n) = D\overline{x}(n-1) + E\bar{u}(n) + Fz(n), \tag{10.58b}$$

$$\overline{q}(n) = D_q\overline{x}(n-1) + E_q\bar{u}(n) + F_q z(n), \tag{10.58c}$$

$$f(q(n)) = 0. \tag{10.58d}$$

We will demonstrate this approach using again the example from the clipping circuit in Fig. 10.3. By ordering the circuit elements as voltage source, resistor, capacitor, first diode, and second diode, the following coefficient matrices can be found:

$$M_v = \begin{pmatrix} 1 & & & & \\ & -1 & & & \\ & & C & & \\ & & 0 & & \\ & & & 1 & \\ & & & 0 & \\ & & & & 1 \\ & & & & 0 \end{pmatrix} M_i = \begin{pmatrix} 0 & & & & \\ & R & & & \\ & & 0 & & \\ & & 1 & & \\ & & & 0 & \\ & & & 1 & \\ & & & & 0 \\ & & & & 1 \end{pmatrix} M_x = \begin{pmatrix} \\ \\ -1 \\ 0 \\ \\ \\ \\ \\ \end{pmatrix}$$

$$M_{\dot{x}} = \begin{pmatrix} \\ \\ 0 \\ -1 \\ \\ \\ \\ \\ \end{pmatrix} M_q = \begin{pmatrix} & & & \\ & & & \\ & & & \\ & & & \\ -1 & 0 & & \\ 0 & -1 & & \\ & & -1 & 0 \\ & & 0 & -1 \end{pmatrix} u = \begin{pmatrix} u_{in} \\ 0 \\ 0 \\ 0 \\ 0 \\ 0 \\ 0 \\ 0 \end{pmatrix}. \tag{10.59}$$

The matrices T_v and T_i can be constructed using $T_v v = 0$ and $T_i i = 0$ yielding

$$T_v = \begin{pmatrix} -1 & 1 & 1 & 0 & 0 \\ -1 & 1 & 0 & 1 & 0 \\ 1 & -1 & 0 & 0 & 1 \end{pmatrix} T_i = \begin{pmatrix} 1 & 0 & 1 & 1 & -1 \\ 0 & 1 & -1 & -1 & 1 \end{pmatrix}. \tag{10.60}$$

After time discretization with a sampling rate of $f_s = 48$ kHz, we can derive

$$\overline{M}'_x = \begin{pmatrix} \text{———} \\ \text{———} \\ -0.5 \\ -48000 \\ \text{———} \\ \\ \text{———} \end{pmatrix} \quad \overline{M}_x = \begin{pmatrix} \text{———} \\ \text{———} \\ 0.5 \\ -48000 \\ \text{———} \\ \\ \text{———} \end{pmatrix} \tag{10.61}$$

In the next step, the equation system from Eq. (10.56a) needs to be solved. The general solution of a system of the form $Ax = b$, which has no unique solution, can be constructed by using the nullspace N of A in conjunction with one particular solution p,

$$x = p + Nz, \tag{10.62}$$

where z is an arbitrary vector. From this general solution, we can directly obtain the system matrices A, B, C, D_q, E_q, F_q. The matrices D, E, and F can be extracted from D_v, E_v, and F_v, i.e. we can take the third row of the matrices, which corresponds to the voltage over the capacitor. The resulting matrices yield

$$A = \begin{pmatrix}-1\end{pmatrix} \quad B = \begin{pmatrix}0\end{pmatrix} \quad C = \begin{pmatrix} 94 \cdot 10^{-3} \\ 0 \end{pmatrix}$$

$$D = \begin{pmatrix}0\end{pmatrix} \quad E = \begin{pmatrix}0\end{pmatrix} \quad F = \begin{pmatrix}1 & 0\end{pmatrix}$$

$$D_q = \begin{pmatrix} 0 \\ 96000 \\ 0 \\ 0 \end{pmatrix} \quad E_q = \begin{pmatrix} 0 \\ 1 \cdot 10^{-3} \\ 0 \\ 0 \end{pmatrix} \quad F_q = \begin{pmatrix} 0 & 1 \\ -5.512 \cdot 10^{-3} & 1 \\ 0 & -1 \\ 1 & 0 \end{pmatrix}. \tag{10.63}$$

The output of the nonlinear state-space system can now be computed by finding a suitable $z(n)$, which is consistent with Eqs. (10.58c) and (10.58d). This vector can then be used to solve the linear part of the system comprising Eqs. (10.58a) and (10.58b). Note that the coefficient matrices might also have different values, owing to the non-uniqueness of the nonlinear equation system from Eq. (10.56a). Consequently, the coefficient matrices depend on the chosen particular solution and the nullspace.

10.6 Exercises

1. Fundamentals

1. What can be said about the output of a memoryless nonlinear system excited with a periodic input?

2. Let $f(x) = |x|$ be a static nonlinear mapping excited with $x(t) = \cos(\omega_0 t)$. Compute the Fourier coefficients of the resulting output $y(t) = |\cos(\omega_0 t)|$. What can be said about the THD?

2. Overdrive, Distortion, Clipping

1. Derive the nonlinear differential equation for the first-order diode clipper assuming identical diodes. Hint: $i_d = I_s \left(e^{\frac{v_d}{\eta v_t}} - 1 \right)$, $\sinh(x) = \frac{e^x - e^{-x}}{2}$.

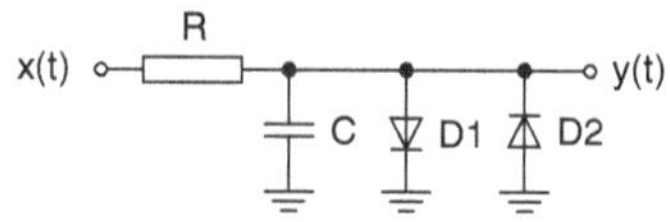

2. What is the difference between soft- and hard clipping nonlinearities? How can they be used to create a distortion or overdrive effect?

3. Nonlinear Filters

1. How can we extend a linear filter design to give it a more natural sound?

2. What is the difference to linear filters regarding the stability?

4. Aliasing and its Mitigation

1. Assume a nonlinear system introducing harmonics rolling off with frequency by approximately $1/f$. When operated at a sampling rate of 44.1 kHz, the aliasing distortion is deemed too high. By what factor, approximately, is the aliasing distortion present below 22.05 kHz reduced when doubling the sampling rate?

2. Apply antiderivative antialiasing to the memoryless systems described by the mapping functions of Eqs. (10.8) and (10.9).

5. Virtual Analog Modeling

1. Give an overview of the three main modeling approaches.

2. How is a circuit element modeled in the wave domain? State the connection between the wave and Kirchhoff domains.

References

[Bil17a] S. Bilbao, F. Esqueda, J.D. Parker, and V. Välimäki: *Antiderivative antialiasing for memoryless nonlinearities*. IEEE Signal Processing Letters, 24(7):1049–1053, 2017.

[Bil17b] S. Bilbao, F. Esqueda, and V. Välimäki: *Antiderivative antialiasing, lagrange interpolation and spectral flatness*. In 2017 IEEE Workshop on Applications of Signal Processing to Audio and Acoustics (WASPAA), pages 141–145, New Paltz, NY, USA, Oct 2017.

[Che04] G. Chen: *Stability of nonlinear systems*. Encyclopedia of RF and Microwave Engineering, pp 4881–4896, 2004.

[Cho20] J. Chowdbury: *Stable structures for nonlinear biquad filters*. In Proceedings of the 23rd International Conference on Digital Audio Effects (DAFx-20), pages 94–100, Vienna, Austria, 2020.

[Eic16] F. Eichas and U. Zölzer: *Virtual analog modeling of guitar amplifiers with Wiener-Hammerstein models*. In 44th Annual Convention on Acoustics, Munich, Germany, 2018.

[Esq15] F. Esqueda, V. Välimäki, and S. Bilbao: *Aliasing reduction in soft-clipping algorithms*. In Proc. 23rd European Signal Process. Conf. (EUSIPCO), pages 2059–2063, Nice, France, 2015.

[Esq16a] F. Esqueda, V. Välimäki, and S. Bilbao: *Antialiased soft clipping using an integrated bandlimited ramp*. In Proceedings of the 24th European Signal Processing Conference (EUSIPCO), pages 1043–1047, Budapest, Hungary, 2016.

[Esq16b] F. Esqueda, S. Bilbao, and V. Välimäki: *Aliasing reduction in clipped signals*. IEEE Transactions on Signal Processing, 64(20):5255–5267, 2016.

[Mul17] R. Muller and Thomas Helie: *Trajectory anti-aliasing on guaranteed-passive simulation of nonlinear physical systems*. In Proceedings of the 20th International Conference on Digital Audio Effects (DAFx-17), pages 87–94, Edinburgh, UK, 2017.

[Par16] J.D. Parker, V. Zavalishin, and E. Le Bivic: *Reducing the aliasing of nonlinear waveshaping using continuous-time convolution*. In Proceedings of the 19th International Conference on Digital Audio Effects (DAFx-16), pages 137–144, Brno, Czech Republic, 2016.

[Par19] J.D. Parker, F. Esqueda, and A. Bergner: *Modelling of nonlinear state-space systems using a deep neural network*. In Proceedings of the 22nd International Conference on Digital Audio Effects (DAFx- 19), Birmingham, UK, 2019.

[Fet86] A. Fettweis: *Wave digital filters: Theory and practice*. In Proceedings of the IEEE, volume 74, 1986.

[Fra13] Fractal Audio Systems: *Multipoint iterative matching and impedance correction technology*, 2013.

[Hol15] M. Holters and U. Zölzer: *A generalized method for the derivation of non-linear state-space models from circuit schematics*. In Proceedings of the 23rd European Signal Processing Conference (EUSIPCO), Nice, France, 2015.

[Hol20] M. Holters: *Antiderivative antialiasing for stateful systems*. Applied Sciences, 10(1), 2020.

[Kem08] C. Kemper: *Musical instrument with acoustic transducer*. "https://www.google.com/patents/US20080134867", 2008.

[Roa79] C. Roads: *A tutorial on Non-Linear Distortion or Waveshaping Synthesis.* Computer Music Journal, Vol. 3, No. 2, pp. 29–34, 1979.

[Ros92] D. Rossum: *Making digital filters sound "analog".* ICMC, pp 30–33, 1980.

[Sche80] M. Schetzen: *The Volterra and Wiener Theories of Nonlinear Systems.* Robert Krieger Publishing, 1980.

[Wer16] K. J. Werner: *Virtual Analog Modeling Of Audio Circuits Using Wave Digital Filters.* PhD thesis, 2016.

[Wri19] A Wright, E. Damskägg, and V. Välimäki: *Real-time black-box mod- elling with recurrent neural networks.* In Proceedings of the 22nd International Conference on Digital Audio Effects (DAFx-19), Birmingham, UK, 2019.

[Yeh10] D.T. Yeh, J.S. Abel, and J.O. Smith: *Automated physical modeling of nonlinear audio circuits for realtime audio effects part I: Theoretical development.* IEEE Trans. Audio, Speech and Language Process., 18(4):728–737, 2010.

Chapter 11

Machine Learning for Audio

P. Bhattacharya, P. Nowak, and U. Zölzer

11.1 Introduction

Machine learning can be described as a set of methods or algorithms that offer the capability to learn from data automatically and develop flexible parameterized models. In classical machine learning, the important features of the input are manually designed and extracted from data, and the system automatically learns to map these features to the requisite outputs. This learning is usually achieved through parameter adaptation and/or hyper-parameter tuning based on a predefined goal or optimization of a cost function. Classical machine learning is used in, and works well for, simple pattern recognition problems. A bulk of the effort or computation would include the design of optimal features for the system. Once the features are handcrafted from a dataset, a generic classification or regression model is used to obtain the output. Some examples of classical machine learning algorithms include linear regression, logistic regression, k-nearest neighbors, and simple decision trees. Representation learning goes a step further and mostly eliminates the need for handcrafted features. Instead, the required features are automatically discovered from the data. With the evolution of graphical processing units, new machine learning methods supporting more complex models have come to the forefront, with deep learning being one of the major topics of research. Deep learning is part of a broader family of machine learning methods, primarily based on artificial neural networks with representation learning. In deep learning, there are multiple levels of feature extraction from raw or processed data. These features are automatically extracted and combined across various levels to produce the output. Each level can be composed of linear and nonlinear, modifiable, and parameterized operators that can extract and represent features from the representations in the preceding level. With the increment of diverse levels inside the model, the complexity of the overall model increases and helps in an improved fine-tuned feature representation. Deep learning architectures such as deep neural network (DNN), recurrent neural network (RNN), and convolutional neural network (CNN) have been applied to

Digital Audio Signal Processing, Third Edition. Udo Zölzer.
© 2022 John Wiley & Sons Ltd. Published 2022 by John Wiley & Sons Ltd.

multiple fields, which include computer vision, audio recognition, natural language processing, social network filtering, machine translation, medical image analysis, material inspection, and gaming. In most of the areas, they have produced much improved results compared to the previous methods.

11.2 Unsupervised and Supervised Learning

Machine learning algorithms can be primarily classified into supervised, unsupervised, and reinforcement learning, of which the first two methods are of particular interest and are used widely in signal processing applications. Figure 11.1 shows a minimal representation of an unsupervised and a supervised system. Unsupervised learning is a task of identifying previously undetected patterns in a dataset with no labels and with a minimum of human supervision. In contrast to supervised learning that usually makes use of human-labeled data, unsupervised learning methods primarily extract underlying statistical or semantic features from the input data and allow for modeling of probability densities over inputs. In audio processing, feature extraction across various domains [Ler12] is very important for applications like audio analysis and fingerprinting, decomposition and separation, and content-based information retrieval [Mül15]. The underlying features or statistics can be analyzed and used to categorize incoming data, identify outliers, or generate latent representations. A clustering method, like the k-means algorithm [Mac67] or a mixture model, is a classical unsupervised learning method that groups unlabeled data into categories. It tries to recognize the common features in data and categorizes a new input based on the presence or absence of such features. Hence, such methods can be used in anomaly detection where an input does not belong to any group [Lu16]. Other popular unsupervised methods like the principal component analysis (PCA), independent component analysis (ICA), and non-negative matrix factorization (NMF) are also used extensively in audio processing and acoustics. ICA is a method which recovers unobserved signals from observed mixtures under the assumption of mutual independence. ICA or its faster variants are used for blind source separation in audio [Chi06] and music classification [Poh06], among others. NMF is another method which has a clustering or grouping property, and is also used for blind source separation [Vir07] and music analysis [Fev09]. Deep unsupervised learning is usually based on deep neural networks without explicit input/label pairs for training. Some models employed in deep unsupervised learning include autoencoder, deep belief network (DBN), and self organizing map (SOM), which are primarily used for dimension reduction or creation of latent variables. SOM is a neural network which produces a low-dimensional representation of a large input space and is useful for visualization. Autoencoders are used to generate latent variables which could be of a smaller size compared to the original high fidelity input and hence could be used in audio compression [Min19]. However,

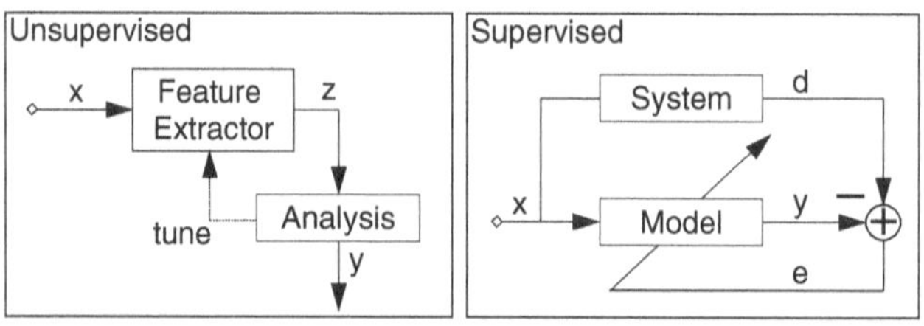

Figure 11.1 A minimal illustration of unsupervised and supervised learning models.

autoencoders, belief networks, or similar networks can learn underlying latent representations or embeddings from the input data, which can be used for classification tasks. An example of a neural network for speaker verification can be found in [Sny17] where the network learns to create embeddings, named as x-vectors.

Supervised learning is the task of learning a generalized mapping function of a system based on example input–label pairs. It models a function from labeled training data consisting of a set of examples. In supervised learning, each example is a pair consisting of an input and a desired output value. Supervised models have a predefined cost function that compares the predicted and desired output values. Based on any difference or similarity between the two, the model parameters are iteratively updated with a goal towards minimizing the difference or maximizing the similarity. Linear and logistic regression methods are two of the basic supervised learning methods based on least square minimization or least absolute deviation, and maximum likelihood estimation. These methods and their improved versions are used to estimate model parameters or coefficients to fit a given data or criteria. Other well-known methods include k-nearest neighbors (k-NN) and decision trees, where the latter is used in multiple audio applications [Lav09, Aka12]. Support vector machine (SVM) is another popular learning method mostly used for classification tasks, and it has been applied to perform audio classification and retrieval [GL03, LL01]. Deep learning methods gradually came to the forefront with DNN, CNN, and generative adversarial network (GAN). Deep learning architectures are gradually being used for a variety of audio applications like audio enhancement [Mee18], music recognition [Han17], pitch estimation [Zha16], virtual analog modeling [Mar20] and more.

This chapter is particularly focused on deep learning methods and it illustrates the concept of backpropagation on simple neural and convolutional network architectures, through useful derivations. Additionally, certain example applications are provided to illustrate the usability and performance of these methods in multiple areas of audio processing.

11.3 Gradient Descent and Backpropagation

Most supervised or semi-supervised machine learning algorithms include an objective function based on a ground truth or a prior, which is usually minimized to find an optimal set of parameters. The consequent derivation from this optimization criteria leads to the method of steepest descent, where the optimal solution is directly proportional to the negative gradient of the objective function. In deep learning, the proposed models are usually multilayered and consist of differentiable parameterized functions. To find the optimal set of parameters, the respective gradients of the objective function with respect to the parameters need to be calculated. The required gradient is eventually broken into a product of multiple local gradients from the hidden layers by the chain rule of derivatives. This method is generally known as backpropagation. The following section describes an artificial neural network (ANN) and exhibits how backpropagation works through the network layers.

11.3.1 Feedforward Artificial Neural Network

An artificial neural network is based on a collection of connected units or nodes called artificial neurons, which loosely model the neurons in a biological brain. Each connection, like the synapses in a biological brain, can transmit a signal to other neurons. An artificial neuron that receives a signal then processes it and can stimulate the connected neurons. In ANN implementations, the signal at a connection is a real number and the output of each neuron is computed by some nonlinear function of the sum of its inputs. Neurons and their connections typically have a weight that is adapted as learning proceeds. The weight increases or decreases the strength of the signal at a connection. In the case of multilayer perceptrons (MLP) [Ros61], neurons may have a threshold such that a signal is sent or the neuron is fired only if the aggregate surpasses the threshold. Typically, neurons are aggregated into layers and different layers may perform different transformations on their inputs. Signals travel from the first layer, or the input layer, to the last layer, or the output layer, possibly after traversing a number of intermediate layers, also referred to as hidden layers. An example neural network with one hidden layer is illustrated in Fig. 11.2. Every node of one layer is connected to every other node of the succeeding layer by scalar weights, which results in a densely connected structure and a matrix of scalar weights. The dimensions of a weight matrix is defined by the dimensions of the connected layers. Given an input element x_i, hidden element y_j, hidden activation a_j, and output element y_k, the operations are given by

$$y_j = \sum_{i=1}^{I} w_{ji} \cdot x_i + b_j, \tag{11.1}$$

$$a_j = \sigma(y_j), \tag{11.2}$$

$$y_k = \sum_{j=1}^{J} w_{kj} \cdot a_j + b_k, \tag{11.3}$$

where w_{ji} and b_j represent the associated scalar weight between x_i and y_j and the bias, respectively, and w_{kj} and b_k represent the associated scalar weight between a_j and y_k and the corresponding bias, respectively. Equation (11.2) refers to an activation operation, where $\sigma(\cdot)$ refers to a sigmoid activation function described in the next section.

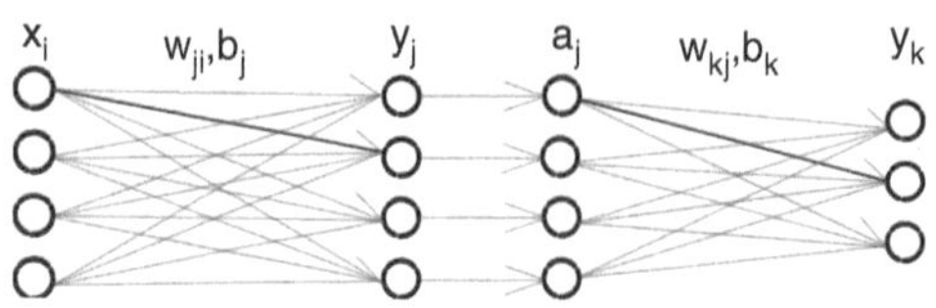

Figure 11.2 A feedforward neural network with one hidden layer.

Activation Function

An activation function performs a nonlinear operation on each neuron individually, retaining the dimensions of that layer. The most commonly used activations in ANN are the sigmoid, Tanh, and the ReLU functions, as illustrated in Fig. 11.3. The sigmoid function is given by

$$\sigma(x) = \frac{1}{1 + e^{-x}}, \tag{11.4}$$

where x is the input to the layer. The derivative of this function is given by

$$\frac{\partial \sigma(x)}{\partial x} = \sigma'(x) = \sigma(x)(1 - \sigma(x)). \tag{11.5}$$

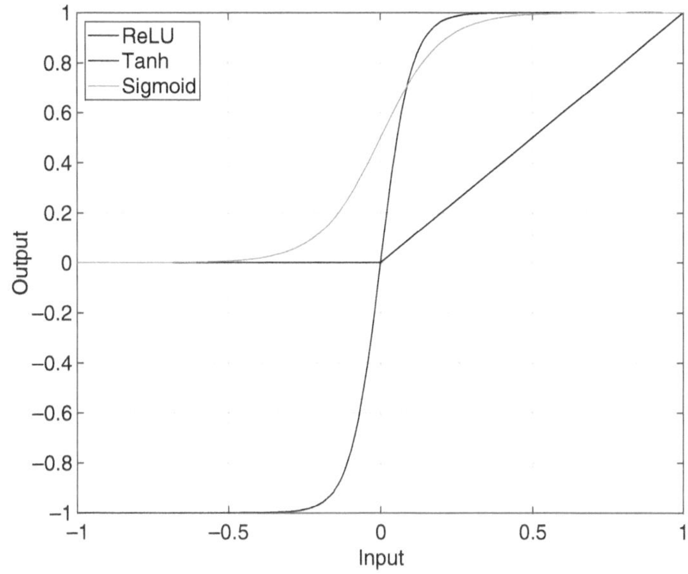

Figure 11.3 Activation functions.

The Tanh function (T) is an activation or gating unit given by

$$T(x) = \frac{e^x - e^{-x}}{e^x + e^{-x}}, \tag{11.6}$$

and its derivative is given by

$$\frac{\partial T(x)}{\partial x} = T'(x) = 1 - T^2(x). \tag{11.7}$$

Rectified linear unit (ReLU) is another activation function, which is given by

$$R(x) = \max(0, x), \tag{11.8}$$

having a point of discontinuity at 0. The corresponding approximate derivative is given by

$$\frac{\partial R(x)}{\partial x} = R'(x) = \begin{cases} 1 & \text{if } x > 0, \\ 0 & \text{if } x <= 0. \end{cases} \tag{11.9}$$

Backpropagation

Based on the network shown in Fig. 11.2, the parameter elements w_{kj}, b_j, w_{ji}, b_k, for any i, j, k, need to be updated and the goal is to calculate the corresponding gradients. The loss function is assumed to be a sum of squared error given by

$$E = \frac{1}{2}\sum_{k=1}^{K}(d_k - y_k)^2, \tag{11.10}$$

where d_k denotes a ground truth element or label. The gradient $\delta_{w_{kj}}$, required for updating the weight parameter w_{kj}, is expressed and extended according to the chain rule of derivatives as

$$\delta_{w_{kj}} = \frac{\partial E}{\partial w_{kj}} = \frac{\partial E}{\partial y_k} \cdot \frac{\partial y_k}{\partial w_{kj}} \tag{11.11}$$

$$= -(d_k - y_k) \cdot \frac{\partial}{\partial w_{kj}} \left(\sum_{j=1}^{J} w_{kj} \cdot a_j + b_k \right) \tag{11.12}$$

$$= -e_k \cdot a_j. \tag{11.13}$$

In a similar way, the gradient δ_{b_k}, to update the bias term b_k, is given by

$$\delta_{b_k} = \frac{\partial E}{\partial b_k} = \frac{\partial E}{\partial y_k} \cdot \frac{\partial y_k}{\partial b_k} = -(d_k - y_k) = -e_k. \tag{11.14}$$

Additionally, the partial gradient of the loss function with respect to the hidden activation δ_{a_j}, which will be backpropagated to the previous layer, is given by

$$\delta_{a_j} = \sum_{k=1}^{K} \frac{\partial E}{\partial a_j} = \sum_{k=1}^{K} \frac{\partial E}{\partial y_k} \cdot \frac{\partial y_k}{\partial a_j} \tag{11.15}$$

$$= \sum_{k=1}^{K} -(d_k - y_k) \cdot \frac{\partial}{\partial a_j} \left(\sum_{j=1}^{J} w_{kj} \cdot a_j + b_k \right) \tag{11.16}$$

$$= \sum_{k=1}^{K} - e_k \cdot w_{kj}. \tag{11.17}$$

In the next step, it is necessary to calculate the gradient $\delta_{w_{ji}}$, required for updating the weight parameter w_{ji}, and it can be expressed and extended according to the chain rule of derivatives. Assuming that a sigmoid activation is used and with the help of the Eqs. (11.17) and (11.3), it can be derived as

$$\delta_{w_{ji}} = \frac{\partial E}{\partial w_{ji}} = \sum_{k=1}^{K} \frac{\partial E}{\partial y_k} \cdot \frac{\partial y_k}{\partial a_j} \cdot \frac{\partial a_j}{\partial y_j} \cdot \frac{\partial y_j}{\partial w_{ji}} \tag{11.18}$$

$$= \sum_{k=1}^{K} - e_k \cdot w_{kj} \cdot \sigma'(y_j) \cdot \frac{\partial}{\partial w_{ji}} \left(\sum_{i=1}^{I} w_{ji} \cdot x_i + b_j \right) \tag{11.19}$$

$$= \sum_{k=1}^{K} - e_k \cdot w_{kj} \cdot \sigma'(y_j) \cdot x_i. \tag{11.20}$$

The expression $\tilde{\sigma}(\cdot)$ in the above equations is the local derivative of the sigmoid activation function, as given by Eq. (11.5). The gradient δ_{b_j} is calculated similarly and is given by

$$\delta_{b_j} = \frac{\partial E}{\partial b_j} = \sum_{k=1}^{K} \frac{\partial E}{\partial y_k} \cdot \frac{\partial y_k}{\partial a_j} \cdot \frac{\partial a_j}{\partial y_j} \cdot \frac{\partial y_j}{\partial b_j} \tag{11.21}$$

$$= \sum_{k=1}^{K} - e_k \cdot w_{kj} \cdot \sigma'(y_j) \cdot \frac{\partial}{\partial b_j} \left(\sum_{i=1}^{I} w_{ji} \cdot x_i + b_j \right) \tag{11.22}$$

$$= \sum_{k=1}^{K} - e_k \cdot w_{kj} \cdot \sigma'(y_j). \tag{11.23}$$

In the next step, the parameters can be iteratively updated with the help of a simple gradient descent, given by

$$w_{kj} = w_{kj} - \eta_w \cdot \delta_{w_{kj}}, \tag{11.24}$$

$$b_k = b_k - \eta_b \cdot \delta_{b_k}, \tag{11.25}$$

$$w_{ji} = w_{ji} - \eta_w \cdot \delta_{w_{ji}}, \tag{11.26}$$

$$b_j = b_j - \eta_b \cdot \delta_{b_j}, \tag{11.27}$$

where η_b and η_w denote the step sizes or learning rates for weights and biases, respectively. The learning rate is usually a small fractional value and η_b is usually lower than η_w. Gradient descent or stochastic gradient descent is usually very slow in convergence when used in a deep learning method across large datasets. Hence, faster gradient descent methods like Nesterov momentum [Nes83], adaptive gradient or adagrad [Duc11], rmsprop [Dau15], and adaptive momentum or adam [Kin15] are primarily used for large-scale deep learning applications.

11.3.2 Convolutional Neural Network

A CNN is an artificial neural network which contains at least one convolutional unit which performs a convolution or cross-correlation between the input and a set of pre-defined filters. The filter coefficients, also referred to as weights, are initialized using one of the widely used initialization methods [Glo10, He15] and are subsequently updated iteratively. The primary advantage of such a network over a feedforward artificial neural network is that a convolutional layer contains fewer parameters compared to a corresponding fully connected layer. A CNN also contains hidden layers and an activation function between each convolutional layer. An example network with one hidden layer is illustrated in Fig. 11.4, which contains a convolutional unit and a fully connected unit. The hidden layer is produced by the filtering operation between the nodes in the input layer and the pre-initialized filters in the convolutional unit. The filtering operation is followed by a ReLU activation function, which is commonly used in a CNN. Finally, the hidden activations are densely connected with the neurons of the output layer. Given an input element x_j, hidden element y_j, hidden activation a_j, and output element y_k, the operations are given by

$$y_j = \sum_{l=-\frac{h-1}{2}}^{\frac{h-1}{2}} w_l \cdot x_{j+l} + b, \tag{11.28}$$

$$a_j = R(y_j), \tag{11.29}$$

$$y_k = \sum_{j=1}^{J} w_{kj} \cdot a_j + b_k, \tag{11.30}$$

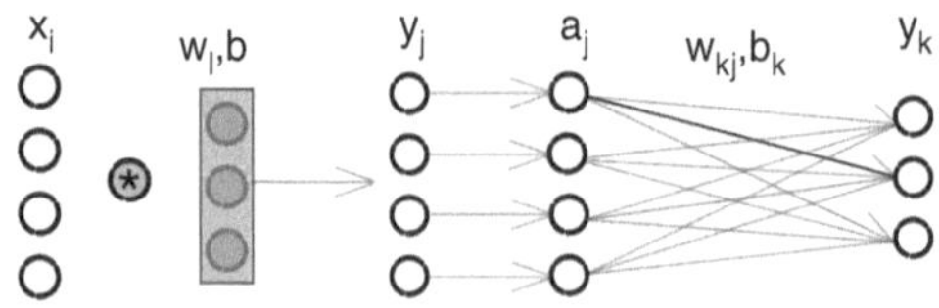

Figure 11.4 A convolutional neural network with one hidden layer.

where w_l and b represent the associated filter coefficient of the filter having a length h and the scalar bias value, respectively, and w_{kj} and b_k represent the associated scalar weights between a_j and y_k and the corresponding bias, respectively. The convolutional unit can also contain multiple filters and bias values resulting in a hidden layer containing multiple stacked vectors instead of one vector.

Based on the network shown in Fig. 11.4, the weight elements w_{kj}, b_j, w_l, and b need to be updated and the goal is to calculate the corresponding gradients. The gradient $\delta_{w_{kj}}$, required for updating the weight parameter w_{kj}, is already given by Eq. (11.13) while the gradient δ_{b_k}, to update the bias term b_k, is given by Eq. (11.14). Additionally, the partial gradient of the loss function with respect to the hidden activation δ_{a_j}, which will be backpropagated to the previous layer, is given by Eq. (11.17). In the next step, it is necessary to calculate the gradient δ_{w_m}, required for updating the filter coefficient w_m, and it can be expressed and extended according to the chain rule of derivatives. With the help of Eqs. (11.17) and (11.30), it can be derived as

$$\delta_{w_m} = \frac{\partial E}{\partial w_m} = \sum_{j=1}^{J}\sum_{k=1}^{K} \frac{\partial E}{\partial y_k} \cdot \frac{\partial y_k}{\partial a_j} \cdot \frac{\partial a_j}{\partial y_j} \cdot \frac{\partial y_j}{\partial w_m} \tag{11.31}$$

$$= \sum_{j=1}^{J}\sum_{k=1}^{K} - e_k \cdot w_{kj} \cdot R'(y_j) \cdot \frac{\partial}{\partial w_m}\left(\sum_{l=-\frac{h-1}{2}}^{\frac{h-1}{2}} w_l \cdot x_{j+l} + b \right) \tag{11.32}$$

$$= \sum_{j=1}^{J}\sum_{k=1}^{K} - e_k \cdot w_{kj} \cdot R'(y_j) \cdot \left(\sum_{l=-\frac{h-1}{2}}^{\frac{h-1}{2}} \frac{\partial w_l}{\partial w_m} \cdot x_{j+l} \right) \tag{11.33}$$

$$= \sum_{j=1}^{J} \delta_{y_j} \cdot x_{j+m}, \tag{11.34}$$

where the expression δ_{y_j} is given by

$$\delta_{y_j} = \sum_{k=1}^{K} - e_k \cdot w_{kj} \cdot R'(y_j), \tag{11.35}$$

and the expression $R'(y_j)$ refers to the local derivative of the ReLU activation function, as given by Eq. (11.9). Equation (11.34) leads to a cross-correlation operation between the incoming gradient and the input vector. The gradient δ_b is calculated similarly and is given by

$$\delta_b = \frac{\partial E}{\partial b} = \sum_{j=1}^{J}\sum_{k=1}^{K} \frac{\partial E}{\partial y_k} \cdot \frac{\partial y_k}{\partial a_j} \cdot \frac{\partial a_j}{\partial y_j} \cdot \frac{\partial y_j}{\partial b} \tag{11.36}$$

$$= \sum_{j=1}^{J} \delta_{y_j}. \tag{11.37}$$

After the calculation of the required gradients, the parameters are updated by stochastic gradient descent or a faster parameter update algorithm.

11.4 Applications

The following sections describe several deep learning applications in the area of audio signal processing. In the first example, the application proposed in [Bha20] is described, where a cascade of parametric peak and shelving filters is adapted to model head-related transfer functions. The iterative optimization method is sample based and uses the instantaneous backpropagation method. In the second example, the sparse feedback delay network (FDN) with Schroeder allpass filters, described in Section 7.3.3, is adapted to model a desired room impulse response or simulate a room impulse response (RIR) based on a desired reverberation time. The final application of the section describes a CNN which is used to reduce Gaussian noise in audio or speech signals.

11.4.1 Parametric Filter Adaptation

Parametric peak and shelving filters, as introduced in Section 6.2.2, can be cascaded to perform audio processing tasks like equalization, spectral shaping, and modeling of complex transfer functions. An independent optimization of the mentioned parameters of each individual filter in this cascaded structure can be performed with the help of a backpropagation algorithm. Earlier research in this area includes the work done in [Gao92], which introduces a backpropagation-based adaptive IIR filter, and in [Ros99], which trained a recursive filter with derivative function to adapt a controller. In the context of neural networks, a cascaded structure of FIR and IIR filters with multilayer perceptrons was used in [Bac91] for time-series modeling based on a simplified instantaneous backpropagation through time (IBPTT). A similar adaptive IIR-multilayer perceptron (IIR-MLP) network was also described in [Cam96] based on causal backpropagation through time (CBPTT). In recent years, backpropagation is extensively used in CNNs as well as recurrent neural networks, the latter being recursive in nature. However, in the aforementioned literature, the adaptation is primarily performed directly on filter coefficients. This work is one of the few methods that adapt the control parameters with deep learning [Ner20, Väl19, Eng20] and the application described in this section is already introduced in [Bha20].

The proposed cascaded structure with M filters is illustrated in Fig. 11.5, where the first and the last filters are shelving filters while the remaining filters are peak filters. Peak and shelving

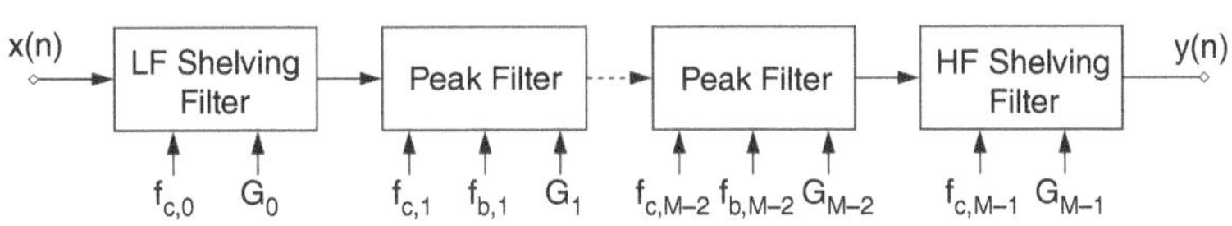

Figure 11.5 A cascaded structure of peak and shelving filters.

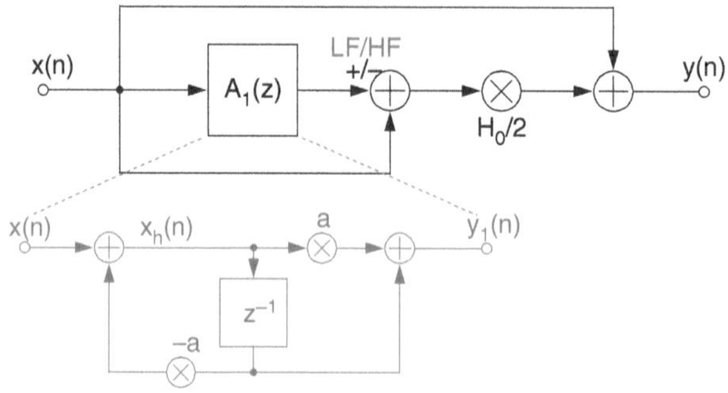

Figure 11.6 Signal flow graph of a LF/HF shelving filter.

filters are described in Section 6.2.2 along with their illustrations and transfer functions. From the transfer functions defining low-frequency and high-frequency shelving filters (LFS and HFS) in Eq. (6.59) and Eq. (6.71), one can derive the signal flow graph in Fig. 11.6 and the corresponding difference equation can be calculated to be

$$y(n) = \frac{H_0}{2}[x(n) \pm y_1(n)] + x(n), \tag{11.38}$$

where $y_1(n)$ defines the output signal of a first-order allpass filter according to

$$y_1(n) = a_{B/C}x_h(n) + x_h(n-1), \tag{11.39}$$

$$x_h(n) = x(n) - a_{B/C}x_h(n-1). \tag{11.40}$$

Similarly, the signal flow graph in Fig. 11.7 and the difference equation of a peak filter can be derived from Eq. (6.81) as

$$y(n) = \frac{H_0}{2}[x(n) - y_2(n)] + x(n), \tag{11.41}$$

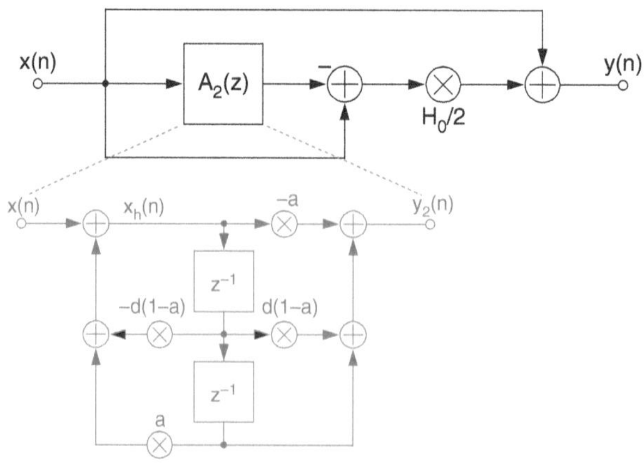

Figure 11.7 Signal flow graph of a peak filter.

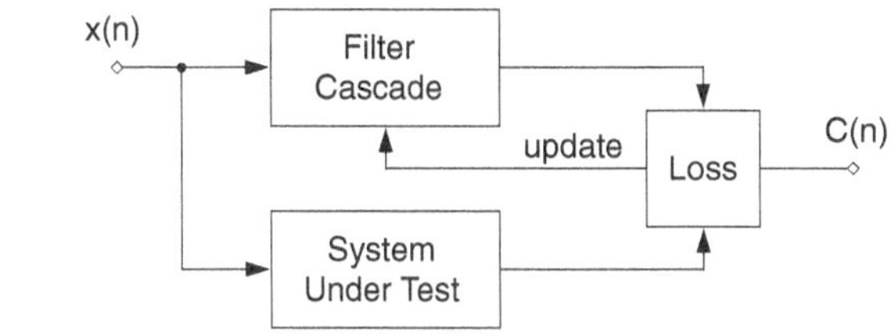

Figure 11.8 Block diagram of the model.

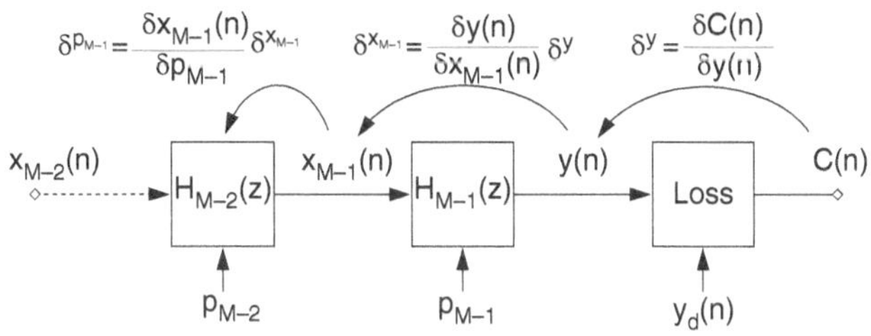

Figure 11.9 Illustration of backpropagation through the cascaded structure.

where $y_2(n)$ defines the output signal of a second-order allpass filter according to

$$y_2(n) = -a_{B/C}x_h(n) + d(1 - a_{B/C})x_h(n-1) + x_h(n-2), \tag{11.42}$$

$$x_h(n) = x(n) - d(1 - a_{B/C})x_h(n-1) + a_{B/C}x_h(n-2). \tag{11.43}$$

The partial derivative of a pre-defined cost function, with respect to the filter parameters in a cascaded structure, needs to be calculated as a product of multiple local derivatives according to the chain rule. Given the cascaded structure and a reference system under test, a global instantaneous cost or loss function $C(n)$ for the nth sample can be defined, as shown in Fig. 11.8. The derivative of the cost function with respect to a parameter p_{M-1} can be written as

$$\frac{\partial C(n)}{\partial p_{M-1}} = \frac{\partial C(n)}{\partial y(n)} \frac{\partial y(n)}{\partial x_{M-1}(n)} \frac{\partial x_{M-1}(n)}{\partial p_{M-1}}, \tag{11.44}$$

according to the chain rule of derivatives, where $y(n)$ represents the predicted output of the cascaded filter structure, $y_d(n)$ represents the desired output of the system under test, $x_{M-1}(n)$ represents the output of the $\{M-1\}$th filter in the cascade, and p_{M-1} represents any control parameter like gain, bandwidth, or center frequency of the $\{M-1\}$th peak filter.

This will result in a simplified instantaneous backpropagation algorithm [Bac91], which is illustrated in Fig. 11.9. To calculate the above derivative, with respect to the instantaneous cost function, it is necessary to calculate the local derivative $\frac{\partial C(n)}{\partial y(n)}$ of the cost function, the local derivative $\frac{\partial y(n)}{\partial x_{M-1}(n)}$ of a filter output with respect to its input, and the local derivative $\frac{\partial x_{M-1}(n)}{\partial p_{M-1}}$ of a filter output with respect to its parameter p_{M-1}. Hence, in general, the above three types of local gradients need to be calculated to adapt the cascaded structure. In the following sections, the local gradients of the filter output against the filter input and the control parameters are derived for shelving and peak filters. Finally, the cascaded structure is used for modeling head-related transfer function (HRTF) magnitudes as an example application.

Shelving Filter

For a shelving filter, local derivatives of the filter output are calculated against the filter input, the gain, and the cutoff frequency. Referring to Eq. (11.38), the derivative of a low-frequency shelving (LFS) and a high-frequency shelving (HFS) filter output $y(n)$, with respect to its input $x(n)$, is calculated as

$$\frac{\partial y(n)}{\partial x(n)} = \frac{H_0}{2}\left[1 \pm a_{B/C}\right] + 1. \tag{11.45}$$

The derivative of the shelving filter output, with respect to the filter gain G, for the boost case is calculated as

$$\frac{\partial y(n)}{\partial G} = \frac{\left[x(n) \pm y_1(n)\right]}{2}\frac{\partial H_0}{\partial G} \tag{11.46}$$

$$= \frac{\left[x(n) \pm y_1(n)\right]}{2}\frac{\partial}{\partial G}\left[10^{\frac{G}{20}} - 1\right] \tag{11.47}$$

$$= \frac{\left[x(n) \pm y_1(n)\right]}{40} 10^{\frac{G}{20}} \ln(10). \tag{11.48}$$

The derivative of the filter output, with respect to the filter gain G, for the cut case is different from the boost case because of the dependence between the gain parameter and the parameter for cutoff frequency that can be seen in Eqs. (6.47) and (6.53). With the help of Eq. (11.38), it can calculated as

$$\frac{\partial y(n)}{\partial G} = \frac{\left[x(n) \pm y_1(n)\right]}{2}\frac{\partial H_0}{\partial G} \pm \frac{H_0}{2}\frac{\partial y_1(n)}{\partial G} \tag{11.49}$$

$$= \frac{\left[x(n) \pm y_1(n)\right]}{40} 10^{\frac{G}{20}} \ln(10) \pm \frac{H_0}{2}\frac{\partial y_1(n)}{\partial G}, \tag{11.50}$$

where the expression $\frac{\partial y_1(n)}{\partial G}$ of Eq. (11.50) can be extended with the help of Eq. (11.39) as

$$\frac{\partial y_1(n)}{\partial G} = \frac{\partial a_C}{\partial G} x_h(n) + a_C \frac{\partial x_h(n)}{\partial G} + \frac{\partial x_h(n-1)}{\partial G}, \tag{11.51}$$

with

$$\frac{\partial a_C}{\partial G} = \frac{-\ln(10) V_0 \tan\left(\pi \frac{f_c}{f_S}\right)}{10\left[\tan\left(\pi \frac{f_c}{f_S}\right) + V_0\right]^2}, \quad \textbf{(LFS)} \tag{11.52}$$

$$\frac{\partial a_C}{\partial G} = \frac{\ln(10) V_0 \tan\left(\pi \frac{f_c}{f_S}\right)}{10\left[V_0 \tan\left(\pi \frac{f_c}{f_S}\right) + 1\right]^2}, \quad \textbf{(HFS)} \tag{11.53}$$

$$\frac{\partial x_h(n)}{\partial G} = -\frac{\partial a_C}{\partial G} x_h(n-1) - a_C \frac{\partial x_h(n-1)}{\partial G}. \tag{11.54}$$

From Eq. (11.54), $\frac{\partial x_h(n-1)}{\partial G}$ is calculated by employing the chain rule of derivatives and IBPTT with the initialization of $\frac{\partial x_h(k)}{\partial G}|_{k=0} = 0$.

Finally, the derivative of the shelving filter output, with respect to the cutoff frequency f_c, is calculated as

$$\frac{\partial y(n)}{\partial f_c} = \pm\frac{H_0}{2}\frac{\partial y_1(n)}{\partial f_c}, \tag{11.55}$$

where

$$\frac{\partial y_1(n)}{\partial f_c} = \frac{\partial a_{B/C}}{\partial f_c}x_h(n) + a_{B/C}\frac{\partial x_h(n)}{\partial f_c} + \frac{\partial x_h(n-1)}{\partial f_c}, \tag{11.56}$$

with

$$\frac{\partial a_B}{\partial f_c} = \frac{2\pi \sec\left(2\pi\frac{f_c}{f_S}\right)\left[\sec\left(2\pi\frac{f_c}{f_S}\right) - \tan\left(2\pi\frac{f_c}{f_S}\right)\right]}{f_S}, \tag{11.57}$$

$$\frac{\partial a_C}{\partial f_c} = \frac{2\pi V_0 \sec\left(\pi\frac{f_c}{f_S}\right)^2}{f_S\left[\tan\left(\pi\frac{f_c}{f_S}\right) + V_0\right]^2}, \quad \textbf{(LFS)} \tag{11.58}$$

$$\frac{\partial a_C}{\partial f_c} = \frac{2\pi V_0 \sec\left(\pi\frac{f_c}{f_S}\right)^2}{f_S\left[V_0\tan\left(\pi\frac{f_c}{f_S}\right) + 1\right]^2}, \quad \textbf{(HFS)} \tag{11.59}$$

$$\frac{\partial x_h(n)}{\partial f_c} = -\frac{\partial a_{B/C}}{\partial f_c}x_h(n-1) - a_{B/C}\frac{\partial x_h(n-1)}{\partial f_c}. \tag{11.60}$$

From Eq. (11.60), $\frac{\partial x_h(n-1)}{\partial f_c}$ is calculated by employing the chain rule of derivatives and IBPTT with the initialization of $\frac{\partial x_h(k)}{\partial f_c}|_{k=0} = 0$.

Peak Filter

For a peak filter, local derivatives of the filter output are calculated against the filter input, the gain, the center frequency, and the bandwidth. Referring to Eq. (11.41), the derivative of a second-order peak filter output $y(n)$, with respect to its input $x(n)$, is calculated as

$$\frac{\partial y(n)}{\partial x(n)} = \frac{H_0}{2}\left[1 + a_{B/C}\right] + 1. \tag{11.61}$$

As done in the case of shelving filters, the derivative of the peak filter output, with respect to the filter gain G, for the boost case will result in

$$\frac{\partial y(n)}{\partial G} = \frac{\left[x(n) - y_2(n)\right]}{40}10^{\frac{G}{20}}\ln(10). \tag{11.62}$$

For the cut case as well, the derivative of the peak filter output, with respect to the filter gain G, is derived in a similar manner. With the help of Eq. (11.41), the derivation leads to

$$\frac{\partial y(n)}{\partial G} = \frac{\left[x(n) - y_2(n)\right]}{40}10^{\frac{G}{20}}\ln(10) - \frac{H_0}{2}\frac{\partial y_2(n)}{\partial G}, \tag{11.63}$$

and the expression $\frac{\partial y_2(n)}{\partial G}$ from the above equation can be extended with the help of Eq. (11.42) as

$$\begin{aligned}\frac{\partial y_2(n)}{\partial G} = -\frac{\partial a_C}{\partial G}x_h(n) - a_C\frac{\partial x_h(n)}{\partial G} - \cdots \\ d\frac{\partial a_C}{\partial G}x_h(n-1) + \frac{\partial x_h(n-2)}{\partial G} + \cdots \\ d(1-a_C)\frac{\partial x_h(n-1)}{\partial G},\end{aligned} \tag{11.64}$$

with

$$\frac{\partial a_C}{\partial G} = \frac{-\ln(10)V_0\tan\left(\pi\frac{f_b}{f_S}\right)}{10\left[\tan\left(\pi\frac{f_b}{f_S}\right)+V_0\right]^2}, \tag{11.65}$$

and

$$\begin{aligned}\frac{\partial x_h(n)}{\partial G} = d\frac{\partial a_C}{\partial G}x_h(n-1) + \frac{\partial a_C}{\partial G}x_h(n-2) + \cdots \\ a_C\frac{\partial x_h(n-2)}{\partial G} - d(1-a_C)\frac{\partial x_h(n-1)}{\partial G}.\end{aligned} \tag{11.66}$$

From Eq. (11.66), the expressions $\frac{\partial x_h(n-1)}{\partial G}$ and $\frac{\partial x_h(n-2)}{\partial G}$ are calculated with the chain rule of derivatives and IBPTT with the initialization of $\frac{\partial x_h(k)}{\partial G}|_{k=0} = 0$ and $\frac{\partial x_h(k)}{\partial G}|_{k=-1} = 0$.

The derivative of the peak filter output, with respect to the cutoff frequency f_c, leads to an expression similar to Eq. (11.55) given by

$$\frac{\partial y(n)}{\partial f_c} = -\frac{H_0}{2}\frac{\partial y_2(n)}{\partial f_c}, \tag{11.67}$$

where

$$\begin{aligned}\frac{\partial y_2(n)}{\partial f_c} = -a_{B/C}\frac{\partial x_h(n)}{\partial f_c} + \frac{\partial d}{\partial f_c}(1-a_{B/C})x_h(n-1) + \cdots \\ d(1-a_{B/C})\frac{\partial x_h(n-1)}{\partial f_c} + \frac{\partial x_h(n-2)}{\partial f_c},\end{aligned} \tag{11.68}$$

with

$$\frac{\partial d}{\partial f_c} = \frac{2\pi\sin\left(2\pi\frac{f_c}{f_S}\right)}{f_S}, \tag{11.69}$$

and

$$\begin{aligned}\frac{\partial x_h(n)}{\partial f_c} = -\frac{\partial d}{\partial f_c}(1-a_{B/C})x_h(n-1) - \cdots \\ d(1-a_{B/C})\frac{\partial x_h(n-1)}{\partial f_c} + a_{B/C}\frac{\partial x_h(n-2)}{\partial f_c}.\end{aligned} \tag{11.70}$$

From Eq. (11.70), the expressions $\frac{\partial x_h(n-1)}{\partial f_c}$ and $\frac{\partial x_h(n-2)}{\partial f_c}$ are calculated with the chain rule of derivatives and IBPTT with the initialization of $\frac{\partial x_h(k)}{\partial f_c}|_{k=0} = 0$ and $\frac{\partial x_h(k)}{\partial f_c}|_{k=-1} = 0$.

Finally, the derivative of the peak filter output, with respect to the bandwidth f_b, is calculated as

$$\frac{\partial y(n)}{\partial f_b} = -\frac{H_0}{2}\frac{\partial y_2(n)}{\partial f_b}. \tag{11.71}$$

From the above equation, $\frac{\partial y_2(n)}{\partial f_b}$ is calculated as

$$\begin{aligned}\frac{\partial y_2(n)}{\partial f_b} &= -\frac{\partial a_{B/C}}{\partial f_b}x_h(n) - a_{B/C}\frac{\partial x_h(n)}{\partial f_b} - \cdots \\ &\quad d\frac{\partial a_{B/C}}{\partial f_b}x_h(n-1) + \frac{\partial x_h(n-2)}{\partial f_b} + \cdots \\ &\quad d(1-a_{B/C})\frac{\partial x_h(n-1)}{\partial f_b},\end{aligned} \tag{11.72}$$

with

$$\frac{\partial a_B}{\partial f_b} = \frac{2\pi \sec\left(2\pi\frac{f_b}{f_S}\right)\left[\sec\left(2\pi\frac{f_b}{f_S}\right) - \tan\left(2\pi\frac{f_b}{f_S}\right)\right]}{f_S}, \tag{11.73}$$

$$\frac{\partial a_C}{\partial f_b} = \frac{2\pi V_0 \sec\left(\pi\frac{f_b}{f_S}\right)^2}{f_S\left[\tan\left(\pi\frac{f_b}{f_S}\right) + V_0\right]^2}, \tag{11.74}$$

and

$$\begin{aligned}\frac{\partial x_h(n)}{\partial f_b} &= d\frac{\partial a_{B/C}}{\partial f_b}x_h(n-1) - d(1-a_{B/C})\frac{\partial x_h(n-1)}{\partial f_b} + \cdots \\ &\quad \frac{\partial a_{B/C}}{\partial f_b}x_h(n-2) + a_{B/C}\frac{\partial x_h(n-2)}{\partial f_b}.\end{aligned} \tag{11.75}$$

From Eq. (11.75), the expressions $\frac{\partial x_h(n-1)}{\partial f_b}$ and $\frac{\partial x_h(n-2)}{\partial f_b}$ are calculated with chain rule of derivatives and IBPTT with the initialization of $\frac{\partial x_h(k)}{\partial f_b}|_{k=0} = 0$ and $\frac{\partial x_h(k)}{\partial f_b}|_{k=-1} = 0$.

Cascaded Structure

With the help of the above formulations, the necessary gradients for parameter update can be derived. As an example, the derivative of the cost function, e.g. the squared-error function given by

$$C(n) = [y_d(n) - y(n)]^2, \tag{11.76}$$

with respect to the gain of the $\{M-1\}$th peak filter G_{M-1} in the filter cascade, as shown in Fig. 11.9, is given by

$$\frac{\partial C(n)}{\partial G_{M-1}} = \frac{\partial C(n)}{\partial y(n)}\frac{\partial y(n)}{\partial x_{M-1}(n)}\frac{\partial x_{M-1}(n)}{\partial G_{M-1}}, \tag{11.77}$$

according to Eq. (11.44). If the boost case is assumed, then with the help of the derivative of the cost function from Eq. (11.76), with respect to $y(n)$ given by

$$\frac{\partial C(n)}{\partial y(n)} = -2[y_d(n) - y(n)], \tag{11.78}$$

Eq. (11.61), and Eq. (11.62), the required gradient for parameter update is derived as

$$\frac{\partial C(n)}{\partial G_{M-1}} = e(n)\left(\frac{H_0}{2}\left[1 + a_{B_{M-1}}\right] + 1\right)\left(K\left[x_{M-2}(n) - y_{1_{M-1}}(n)\right]\right), \tag{11.79}$$

where

$$e(n) = \frac{\partial C(n)}{\partial y(n)}, \tag{11.80}$$

$$K = \frac{10^{\frac{G_{M-1}}{20}}\ln(10)}{40} = \frac{V_{0_{M-1}}\ln(10)}{40}. \tag{11.81}$$

The gain G_{M-1} is subsequently updated by the above expression, preferably with the adaptive momentum update for faster convergence.

Head-related Transfer Function Modeling

An HRTF is a direction dependent transfer function between an external sound source and the human ear. The inverse Fourier transform of the HRTF is the head-related impulse response (HRIR). In spatial audio through headphones, mono signals are filtered with the corresponding HRIRs to create a virtual sound that is localized in a certain direction. To achieve a good resolution of the 3-D space, HRIRs have to be saved for a high number of directions, which results in a large amount of stored data. Hence, parametric IIR filters can be used to model the HRTF magnitudes with a lower number of saved parameters.

For the initialization of the cascaded filter structure, an approximated number of required peak filters needs to be determined. Afterward, the allocation of the initial parameter values is done, as illustrated in Fig. 11.10. Initially, the magnitude response is smoothed and the mean of the magnitude response is subtracted. Based on the prominence and proximity to one another, a finite number of peaks and notches are selected. The initial center frequency of a peak filter is determined based on the position of a peak or notch while the initial cutoff frequencies of the shelving filters are determined based on the slopes in the magnitude response. The gain of a filter is initialized by the magnitude of the transfer function at the position of the peak or notch. To reduce the summation effect arising from the cascade, the gain of every peak filter is scaled by a fractional factor depending on the magnitude. Gains of notches in the positive half and peaks

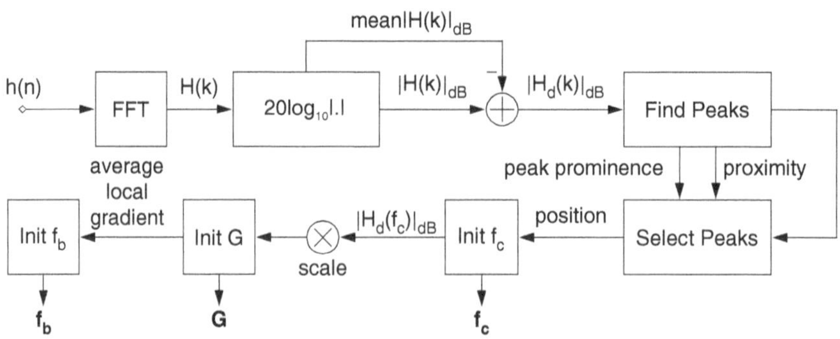

Figure 11.10 Block diagram illustrating the initialization method for HRTF modeling.

in the negative half are converted to small negative and positive values, respectively. Finally, the bandwidth of a peak filter is initialized based on the average local gradient of the magnitude response around the peak's position. The number of filters can be reduced or increased based on the third step. A major drawback of such an initialization is that more than an optimal number of filters will usually be proposed. Additionally, the peak picking method will be insufficient in flat regions. However, a direct initialization is simple and the simultaneous filter update improves the run-time.

To adapt the cascaded structure, a log-spectral distance between the estimated and the desired magnitude responses is considered as the objective function, which is given by

$$C(k) = \left(|Y_d(k)|_{\mathrm{dB}} - |Y(k)|_{\mathrm{dB}}\right)^2, \tag{11.82}$$

where $Y_d(k)$ and $Y(k)$ denote the magnitude responses of the desired and estimated output signals, respectively, in dB, and k denotes a frequency bin. The derivative of the Fourier transform between time and frequency domain can be performed with the help of Wirtinger calculus, as demonstrated in [Car17] and [Wan20].

For the evaluation, a subset of HRIRs from the CIPIC database [Alg01] is chosen to evaluate the HRTF magnitude approximation. In the first step, the HRIRs are converted to HRTFs and the magnitude responses are treated as the desired signal for the filter cascade. Nevertheless, before performing the discrete Fourier transform, the HRIRs are padded with zeros to a length of 1024 to achieve a better frequency resolution. Afterward, the aforementioned initialization is performed. Owing to differences in the HRTFs between subjects and directions, every transfer function needs a unique initialization, which can result in a different number of peak filters and initial parameter values. After the initialization of the cascaded structure, the filters are trained and updated for 100 epochs with the adam method [Kin15]. However, in the cases of most HRTFs, a smaller number of epochs is sufficient to achieve a good approximation. The learning rate during the update method is selected as $\eta = 10^{-1}$. Additionally, there is a learning rate drop factor of 0.99 for every time the error in a particular epoch is higher than the error in the previous epoch. The aforementioned hyper-parameters might change for a few exceptional cases where convergence requires different values.

In Fig. 11.11, the magnitude responses of the desired HRTF, the initial HRTF estimate, and the final approximation of the right ear of 'Subject_008', for an azimuth $\varphi = 20°$ and elevation $\theta = 0°$, are plotted. As can be seen, the algorithm is able to reproduce the desired magnitude response with 17 peaks and two shelving filters. A more detailed evaluation and discussion can be found in [Bha20].

11.4.2 Room Simulation

In the following application, a sparse FDN, introduced in Section 7.3.3, is adapted to simulate a desired RIR. This section provides some derivations related to the proposed FDN with a cascaded allpass filter and delay line in each of the K branches having a sparse diagonal feedback matrix. Figure 11.12 shows an overview of the proposed FDN. The corresponding difference equations are given as

$$y_k(n) = s_k(n - M_k) - m \cdot s_k(n), \tag{11.83}$$

$$s_k(n) = x_k(n) + m \cdot s_k(n - M_k), \tag{11.84}$$

$$x_k(n) = x(n) + y_{k-1}(n - D_{k-1}), 1 < k \leq \mathrm{K} \tag{11.85}$$

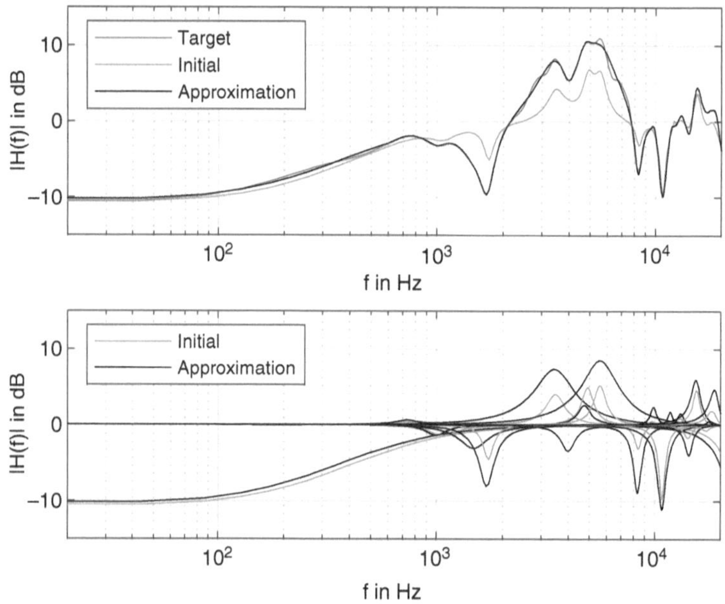

Figure 11.11 Magnitude responses: in the top plot, the whole filter cascade; and in the bottom plot, all individual filter stages of the desired HRTF, the initial HRTF estimate, and the final approximation of the right ear of ‘Subject-008’, for an azimuth $\phi = 20^\circ$ and elevation $\theta = 0^\circ$.

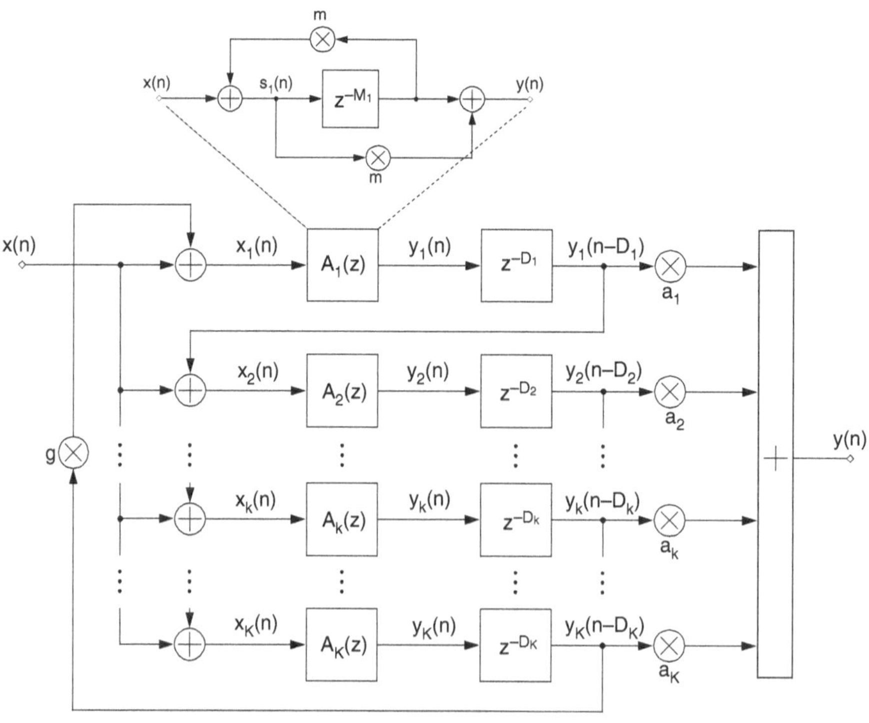

Figure 11.12 Feedback delay network.

$$x_1(n) = x(n) + g \cdot y_{\mathrm{K}}(n - D_{\mathrm{K}}), \tag{11.86}$$

$$y(n) = \sum_{k=1}^{\mathrm{K}} a_k \cdot y_k(n - D_k), \tag{11.87}$$

where $y(n)$ denotes the overall output of the network, a_k denotes the kth coefficient of a mixing vector, $y_{\mathrm{k}}(n)$ denotes the output of the kth Schroeder allpass, M_k and D_k denote the associated delays inside the allpass and the subsequent delay line of the kth branch, respectively, $s_k(n)$ denotes a state of the kth Schroeder allpass, $x_k(n)$ denotes the input to the kth Schroeder allpass, $x(n)$ denotes the input signal to the overall network, and g, a_k are the control parameters. The output of the FDN is sent to a loss function given by

$$E = \sum_{n=1}^{N} \left[d(n) - y(n)\right]^2, \tag{11.88}$$

where d denotes the desired signal that works as a ground truth or label. To adapt g, the expression $\frac{\partial E}{\partial g}$ should be calculated with chain rule of derivatives and backpropagation as given by

$$\frac{\partial E}{\partial g} = \sum_{n=1}^{N} \frac{\partial E}{\partial y(n)} \cdot \frac{\partial y(n)}{\partial g} \tag{11.89}$$

$$= \sum_{n=1}^{N} \sum_{k=1}^{\mathrm{K}} e(n) \cdot a_k \cdot \frac{\partial y_k(n - D_k)}{\partial g} \tag{11.90}$$

$$= \sum_{n=1}^{N} \sum_{k=1}^{\mathrm{K}} e(n) \cdot a_k \cdot \frac{\partial y_k(n - D_k)}{\partial s_k(n - D_k)} \cdot \frac{\partial s_k(n - D_k)}{\partial x_k(n - D_k)} \cdot \frac{\partial x_k(n - D_k)}{\partial g} \tag{11.91}$$

$$= \sum_{n=1}^{N} \sum_{k=1}^{\mathrm{K}} e(n) \cdot a_k \cdot m \cdot \frac{\partial x_k(n - D_k)}{\partial g}. \tag{11.92}$$

In Eq. (11.89), the following expressions are substituted as

$$\frac{\partial E}{\partial y(n)} = -2\left[d(n) - y(n)\right] = e(n), \tag{11.93}$$

$$\frac{\partial y(n)}{\partial g} = \sum_{k=1}^{\mathrm{K}} a_k \cdot \frac{\partial y_k(n - D_k)}{\partial g}, \tag{11.94}$$

while the following expressions from Eq. (11.91) are given by

$$\frac{\partial y_k(n - D_k)}{\partial s_k(n - D_k)} = \frac{\partial s_k(n - D_k - M_k)}{\partial s_k(n - D_k)} - m \cdot \frac{\partial s_k(n - D_k)}{\partial s_k(n - D_k)} = -m, \tag{11.95}$$

$$\frac{\partial s_k(n - D_k)}{\partial x_k(n - D_k)} = \frac{\partial x_k(n - D_k)}{\partial x_k(n - D_k)} + m \cdot \frac{\partial s_k(n - D_k - M_k)}{\partial x_k(n - D_k)} = 1. \tag{11.96}$$

From Eq. (11.92), the expression $\frac{\partial x_k(n-D_k)}{\partial g}$ can be derived further and has to be expressed in the form of $\frac{\partial x_1(\cdot)}{\partial g}$, because $x_1(\cdot)$ is a function of g. With the help of Eq. (11.85) the derivation can be written as

$$\frac{\partial x_k(n-D_k)}{\partial g} = \frac{\partial y_{k-1}(n-D_k-D_{k-1})}{\partial g} \tag{11.97}$$

$$= -m \cdot \frac{\partial x_{k-1}(n-D_k-D_{k-1})}{\partial g} \tag{11.98}$$

$$= -m \cdot \frac{\partial y_{k-2}(n-D_k-D_{k-1}-D_{k-2})}{\partial g} \tag{11.99}$$

$$= (-m)^2 \cdot \frac{\partial x_{k-2}(n-D_k-D_{k-1}-D_{k-2})}{\partial g}. \tag{11.100}$$

The expression in Eq. (11.100) can be extended and expressed in a more general form given by

$$\frac{\partial x_k(n-D_k)}{\partial g} = (-m)^p \cdot \frac{\partial x_{k-p}(n-D_k-\cdots-D_{k-p})}{\partial g}. \tag{11.101}$$

With $p = k-1$, the expression in the above equations can be expressed as a partial derivative function of x_1, differentiable with respect to g, and is given by

$$\frac{\partial x_k(n-D_k)}{\partial g} = (-m)^{k-1} \cdot \frac{\partial x_1(n-D_k-\cdots-D_1)}{\partial g} \tag{11.102}$$

$$= (-m)^{k-1} \cdot \left(y_K(n-\sum_{i=1}^{k} D_i - D_K) - \cdots \right.$$
$$\left. g \cdot \frac{\partial y_K(n-\sum_{i=1}^{k} D_i - D_K)}{\partial g} \right). \tag{11.103}$$

The second expression in Eq. (11.103) can be extended further and can be neglected for a large K because $m = 0.5$ is usually a small fraction. Combining Eq. (11.92) and Eq. (11.103) will lead to the approximate expression given by

$$\frac{\partial E}{\partial g} = -\sum_{n=1}^{N}\sum_{k=1}^{K} e(n) \cdot a_k \cdot m \cdot \frac{\partial x_k(n-D_k)}{\partial g} \tag{11.104}$$

$$= \sum_{n=1}^{N}\sum_{k=1}^{K} e(n) \cdot a_k \cdot (-m)^k \cdot y_K(n-\sum_{i=1}^{k} D_i - D_K). \tag{11.105}$$

In addition to controlling g, the parameters a_k of the mixing vectors can also be adapted and the expression $\frac{\partial E}{\partial a_k}$ needs to be derived. This derivation is given by

$$\frac{\partial E}{\partial a_k} = \sum_{n=1}^{N} \frac{\partial E}{\partial y(n)} \cdot \frac{\partial y(n)}{\partial a_k} \tag{11.106}$$

$$= \sum_{n=1}^{N} e(n) \cdot y_k(n-D_k). \tag{11.107}$$

Finally, the respective updates could be performed with gradient descent as

$$g = g - \eta_g \cdot \frac{\partial E}{\partial g}, \tag{11.108}$$

$$a_k = a_k - \eta_a \cdot \frac{\partial E}{\partial a_k}, \tag{11.109}$$

where η_g and η_a are the respective learning rates and η_a is usually chosen as a very small fraction compared with η_g.

The update can occur for a finite number of iterations or it can stop based on a minimum error threshold. Instead of a desired RIR, the optimization can also be driven by an approximate desired energy decay curve (EDC) of an impulse response. A normalized EDC for an impulse response $h(n)$ in a discrete case is defined as

$$\text{EDC}(n) = 10 \cdot \log_{10}\left(\frac{\sum_{k=n}^{N} h(k)^2}{\sum_{k=1}^{N} h(k)^2}\right). \tag{11.110}$$

The EDC is used for estimating the reverberation times such as T_{60}, and for a long RIR, it decays almost linearly before it subsides rapidly near the end. Hence, in the absence of a desired RIR, a normalized EDC can be constructed based on a desired reverberation time with the assumption of linearity, and this approximate EDC can be used as a ground truth. In this simulation, the input $x(n)$ is an impulse and therefore the estimated output $y(n)$ is an impulse response. The EDC of $y(n)$ is then calculated and its deviation from the approximate ground truth is used as an error to drive the FDN parameter adaptation. The mixing coefficients a_k are initialized randomly and scaled, while the initial value of the parameter g is selected as 0.3.

Figure 11.13 shows a simple example of a simulation based on a desired RIR as the ground truth. The first plot shows the normalized decay curves for the final estimated impulse response after 20 iterations, the desired or ground truth impulse response, and the initial impulse response. The second plot shows the final RIR, which is broadband in nature. The EDC indicates a T_{60} reverberation time of approximately 0.67 seconds and the density of the impulse response depends on the number of branches in the network. In this simulation, the respective delays inside the

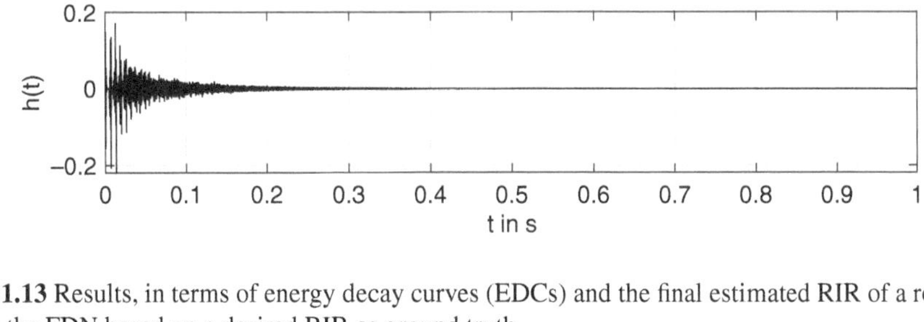

Figure 11.13 Results, in terms of energy decay curves (EDCs) and the final estimated RIR of a room simulation by the FDN based on a desired RIR as ground truth.

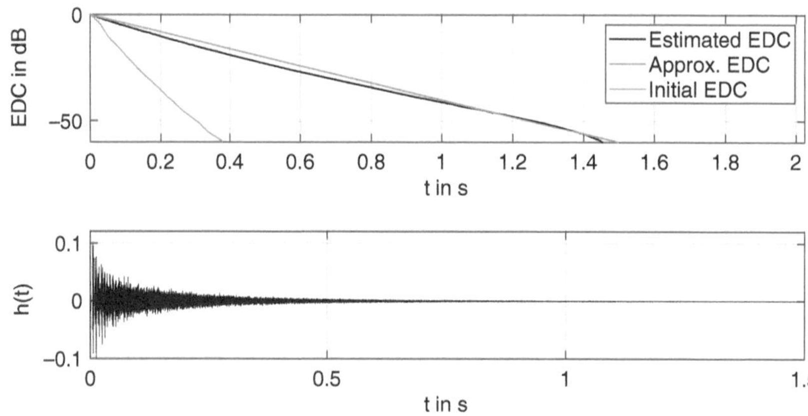

Figure 11.14 Results in terms of EDCs and the final estimated RIR of a room simulation by the FDN based on a desired reverberation time T_{60} as ground truth.

allpass filters are purely prime numbers, in terms of number of samples, and they are selected as $M_1 > 200$ and $M_1 < M_k < 1.5 \cdot M_1$. If the value of M_k is very high, a subset of prime values between M_1 and M_k is selected to reduce the number of branches and accelerate convergence. However, reducing the number of branches leads to a reduction of impulse response density. The delay after an allpass filter is selected as $D_k \leq \frac{M_k}{10}$.

Figure 11.14 shows an example of a simulation based on a desired T_{60}. The ground truth is created by a linear interpolation on the basis of the proposed T_{60} as 1.5 s and is treated as an approximation of the expected EDC. It is however noteworthy that such an approximation of the ground truth is not entirely well posed, owing to its dissimilarity with real EDCs near the tail end. Through the iterations, the EDC of the estimated impulse response gradually drifts towards the approximated ground truth and can eventually improve further. The corresponding final impulse response after 10 iterations is also shown in the plot. The method can be extended to stereo by adding an additional mixing vector of coefficients. A pseudo-random initialization of each mixing vector ensures that the individual impulse responses are decorrelated to avoid a narrow sound field while rendering an audio or they can be decorrelated during adaptation, if required.

11.4.3 Audio Denoising

Noise reduction is one of the most widely performed low level applications in audio signal processing. One of the earliest approaches in noise reduction is the method of spectral subtraction [Bol79], which is quite fast and is suitable for reducing stationary noise. Another popular approach to noise reduction is Wiener filtering based on minimum mean squared error, where an estimation of the *a priori* signal-to-noise ratio (SNR) is required [EM84]. Other methods based on adaptive filtering [Tak03], wavelets [Soo97], adaptive time-frequency block thresholding [Yu08], and statistical modeling [God01] are proposed for efficient audio restoration. Additionally, companding based noise reduction employed by Dolby and dynamic noise limiters introduced by Philips are some of the commercially used denoising systems. In recent years, deep learning methods have been successfully used in audio denoising and notable deep neural network architectures like CNN [Par17], wavenet [Ret18], and recurrent neural network (RNN) [Val18] have shown very promising results. In this section, a convolutional neural network model is described for an offline audio denoising application against additive Gaussian noise.

Denoising Model

Figure 11.15 illustrates the proposed denoising model. In the training phase, Gaussian noise, denoted by $r(n)$, is initially added to the clean audio signal sampled at 48 kHz and denoted by $x(n)$. It is ensured that the overall SNR in each audio file is 5 dB. The noisy audio is transformed with a short-time Fourier transform (STFT) using a hamming window of 1024 samples, 75% overlap, and a frequency resolution of 46.875 Hz. After the transform, its magnitude response is calculated, and the coefficients are normalized between 0 and 1. With a unit stride, consecutive coefficient vectors of the magnitude response, denoted by $|X_N(m,k)|$ and centered around the mth frame, are concatenated along the depth to produce a sequential input for the CNN. The model predicts the denoised coefficient vector of the magnitude response, denoted by $|Y(m,k)|$, while the corresponding magnitude response from the clean audio, denoted by $|X(m,k)|$, is used as the ground truth. The CNN is trained for several epochs with a combination of quadratic loss functions. In the testing phase, the aforementioned data pre-processing is performed for the test audio and the trained model is used to estimate the denoised magnitude response. The phase response of the noisy audio is then combined with this magnitude response and the inverse STFT is performed to reconstruct the denoised audio signal.

Network Architecture

The CNN has a feedforward architecture with inception and attention modules, as illustrated in Fig. 11.16. The input structure contains five consecutive vectors of 513 coefficients each, centered

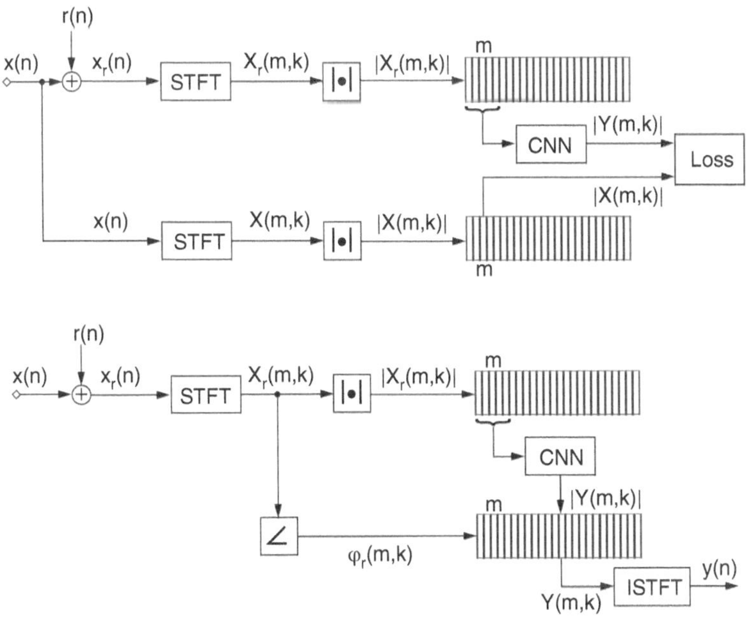

Figure 11.15 Illustration of the denoising model during training phase (top plot) and the testing phase (bottom plot).

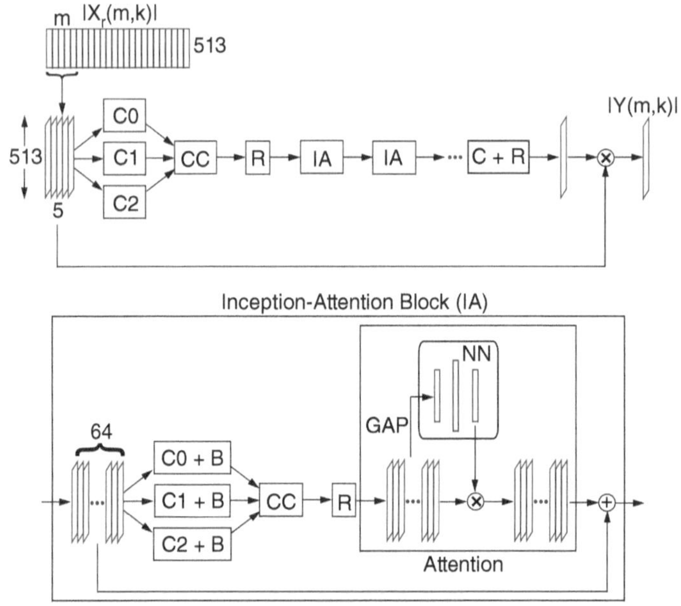

Figure 11.16 Illustration of the CNN architecture, where C0, C1, and C2 denote convolution layers with filter sizes of $3 \times 1 \times d_{in} \times 32$, $3 \times 1 \times d_{in} \times 16$, and $3 \times 1 \times d_{in} \times 16$ and dilation factors of 0, 1, and 2, respectively, C denotes a convolution layer with filter size of $3 \times 1 \times 64 \times 1$, R denotes a ReLU activation, CC denotes concatenation, BN denotes batch normalization, GAP denotes global average pooling, and NN denotes a neural network.

on the *m*th frame index and it is sent to an inception module [Sze15] containing three parallel convolution layers. The filter groups in the layers have a spatial dimension of 3×1 but different dilation factors of 0, 1, and 2, and they produce output feature maps with depths of 32, 16, and 16, respectively. Dilation of a filter refers to the insertion of zeros in between filter coefficients. This ensures that there is an increment in its receptive field but not in the number of training parameters. Multiple dilation factors help in aggregating multiresolution features, which improves the network performance. The padding is adjusted according to the dilation factors and filter sizes so that the output feature maps have the same spatial dimensions. The features generated by the parallel convolutional layers are concatenated along depth to create an output of 64 feature maps, and it is rectified by the ReLU layer. These feature maps are sent to a residual inception-attention block which features the aforementioned inception module containing an additional batch normalization operation after each convolution layer, and the inception module is followed by an attention module. The attention mechanism was introduced as an improvement in neural machine translation system [Bah15]. Although the mechanism is primarily used in RNNs or LSTMs for natural language processing [Che16], attention has emerged as an important module in computer vision [Xu15] and audio processing [Hao19].

In the current CNN, a simple self-attention mechanism is used. At the beginning of this module, a global average pooling is performed across each feature map of the incoming three-dimensional input, and the operation is given by

$$x_j = \sum_{i=1}^{H} X_{ij}, \tag{11.111}$$

where X_{ij} denotes the ith element in the jth feature map, x_j denotes the corresponding scalar output, and H denotes the height of the feature map while its width is 1. Pooling on every feature map of the tensor results in an output vector which is processed by a simple neural network with 1 hidden layer and ReLU activation function. At the output of the neural network, a sigmoid function, given by Eq. (11.4), is used to limit the values between 0 and 1. As an alternative, a softmax function can also be used to get a normalized output vector. Finally, each coefficient of this vector acts as an independent multiplier to tune each feature map of the incoming three-dimensional structure.

A series of inception–attention blocks are cascaded before a final convolution and ReLU layer which estimates a spectral mask, and the mask is multiplied by the input coefficient vector corresponding to the mth frame to suppress noise. The result is sent to the loss layer where the error is calculated with respect to the ground truth coefficient vector.

Model Evaluation

To train the model, a combination of losses is used. The first loss is the mean squared error between the magnitude coefficients given by

$$L_{\text{mse}}(m) = \frac{1}{K}\sum_k (|X(m,k)| - |Y(m,k)|)^2, \tag{11.112}$$

where $|X(m,k)|$ and $|Y(m,k)|$ denote the magnitude response vectors of the desired and estimated output signal, respectively, in dB, k denotes a frequency bin index, m denotes the frame index, and K denotes the number of bins. The second loss function is the absolute error between the absolute values of the transformed magnitude response coefficients corresponding to the estimation and ground truth. This loss function is given by

$$L_{\text{dft}}(m) = \frac{1}{K}\sum_k ||\text{DFT}(|X(m,k)|)| - |\text{DFT}(|Y(m,k)|)||, \tag{11.113}$$

where $|\cdot|$ denotes the absolute value. This loss function shows an improved noise suppression capability, particularly in unvoiced or silent regions to reduce unpleasant artifacts.

To train the denoising model, a dataset is built with audio files containing both high fidelity speech and music sampled at 48 kHz. Short speech files are collected from the PTDB-TUG [Pir11] dataset, and music files of longer duration are collected from multiple classical music datasets including the Bach10 [Dua10] and Mirex-Su [Su16] datasets. Two networks are trained with speech and music files, respectively. To train the first network, 200 speech files are selected and some examples from the remaining speech files are selected for testing. The second network is trained with 20 music files while the remaining music files are selected for testing. In this example, a CNN model with eight inception–attention blocks is used for speech denoising and a model with twelve inception–attention blocks for denoising music signals, based on experiments. The models are trained for 70 epochs with adaptive momentum update. The performance of the models can be evaluated by different objective metrics. Improvement in SNR is an indicator of denoising performance, while perceptual evaluation of audio quality (PEAQ) [Thi00] and speech quality (PESQ)

Table 11.1 Average performance of the denoising CNNs on noisy speech and music

Audio	SNR (dB)	ODG	MOS
Noisy Speech	5	−3.91	1.48
Denoised Speech	18.8	−3.29	2.49
Noisy Music	5	−3.91	–
Denoised Music	16.3	−3.64	–

[ITU01] are important methods to evaluate audio and speech quality. The PEAQ method provides an objective difference grade (ODG) which rates an audio between −4 and 0. An ODG score of 0 indicates imperceptible differences compared with the reference and a score of −4 indicates an audio with annoying artifacts. Similarly, PESQ provides a mean opinion score (MOS) that ranges from 1, which indicates a poor speech quality, to 5, which indicates an excellent speech quality. The test dataset contains 54 audio files, out of which the first 50 files contain short speech and the remaining files contain music of relatively longer duration. Table 11.1 illustrates the performance of the CNNs for speech and music signals. The average SNR of the denoised speech signals produced by the CNN indicates an improvement of 13.8 dB. Similarly, the average ODG score of the speech after noise suppression is improved by 0.62. To perform the wideband PESQ evaluation, the speech signals are resampled to 16 kHz. The average MOS of the denoised signal is nearly improved by 1. The average SNR of the denoised music signals produced by the CNN shows an improvement of approximately 11 dB. However, the average ODG score of the music signals after noise suppression does not show an improvement similar to that of speech. This indicates that music is relatively more difficult to denoise compared to speech because of a low number of pauses and the presence of more high-frequency content which gets suppressed along with noise and affects the perceptual quality. Figure 11.17 illustrates an example of denoising on a female speech signal corrupted with noise. The SNR of the denoised speech is 20.9 dB, its ODG score is −2.91 compared with a score of −3.91 for the noisy speech, and its MOS score improves from 1.28 to 2.44. Figure 11.18 illustrates an example of denoising on a musical piece from the Mirex-Su dataset corrupted with noise. The overall improvement in SNR is approximately 10.8 dB, while the ODG score of the denoised signal is −3.45 compared with a score of −3.9 for the noisy signal. Figure 11.19 shows the corresponding spectral representations of the noisy, denoised, and original signal.

The applications described in this chapter are a few of the many audio processing applications which can be categorized as supervised regression problems. Similar applications with deep learning include audio enhancement in terms of super-resolution [SD21], supervised audio source separation [LM19], and neural source modeling [Wan20]. Supervised audio classification problems, however, categorize audio signals or inherent features within audio signals into distinct classes which can be used for further processing. Classification problems are also extensively researched and some of the notable applications include pitch detection [SZZ16], music genre classification [Ora17], and speaker recognition [Sny18], among others.

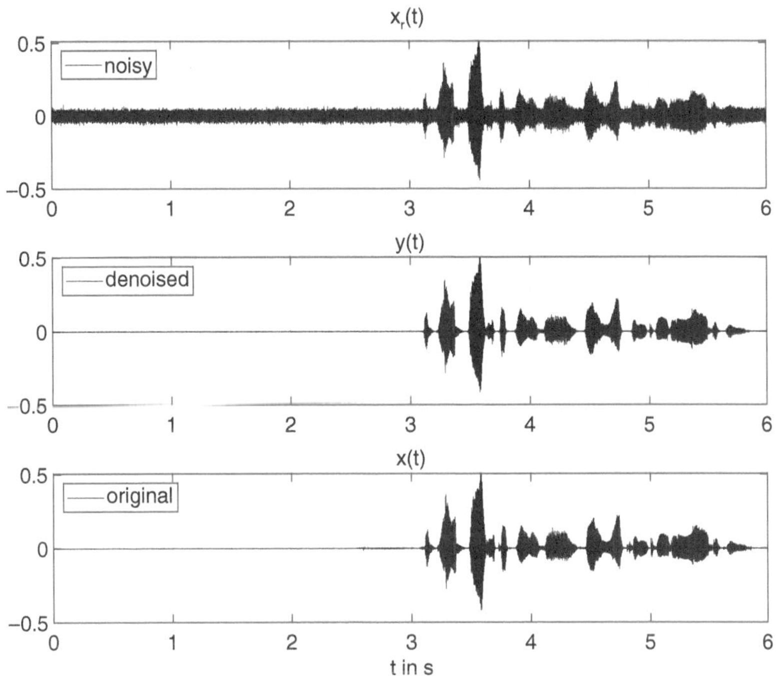

Figure 11.17 Example of noise suppression in an audio file from the PTDB-TUG speech dataset.

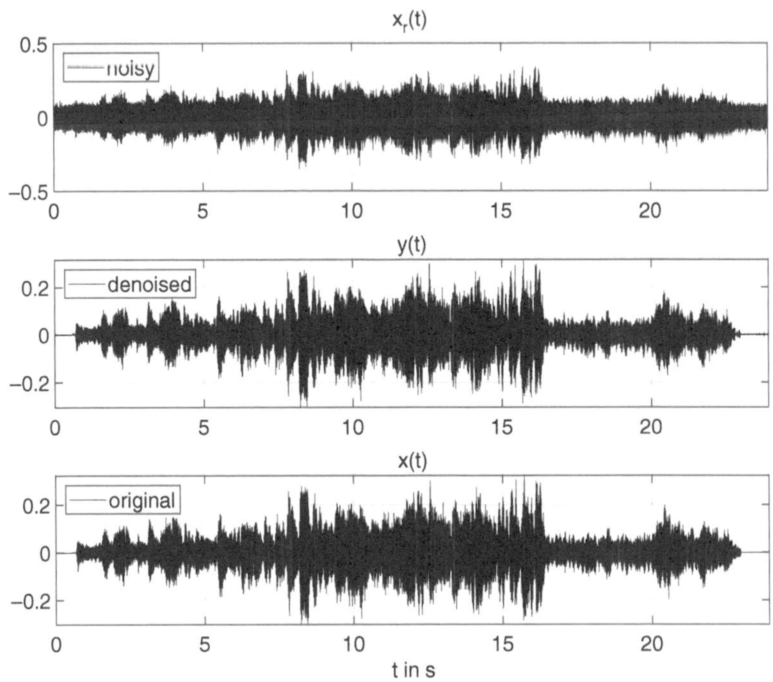

Figure 11.18 Example of noise suppression in an audio file from the Mirex-Su dataset.

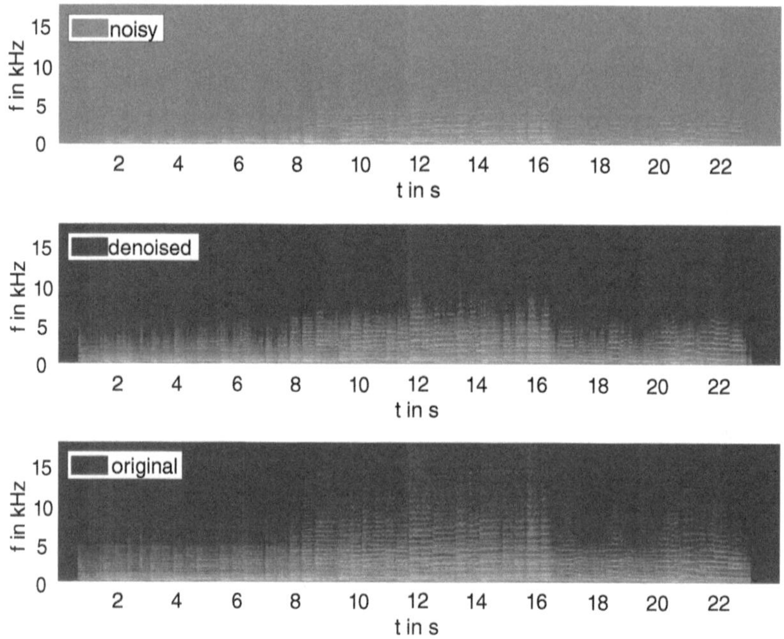

Figure 11.19 Spectrogram of the example noisy, denoised, and original audio file from the Mirex-Su dataset.

11.5 Exercises

1. Write a script for a simple feedforward neural network with an input layer of dimension 1, an output layer of dimension 1, a hidden layer of dimension 5, and an activation and a loss function.

   (a) Train the network with stochastic gradient descent for 50 epochs to fit the function $y = max\,(\sin(x), -0.2)\,, -1 \leq x \leq 1$, monitor the error, and test the network.

   (b) Perform the training with a higher number of epochs.

   (c) Perform the training with multiple hidden layers.

2. Replace a fully connected layer with a convolution layer and train the network for the previous example.

3. Perform the above exercise and the examples from the chapter using a deep learning toolbox in a framework of your choice (Matlab, Pytorch, Tensorflow).

References

[Aka12] M. Akamine and J. Ajmera: *A decision-tree-based algorithm for speech/music classification and segmentation.* EURASIP Journal on Audio, Speech, and Music Processing, 10, Feb 2012.

[Alg01] V. R. Algazi, R. O. Duda, D. M. Thompson, and C. Avendano: *The CIPIC HRTF database*. Proceedings of the 2001 IEEE Workshop on the Applications of Signal Processing to Audio and Acoustics, pages 99–102, Oct 2001.

[Bac91] A.D. Back and A. C. Tsoi: *FIR and IIR synapses, a new neural network architecture for time series modeling*. Neural Computation, 3(3):375–385, 1991.

[Bah15] D. Bahdanau, K. Cho, and Y. Bengio: *Neural machine translation by jointly learning to align and translate*. In 3rd International Conference on Learning Representations, May 2015.

[Bha20] P. Bhattacharya, P. Nowak, and U. Zölzer: *Optimization of cascaded parametric peak and shelving filters with backpropagation algorithm*. Digital Audio Effects 2020 (DaFX), 2020.

[Bol79] S. F. Boll: *Suppression of acoustic noise in speech using spectral subtraction*. IEEE Transactions on Acoustics, Speech, Signal Processing, ASSP-27(2):113–120, Apr 1979.

[Chi06] J-T. Chien and B-C. Chen: *A new independent component analysis for speech recognition and separation*. IEEE Transactions on Audio, Speech, and Language Processing, 14(4):1245–1254, 2006.

[Che16] J. Cheng, L. Dong, and M. Lapata: *Long short-term memory networks for machine reading*. In Proceedings of the 2016 Conference on Empirical Methods in Natural Language Processing, pages 551–561, Nov 2016.

[Car17] H. Caracalla and A. Roebel: *Gradient conversion between time and frequency domains using Wirtinger calculus*. In Digital Audio Effects 2017 (DaFX), Edinburgh, United Kingdom, Sep 2017.

[Cam96] P. Campolucci, A. Uncini, and F. Piazza: *Fast adaptive IIR-MLP neural networks for signal processing application*. Acoustics, Speech, and Signal Processing, IEEE International Conference on, 6:3529–3532, Jun 1996.

[Dau15] Y. N. Dauphin, H. de Vries, J. Chung, and Y. Bengio: *Rmsprop and equilibrated adaptive learning rates for non-convex optimization*. CoRR, abs/1502.04390, 2015.

[Duc11] J. Ducji, E. Hazan, and Y. Singer: *Adaptive subgradient methods for online learning and stochastic optimization*. Journal of Machine Learning Research, 12:2121–2159, Jul 2011.

[Dua10] Z. Duan, B. Pardo, and C. Zhang: *Multiple fundamental frequency estimation by modeling spectral peaks and non-peak regions*. IEEE Trans. Audio Speech Language Process., 18(8):2121–2133, 2010.

[EM84] Y. Ephra–m and D. Malah: *Speech enhancement using a minimum mean-square error short-time spectral amplitude estimator*. IEEE Transactions on Acoustics, Speech, Signal Processing, ASSP- 32(6):1109–1121, Dec 1984.

[Eng20] J. Engel, L. Hantrakul, C. Gu, and A. Roberts: *DDSP: Differentiable Digital Signal Processing*. International Conference on Learning Representation, pages 1–19, 2020.

[Fev09] C. Fevotte, N. Bertin, and J-L. Durrieu: *Nonnegative matrix factorization with the itakura-saito divergence: With application to music analysis*. Neural Computation, 21(3):793–830, 2009.

[Glo10] X. Glorot and Y. Bengio: *Understanding the difficulty of training deep feedforward neural networks*. Journal of Machine Learning Research, 9:249–256, 2010.

[GL03] G. Guodong and S.Z. Li: *Content-based audio classification and retrieval by support vector machines*. IEEE Transactions on Neural Networks, 14(1):209–215, 2003.

[Gao92] F.X.Y. Gao and W.M. Snelgrove: *An adaptive backpropagation cascade IIR filter*. IEEE Transactions on Circuits and Systems II: Analog and Digital Signal Processing, 39(9):606–610, Sep 1992.

[God01] S.J. Godsill, P.J. Wolfe, and W.N.W. Fong: *Statistical model-based approaches to audio restoration and analysis*. Journal of New Music Research, 30(4):323–338, 2001.

[Han17] Y. Han, J. Kim, and K. Lee: *Deep convolutional neural networks for predominant instrument recognition in polyphonic music*. IEEE/ACM Transactions on Audio, Speech, and Language Processing, 25(1):208–221, 2017.

[Hao19] X. Hao, C. Shan, Y. Xu, S. Sun, and L. Xie: *An attention-based neural network approach for single channel speech enhancement*. In ICASSP 2019 - 2019 IEEE International Conference on Acoustics, Speech and Signal Processing, pages 6895–6899, 2019.

[He15] K. He, X. Zhang, S. Ren, and J. Sun: *Delving deep into rectifiers: Surpassing human-level performance on imagenet classification*, 2015.

[ITU01] ITU Perceptual evaluation of speech quality (PESQ): *An objective method for end-to-end speech quality assessment of narrow-band telephone networks and speech codecs*, 2001.

[Kin15] D.P. Kingma and J.Ba. Adam: *A method for stochastic optimization*. In 3rd International Conference on Learning Representations, ICLR, May 2015.

[Lav09] Y. Lavner and M.E.P. Davies: *A decision-tree-based algorithm for speech/music classification and segmentation*. EURASIP Journal on Audio, Speech, and Music Processing, 239892, Jun 2009.

[Ler12] A. Lerch: *An introduction to audio content analysis: applications in signal processing and music informatics*. John Wiley & Sons, Ltd., 2012.

[LL01] L. Lu, S.Z. Li, and H-J. Zhang: *Content-based audio segmentation using support vector machines*. In IEEE International Conference on Multimedia and Expo, pages 749–752, 2001.

[LM19] Y. Luo and N. Mesgarani. *Conv-tasnet: Surpassing ideal timefrequency magnitude masking for speech separation*. IEEE/ACM Trans. Audio, Speech and Lang. Proc., 27(8):1256–1266, Aug. 2019.

[Lu16] Y-C. Lu, C-W. Wu, A. Lerch, and C-T. Lu: *Automatic outlier detection in music genre datasets*. In Proceedings of the 17th International Society for Music Information Retrieval Conference, ISMIR, pages 101–107, Aug 2016.

[Mül15] M. Müller: *Fundamentals of Music Processing: Audio, Analysis, Algorithms, Applications*. Springer Publishing Company, Incorporated, 2015.

[Mac67] J. MacQueen: *Some methods for classification and analysis of multivariate observations*. In Proceedings of the 5th Berkeley Symposium on Math, Statistics, and Probability, pages 281–297, 1967.

[Mar20] M.A.R. Martinez, E. Benetos, and J.D. Reiss: *Deep learning for black-box modeling of audio effects*. Applied Sciences, 10(2), 2020.

[Mee18] H. S. Meet, N. Shah, and A. H. Patil: *Time-frequency masking-based speech enhancement using generative adversarial network*. In 2018 IEEE International Conference on Acoustics, Speech and Signal Processing (ICASSP), pages 5039–5043, 2018.

[Min19] G. Min, C. Zhang, X. Zhang, and W. Tan: *Deep vocoder: Low bit rate compression of speech with deep autoencoder*. In 2019 IEEE International Conference on Multimedia Expo Workshops (ICMEW), pages 372–377, 2019.

[Ner20] S. Nercessian: *Neural parametric equalizer matching using differentiable biquads*. Digital Audio Effects 2020 (DaFX), 2020.

[Nes83] Y. Nesterov: *A method for unconstrained convex minimization problem with the rate of convergence* $O(1/k^2)$. Doklady ANSSSR, pages 543–547, 1983.

[Ora17] S. Oramas, O. Nieto, F. Barbieri, and X. Serra: *Multi-label music genre classification from audio, text, and images using deep features*. CoRR, abs/1707.04916, 2017.

[Poh06] T. Pohle, P. Knees, M. Schedl, and G. Widmer: *Independent component analysis for music similarity computation*. In ISMIR 2006, 7th International Conference on Music Information Retrieval, pages 228–233, Oct 2006.

[Par17] Se Rim Park and Jin Won Lee: *A fully convolutional neural network for speech enhancement*. In Proc. Interspeech 2017, pages 1993–1997, 2017.

[Pir11] G. Pirker, M. Wohlmayr, S. Petrik, and F. Pernkopf: *A pitch tracking corpus with evaluation on multipitch tracking scenario*. In Interspeech 2011, pages 1509–1512, 2011.

[Ros61] F. Rosenblatt: *Principles of Neurodynamics. Perceptrons and the Theory of Brain Mechanisms*. Defense Technical Information Center, 1961.

[Ros99] R.E. Rose: *Training a recursive filter by use of derivative function*, US patent US5905659A, May 1999.

[Ret18] D. Rethage, J. Pons, and X. Serra: *A wavenet for speech denoising*. In 2018 IEEE International Conference on Acoustics, Speech and Signal Processing (ICASSP), pages 5069–5073, 2018.

[SD21] S. Sulun and M.E.P. Davies: *On filter generalization for music bandwidth extension using deep neural networks*. IEEE Journal of Selected Topics in Signal Processing, 15(1):132–142, 2021.

[Sny17] D. Snyder, D. Garcia-Romero, D. Povey, and S. Khudanpur: *Deep neural network embeddings for text-independent speaker verification*. In Proc. Interspeech 2017, pages 999–1003, 2017.

[Sny18] D. Snyder, D. Garcia-Romero, G. Sell, D. Povey, and S. Khudanpur: *X-vectors: Robust dnn embeddings for speaker recognition*. In IEEE International Conference on Acoustics, Speech and Signal Processing, pages 5329–5333, 2018.

[Soo97] I. Y. Soon, S. N. Koh, and C. K Yeo: *Wavelet for speech denoising*. In Proceedings of IEEE TENCON –97, IEEE Region 10 Annual Conference of Speech and Image Technologies for Computing and Telecommunications, volume 2, pages 479–482, 1997.

[Sze15] C. Szegedy, L. Wei, J. Yangqing, P. Sermanet, S. Reed, D. Anguelov, D. Erhan, V. Vanhoucke, and A. Rabinovich: *Going deeper with convolutions*. In 2015 IEEE Conference on Computer Vision and Pattern Recognition, pages 1–9, 2015.

[Su16] L. Su and Y-H. Yang: *Escaping from the abyss of manual annotation: New methodology of building polyphonic datasets for automatic music transcription*. In Music, Mind, and Embodiment, pages 309–321. Springer International Publishing, 2016.

[SZZ16] H. Su, H. Zhang, X. Zhang, and G. Gao. *Convolutional neural network for robust pitch determination*. In IEEE International Conference on Acoustics, Speech and Signal Processing, pages 579–583, 2016.

[Thi00] T. Thiede, W. C. Treurniet, R. Bitto, C. Schmidmer, T. Sporer, J. G. Beerends, and C. Colomes: *PEAQ - the ITU standard for objective measurement of perceived audio quality*. Journal of the audio engineering society, 48(1/2):3–29, Feb 2000.

[Tak03] H. Takeshi, M. Takahiro, I. Yoshihisa, and H. Tetsuya: *Musical noise reduction using an adaptive filter*. Acoustical Society of America Journal, 114(4), Oct 2003.

[Val18] J-M. Valin: *A hybrid dsp/deep learning approach to real-time full-band speech enhancement*. In 2018 IEEE 20th International Workshop on Multimedia Signal Processing (MMSP), pages 1–5, 2018.

[Väl19] V. Välimäki and J. Rämö: *Neurally controlled graphic equalizer*. IEEE/ACM Transactions on Audio, Speech, and Language Processing, 27(12), Dec 2019.

[Vir07] T. Virtanen: *Monaural sound source separation by nonnegative matrix factorization with temporal continuity and sparseness criteria*. IEEE Transactions on Audio, Speech, and Language Processing, 15(3):1066–1074, 2007.

[Wan20] X. Wang, S. Takaki, and J. Yamagishi: *Neural source-filter waveform models for statistical parametric speech synthesis*. IEEE/ACM Transactions on Audio, Speech, and Language Processing, 28:402–415, 2020.

[Xu15] K. Xu, J. Ba, R. Kiros, K. Cho, A. C. Courville, R. Salakhutdinov, R. S. Zemel, and Y. Bengio: *Show, attend and tell: Neural image caption generation with visual attention*.

In Proceedings of the 32nd International Conference on Machine Learning, volume 37, pages 2048–2057, Jul 2015.

[Yu08] G. Yu, S. Mallat, and E. Bacry: *Audio denoising by time-frequency block thresholding*. IEEE Transactions on Signal Processing, 56(5):1830–1839, 2008.

[Zha16] J. Zhang, J. Tang, and Li-R. Dai: *RNN-BLSTM based multi-pitch estimation*. In Interspeech 2016, pages 1785–1789, 2016.

Index

Digital Audio Signal Processing, Third Edition. Udo Zölzer.

© 2022 John Wiley & Sons Ltd. Published 2022 by John Wiley & Sons Ltd.